A. Brunn L. Baehr H.-J. Karpe (Eds.)

# Conversion

## Opportunities for Development and Environment

Springer-Verlag
Berlin Heidelberg New York
London Paris Tokyo
Hong Kong Barcelona Budapest

Anke Brunn
Ministerin für Wissenschaft
und Forschung des Landes NRW
Völklinger Straße 49
4000 Düsseldorf
Germany

Lutz Baehr
Director
Department for Economic
and Social Development (DESD)
United Nations
New York
USA

Hans-Jürgen Karpe
Institut für Umweltschutz
Universität Dortmund
Postfach 50 05 00
4600 Dortmund 50
Germany

ISBN-13: 978-3-642-95703-1 e-ISBN-13: 978-3-642-95701-7
DOI: 10.1007/978-3-642-95701-7


Softcover reprint of the hardcover 1st edition 1992



Typesetting: Camera ready by author

60/3020-5 4 3 2 1 0 - Printed on acid -free paper

# CONTENTS

**ANNEX**

## Abbreviations and accronyms

| | |
|---|---|
| CFE | Conventional Force Reduction in Europe |
| CIS | Commonwealth of Independent States |
| CSTD | Centre for Science and Technology for Development |
| DMSP | US Defence Meteorological Satellite Program |
| DoD | Departement of Defence |
| GDR | German Democratic Republic |
| GNP | Gross National Product |
| IDNDR | International Decade for Natural Disaster Reduction |
| IMF | International Monetary Fund |
| INFU | Institut für Umweltschutz |
| IUCN | International Union for the Conservation of Nature and Natural Resources |
| MWF | Ministerium für Wissenschaft und Forschung |
| NASA | National Aeronautics and Space Administration |
| NATO | North Atlantic Treaty Organization |
| NGO | Non-governmental Organization |
| NVA | Nationale Volksarmee |
| R & D | Research and Development |
| S & T | Science and Technology |
| UNCED | United Nations Conference on Environment and Development |
| UNCTAD | United Nations Conference on Trade and Development |
| UNHA | United Nations Humanitarian Assistance Team |
| UNIDO | United Nations Industrial Development Organization |
| UNO | United Nations Organization |
| WMO | World Meteorological Organization |

# PREFACE

This Conference Proceedings on 'Conversion Opportunities for Development and Environment', provides the edited version of the papers presented by the participants to the conference. This proceedings also provides a summary of the deliberations among distinguished groups of eminent scientists, professionals, policy makers and industrialists, dealing specifically with the opportunities and constraints in redirecting the savings from military activities for the benefit of socio-economic development and protection of environment.

We hope that this proceedings would be useful not only in generating an awareness of the magnitude of the opportunities available to mankind but also provoke scientists, technologists and policy makers to initiate specific programmes and promote investments which go beyond the recommendations of the conference.

We gratefully acknowledge stimulating contributions of the participants from around the world and are very pleased with the large attendance of high level participants from the developing countries and Eastern Europe.

The conference was basically sponsored by the State of North Rhine-Westphalia.

We are particularly grateful to the additional support provided by the Initiativkreis Ruhrgebiet, the Bundesstiftung Umwelt, and the City of Dortmund.

Anke BRUNN
Minister for
Higher Education and
Research
NORTH RHINE-
WESTPHALIA

Dr.Lutz BAEHR
Director
UNCSTD
NEW YORK

Dr.Hans J.KARPE
Director
INFU
DORTMUND

# EXECUTIVE SUMMARY

M. Anandakrishnan, Rapporteur

## OVERVIEW

The International Conference on Conversion - Opportunities for Development and Environment, jointly organized by the United Nations Centre for Science and Technology for Development (New York), and the Ministry of Higher Education and Research, State of North Rhine Westphalia, the Institute of Environmental Protection Policy (INFU) of the University of Dortmund, Germany, has attracted the participation of over 200 senior political leaders, industrialists, scientists and technologists from 35 countries, and representatives of UN organizations and specialized agencies which are currently engaged in exploring practical measures for promoting the conversion of military facilities and capabilities for sustainable human benefits.

The Conference was the third in a series of international gatherings on CONVERSION following the earlier meetings in Beijing and Moscow. Going beyond the feasibility and technical aspects of Conversion, the Dortmund Conference focussed on concrete approaches to redirecting the resources specifically for socio-economic development and environmental protection.

Welcoming the participants, the Lord Mayor of the City of Dortmund, Mr. Günter Samtlebe, referred to the successful example of the economic conversion of the Ruhr area from a military economy during World War II into a peaceful and flourishing market economy, and the establishment of technical institutes in the vacated army barracks.

In his Opening Address, the Minister President of the State of North Rhine Westphalia, Dr. Johannes Rau, emphasized the need for serious political and policy dialogues on conversion, beyond academic discussions, inspite of the many new uncertainties and imponderables. He cited the experience of the State in the restructuring process, resulting in establishment of several thousand high technology industries, many of which were in a leading position in the development of highly advanced environmental technologies. He stressed the importance of global attention in eliminating weapons of more destruction, eradication of hunger and poverty and the establishment of a UN framework to enlarge the term "security" to include protection against natural and environment disasters.

In his keynote address, the German Federal Minister for the Environment, Natural Protection and Nuclear Safety, Prof. Dr. Klaus Töpfer pointed out the vast resources released on account of the end to East-West confrontation, which at the same time led to elimination of lots of jobs, highlighting the importance of attention to be paid to the Conversion issues. He called for Conversion of mental attitudes as an important first step leading to new initiatives towards sustainable global environment and suitable monitoring systems under United Nations auspices.
Addressing the participants, the Chairperson of the Conference, Mrs. Anke Brunn, North Rhine Westphalia Minister of Higher Education and Research, called for elimination of investments in arms technology. She outlined her

government's policy of not supporting any research activity for military purposes and describes the programmes of the State's Scientific Council on Conversion.

Mr. Kofi Annan, Assistant Secretary-General, Office of Peace Keeping Operations, United Nations, mentioned in his opening address that the end of the Cold War has not put an end to regional conflicts. He emphasized the need to clearly articulate the links between peace and sustainable development, and the importance of eliminating the roots of regional conflicts and local strifes. He said that the world body should move from peace-keeping to peace-making and peace-building. He called for a creative approach to conversion, to enable the United Nations to respond in a timely and effective manner to unanticipated humanitarian emergencies along the lines proposed for the U.N. humanitarian assistance teams.

Prof. Essam Eldin Galal, in his key-note address, drew attention to the contradictions in national and international behaviour which hampers the definition and effective implementation of the Conversion process. He pleaded for restoration of international attention to the miseries faced by developing countries and the aggravating environmental degradation.

The deliberations reflected the growing awareness about the vast economic and environmental benefits of Conversion at the highest levels of government, industries and research establishments. The elimination of the cold war conditions of the past decades and the likely prospects of reducing local conflicts around the world present an unique climate to articulate methodically and intensely the immeasurable benefits of Conversion to mankind. Several immeasurable obstacles to these efforts were also recognised.

**THE IMPERATIVES OF CONVERSION**

The existence of high-level political will is a pre-condition to the promotion of the Conversion process, even though the economic imperatives and environmental compulsions could hasten this process.

Conversion should be viewed as a simultaneous and integral part of the arms reduction efforts so that the employment creation potentials of Conversion are appreciated as against the anxiety of unemployment due to arms reduction. Similarly the export potentials for civilian goods and services created by Conversion should outweigh the economic returns of the arms trade. Many new fields of industrial developments could arise.

The specific scope for successful initiation of the Conversion process will have to be defined in terms of the current economic and political situation in different countries such as those prevailing in Eastern Europe, especially in the Commonwealth of Independent States, developing countries and the industrialized countries.

Conversion should be perceived as an effort of international disarmament, and should be clearly distinguished from activities which encompass the restructuring, modernization of existing military contingents.

The opportunities for the environmental protection arising out of the Conversion efforts are seen both from the point of view of better use of resources for development of sustainable environment, as well as for the clean-up of the environments already ruined by the military activities. The fact that man has become more dangerous to nature than nature to man has to be hgihlighted in this context.

The socio-economic development including opportunities for education, health care and employment generated by the Conversion efforts should help to alleviate the persistant miseries of large populations around the world.

## NATURE OF PROBLEMS TO BE ADDRESSED

It is necessary to bring about high level consciousness on the costs and benefits of continuing the current levels of military systems vis-a-vis diversion of resources for civilian benefits. A reorientation of value system currently distorted by military power is essential.

Conversion should be supported and reflected by all societal groups which are actively involved in conversion activities or socially affected by decisions related to those activities.
Though the conversion issues are largely country and region-specific, their successful implementation calls for a coordinated global strategy under the leadership of the United Nations.

Over exploitation of natural resources as a result of emphasis on militarization of economies should be reversed. Some technologies of mass destruction such as nuclear weapons and missiles technologies have contributed to diversion of scace resources from developmental efforts. This situation needs to be eliminated.

The fixed and working capital for military activities should be drastically reduced compared to those for civilian activities.

The national laws should be redrafted to bring military activities under the same environmental legal framework as civilian activities.

Even as the World Community strives to move towards elimination of weapons of destruction, which may take many more decades to succeed, the progressive steps in arms reduction should be combined with the benefits of conversion in order to provide the evidence of positive benefits of arms reduction process.

The evolving concept of security should embrace not only defence and foreign policy but also macroeconomic and environmental policy issues. At national level interdisciplinary study group involving political scientists, economists, environmentalists and technolgists should be contributed to evolve practical steps towards specific Conversion schemes.

Verification schemes should be developed not on the basis of mistrust, but as a means of information sharing for improving conversion process.

## RECOMMENDATIONS

During the discussions in the plenary sessions, working group meetings and written presentations, a number of conrete proposals emerged for enhancing the prospects of implementing conversion schemes. These are in the nature of actions that can be taken at the international as well as national levels. The proposals mainly fall under the categories of a set of studies, informatiion exchange and clearing house systems, major publications, follow-up conferences on specific issues, a variety of technological converison schemes and innovative institutional designs.

### At International Level

The United Nations should ensure that the modern technologies such as biotechnologies, electronics, lasers, new materials and new energy technologies are not diverted for military application.

The monitoring systems created in the work of many arms reduction agreements among countries should develop promotional system of economic conversion and environmental restoration. The role of the Security Council should be redefined to include economic and environmental security as essential components of international security. Environmental terrorism, should be explicitly defined and penalised.

All International financial facilities, existing or newly proposed, meant for economic development and environmental protection should give high priority to those initiatives involving Conversion process.

Global monitoring satellites currently used exclusively for military purposes should be brought under the purview of an International Organization to enable their utilization for civilian benefits especially for rapid deployment purposes in case of disasters and emergencies and for sharing of global data among all countries for resource development.

Technologies developed for decommissioning and denaturization of military hardware and amunitions should be made widely accessible to any interested countries through an international technology transfer regime. The protection of environment from dangers associated with the denaturization schemes should become an essential part of the Conversion programmes.

The proposal to constitute a United Nations Humanitarian Assistance Team (UNHAT) designed as an intenational corps consisting of military and civilian personnel for emergency rapid deployment under UN auspices, should be supported by the international community.

**At National level**

Levels of R&D expenditures on military as compared to civil use should be documented and the benefits analyzed. A deliberate shifting of military R&D for civilian purposes should be encouraged.

Those countries which attempt to reduce arms expenditure do not explicitly reallocate savings (peace dividends) for socio-economic or environmental development needs. The national accounting systems should be restructured to distinctly allocate military savings to conversion process. More than the money, the real "peace dividends" should be viewed in terms of resources of trained personnel and precious assets freed from military for civilian use.

Institutional restructuring of military establishments and associated industrial production systems should be brought under direct civilian legislative system.

Diffusion of knowledge from military to civilian entities in R&D as well as production should be encouraged through appropriate incentive schemes and legislation. Special institutes should be established at the national level by converting existing military institutions to facilitate such diffusions. These institutes could evolve operational programmes dealing with conversion of military hardware, military facilities (such as barracks), personnel retraining, environmental recovery of polluted land and water resources and existing R&D facilities.
A realignment of national budget priorities should be made to stimulate Conversion efforts. The budgets should provide for economic incentives for reorganizing and retooling of production enterprises, retraining of labour, assistance to communities affected by closures of military plants etc.

Industries such as aerospace, computers, electronics, instruments, machine tools and advanced materials which tend to seek primarily military markets should be encouraged to develop civilian products for the national and global markets.

Some developing countries engage in arms production for export purposes as a means of earning foreign exchange to pay off the foreign debts. Therefore, debt forgiveness in exchange for conversion of arms factories to civilian production would pave the way to encourage civilian exports instead of arms production.

Many industrialized countries encourage arms production and sale to other developed as well as to a large number of developing countries as a part of trade strategies. National legislation to curb export of arms to developing countries and withdrawal of subsidised facilities for such purposes should be enacted.

Arms exports should be regulated by law to prevent exports to countries engaged in war or countries not complying with human rights and countries under United Nations embargo.

The use of military personnel during peace time in civilian projects and use of military facilities (such as airports) for civilian use should be planned as a dual-use strategy.

**FOLLOW UP MEASURES**

Considering the need to bring about clarity of ideas to the various issues embodied in the above recommendations, and the need for sensilization of various concerned groups to the Conversion issues, the Conference sugggested that:

1. The United Nations should explore the possibility together with the Government of Russia and Cyprus to convene a follow-up conference of experts on Conversion of Military Areo-space technologies in Moscow, preferably before end of 1992; and another conference of experts on Conversion Issues of the Middle East in Nicosia, preferably in 1993. The Asian Institute of Technology in Bangkok has also expressed its interest.

2. The United Nations should support the incorporation and dissemination of available knowledge and experiences on Conversion issues into one of the Information Exchange Networks of Universities. An active network for information training and research on conversion should be initiated for collection of information and data on conversion, provide referral service to academic and training institutions, as well as to policy making organizations. Initially, such a network may be started by utilizing the existing network of University Information System, located in the University of Dortmund.
3. A feasibility study on the establishment of an effective mechanism at national level to convert military production and technologies for peaceful use should be undertaken, initially on a pilot basis in China, with posible support of the United Nations and Germany.

4. A Conversion Management and Training MANUAL utilizing experience within the UN family will be established (by the Treuhandanstalt) after the Conference. The MANUAL will be organized as a computer mail service as well.

5. The need was stressed to strengthen the research in the field of conversion. Therefore it was proposed by the Minister and Chair of the Conference, Mrs. Anke Brunn, to establish an International Conversion Center, which could be located in Bonn as a future center

for international development and cooperation. This "Clearing House" should provide the newly emerging scientific conversion community in cooperation with the United Nations with the exchange of information, activities on research and professional retraining for military personell.

# The Dortmund Declaration

The current status of military production and expenditures continues to be a source of deep concern in a world in which the majority of countries face severe economic, social and financial problems compounded with an increasing environmental degradation, despite an evident decline of a major military threat in the recent years resulting from the East/West detente.

This situation underscores the need for the conversion of military production and technologies as an effective response to national and international disarmament efforts, and as a contribution to a new perception of an international strategy of security and development in harmony with the natural carrying capacity of our planet.

In order to overcome the potential obstacles for implementing Conversion efforts, it has become necessary to discuss and document the successful Conversion initiatives and thereby identify technological options, policy issues and choices for actions at national and international levels.

With this purpose in view, a representative group of scientists, technologists, industrialists, trade union leaders, political leaders, administrators and members of concerned non-governmental organizations met in Dortmund, Germany, from 24-27 February 1992, for a conference on "Conversion: Opportunities for Development and Environment".
The Conference was jointly organized by the Ministry for Science and Research of the State of North Rhine Westphalia, the Institute of Environmental Protection Policy of the University of Dortmund and the United Nations Centre for Science and Technology for Development, under the patronage of the Prime Minister of the North-Rhine Westphalia, Mr. Johannes Rau, with Mrs. Anke Brunn, North-Rhine Westphalia Minister of Higher Education and Research, as Chairperson.

The discussion of this conference resulted in a broad understanding among participants on Conversion and related environment issues, constituting the "Dortmund Declaration".

1. Disarmament is an essential pre-condition to respond to increasing economic, social and financial needs and to reverse trends of environmental degradation at local, national and international levels. However, to implement disarmament efforts and to reduce the international proliferation of arms, it is indispensable to introduce policies and programmes of conversion of military production and technology for its civil use, preferably under the governance of the United Nations.

2. Policies and actions in this regard have to be based on an increased public awareness of the inter-relationship of disarmament and conversion of military production.

3. The integration of military production into the civil economy requires a public transparency of this sector and its related policy decision-making. This also applies to the sale and trade of arms precondition to

successfully monitor the flow of arms.

4. The gradual reduction of subsidies and privileges inherent to military production in many countries has to acccompany conversion efforts.

5. Conversion policies have also to address the gradual transformation of military forces and their equipment into mobile civil units to prevent and combat natural disasters under the governance of the United Nations and following the concept of an international "green helmet" army.

6. Policies of military conversion should address environmental issues. Specific conversion activities should be based on technology assessment exercises taking into account environmental concerns.

7. Restructuring and modernization efforts involving discard of obsolete military equipment should not be counted as a conversion process.

8. The transformation of military related research and science and technology activities has to be addressed as a central area of conversion policy.

9. Scientific institutions should be supported worldwide in their efforts to give special attention to the research on economic, technological, sociological, organizational and environmental aspects of conversions and to seek new forms of international research partnership, and in particular, in co-operating with institutions in developing countries and countries with economies in transition.

10. The 'peace dividends' result from the end of the cold war. However, a considerable amount of resources have to be allocated immediately for the safe disposal of arms and other related conversion tasks. Principally 'peace dividends' should be made explicitly and increasingly available to finance development co-operation and international programmes to respond to global environmental threats.

11. International development co-operation should recognize the legitimacy of conversion efforts and support related activities in developing countries by providing technical expertise and financing on concessionary terms.

12. The United Nations should undertake follow-up efforts to disaggregate sector - specific and region-specific issues for further clarification and new initiatives in conversion process along with recommendations of the Dortmund Conference. All constituencies either involved in conversion process or affected by it should be encouraged to participate in the coordinated efforts under the United Nations auspices.

# SPECIAL ADDRESSES

# WELCOME ADDRESS
# BY

**Anke Brunn**
*North Rhine Westphalian Minister of Higher Education and Research*

Only a few months ago, "conversion" - the transformation of military into civil potential - seemed to be a mere academic concept, a utopia in the face of numerous bloody conflicts all over the world.

Nevertheless, at a time when the Gulf crisis was becoming a real threat to international peace, I supported the idea of a UN conference on the theme of conversion. During preparations to the conference, the Gulf War broke out and eventually ended. Tens of thousands of human beings lost their lives because of this conflict; the environmental damage is immense; the cost runs into many billions of dollars.

Given the facts of world politics, conversion seemed to be at the bottom of the political agenda.

Which is why I was all the more pleased that we were able to make North-Rhine Westphalia and the City of Dortmund the venue for our UN conference and thus the centre of conversion. Well over 200 experts from more than forty countries all over the world came to the Ruhr to discuss how to put war materials to peaceful use.

Scientists, politicians and representatives from the field of industry and commerce sat at one table. The conversion of aviation and space technology, the cleaning up and rehabilitation of former military sites, newcareer concepts for military personnel, the special problems of the underdeveloped countries... all these were central topics for discussion.

The great commitment on the part of the experts, the vitality of the talks and the encouraging feedback from both media and public only go to prove what we had already surmised: conversion has become a topic of world wide significance -- one that will be crucial to the future of mankind.

A theme that will not remain in the realm of worlds, I trust, but one that will be transferred into action. The basis for such action is the Dortmund Declaration, a compilation of the results of the UN Conference results formulated as demands on the growth market "conversion", scientific community conversion and the responsibility of politicians throughout the world.

Conversion has become a self-evident topic for discussion. It must now become a reality, world wide.

# WELCOME ADDRESS BY

**Günther Samtlebe**
*Lord Mayor of Dortmund*

Right honorable Prime Minister, esteemed Minister, right honorable Chancellor Brandt*, your Excellencies, Ladies and Gentlemen...

Only recently, I read a sentence which perfectly sums up both the general situation today and this conference:

"History is a prologue".

We are presently engaged in writing a new book of international relations between countries, states and people. Let us consider this conference as the beginning of the first chapter of a new, joint work.

I am very pleased, Ladies and Gentlemen, to be able to welcome you today to this International Conversion Conference organised by the United Nations and held here in Dortmund.

On behalf of all the people of this town and of this region, I bid you a very warm welcome. Along with the conference organisers, the Institute for Environmental Protection at our university, we are very proud that this conference has aroused such great interest worldwide, and that so many persons of international repute from so many different countries are participating in it.

We consider it a special honour and a great distinction that the Prime Minister of North-Rhine Westphalia, Dr.h.c.JohannesRau, has agreed to act as patron to the Conference. Mr. right honourable Prime Minister, I extend to you my whole-hearted welcome.

Ladies and Gentlemen, it is also my special pleasure to be able to greet in our midst Chancellor Willy Brandt, Deputy President of the North-South Commission and one of the most prominent mentors and forerunners in the process of reapproachment.

# GREETING ADDRESS

Внешнеполитическая ассоциация

103064, Москва, ул. Елизаровой, 10
Тел. (0-95) 975-21-67
Факс (0-95) 975-21-90

Foreign Policy Association

103064 ul. Elizarovoy 10 Moscow
Tel. (0-95) 975-21-67
Fax (0-95) 975-21-90

*Dear Friends!*

*I realise the importance of your conference which is dedicated to one of the most vital issues confronting the word community today. I share whole heartedly the aim of the conference -- to direct conversion to the course of social and economic development and the protection of the environment. And I shall do my best to contribute to the achievement of this aim.*

*It is not a secret that in different states, including the members of the Commonwealth of Independent States, a huge quantity of lethal weapons capable of exploding our planet many times, has been accumulated.*

*The process which took place in the former Soviet Union and in the countries of Eastern Europe opened prodigious opportunities for international humanitarian cooperation including cooperation in the field of conversion. To bring local, current and temporary political and economic interests into the strategical direction of the development of world community is, from my point of view, the aim of all social and political forces.*

*However, disarmament, and especially conversion, is a very complicated, laborious and expensive process. And one state alone is not likely to carry it out without international cooperation and aid.*

*Hence there is no alternative to the multilateral collaboration in the field of conversion if we want to include it in the interests of development and protection of environment.*

*Let me wish you real success in your work, as well as health and prosperity. I avail of this opportunity to express my deep gratitude to the leadership of Germany and the Government of the Land of West-Vallen for the Organisation of this remarkable conference and for the invitation to take part in its work. I regret that due to several unavoidable reasons I am unable to participate in it.*

***Eduard Shevardnadze***
*Moscow*
*02 22 92*

# OPENING ADDRESS BY

**Dr.h.c.Johannes Rau**
*Prime Minister of North-Rhine Westphalia*

## I

How will this century close? As it began, at war? Or in a state of peace recognised on all sides?

A most difficult question to answer. Even the old certainties of the Cold War have vanished, making way for new uncertainty and confusion. And as to whether the situation will turn out for the good, we vacillate between reasonable hope and justified doubt. We are truly on a roller-coaster ride of the emotions. Far too much is happening in far too short a time. For example:

- joy at the downfall of totalitarian dictatorships, at the victory of liberty and democracy in East Europe;
- relief at the end of the East-West Conflict and at the fading away of an atomic threat that, over the years, had seemed to be one of Nature's immutable laws;
- shock at the military dispute in the Persian Gulf and deep concern about the threat to Israel's very existence;
- and grief at the civil war on the Balkan Peninsula, at the ethnic and national conflicts in the Caucasus and elsewhere.

We have been forced to realise that the end of the East-West Conflict would not automatically mean the end of war and violence inside Europe or outside it.

Nevertheless, expectations have recently been growing for peace, security and the world-wide implementation of human rights.

"Swords to ploughshares" -- today, this biblical theme is no mere dream but a justified hope for many people. With the end of the Cold War, we may assume - in East and West, North and South - that it will be possible in future to use more sensibly the finance that has been wasted on weapons and armies in the past... to fight starvation, sickness and need, to safeguard the natural basics of life.

## II

Conversion -- originally a concept by experts for experts -- might well become a keyword of the decade. Indeed, as a central task, conversion is already an item on the agenda of world politics.

We know today that the success of international detente depends, to a very great extent, on what means and at what cost disarmament will be translated into practice.

Of increasing importance is the question:

How can we - on a global basis - utilise military capability for humane purposes, dismantle military research and persuade arms manufacturers not to make weapons that are no longer needed but, instead, to produce those civil goods that we lack?

That is the theme of conversion, the theme of the international conference that we are opening today in Dortmund.

It is the third event of its kind, after Moscow and Peking, with the participation of the UN. It is the first of its kind in whichenvironmental issues have been given so much attention. And it is the very first time that the United Nations have organised an internal expert conference in conjunction with a German federal state.

This fills me with great pleasure. It makes me proud. I am convinced that those of you who have come from more than thirty different countries to the Ruhr Area to be our guests will take a positive impression of North-Rhine Westphalia back with you.

The people here in the "Revier", the Ruhr Area, are very distinctive -- as you will experience for yourselves. If less urbane than the people in other metropolitan areas, they are more down to earth, pragmatic and reserved. Once they've opened up to you, you can count on their goodwill and friendship -- unlimitedly so. Solidarity and consideration for others have always been a long and powerful tradition in the Ruhr and Rhine areas.

## III

It is also my opinion that the UN choosing North-Rhine Westphalia was not coincidental and that there are very good reasons for holding an international conversion conference in the Ruhr Area and in Dortmund in particular.

This region has, in a matter of only a few decades, undergone a double structural change successfully. A modern industry and service sector region has emerged from the former armaments forge that supplied two world wars and from an area that for decades was characterised by coal and steel. Now the International market leader in the development of state-of-the-art technologies for the environment and the future, the Ruhr Area today - as I never tire of saying - hasthe highest number of universities and polytechnics in Europe.

As conversion experts, you will be interested to know that the universities of North - Rhine Westphalia have no place for military research -- only conversion research focused at various institutes throughout the state. For example,

- the Rhine-Westphalian Institute for Economic Research in Essen;
- the Institute for Development and Peace at the Combined University and Polytechnic of Duisburg;
- the Institute for the Law of Peace Maintenance and Humanitarian Law at the Ruhr University of Bochum;
- the Institute for the Aerospace Law at the University of Cologne;
- and, of course, the Institute for Environmental Research at the University of Dortmund which - under the directorship of Professor Karpe - has excelled itself with the idea and the implementation of this conference.

Conversion in North-Rhine Westphalia is not just a research issue. It is a current aspect of state policy since the withdrawal of troops means the loss of jobs and the release of land which, more often than not, is ecologically polluted. The realignment from military to civil production hits hard at companies and employees, sometimes affecting the lifeblood of entire regions.

The State Government of North-Rhine Westphalia is tackling these problems head-on within the overall framework of its structural policy for economic and ecological renewal.

Our foreign guests will be able to see this for themselves during and after theconference when they visit research and technology centres as well as private companies.

## IV

Where global threats vanish, money for armaments can be saved and given over to better purposes: the peace dividend. Measured against military budgets whose size makes you dizzy to contemplate, even small reallocations can have a great effect.

In a survey conducted in October 1991, the famous Stockholm Peace Research Institute calculated that global expenditure on armaments ran to the staggering sum of 1,000 billion dollars.

A mere five days of that amount would be sufficient to provide the population of the Third World with sanitation facilities and a clean water supply. Nine hours of armaments spending would finance the vaccination of all children and so prevent a million deaths a year.

Both Willy Brandt and Richard von Weizsäcker have pleaded that a third of the peace dividend be put into the development budget. In terms of target and long-term perspective, this is a sensible proposal even if there is still a lot of persuasion to be done before realisation of the proposal nationally and internationally.

After all, world peace will not become a reality without the development of the Third and Fourth World. And without an agreement between North and South as to the use of nature - as exemplified by the tropical rain forests - the environment will suffer. There can be no doubt (and this will be shown by the UN Earth Summit to be held in Rio deJaneiro this June) that peace, development and environment policies on a global scale form the magical triangle of world security policy.

These objectives can only be reached jointly; on our own, we will all fail.

## V

As for the state of the world today, it may safely be said that the old threats have been replaced by new dangers:

- In Eastern and Southern Europe, nationalism, chauvinism and racism have flared up again, threatening to fill the vacuum that communism has left behind.
- On the Balkan Peninsula, in the Caucasus and in other regions of Eastern and Southern Europe, ethnic and national conflicts have broken out, with much loss of blood.
- There is the justified concern about weapons of mass destruction being passed on to an increasing number of states, possibly with the assistance of unemployed weapon producers and atomic scientists from the former Soviet Union.
- Waves of migration all over the world are taking place for reasons of hunger and needs as well as political, ethnic and religious prosecution.
- Militant Islamic fundamentalism is spreading like a forest fire, usually caused by material need and desperation.
- The world is being inundated with drugs, accompanied by a growth in organised crime.
- Virtually all the ecological data available to us (not just the ozone values) and the alarming number of famines and natural disasters indicate that the limits ofenvironmental pollution have already been reached.

## VI

No single person can solve the world's problems overnight. Nevertheless, it is our duty to seek solutions, for I believe that it still isn't too late if we act quickly -- in fact, quicker than some people would have us believe.

The rich industrialised countries of the West are particularly called on for active responsibility and solidarity. For those of us in the Community, European policies may no longer be confined to mere EC policies. Which is why we must ensure that freedom and democracy flourish in Eastern Europe. We can best do this by helping to improve the environmental and living conditions there and by securing the livelihoods of the people. We would, in doing so, build solid and dependable foundations for our shared houses as well as acting with a sense of political responsibility and in our own interest

-- as those people with vision have already perceived.

A strong region in the heart of Europe, North-Rhine Westphalia accepts its Pan-European responsibility. The State Government of North-Rhine Westphalia is committed to aid in Poland, Romania, Hungary, Czechoslovakia, Bulgaria and, of course, in the Community of International States. At the forefront here is humanitarian, immediate relief -- above all, an improvement to the often disastrous medical provision at grass-roots level.

North-Rhine Westphalia is also helping with the cultivation of business and economic contacts. It is promoting practical reconciliation and understanding with thosepeoples in Eastern and Southern Europe who were subjugated and enslaved by the Germans during the Second World War.

Having said that, we in the West should not forget the South in helping the East. We must show the peoples of the South by our deeds that we do not want a Europe that confines itself at best to inner-European balancing out. The people of the South do not expect us to give more; they expect us to learn to take less. This applies in particular to our excessive demands on nature.

One thing is clear: the world, including the industrialised North, cannot afford the loss of another decade for the South.

## VII

I am reminded of a saying made many years ago by the American dramatist Thornton Wilder: "The things common to all people are beginning to take on much more importance than those things that separate them".

This is true and because it is so, because politics, economics, technology and science are all globally interwoven, we need a New World Order -- an order that could be founded in the United Nations. The peacemaking and peacekeeping instrumentation of the United Nations must, therefore, be considerably strengthened.

The summit meeting of the UN Security Council at the end of January 1992 was a big step forward with its emphasis on preventive diplomacy, on joint and prompt arbitration in regional conflicts and on the resolution to campaign against the further spread of atomic, biological and chemical weapons. It was right to expand the definition of security so that now, under the new definition, UNaction is possible in the event of human rights violation, environmental disaster and refugee movement.

That which has already been achieved must be safeguarded and built upon. All Europeans - and, of course, we Germans - are called on to make the UN more capable of acting so that it can secure and promote peace and development throughout the world. On our own continent, we can do this by strengthening the role of the Conference for Security and Cooperation in Europe.

I most heartily applaud the acceptance of the CIS states, the former Soviet Union, under the roof of the CSCE.

What we now need - and here I agree with Foreign Minister Hans-Dietrich Genscher - is the better coordination of measures for avoiding conflict and for handling crises and the institution of a Pan-European Court of Arbitration for the peaceful settlement of disputes.

I know that CSCE blue helmets have been proposed and I appreciate the thinking behind such proposals. But, at the moment, I would give greater priority to a European Peace Corps precisely because peace and development are so inextricably linked. We now need - and I have said this persistently - rapid deployment of troops not only for military purposes but also so that we can help - in word and in deed - the countries of Central, Eastern and Southern Europe with their process of development in the fields of administration, law, education, health and business.

The end of the East-West conflict has been of foremost benefit to Germany -- unexpectedly, it made German Unification possible. The cessation of the Cold War has,however, also presented us Germans with a new and greater responsibility for Europe and the world. "Many non-Germans" wrote the Swiss author Adolf Muschg recently "view the enlarged Germany with a feeling of concerned but great hope". I say we should not disappoint such hopes. For this reason, we Germans must begin with the disarming of the *Bundeswehr* and, as far as is feasible, stop the export of German weapons.

The State Government of North-Rhine Westphalia has repeatedly pointed to the urgency of using all the legal possibilities open in order to prevent the trade in death.

**VIII**

Here the wheel turns full circle. To ensure that disarmament is not only politically but technologically possible, and national economics no longer have to rely on weapon exports, we need conversion experts. In addition to the politicians' responsibility, the competence of scientists with their wealth of ideas is required.

In this way, the yield from the peace dividend will be wisely spent on necessary and correct measures such as dismantling of weapon arsenals, the cleaning up of militarily caused damage to the environment and the limitation of negative employment effects.

Above all, conversion is a chance for development and the environment if we succeed in deploying the highly advanced technological and logistic potential of the national armies more efficiently than in the past to fight starvation and natural disasters, or if we succeed in feeding the results of military research into the development of civil information technologies, into medical analysis and into biotechnology.

Over the next few days, here in Dortmund, scientists from all over the world will discuss these and other possibilities - as well as the basic conditions for conversion - with representatives from the fields of politics and business in an attempt to draw up suitable strategies for concerted action. I wish you all good fortune and great success.

I wish you this all the more since - apart from specialist detail - nothing less than the global implementation of the human right to an existence in peace and liberty, the human right to a life free of hunger and the human right for a clean and non-poisoned environment is, in fact, your unwritten agenda.

I am glad to say that your conference has already "fallen on fertile ground" in North-Rhine Westphalia. As soon as this conference is over, an interdisciplinary scientific Conversion Study Group will commence work. We in North-Rhine Westphalia link the setting up of this study group with the hope that we will regularly be able to welcome the United Nations and conversion experts from all over the world as guests to our state. Our hail of welcome in their coalmining tradition is *GLUCKAUF* or good luck!

# OPENING ADDRESS BY

**Mr.Kofi A. Annan**
Assistant Secretary - General
*Office for Peace-Keeping Operations, United Nations*

It is an honour for me to represent the United Nations at this conference which will discuss one of the most important and topical issues of the day, namely "the process of transition from military to civilian economies".

This is a very timely opportunity to take stock and to examine ways in which the efforts of the international community may be reoriented as we move away from the militarism and threats of the cold war.

The world is looking forward with hope and apprehension towards the establishment of a new system, perhaps of a new order, in which peace, co-operation and the rule of law will prevail.

On the other hand, as we are all aware, the end of the cold war has not put an end to regional and ethnic conflicts, to the violation of human rights and to the misery inflicted by poverty and under-development.

There is a vast agenda ahead of us, the outline of which we are only beginning to discern. This conference may well become one of the pioneering efforts in reorienting our perspective on some of the major problems of the world and the shape of the future. In so doing, this conference may also help to define the new emerging role of the United Nations.

We are now in a position to decrease military budgets for the benefit of development. The linkage between peace,political stability and sustainable economic development cannot be over-emphasized. As the Prime Minister of Japan, His Excellency Kiichi Miyazawa, stated at the summit meeting of the Security Council on 31 January 1992 in New York "The trend toward world peace that we are witnessing today will not endure unless the *dividends of peace* are shared by all, but especially by the peoples of the South who are beset with famine, poverty, disease and other hardship."

The task of converting the worldwide military industrial complex to civilian uses can only be achieved in a climate of stability and peace where conflicts between and within countries can be effectively controlled. In this context, it will also be necessary to reduce the depressingly large proportion of governmental and human resources, especially among the developing countries, that is devoted to the production or the purchase of arms. It is not possible to abstract away the emergence of an economically and ecologically sustainable development process from the maintenance of peaceful conditions and the control of international conflicts. In fact, the latter is a necessary condition for the former.

Instead of entering a post-nuclear age, we are now witnessing the possible beginning of an age of dangerous and uncontrolled proliferation of regional conflicts and civil strife in several areas. Such strife tends to escalate into regional and international threats to peace and security. Moreover, theweak economies of the South remain in the grip of deepening crisis, while growing poverty -- a tragically potent source of instability -- is becoming a menace to recent gains in democracy.

Accordingly, when all is said and done, we must return to the primary purpose of the United Nations, namely the maintenance of international peace and security.

We are rethinking the ways in which the United Nations can approach this task. By all accounts, the most innovative contribution of the United Nations in this realm has been the development of peace-keeping operations. One of the first things that the transformed international system has been required to do, is to broaden the scope of this instrument of peace.

Instead of peace-keeping limited to military observation and interposition, we are now thinking in terms of peace-making and peace-building, the supervision of free and fair elections, the monitoring of the observance of human rights and in certain cases, the assumption of quasi-governmental responsibilities.

There seems to be every reason to believe that the increasing demand for launching unanticipated peace-keeping and related operations will continue.

Furthermore, the United Nations has been called upon to give top priority to preventive diplomacy to help avert future conflicts and to strengthen its machinery for large-scale humanitarian assistance.

As the leaders of a prominent group of countries phrased it "the conditions now exist for the United Nations to fulfill completely the promise and vision of its founders".

In the last year or so, the United Nations has been entrusted with a multitude of tasks. It has launched no less than seven new operations: along the Iraq-Kuwait border (UNIKOM), in El Salvador (ONUSAL), in Angola (UNAVEM II), in Central America (ONUCA), in Western Sahara (MINURSO) and in Cambodia (UNAMIC). As you all know, a major peace-keeping operation is about to be launched in Yugoslavia in dauntingly difficult circumstances. And let us not forget, of course, the five other peace-keeping operations already in place (UNDOF) in Syria, (UNIFIL) in Lebanon, (UNTSO) in Israel, (UNMOGIP) in India/Pakistan and (UNFICYP) in Cyprus.

In addition, a highly complex set of tasks is being undertaken by the United Nations in Iraq pursuant to resolutions of the Security Council. This includes the destruction under United Nations supervision of Iraq's weapons of mass destruction, a boundary demarcation between Iraq and Kuwait, the return of all Kuwait property seized by Iraq and the setting up of a Compensation Fund for the payment of claims for loss and injury resulting from the invasion of Kuwait.

Finally, the Organization must pursue actively its efforts in areas such as the environment, drugs control, human rights and humanitarian relief. And, of course, there is the continuing preoccupation with the traditional economic and social development issues.

Turning once again to the main issue of maintenance of peace, recent developments have again highlighted the difficulties that the organization faces in mobilizing resources, personnel, organizational capacity, logistical and security arrangements to deal withpeace-keeping, humanitarian and other operational activities.

This situation has at times made it difficult for the United Nations to respond in a timely and effective manner to unanticipated humanitarian emergencies. I would suggest that what is needed in this regard is a creative approach to conversion; not only the conversion of industrial plants but the reorientation of the purposes of organized military manpower. What is also needed for this purpose is to find ways to attract the participation of governments which have not hitherto been in a position to do so for constitutional, political or financial reasons.

In many cases, the one mechanism which most governments have in a state of readiness is the military, although certain non-military governmental and/or non-governmental entities may likewise be available in certain countries in a state of constant readiness. Even some of the major countries, when confronted with severe disasters, have found it necessary, in providing assistance, to resort to military personnel and logistic facilities in order to respond with the speed that the event calls for. In situations of this kind, the United Nations, despite its substantial achievements, has found itself handicapped as it does not have adequate capability to draw upon.

What is needed when disaster strikes is an enhanced capability for the immediate deployment of organized personnel, logistic support for such deployment and the high speed delivery of humanitarian relief supplies in the field.

Heretofore, the United Nations has used military and related civilian units of Member States only in the context of peace-keeping. A complementary idea should be explored. This would involve the setting up of the "United Nations Humanitarian Assistance Teams" (See Attachment I & II).

These teams would be designed as an international corps, consisting of military and civilian personnel, to be called upon for emergency rapid deployment, under United Nations auspices.

Governments would be invited to participate in this scheme, of course within the limits of their capabilities, by earmarking trained and equipped component or components of their military and/or other standing organized governmental bodies, including national search and rescue teams, to be available on a continuing year-round basis for immediate intervention at the request of the Secretary-General.

This brings me inevitably to the problem area which has been the cause of profound concern to the Secretary-General and to his predecessor, namely, the financial situation of the United Nations. None of the activities that we are thinking of, no matter how indispensable, can go forward without the necessary resources, human, financial or material.

It is particularly ironical that, at a time when the international community is turning increasingly to the United Nations, the difficult financial situation of the organization should be exacerbated by a lack of commitment to provide the Organization with the resources necessary to fulfill its responsibilities.

As of 31 December 1991, 102 Member States had yet to pay their 1991 contributions. (Unpaid assessments totalled $ 816,558,000). It is in that context, the then Secretary-General, Mr.Perez de Cuellar, put forward to the General Assembly a set ofproposals designed to induce Member States to pay their contributions on time and in full. One of the Secretary-General's proposals was to apply interest charges to outstanding contributions of all Member States in arrears.

The most innovative idea was to establish a multi-billion Peace Endowment Fund which would cover the costs of authorized peace-keeping operations, conflict resolutions and related operational activities designed to contain and control developments that may threaten international peace and security.

I hope that the economically advanced Member States, especially those that bear a particular responsibility because of historical and political reasons for the success of the Organization, would not only pay their assessed contributions but also make generous voluntary contributions to the endowment fund.

As this conference is about to embark upon the consideration of the major aspects of post cold-war conversion, we should bear in mind the significance of the present historic moment, which represents the fulfillment of so many dreams. It is now up to the international community to seize the moment and find ways and means of meeting the challenges ahead.

I am pleased that this conference is taking place in Germany, which has a long-standing experience in dealing with humanitarian emergencies and has consistently been responsive to disaster relief needs of countries around the world. I am confident that Germany will persevere in its generous and dynamic efforts and will maintain its leadership role in the areas of humanitarian relief, economic development, industrial and organizational conversion.

# ATTACHMENT I

**A Proposal for the Establishment of the United Nations Humanitarian Assistance Teams, (UNHAT)**

**Introduction**

Recent developments leading to the proliferation of demands for United Nations humanitarian, peace-keeping and other operational activities have again highlighted the difficulties that the United Nations faces in mobilizing resources, personnel, organizational capacity, logistical and security arrangements to deal with such emergencies.

On the one hand there is the problem of raising the necessary voluntary contributions from Member states that are faced with increasing financial demands and, given their own international pressures and other constraints, are not always able to respond promptly to those demands. This situation has at times made it difficult for United Nations to respond in a timely and effective manner to unanticipated humanitarian emergencies. Consequently, there is an imperative need to find creative ways of raising funds and eliciting other contributions.

There is also an urgent need to find creative means of attracting the participation in these activities of Governments which have not hitherto been in a position to do so for constitutional, political or financial reasons.

In many cases, the one mechanism which most Governments have in a state of readiness is the military, although certain non-military governmental and/or non-governmental entities may likewise be available in certain countries in a state of constant readiness. Even some of the major countries, when confronted with severe disasters, have found it necessary, inproviding assistance, to resort to military personnel and logistic facilities in order to respond with the speed that the events called for. In situations of this kind, the United Nations, despite its substantial achievements, has found itself handicapped as it does not have adequate capability to draw upon. This shortcoming has been the subject of criticism in some quarters.

However, the United Nations system has substantial operational capability for emergency disaster relief, including close co-operation with the activities of Non-Governmental Organizations (NGOs) in the field. This capability should be built upon and, as necessary, supplemented and improved, chiefly for the purpose of high-speed initial intervention. More specifically, that is needed when disaster strikes, is an enhanced capability for the immediate deployment of organized personnel, logistics support for such deployment and the high speed delivery of humanitarian relief supplies in the field.

**New Approach**

Heretofore, the United Nations has used military and related civilian (e.g. civilian police) units of Member States only in the context of peace-keeping. However, a complementary idea should be explored. This would involve the setting up of "United Bands" (because of the blue arm bands that they would wear). Briefly, this new concept will work in the following manner:

1. United Nations Humanitarian Assistance Teams (UNHAT) will be designed as an international corps, consisting of military and civilian personnel, to be called upon foremergency rapid deployment, under United Nations auspices. UNHAT will be so organized as to enable the United Nations system to provide humanitarian relief in any

part of the globe within twenty-four hours in the event of major natural or man-made disasters or any other life-threatening situation.

2. All Governments will be invited to participate in the UNHAT system by earmarking trained and equipped component or components of their military forces and/or other standing organized governmental bodies, including national search and rescue teams, to be available on a continuing year-round basis for immediate intervention at the request of the Secretary-General. Similar arrangements could be initiated directly with certain NGOs which have a well established capacity for rapid disaster relief intervention.

3. UNHAT will constitute a permanent element of the United Nations system for humanitarian assistance.

4. UNHAT will, in principle, be unarmed.

5. Extra costs incurred by Governments that are directly attributable to their participation in the UNHAT system, will be treated for budgetary purposes as voluntary contributions and should be in addition to -- and not in lieu of those which they currently provide for humanitarian relief. In this way UNHAT can provide the opportunity for Governments who have hitherto not been able to participate in such humanitarian activities, to join in the collective burden sharing.

6. Experience indicates that a certain amount of cash is needed to meet the cost of the most immediate and urgent relief supplies required on the ground, (e.g. blankets,shelter, tools etc. and other start-up costs). For this purpose, it is proposed that a humanitarian fund be established at a reasonable level to provide for such items, possibly based on a one time assessment and subject to replenishment as necessary.

7. The United Nations will provide permanent central machinery to ensure international co-ordination of these components and related arrangements. Such co-ordination will be designed to ensure that, when deployed, the UNHAT components will be able to work together smoothly. This may be achieved by way of training, visits by briefing officers, distribution of briefing kits and practice in the use of compatible equipment and communication and transport facilities etc.

8. A suitable co-ordinating unit will be designated by the Secretary-General. It could be located in UNDRO or be part of a possible oversight office of a United Nations Humanitarian Operations Co-ordinator.

9. In case of major disasters, it should be understood that the deployment of UNHAT will be among the first in a series of measures undertaken by the United Nations in the field. This humanitarian corps will be called upon in the very first phase of the emergency, which will be defined as the time span during which human lives are at risk in the immediate aftermath of a disaster.

10. UNHAT's intervention will cease after the first phase of the emergency has been addressed, that is after the necessary steps for the longer-term co-ordinated involvement of the appropriate agencies and programmes of the United Nations system have been taken. Its responsibilities and facilities will then be handed over to the appropriate agencies and programmes of the UnitedNations system, as designed by the Secretary-General. A smooth transfer of responsibilities and exchange of information must be ensured.

11. The co-ordinating office mentioned in paragraph 7 above will maintain a central up-to-date register of the personnel, equipment, supplies, logistics and transport support that could be made available for immediate deployment by Governments participating in this programme.

12. As soon as UNHAT's involvement has been decided upon, an initial list of requirements will be immediately established on the basis of existing UNDRO practice. These requirements will be matched with the

above-mentioned register with a view to securing rapid availability of required personnel, facilities and supplies.

13. When a disaster occurs, the Secretary-General will designate the contributing Governments to be requested to provide the assistance required.

14. The Secretary-General may also wish to consider possible security aspects, including the relationship between humanitarian and peace-keeping operations.

15. The Secretary-General will, in due course, request the inclusion of an additional item in the agenda of the General Assembly, with attached explanatory memorandum, outlining the substance of the UNHAT's proposal.

16. Certain principles and guidelines applicable to the structure, operations and status of UNHAT are set forth in the attached annexure.

**Observations**

This arrangement will present a number of advantages:

a. These units will be available as soon as needed and will have the capability of operating in an organized fashion. This will *inter alia* avoid the difficulties of building up new organizational structures, so to say from scratch, and in the worst conceivable circumstances.

b. Countries may find it feasible financially to make available existing elements of their military and related civilian effective. By contrast, recent experience, especially with regard to man-made emergencies, has highlighted the difficulties encountered by many Governments in providing voluntary contributions in kind or in cash to the United Nations.

c. It will also facilitate, in some cases, the involvement of certain countries which, in the absence of a United Nations umbrella, would find it difficult, for various reasons, to play a suitable role in dealing with humanitarian emergencies abroad.

d. It will make it possible for certain countries to employ unarmed military and related civilian units under United Nations auspices while avoiding the constitutional and political difficulties of involvement in United Nations peace-keeping operations.

e. Their presence will provide a sense of security among affected populations and thus serve as a confidence-building measure that could, in the event of man-made disasters involving international orcivil conflicts, facilitate appropriate peace-making efforts.

It is evident that this proposal would serve no useful purpose if it were to result in Governments that contribute components to the "humanitarian assistance teams" reducing their contributions in cash or kind. Therefore, the earmarking and deployment of the humanitarian teams should in no way be construed as a substitute for the other contributions that Governments have heretofore and will in the future make in connection with emergency humanitarian situations.

The scheme suggested above will, of course, have to be properly integrated in the overall United Nations system machinery for dealing with humanitarian problems.

In the present circumstances, when the demands for emergency humanitarian actions seem to overwhelm the capacity of the international community to respond, the introduction of the new approach would undoubtedly reflect favourably on the ability of the United Nations to address this problem in a new and imaginative manner.

# ATTACHMENT II

### Principles and guidelines for UNHAT

a. UNHAT must have at all times the full confidence and backing of the General Assembly and (as appropriate in the case of operations involving security aspects, of the Security Council). It will expect to operate with the full co-operation of the host governments, which will also ensure that the full co-operation of the relevant local authorities will be forthcoming.

b. UNHAT will be under the command of the United Nations vested in the Secretary-General, under the authority of the General Assembly (and, as appropriate, of the Security Council). The command in the field will be exercised by a Team commander appointed by the Secretary-General according to reporting arrangements to be determined in due course. The Secretary-General will keep the General Assembly (and, as appropriate, the Security Council) fully informed of developments relating to the functioning of each operation and each UNHAT team. All matters which could affect the nature and continued effective functioning of any UNHAT operation will be referred to the General Assembly (and/or, as appropriate to the Security Council) for appropriate action.

c. UNHAT and each of its teams will enjoy the freedom of movement and communication (including code) and other facilities necessary for the performance of their tasks. UNHAT and its personnel should be granted all the relevant privileges and immunities provided for by the Convention on the Privileges and Immunities of the United Nations.

d. UNHAT and its teams shall operate at all times separately from the armed and unarmed agencies of the government(s) concerned, but will co-operate closely with such agencies in the performance of disaster relief functions. UNHAT will carry out the instructions of, and will report to, the Secretary-General only and will at no time receive or carry out the instructions of any government.

e. In performing its functions UNHAT will act with complete impartiality and in conformity with the principles of the Charter, of the Universal Declaration on Human Rights, and of the Geneva Conventions. It will in no circumstances permit its functioning to result in discriminatory treatment based on racial, national, ethnic, religious, sexual, party or communal criteria.

# KEYNOTE ADDRESS

# THE NEW WORLD ORDER - THE IMPACT OF ITS SECURITY AND ECONOMIC ATTRIBUTES

**Prof. Essam E. GALAL**
**Egypt**

May I express my gratitude for the honour bestowed on me by the organizers of this conference, in addressing this distinguished gathering.

The subject of this conference reflects the realities of the unique opportunities and challenges facing our world at this historical juncture. While our present knowledge and capacity for prediction may not be enough to define precisely the scope of the challenges or the feasibility of the opportunities, it is certain, that our profound understanding and determined concern, are going to be decisive factors in warding off the threats and seizing on the benefits. Current international change has no built-in guarantee of progress. This is the predicament of the New World Order.

The focus of this conference should be the challenge that Twin MADs namely, the MAD (Mutual Assured Nuclear Destruction) and the MAD (Mutual Assured Environmental Devastation), and the associated recklessness, instruments and practices are still very much in existence, inspite of the mounting justification and motivation for their demise.

With the disintegration of the Warsaw Pact, the fragmentation of the USSR and their ideological, political and economic transformation, the very raison d'etre of the nuclear confrontation has been erased. General Colin Powell, the US chairman of the Joint Chiefs of Staff puts it succinctly "I'm running out of demone. I'm running outof villains. I am down to Castro and Kim Il Gung". Even Mr.Il Sung has cut the villains by half by agreeing to nuclear control. Mr.Yeltsin has undertaken to point his 13,600 nuclear heads away from the US targets. It will certainly stretch his ingenuity to find alternative targets on this globe.

Robert McNamara the former US Defence Secretary in answering the question "How much is enough"? suggested in the early 60's as an upper limit that only 100 one-megaton nuclear bombs, can destroy 20-25% of the soviet population and 50% of its industrial capacity. Herbert York, former Director of Defence Research & Engineering recently concurred with this estimate of a minimum deterrent of 100 nuclear warheads.

Buttressing all these arguments for change, there are still two mundane considerations that should be within the reach of politicians' grasp. These are the economic crises and the public glamour for the "Peace dividends" including growth and environment protection.

It is both edifying as well as a salutary warning for this conference to contemplate the responses to all these irresistible pressures for change.

In his State of the Union address, Mr.Bush indicated his readiness to cut by more than half all strategic warheads, that is beyond the 30% reduction agreed by START, thus bringing down the existingarsenals of around 12,000 each to around 4,500. Mr.Yeltsin, perhaps because of a much more crushing economic crisis and internal political turmoil had no hesitancy to go beyond the suggestion of Mr.Bush, by proposing a ceiling of 2,500 warheads apiece. This proposal was rapidly dampened by the US Secretary of Defence although the START 30% reduction took several years of negotiation and will take several years to implement and the "Alarming" Yeltsin ceiling is well within the scope of perpetuating the nuclear bipolar hegemony and its MAD doctrine i.e. sustaining the Old Order.

Mr.Yeltsin was not about to forsake his claim for supremacy either. He proposed a USA-USSR partnership in the Star War programme to transform it into a "global" missile defence system, a timely reminder of the "parity" as the foundation of the cherished bipolarity.

Inspite of the credible claims of amity and of enmity, the mundane attributes of the rivalry surprisingly seem still to have a volition of their own. Mr.Bush targeting the landbased Urved missile, the basis of USSR nuclear strategic forces for the major cuts and Mr.Yeltsin aiming at across the board cuts targeting the sea-based advantage of the USA.

The disorientation and confusion of the transition, whatever its ultimate outcome, manifests itself in some amazing and possibly disquieting trends. As an example of the concerted manipulations by the nuclear super powers one may point out the efforts to ensure the Russian Federal Republic nuclear hegemony over the commonwealth in order to contribute to regional and non-proliferation stability. In the longer term, however, it will revivenuclear bipolar rivalry and hegemony. U.K. and France too, still adamant on keeping their nuclear arsenals (to "eat at the top table" in MacMillan's words) are equally keen to safeguard their foothold in the order, inspite of their difference on the future role of USA in Europe. Mr.Mitterand on the 10th of Jan.1992 noting that two of the twelve members of E.C. have their own nuclear forces, he stated "they have a clear doctrine for their national defence. Is it possible to imagine a European doctrine? That question will very quickly become one of the major issues in the construction of European defence". Was he questioning the bipolarity of the New Order? May be. But certainly not challenging its doctrines of hegemony.

To Sum Up: The excesses of the old Global Security Order currently looks beyond perpetuation. The propagation of its doctrine and attributes, however seem to be backed by determined and effective stakeholders.

Moving to the second parameter of World Security Order - global security management, similarly, has produced some promising "Peace Dividends". The phasing out of the massive military and economic backing has significantly contributed to ending the wars by proxy in Afghanistan, Namibia, Ethiopia, Cambodia and El Salvador. Equally promising was the permission to allow the UN and International Law to play a long denied role in resolving security threats in Namibia, Cambodia, El Salvador and Somalia. Even in the Gulf War the UN and its International Law mandate was assigned the useful role of a respected platform for an operation under full American command, but the UN role is being upgraded in the unrewarding Yugoslav disputes. On the other hand, therole of UN, Western Europe and most of the world community is being shamefully demoted in the Arab Israeli negotiation. What is even more disquieting from the prospects of the New World Order is that the self appointed guardian of the Order in this critical dispute assures that no settlement (International Law) will be reinforced a far cry from the Gulf situation and the sanctity of international law and UN resolutions. "Benign" neglect is sedately being practiced towards the worst of oppressor, long cherished and sustained, in Africa and Latin America. A good beginning.

On restriction of arms transfer, there is much talk and selective enforcement, disturbingly outside the jurisdiction of the UN and International Law. On the other hand, thee is much talk of massive transfers to U.A.E. said to amount to 30 billion. Already $19 billion are contracted.

A lot of pressure is being applied to China to end its missile trade parallel to continuing the finance of Israel advanced anti-missile system and flow of significant technology under the guise of the strategic USA-Israel alliance which certainly contribute to Israel most advanced missile arsenal outside the major powers and reportedly shared with South Africa.

The German secret service has been at last caught practicing its illegitimate transfer of vintage Soviet to Israel. On the other hand, the first legitimate transfer of the Czechoslovakian arms to Syria did not escape the vigilance of secret service and the German navy. Hopefully, a lapse in the practices of the most self restrained major power.

To sum up, in the global security field like the nuclear disarmament field, a window of great opportunity has been opened by the drastic changes of the last few years. Some promising progress has been achieved; on the other hand, some ingrained excesses and coveted privileges may hinder the emergence of a New Order or mutilate its attributes.

Only concerted and sustained vigilance can help overcome these hindrances. Their implications for development and the environment remain, upto today at best, dubious.

In view of the these disturbing contradictions, conversion in the terms of this conference has to be well defined. It should not mean the more divergence of military productive capacity made redundant by incidental reductions of military expenditure due to economic or political constraints. What should concern us here, is the reallocation of squandered human resources, assets, natural and financial resources away from threatening and destructive tools of violence, towards socially and environmentally neglected vital needs, nationally, regionally and internationally. In the final analysis, we are concerned here with conversion of orientations, priorities and value systems.

Admittedly, this is not an easy task. I propose however it is the only lasting way out of our contemporary dilemmas. Let us first look at the nature of the challenge. SIPRI estimates the number of the large arms producers of more than $300 million, sales to be 100 internationally -- outside the East European countries. The USA has 48 of these. Developing countries has three in the lowest rank. Together with those oflsrael and South Africa they have sales amounting to 1.7% of the total.

Since 1985 the US procurement budget has been reduced by half to $ 65 billion and real defence spending reduced by 25%. The USSR, or whatever is replacing it nowadays, is also undergoing some drastic cuts; in 1980 it amounted to 8.2% of the previous year. While military expenditure still represented one third of the state budget in 1991, an alarmingly declining GNP estimated at 30% in 1992, will enforce more severe cuts.

Yet the pivotal issue remains - how do these cuts get incorporated into a lasting restructured economy oriented towards redeployment of spared resources in socially beneficial production.

Judith Reppy in studying the adjustments that followed the second World War and the Korean and Vietnam wars concluded "When the economy is strong, conversion can take place without requiring special provisions for the defence labour force, but without the background of glowing economy there can be real hardships". No one can characterize the economies of the super powers as being particularly strong or growing at this juncture.

An additional difficulty relates to the interest, mechanisms and practice that managed successfully the arms race i.e. the military - industrial complexes. The development of a counter weight of a cohesive network of socially committed actors to balance and ultimately supersede these complexes is not a trivial task.

It has to be taken into consideration that the first reflex reaction of the militaryindustry to procurement cuts, is likely to be more adjustments within the framework of its specialised function. United Technologies responded by eliminating 13400 jobs and closing or consolidating 100 facilities. General dynamics boosted its cash reserve by selling its Cessena light aircraft subsidiary. some major military producers still depend on plentiful funds hoarded from past large budgets that can tide over a couple of years, with saving in costs, moving to cheaper locations, concentrating on areas of pre-eminence and trimming of unprofitable operations. In short it is entrenchment and not conversion. Michael Rich of Rand predicts, big defence companies will remain defence companies". One element of Government policy is to encourage this trend; the funds for R & D remains protected from cuts and stand at $ 40 billion for USA and government hesitates to formulate a specific policy towards conversion.

In USSR, conversion is a more urgent economic and political necessity but is nevertheless handicapped by even more formidable burdens. In addition to the national fragmentation, political upheaval and economic collapse, there are immense problems of finance, restructuring, inefficiency and mismanagement. The need for reinventing the instruments of transition in central command economy do not stop at the command machinery but extends to the very roots of the productive and management structures.

As previously indicated conversion of military R & D is an important requirement of overall military production conversion. There are many reasons for this. The size of R & D military resources is huge - $ 100 billion internationally or about 10% of total military expenditure and it represents 25%of global international R & D financial and personnel resources. Moreover, 80% of it concentrated within the super powers domineering structure. It acts as a decisive "pull" factor on future demands and procurement at the level of the superpower and percolates its pull effect down the whole global power structure. In addition, it has a very destabilising potential through its capacity to neutralise the benefits of arms limitation agreement through upgrading the efficacy of the unit and expanding its impact. In addition, it paves the way for draining valuable resources, rekindles arms production and undermines mutual confidence. Unlike production and deployment, it is difficult to monitor or control; and it manifests a tendency to lag behind cuts in other military expenditures. Clearly a "special regimen" of military R & D conversion must be enforced and controlled by an international convention and machinery.

To sum up: Political and economic change open a window of opportunity for conversion. This no doubt would release resources for socially urgent needs as well as environmental protection and improvement. For these peace dividends eventually to spread to the development needs of the south however, it needs mush more than the difficult management of conversion of the military industry. An equally difficult management of conversion of international value system and partnership would be necessary. This should be one of the main tasks of this conference. Let me move to the third dimension front the New Order seems to be pregnant with opportunities for far reaching changes very much related to our concerns here in this conference. They are not as yet precisely definable but can reasonably be extrapolated. However obdurate nuclearpower may prove to be, it is reasonable to predict it will be operating in a new political and economic milieu that will permit it a much less commanding role than it has hither to enjoyed. One of the most promising developments in this direction was modestly initiated in 1952 by institutionalising the European Coal and Steel Community with its six members. Probably nudged by the upheaval in Eastern it took, and probably the World with it, a giant step in shaping the future of the World Order in Maastricht, Dec. 1991.

On the economic front with the single European Act already setting a deadline for the Community Common Market, Dec.1992 at Maastricht, it set an epic programme and pattern of future cooperation and unity. Most outstanding is the European Monetary Union with an Institute to coordinate national policies by 1994, and develop into a European Central Bank, Eurofed, by July 1998 and a common European currency, ECU by Jan. 1999 if convergence would be achieved among member states. With a GNP, labour force and potential market exceeding that of rivalry units it could contribute to regional and international stability and prosperity. The trend setting agreement to pool much of the management of European economy in Eurofed including interest rates, money supply and indebtedness if implemented successfully, will be a salutary model for future international evolution.

On the political front, the call for "an even closer union among peoples of Europe" has not been neutralised by omitting the reference to Europe "federal goal" from the preamble, as the instruments and trends towards such potential future goal indicates.

A European police force at least the nucleus of a future European Army, even if it at this stage a pillar of NATO, are such instruments. Even more inspiring is the move towards supernational sovereignty exemplified by the impressive responsibilities conceded to EC in the fields of culture, environment, consumer protection, labour, interstate highway systems, regional development programmes to assist the less developed in the present and possibly future extended membership. Even more relevant to the New Order, Maastricht furnishes a basis for future progress towards mechanism to harmonise and coordinate European Foreign Policy, hopefully independent policy.

Considering the collapse, near chaos and clear unpredictability of the change and structures on East European borders, which was not seriously dealt with, and the economic, social and political calamities in the South "benignly" neglected, it is encouraging to see scope and depth of the change indicating how national and regional efforts can contribute significantly in deciding the attributes of the New World Order and pave the way in facing up earnestly up to the challenges we are gathered here to discuss. Including the challenges of the Twin MADs.

The fourth dimension of the World Order relates to "The South". If the changes in Europe took half a century to materialise, then we are justified to give weighty consideration to the omens of change in the South, however modest, they may have.

The wave of tentative democratic vitalization in some Latin American, Middle East, South -Eastern Asian and latelyAfrican countries is noteworthy. Predictably, it is not uniformly deep rooted or far reaching but certainly promising.

Like Eastern Europe it is too early to predict how the violent changes will settle down and what forms, structures and trends will finally materialise. Even in the transition, tremendous economic, political, social and ethnic handicaps are likely to confuse and dislocate the operation. Considering the long-term decisive impact of the outcome of this transition on the future World Order, the so-called "Benign" neglect by the north has to be perceived nowadays as vicious disregard it disrupts the trends towards reform as it seems to be doing.

Taking the worst case scenario, regarding this neglect there are some sobering reflections that should engage the concern of this conference now that some strenuous political and economic reforms are emerging in some developing countries often at unbearable social and human costs. The Sough may be destined to endure the maximum of the calamities of its poverty, ignorance and disease on its own. The calamities, however, of its environmental degradation, population explosion and social turmoil inevitably will spill over the rest of the world.

Considering the crushing burdens of survival, may I sadly propose that it is almost tragically comic to expect most of the LDCs to be able to mobilize the resources urgently needed for global environmental and climatic protection. Failure to heed our common responsibilities may one day be soon realised for what is its criminal negligence. That is all what this conference is about.

# KEYNOTE ADDRESS
# BY

**Prof. Dr. Klaus TÖPFER**
*German Federal Minister for the Environment and Nuclear Safety*

The situation which Europe and the whole world are confronted with today involves both threats and a number of possibilities for positive development.

Let me first address the framework conditions of international politics:

1. The ice age of the cold war between east and west is over. Creating peace with fewer weapons is no longer an utopian vision but may again become true. This is the result of the break-down of an ideology,

- an ideology by means of which one of the last colonial empires was kept in a suffocating stranglehold,

- an ideology which placed the individual personality under the reign of the collective and which therefore had to rely on central planning of the economy.

Thus, a decisive confrontation has been ended which, for keeping up the antagonism of systems, was forced to consume and waste enormous energy: human creativity, scarce material resources, limited energy sources and limited human work force.

Overcoming the east-west-confrontation meant that the co-ordinates of international politics have undergone such profound changes that we have to conceive new ideas for shaping our future, ideas which are so fundamentally different that they bring about consequences which many of us are not yet aware of.

However, there is cause for concern that the stability created by the balance of horror might be replaced by the instability of an intellectual vacuum.

2. This break-down of the communist-leninist ideology as the basis of a what the communists termed "real existing order for the state and the society" allowed liberation - liberation for the individual person who, very soon realized that his newly won freedom was limited by the continuing constraints of deeply rooted and rigid structural patterns of both the economy and the society. The loss of jobs in companies which were highly subsidised in economic and ecological aspects and therefore could not meet the competitive challenges of the market economy meant a blow to the individual person who, very quickly did no longer attribute this break-down of the economy to the true reason for it, that is, the mercyless

person who, very quickly did no longer attribute this break-down of the economy to the true reason for it, that is, the mercyless exploitation at the expense of human health and the environment, the waste of countless resources for the sake of securing the own ideological coercion system nationally and internationally.

To a higher or lower degree many people became aware of what a Polish philosopher summed up in the following sentence: "Freedom itself never lives up to the dream of freedom!"

Most important, however, is the following fact:

The old order of unfreedom and constraints has been overcome. However, what is left behind is an intellectual vacuum. The question as to how this intellectual conversion can be accomplished, will be of crucial importance.

Alarmed, we realize that this intellectual vacuum which came about by "real existing communism" may very quickly and agressively be filled by national, and sometimes even regional chauvinism.

However, we also observe with great concern how islamic fundamentalism may very quickly put intolerant restrictions by religious mania on the newly won freedoms. Two development trends represent realistic threats and challenges for the intellectual conversion: re-nationalisation on the one hand re-ideologization on the other.

The former Soviet Union was the last big colonial empire - held together merely be the ruling ideology. After the break-down of the ideology the colonial empire too was bound to collapse. The search for a new national identity becomes top priority. The dramatic development of this process is brought home to us every day by terrifying television pictures on new wars: in Armenia and Azerbaijan, in desintegrating Jugoslavia, in Moldavia .... More examples can be given, others must be feared to follow.

What can be our response? In western Europe we have embedded nationalism into the process of European integration and in Germany the creation of a Federal state was our answer to the question of regional identity. Both these solutions also should be taken into account as models by the new democratic national states in central and eastern Europe. Europe needs the strength to go beyond the integration of western Europe - it was particularly for this reason that the Maastricht summit was so important on our way towards political union in Europa as the Europe of western integration has to prove itself capable, also capable of integrating the nations of central and eastern Europe. And we must continue to strengthen our federative structure and incorporate it into Europe instead of giving it up for Europe.

Our other concern is the re-ideologization - a most realistic concern, when one looks at the developments of the islamic fundamentalism, in particular in the newly formed republics in the southern areas of the former Soviet Union. These developments have very direct and dangerous consequences such as the threat of so-called vagabonding traders of nuclear power and chemical substances proliferating fissionable material like plutonium or chemical warfare agents which are not included into a monitored disarmament process but very quickly might reach the wrong hands of new ideologists.

The time of detente which we urgently need as a result of the disarmament process must contribute to reconstructing an efficient economy and to rehabilitating an ecological legacy of the former "real existing socialism" the extent of which has been unconceivable so far. The ideological rift of the world, the cold war between east and west, has been overcome. However, the consequences of the ensuing collapse, in particular in central and eastern Europe which we Germans experience most directly in our own reunited "Fatherland" will now present an enormous challenge. It means is a great opportunity, the acid test for this generation: to eliminate the threat of a new division, of a new cold war, to overcome the division between north and south, between the have and the have-nots.

If the process of transition from military to civil economic activities is not to be interrupted again very soon we have to accomplish the intellectual conversion. If we fail to do this we will soon have to waste considerable energy for curbing ideological intolerance, energy which we are in urgent need of in order to achieve disarmament in the war between the human race and nature.

What will be required for this purpose are less missionary crusades for democracy, social market economy and tolerance, as important a large number of human contacts as practical proof of these values may be. This also requires a new form of creativity in foreign and safety policy - above all in creating new, regionally balanced security systems, both for the Near East as well as for central and eastern Europe as long as these countries cannot yet be incorporated directly into the European Community. Also, follow-up measures to the CSCE process will have to be considered in order to make use of them for securing the newly won freedoms and democratic possibilities for development.

The cold war between east and west has been overcome within the framework of the CSCE process.

The Final Act of Helsinki provided the basis upon which large-scale confidence-building measures were gradually undertaken, barriers were removed and wide gaps were bridged. The threat of a new confrontation, now between north and south, between poor and rich must also be avoided or eliminated within the framework of a process. Therefore it is our aim to embark on a new "process of Rio" at the Rio conference, to adopt a final act of Rio and on this basis to enable confidence-building measures of cooperation in development and environment policy between north and south, thus preventing poor and rich from continuing to drift apart.

Does such a comparison indeed reflect reality? Does it not dramatize the depiction of a threat? This threat is not exaggerated at all. On the contrary: particularly in the preparation phase for Rio it becomes clear that the developing countries do no longer want to content themselves with a division of labour which enables the industrialised north to enjoy the benefits of technological progress and burdens the poor developing countries in the south with the major share of direct and indirect costs of this prosperity. To an increasing extent attention is drawn to the intolerable fact that the benefits of modern technologies are regionalsed, their drawbacks however globalised. CFCs contributed towards improved economic processes in the highly industrialised north and towards ensuring a higher level of prosperity - the negative side of the coin however, the ozon depletion, is the consequence felt in particular in the underdeveloped south. Let me add just a few of a large number of examples that might be given here:

- Enormous $CO_2$ emissions in the industrialised north and the rise of the sea level in the south. Thus, $CO_2$ emissions represent a very effective from of agression against Bangladesh or small island states in the Pacific which can no longer protect themselves against the rise of the sea level.

- Safeguard of the genetic potential and the $CO_2$ sinks in the tropical rain forests and the transformation of forests in the industrialised north into "lumber factories" where, on the basis of economic principles, entire monocultures were created by mere maximisation of growth of the biomass.

Developing countries increasingly consider these and other examples proof of the fact that we pursue a kind of neocolonialism by means of high environmental standards which we try to impose on them: their own individual locational advantages cannot be made use of. At the first UNCED conference in Stockholm Indira Ghandi put it into a nutshell:

"The countries of the north want to save us from committing the same errors which they themselves committed in the past. However they forget that it was exactly these errors which provided the basis for their own wealth."

Rio de Janeiro has been given the motto "We are one world". We must be aware of the fact, however, that this unity also will involve costs. I am afraid that we have not yet made this sufficiently clear to our fellow-citizens. But if we continue to ostentatiously and often excessively praise our (material) wealth all over the world migration movements will further increase. This also applies to environmental refugees who are left with no other alternative than to flee to escape the collapse of their natural eco-systems. The dramatic pictures of the results of continuing desertification in Africa are brought home to us every day - and their repercussions are made worse by regional and national chauvinism also to be found in that region.

These words contain only a few characteristics depicting the threat of a now cold war, of a conflict between north and south, between poor and rich. Here too, a process clearly geared towards detente with confidence-building measures must be implemented.

In doing so, it is necessary not only to harmonie analyses but also to look for specific approaches towards solutions and for packages of measures. I should like to emphasize above all the role which the further development of technology has to play in overcoming these problems: not opting out but responsibly handling complex techniques can and may be the answer of industrialised, technologically further developed states to the situation depicted above. In this context, I am increasingly worried by unacceptable double standards prevailing in many industrialised countries of the west and the north and which are obviously particularly manifest in Germany: people seem to believe that, by opting out of certain technologies in their own country they have already met their responsibility. By increasing standards and requirements, combined with a reluctance to take decisions on environmentally-sound installations which might almost be called inertia, these technologies are shifted to neighbours near and for who, in the best case, are able to manage the technical features as well as we do but, in the very probable case, do not have the same safety culture as the one which took many decades to grow here in Germany. In this context, too, numerous examples may be mentioned.

Export of wastes must be mentioned here. We continue to observe such exports, which even have been made the business of mafia-like systems, not least because of the fact that we ourselves do not shoulder anymore the responsibility at home by constructing disposal facilities with a maximum of technical standards. It is exactly these facilities which have been developed by our engineers who have developed such facilities and it is exactly us who have developed the expertise to handle the hazards of our wealth without negative impacts on human health. However, we transport these wastes across borders where they end up in facilities which, in many cases, are by far less well equipped they would be required to in our country. In this context we could also speak of waste reprocessing plants instead of waste incineration plants - many more examples could be given that are an expression of the above-mentioned double standards.

Robert Jungk once said:

"I do not want to be matter of fact. I am concerned." In my opinion this sentence has to be worded the other way round if one is truly willing to responsibly listen to the message of Rio."

Especially since I am concerned I have to be matter of fact. Not opting out, but responsibility in assessing and using modern technologies, responsibly developing further our knowledge of the elements of nature and life thus represent the answers to Rio. This does not at all mean "yes" for using a technology and it may mean "no" for not applying a technology. If so, however, this should not be done because of the superficial position of "not in my country", but because of an overall ethical responsibility which provides the basis.

This may bring about new and important consequences also for politics. Should we not embark on new considerations on proliferation? Is it not too an unilateral step when we - very rightly so - sign a new treaty on the non-proliferation in the field of nuclear energy and nuclear fission but do not think about such "non-proliferation treaties" for genetic technology and for certain developments in bacteriology?

The requirement of technological change and its responsibility, however, must be further specified. It must be emphasized again and again that technical progress is not a free gift from heaven, that is to say, it does not come about by pure chance. Technical progress rather is made in situations in which bottlenecks have to be overcome. In the past technical progress therefore always meant labour-saving technical progress, meant rationalisation and automation. This will certainly no longer be the case when it comes to overcoming the global problems and development trends. Every day there are 250,000 people more on this planet which means, at least in terms of quantity, a new distribution of scarcities.

What is becoming scarce in an almost dramatical form everywhere in the world, however, are the so-called environmental assets, that is air, water, soil and the resources we have available. If we do not assess their scarcity with the right prices, we must not be surprised that no or not sufficient technical progress for overcoming these scarcities, that is for a more economical handling of air, water, soil, energy and resources is made. Therefore, preventative environmental policy does not depend on economic cycles, must follow a steady course and must be given medium-term objectives in order to provide an incentive of technology to develop towards this direction.

Changes in scarcity stimulate technology - these changes as well as prices truly reflecting their ecological value also lead to a change in behaviour. Even though the development of technological solutions must be particularly emphasised, we must not follow the old German proverb according to which the engineer can cope with everything. In this context too, a change of behaviour is imperative - not by making moralising demands for askesis, but by taking into account the true scarcities which we are confronted with in our economies and thus also in our consumption behaviour.

It is exactly because I am concerned, I have to be matter of fact: this matter of fact attitude will also in future have to characterise also in future technological change and necessary changes in behavior. I know how difficult it is to reconcile matter of factness and emotions in a constructive way. However, I am convinced that those responsible in science, the economy and politics all over the world are willing to tackle tjis task resolutely.

# ARGENTINA

# CONVERSION AND ENVIRONMENTAL CONFLICT

**Arturo Abriani**
*Fundacion Instituo de Estudios Estrategicos de Buenos Aires, Argentina*

## Introduction

The aim of this paper is to discuss how conversion and the notion of environmental conflict might interact with each other. It first considers the idea of conversion as supplementary to the concept of arms reductions, after identifying some strategic prerequisites. In the following sections the discussion centres on the ideas of environmental security and environmental conflict, approaching the subject from the perspective of the interaction between states. Finally, the paper explores areas of linkage between the conversion process and environmental conflicts.

## The Strategic Meaning of Conversion

With the easing of East-West military tensions that followed the political developments which occurred in the Soviet Union and Eastern Europe at the end of the 1980s, the notion of **conversion** suddenly became a key concept for analysts and decision-makers dealing with economic development as well as for those engaged in environmental protection matters.

Indeed there was a rapid succession of decisions regarding arms limitation and reduction either by unilateral actions or agreements. Those in charge of implementing those decisions had to do so without having comprehensive contingency plans for conversion. The political, macro-economic, social, technological, and environmental challenges posed by the urgent need to convert armed forces and military assets to civilian usages captured most of the attention of those directly concerned, while the strategic prerequisites seemed to remain in the background.

The emphasis on socio-economic and engineering aspects of the subject was evident in meetings conducted between states and in international conferences and seminars. Only issues such as the possibility that arms made redundant by the end of the Cold War and the consequent reductions in defence spending might be traded in the Third World market, or the proliferation potential of migration of experts on military technology, seemed to drive the discussion onto strategic grounds.

From a broad analytical perspective, however, its seems convenient that attention should be paid to the strategic context of the subject, in order to provide realistic and sound bases for those involved in endeavours aimed at the exploitation of the `peace dividends', such as environmental protection activities.

As a concept, **conversion** is subsidiary to the idea of **disarmament**. Conversion is a consequence of **disarmament**; more precisely, it forms a part of the implementation of defence and foreign-policy decisions aiming at reducing the levels of military force. This reduction might be considered as appropriate by policy-makers as a result of the changes in the international situation, allowing the achievement of acceptable levels of security with lower levels of military force-in-being, or political-economic domestic pressures determining the reallocation of resources between sectors of the national economy, or by a mix of these factors.

The idea of **disarmament** pertains to the set of main concepts of Strategic Studies, which also includes the notions of deterrence, defence, limited war, crisis management,arms racing, proliferation, and arms control. In this context, as Barry Buzan points out, **disarmament** relates to one of the fundamental variables affecting the international system: the nature of the prevailing technology, i.e., the instruments of force available to political actors.[1]

In fact, **disarmament**, as an approach to peace and security based upon the reduction or abolition of weapons, has traditionally deserved particular attention in the Strategic Studies literature. But due in part to various military, economic, and political flaws in its logic, and also due to the particular nature of the strategic arms race between the USA and the USSR in the context of the Cold War, the idea lost ground in favour of arms control, which comprises"... restraint internationally exercised upon armaments policy, whether in respect of the level of armaments, their character, deployment or use".[2]

Disarmament and conversion plans defining the degree of reduction of forces -- types and numbers of armaments, personnel, and production plants -- should match, and follow the reappraisal of military strategy, in the light of the evolving international context. Strategy deals with the use of military force as a means to political ends, i.e., it deals with the implementation of defence and foreign policies. These are shaped by threat perceptions and security needs, and constrained by the resources available to the state.

Indeed, disarmament and conversion, as particular aspects in the process of force planning, imply the integration of defence planning and macro-economic planning. They involve costs, not only those related to the economic and social impact of defence cuts, but also the investment costs to be incurred to initiate the implementation of the relevant decisions. And they take a length of time related to the pace of evolution of nationalsecurity interests and threats and to the responsiveness of the force planning system.

The present disarmament and conversion trends are peculiar in various aspects.

First, they are to be conducted after a major and long-lasting arms race -- that of the Cold War -- which in its final stage has been primarily qualitative and shaped by arms control agreements, implying complex quantitative/qualitative balances of armaments and forces, and verification processes --aimed at monitoring and evaluating the parties' compliance with the rules of the agreements. Second, as a consequence of the `cold´ nature of the ended war, the armed forces and military industry of the losing side are intact, as are those of the winners. Third, although military power is declining in importance in the post-Cold War world, it remains a relevant factor under likely scenarios of regional instability; thus disarmament is not expected to be either general or complete.

These features imply additional military, political, and economic problems for the present disarmament and conversion efforts, both in terms of scale -- the numbers of forces involved -- and in terms of the uncertainty regarding the kind and sizes of military force required in the emerging world order. In the background, the economic troubles currently faced by the main actors undergoing the process of arms reduction -- particularly, by the former Soviet Union -- further complicate the issue.

## Environmental Conflict

As a component of the wide spectrum of contemporary non-military threats to security, environmental issues are emerging as a new conflictual element in the relations between states and groups of states. Concepts suchas **environmental security and environmental conflict** are thus being given a central place in international politics.

Both ideas are not new if considered from the traditional perspective of ecology. But the further, broader and indeed complementary meaning results when recognizing environmental effects as a form of interaction among the components of the international system.

The traditional ecological perspective is obviously concerned with man's interaction with nature, focussing primarily on issues such as air, water and soil pollution, depletion of resources, and environmental hazards generated by industrial activities. Its main preoccupation is the interdependence between economic development and environmental protection. From this perspective, the idea of sustainable development is essential to the achievement of environmental security. Environmental degradation caused as a result of war also pertains to the scope of traditional ecological concern.

From the other, complementary perspective -- that of international politics or international relations -- the focus is on the new type of interaction between states, stemming from the generation of trans-border, regional or global environmentally harmful effects. This view is concerned with environmental problems such as issues affecting the vital security interests of the state, and as a factor eventually leading to organized violence.

Once the environmental dimension is added to the identification of national security objectives, threat perceptions regarding environmental security may lead to conflictual interaction. Thus environment becomes a debatable issue in which the control of environmental behaviour perceived as hostileis the subject of contention between states or groups of states.

In the above context, the expression **environmental conflict** may thus convey the idea of armed conflict caused by conflicting interests of states regarding the control of environmental effects.

The various environmental issues which may lead to international conflicts comprise of a number of global effects and a wide spectrum of regional effects. Although trans-frontier ecological impacts and -- rather obviously -- disputes about natural resources have been recognized as a source of international conflict in the past, the realization that there are environmental problems of global implications which may lead to conflicts is relatively new.

From the perspective of international law there is a basic principle according to which trans-frontier damage is prohibited. As a result of a process developed in recent years, the existing international law related to the protection of the environment has widened in scope and includes -- in addition to norms regarding trans-border damage to components of the environment -- regulations in respect to:

> ... exploitation and protection of those natural resources which are commonly shared by several States, the use and protection of areas beyond national jurisdiction, such as high seas and outer space, as well as the managing of environmental problems of global relevance. (...) There exists, however, no consensus upon the exact meaning of the notion of global relevance (...) and, first of all, on the question of who identifies environmental problems being of global relevance. Finally, it is disputed who has ... to shoulder the economic burden of measures to counter damagesor hazards to the environment of global relevance.[3]

A comprehensive list of environmental global issues -- as identified from a Latin American perspective -- includes the following problems: nuclear risk; global warming; drug-production, consumption, trafficking; loss of bio-diversity; depletion of the ozone layer; pollution of the oceans and exploitation of oceanic resources; use of resources from the Antarctic; and use of the outer space.[4]

Most of these conflicting issues share with other contemporary global problems the feature of having a North-South dimension, as explained in a recent paper by Virgina Gamba-Stonehouse, discussing whether environmental crises are the causes or the consequences of international conflict, given the asymmetries in patterns of development of the First and Third World:

> Ever since the reduction of the East-West dimension of international conflict, there has been a new concentration on the North-South dimension of potential and present security concerns. (...) Issues such as immigration control, drug-trafficking, external debt, transfer of technology, weapon trafficking, and international competition for trade and resources, plus issues related to population growth and energy needs of many countries have originated new kinds of conflict. Nations see these issues not as common problems but as individual threats.[5]

This North-South cleavage of most global environmental issues is also stressed in the above quoted paper by R.Wolfrum, when discussing the restrictions on trade implied in the international agreements on the depletion of the ozone layer:

> ... the industrial countries are in practice in the position to dictate their environmental policy upon the developing countries. Therefore, complaints of the developing countries of an upcoming "environmental imperialism" is not without reason if such mechanism is applied without a sound basis in treaty law having been accepted from among members of all groups of States.[6]

Most global environmental issues are related to the energy problem and consequently to the international politics of resources such as oil, having an old and considerable record as a cause of international disputes. That relationship is direct when considering the problem of climate change. And again as in the case of global energy transactions, the notion of interdependence provides a valid approach to the study of international environmental interactions: given the global character of environmental phenomena involved, the developed countries cannot solve the greenhouse problem by themselves.

This North-South dimension of the issue is directly related to its potential as a source of conflict, since there is substantial asymmetry in perceptions about the matter. For countries in the South, facing the problems of underdevelopment, the environmental challenge is seen, indeed, as a global threat. But in their perceptions, remedial action seems to be the primary responsibility of developed nations: the costs of restoring environmental balance should be shared by the different countries in proportion to their present and past contributions to the damages caused.

**Environmental conflicts** may also result from regional disputes over ecological matters. There are many examples in recent years comprising of competition for resources,energy and development. Theabove quoted paper by V. Gamba-Stonehouse discusses the following four cases: the issue of the Gulf of Paria between Trinidad and Venezuela in the Caribbean; the potential resource conflict in the exploration and exploitation of resources of the continental shelf of the Western South Atlantic, between Argentina, Uruguay, Brazil, Chile and Great Britain; the case of Turkey's Ataturk dam and other irrigation and electric power projects, affecting downstream countries such as Syria, Iraq and the Gulf ecosystem; and the conflicts of the late

1970s between Argentina, Brazil, and Paraguay due to the construction of hydro-electric projects in the River Plate Basin system.[7]

**The Links Between Conversion and Environmental Security**

The general discussion made in previous sections regarding the meaning of conversion and the environmental facets of security and conflict aimed at providing a conceptual framework to consider whether there are links between both subjects. There seem to be at least three different areas of linkage: i. the reallocation of resources, essential to the idea of conversion; ii. the implementation of conversion; iii. threat perceptions in terms of environmental security as a restraint for disarmament -- thus as an obstacle for conversion.

The first area relates the possibility of re-orienting the physical, human, financial and material resources from military uses towards civilian purposes pertinent to the protection of the environment. Given the nature of the challenges against environmental protection, the technical skills and R&D potential of the military sector are assumed to perform an important role in fields crucial to global environmental monitoring and control, such as land surface remote control, chemical analysis, cartography, medicine,microbiology, and radiology.

Although perhaps not covered by a strict definition of conversion, the redeployment of military units provide another opportunity in this first area of linkage between disarmament trends and environmental issues. It comprises of the various options for the short or medium-term use of existing military structures for civilian areas of responsibility, and includes disaster relief operations such as the case of aid for refugees or environmental disasters, and the use of intelligence and communications military systems in operations related to the interdiction of drug trafficking.

If the disarmament process -- and the consequent conversion -- made possible by the reduction of the East-West dimension of international conflict allows the allocation of additional means for environmental protection projects and sustainable development, then the emerging North-South conflict on global environmental issues could be more easily prevented and resolved. Such a trend would also contribute to diminish the South-South disputes which originated with the environmental degradation or competition over resources.

The second area of linkage, the implementation of conversion -- more generally, of disarmament -- might generate environmental risks and eventually lead to international conflict, if ecologically sound procedures are not used.

The main problem to be faced in this regard is that of military hardware destruction and disposal; i.e., the elimination of military equipment, weapons systems and ammunition having no civilian applications. This includes the destruction and disposal of elements such as missiles, warheads, shells, fissionable, chemical and biological materials, and certain fuels. In also relates to the question of solving technological and planning problems related to former military sites --bases, firing and training areas, and launching sites.

The main challenges regarding the military hardware issue are posed by the destruction and disposal of chemical weapons and radioactive waste from nuclear weapons. The problems are two fold: technological and economic. There is a lack of environmentally safe technical experience in the handling and disposing of such kinds of highly hazardous wastes; the amounts of stockpiles involved are large and the processes implied are costly, conspiring against the adoption of ecologically safe techniques in the context of economic crisis, since these add important costs to the burden of conversion.

The additional difficulties posed by the requirements of environmentally safe disarmament are clearly described in the following paragraph quoted from an article by K.Hayward and T.Taylor:

> ... there must be no illusions about the difficulties involved in rendering safe, or finding alternative use for, some of the most lethal materials and indestructible machines ever produced. (...) We must also face the blunt truth that some of these items are often more expensive to convert than to construct in the first place. Some hazardous materials are actually cheaper to make than to dispose safely.[8]

The above difficulties imply that the prevention of conflicts eventually generated from hazardous disarmament/conversion activities will not be easy. Available experience regarding private business handling of international transport and disposal of toxic and dangerous industrial waste is not satisfactory. In recent years there have been numerous cases of illegal traffic of hazardous waste -- generated in andexported from industrialized countries and shipped to developing countries.[9]

Directly related to conversion, the issue of chemical and nuclear weapons hazardous waste and its disposal is becoming a matter of dispute among the independent states emerging from the former Soviet Union, and there seem to be perceptions among certain of the smaller republics that a North-South type relationship could develop about the disposal of waste resulting from disarmament plans centrally established.

The problem of misuse, theft and clandestine exports of weapons also pertains to this area of linkage between disarmament and environmental conflict. Under conditions of economic emergency and domestic political disarray, there are risks of illegal arms transfers. Should these include weapons of mass destruction, there is the danger of environmentally dangerous material reaching unskilled hands.

From another perspective, socio-economic disruptions generated by the closing of military industries may also create problems at the domestic level, eventually leading to local international conflicts as a consequence of disputes over trade and resources, immigration control, and other environment-related non-military threats to security.

The third and last area of contact between the two sets of ideas considered in this paper is concerned with the eventual influence of environmental threat-perceptions on the level of armaments regarded as necessary to provide security -- environmental security. Thus the point here is whether environmental conflict as a source of tension could lead to additional arming or counter the disarmament trends generated by the easing of tensions in the East-West conflict.

At the global level it remains to be seen what combination of agreements and unilateral actions comes out of the ongoing developments and international efforts. Perhaps the forthcoming United Nations Conference on Environment and Development will provide a measure of the mix of collaboration and conflict foreseeable in the near future regarding the international politics of the global environment.

At the regional level, the post - Cold War world is the subject of speculations about the possible scenarios regarding economic and political perspectives in the 1990s. Some of these foresee increasing regional conflicts in the absence of the stabilizing influence of the deterrence of relation between the USA and the USSR. Environmental conflicts are not expected to be the exception to this rule.

## Concluding Remarks

Rather paradoxically, parallel with the easing of East-West conflict and the consequent arms reduction processes which are expected to produce `peace dividends', environmental conflicts are emerging as a menace to peace and security. This emerging perception should result in decisions aimed at reassigning resources towards environmental protection, both at global and regional levels. As well as the notion of sustainable development as a key concept to tackle the issue of the relationship between development and environment, the idea of environmental security as a common value should lead to collaborative interaction at the international level, regarding the control of transborder and globally harmful environmental effects. As in most forms of interaction between states, the avoidance of misperceptions is crucial in this respect.

## Endnotes

1. Buzan, Barry, *An Introduction to Strategic Studies Military Technology and International Relations*. Macmillan, London 1989 (1987), p.6. The other fundamental variable affecting the international political system, according to Buzan, is the system's political structure.

2. Bull, Hedley, `The Objectives of Arms Control', in Art, R. and Waltz, K. (eds), *The Use of Force : International Politics and Foreign Policy*, Little, Brown & Co., Boston, 1971, p.349.

3. Wolfrum, Rudiger, "International Responsibilities and Conditions in the Preservation of Environment: Impact upon Developing Countries," paper contributed to Pugwash Meeting No.184: 3rd Workshop on Non-Military Dimensions of Global Security, Environment and Development in Latin America, Lima, Peru, November, 1991, p.3.

4. Comision de Desarrolly Medio Ambiente de America Latina y el Caribe, Nuestra propia Agenda. Banco Interamericano de Desarrollo, Washington, DC, 1991, p.40.

5. Gamba-Stonehouse, Virginia, "Environmental Crises: Cause or Consequence of International Conflict?", paper presented to the PRIO/UNEP symposium "Conversion and the Environment", Polytechnical Institute, Perm, Soviet Union, November, 1991, p.3.

6. Wolfrum, R. op.cit., p.24.

7. Gamba-Stonehouse, V., op.cit., pp.5-9.

8. Hayward, Keith and Taylor, Trevor, "Military Hardware," Paper presented at UN Conference "Conversion: Economic Adjustment in an Era of Arms Reduction", Moscow, USSR, August, 1990, published in *Disarmament*, Topical Papers 5, UN, New York, 1991, pp.262-3.

9. This issue is discussed in Charlin, Marcelo and Varas, Augusto, "Conversion and Environment: The Role of Local Government in the Third World", paper contributed to the PRIO/UNEP symposium "Conversion and the Environment", Polytechnical Institute, Perm, Soviet Union, November, 1991.

# THE CONVERSION OF THE MILITARY-INDUSTRIAL COMPLEX IN PROGRESS IN ARGENTINA

**Alfredo Luzuriaga**
*Argentina*

The end of the cold war has provided the world with a great opportunity for achieving important goals in terms of peace keeping and disarmament, which would have been hard to believe just a few years ago.

Of course, this situation is particularly important to the central powers, who have been principal actors in this expensive struggle for almost half a century. But developing countries should also make their own effort to make good use of the same opportunity at the regional or individual level.

Reduction of military expenditures is mandatory if they wish to solve the present economic crisis that they are passing through.

This reduction can be achieved using new concepts on security and introducing regional agreements and confidence building measures.

South American leaders have foreseen this situation and made an important progress in regional distension. Also, an ambitious economic integration program was set up the during last years.

Argentina has promoted peace-keeping policies and, with that purpose in mind, has fostered agreements to solve conflicts arising from territorial claims or power struggle issues. So, she came to a solution about ancient boundary conflicts with Chile, signed an agreement with the United Kingdom to avoid problems in the South Atlantic and also signed another agreement with Brazil not to develop any type of nuclear weapons, and another one for keeping the region free of chemical weapons.

The present government is conducting a severe adjustment policy to redress its mined economy and to modernize the structure of the state, whose constant deficit was the main cause of a crisis that reached hyper-inflation.

The adjustment includes, among other things, the transformation, liquidation or privatization of state owned enterprises, one of the main sources of the fiscal deficit.

As part of this program an important re-conversion process of the military-industrial complex is included. Overdimensioned and ill-managed, it represents a significant percentage in that deficit. The global losses of 1991 reached 700 millions of dollars, an important figure for the national economy.

It is important to give a brief account of how it was formed, in order to reach a better understanding of its dimensions and the multiplicity of products involved.

About the late thirties, following ideas that were popular in Europe at that time, the Argentine military considered it convenient to become independent in the production of materials necessary for the arms industry. So, the concept of "Strategic Industries", still used, was born.

In a country like Argentina, which was not industrialized at that time, this idea was immediately associated with heavy industries and petrochemicals, i.e., basically steel and fuels.

At the same time factories were developed to produce explosives, ammunition and portable arms.

Also, an aircraft industry and an important shipyard were established. By October 9th, 1941 the General Directorate for Military Industries (Direction General de Fabricaciones Militares) was created and until now has been conducting the military industries and the task related to industrial mobilization. later on, the aircraft factory and the shipyard were transferred to the air force and the Navy.

At that time the situation of the national economy was very different from what it is nowadays and the State was able to afford the high cost of those industrial developments which were undoubtedly beneficial to the growth of the country.

In this context, the military industrial complex of Argentina has been growing since then, reaching its present dimensions. It is important to highlight that the produced goods tended to shift more to civilian than to military ones.

By the late seventies, under a military rule, an ambitious plan was drawn up to incorporate modern technology into the arms industry. Towards this end, and associated with European companies, the Navy raised a shipyard for building submarines, the army a factory to produce a combat tank and the Air Force an advanced jet trainer aircraft, while developing with its own technology a medium range missile.

Financial problems, management failures and difficulties with a restricted and highly competitive international market, drove these industries to bankruptcy.

Out of four submarines intended to be constructed, none was finished and neither tanks nor aircrafts could be sold to other countries. The missile project, known as "Condor II", was deactivated and finallydismantled by government orders on February 11th, 1992.

Since the day that the present government took office, it had to face a severe economic crisis resulting in hyper-inflation, which has deeply affected the national mentality.

The decision was to adopt a rigid adjustment and disregulation program, trying to cut out the share of the State in the national economy, as it is the main source of fiscal deficit.

To this effect, Act No.23, 696, called "ADMINISTRATIVE EMERGENCY AND STATE RESTRUCTURING" was passed.

The privatization of state companies is carried out by means of Public Biddings and the bidding specifications containing the offer basis. Conditions are sent to the different areas of the Government so as to incorporate suggestions, corrections and observations wherever necessary. In all cases, the bidding forms must be approved by a Commission of the National Congress.

The Ministry of Defense has made significant advances in the Restructuring and Privatization Program of the enterprises under its sphere of influence. The objectives and strategies to follow for their transformation or sale were developed by a group of technical experts with the advice of the World Bank.

A program has been drawn up bearing in mind the National government policies, those of the Armed Forces (regarding critical military production material) and the situation of the state companies. The program will allow the privatization, throughout 1992, of those companies that are under the Ministry's rule today.

The aims of the privatization program include :

The reduction of the fiscal deficit by privatizing industrial plants which, under State management, are inefficient or show losses.

To eliminate the participation of the Ministry of Defence in profitable enterprises.

An increase in the participation of the private sector in the National Economy.

An improvement of the productivity and efficiency of the enterprises to be privatized.

The technical team has performed the following tasks:

Economic and legal analysis of the assets to be transferred to private hands.

Study and preparation of the basis and conditions for the sales offer.

**Assets Appraisals**

In order to succeed in selling the state companies it is necessary to determine different degrees of assessments based on various methodologies. Thus a parameter is obtained, which will permit the analysis of the reasonableness of the possible offers made by potential buyers.

The appraisals are carried out by government agencies which apply the following valuation methodologies:

Technical appraisal (applied only to Fix Assets)

Determination of value according to comparable transactions.

Determination of net worth by discount cash flows.

**Market Value**

The study of each company's books had also to be considered. It was important to avoid any possible mistake which could hinder the transactions; thus, for example, the stocks had to be analyzed.

The program scheduled for 1992 is shown in Appendix 1.

The defense complex, described in Appendix I, accounts for 27469 employees distributed as follows :

STEEL MILLS : 14, 514
CHEMICAL AND PETROCHEMICAL COMPLEXES : 3, 240
SHIPYARDS AND DOCKYARDS: 2,712
CIVIL PRODUCTION : 3, 906
MILITARY PRODUCTION : 3,097

Within these sectors, those of higher incidence are: SOMISA in steel production (10,087) and AREA MATERIAL CORDOBA in civil production (2,950).

In order to reduce the number of employees, a condition necessary before privatization, a voluntary retirement program was scheduled. To that purpose, a loan from the World Bank of 300 million dollars was granted to compensate the workers on retirement.

The total assets of the enterprises under analysis amount to US$ 5,679 million approximately. Within this total, the steel related sector is that of greatest incidence with almost 50 per cent of the total amount.

The total sales amount of the state-run companies for the year 1990 reached US$ 1,446 million. It is also the only steel producing sector which concentrates the higher share in the total sales with about 65% approximately.

SOMISA's high incidence, with a total sales participation of 50% approximately, is to be highlighted.

Leaving aside six enterprises already awarded to the private sector and two others open for bidding, there still remain more than 28 firms in the defence area. The main activities of these enterprises include, besides the war industry, the steel mills, the chemical and petrochemical industry, the steel mills, the chemical and petrochemical industries, the shipyards and dockyards (for ship building and repairs) and a series of enterprises committed to the civil production of a diversity of items ranging from ploughing disks and railroad material to different types of services. Some of the them work as corporations, others as joint stock companies and others as entities without legal autonomy but dependant on the Direccion General de Fabricaciones Militares (Directorate of Military Factories) or on the Air Force.

This group of industries, spread throughout the territory, represents the efforts carried out by the National State at a moment when it was necessary to supply the Country with technological capacity to meet the demands of those times. Today, they constitute a graphic representation which shows the magnitude of the restructuring to be undertaken.

Within this framework of privatization, there were several calls for biddings and the following enterprises were awarded to the private sector:

**A petrochemical Complex (POLISUR S.A.M. MONOMEROS VINILICOS S.A.M., PETROPOL S.A.M. and INDUCLOR S.A.M.)**

The prices obtained from the sales reached on the whole the amount of US$ 178 millions; 73% to be paid in Government bonds and the rest in cash.

**A Dockyard (TANDANOR S.A.)**

On 10/31/91, Tandanor was awarded to a consortium composed by the BANCO HOLANDES UNIDO, CIAMAR and SUD MARINE ENTERPRISES, for US$ 59,760,000; US$ 7,200,000 to be paid in cash and the balance due, guaranteed and with interests, in 9 years with 3 years of grace.

**A Steel Company (FORJA ARGENTINA S.A.)**

On 01.17.91, 80% of the capital stock was awarded to the FORCOR S.A. company for US$ 742,000, payable in cash.

The following enterprises are undergoing the process of acceptance of bids.

**Chemical Plants**

- Carboquimica Argentina S.M.
- Petroquimica Rio Tercero S.A.
- Ex METEOR Industrial Plant.

Due to the isolated commercial operations reported and the generalized failure that obstruct the fulfillment of their corporation objectives, the liquidation of the following enterprises was set forward or accelerated:

**Miscellaneous Enterprises**

- E.D.E.S.A.
- Aceros Ohler S.A.
- Altos Hornos Zapla Construcciones S.A.
- Consultara S.A.
- Industrials Mecanicas del Estado S.A.
- Sidinsa S.A.

In the short run, the process for the liquidation of the six state enterprises in this area will be over.

This profound change intended by privatization will be complemented by the conversion of some of the factories to civilian production, trying to keep some capabilities for military items to be produced. It is considered that a private owned enterprise with a production of 80% civilian items and 20% of military ones is the best combination obtainable.

The program will have also an important social cost. Many people will lose their jobs. It is important for the reactivation of the economy to lower disemployment.

Such a modification of a traditional part of the Argentine strategic thought should be considered as a confidence-building measure that has been taken by their own decision, and one that will contribute to peace-keeping in the region.

## APPENDIX I

## DEFENSE RELATED ENTERPRISES TO BE RESTRUCTURED AND/OR PRIVATIZED

### A STEEL MILLS

(1) D.G.F.M.ALTOS HORNOSN ZAPLA

| | | |
|---|---|---|
| Geographical Situation | : | PALPALA - PCIA DE JUJUY |
| Main Products | : | THOMAS ROLLED SHEETS, ELECTRICAL ROLLED SHEETS, ELECTRICAL STEEL BILLETS, ELECTRICAL FURNACE LINGOTS. |
| Personnel | : | 3,071 employees. |
| Assets | : | 357,807 (in thousands of US$) |
| Sales | : | 72,335 (in thousands of US$) |

(2) SOCIEDAD MIXTA SIDERURGICA ARGENTINA (SOMISA)

| | | |
|---|---|---|
| Geographical Situation | : | SAN NICOLAS-PCIA. DE BUENOS AIRES |
| Principal Products | : | SINTER, METALLIC COKE, CAST IRON, IRON PIGS, ROUGH PLANES, STEEL BILLETS, METAL PLATES (OBTAINED BY COLD OR HOT PROCESSING), TIN PLATES, RAILS, GIRDERS. |
| Personnel | : | 10,087 employees. |
| Assets | : | 2,185,634 (in thousands of US$) |
| Sales | : | 842,627 (in thousands of US$) |

(3) HIERRO PATAGONICO S.A.M. (HIPASAM)

| | | |
|---|---|---|
| Geographical Situation | : | SIERRA GRANDE-PCIA. DE RIO NEGRO |
| Main Products | : | IRON ORE PELLETS |
| Personnel | : | 1,370 employees. |
| Assets | : | 226,191 (in thousands of US$) |
| Sales | : | 27,821 (in thousands of US$) |

## B CHEMICAL AND PETROCHEMICAL INDUSTRIES

(1) PETROQUIMICA GENERAL MOSCONI S.A.

| | | |
|---|---|---|
| Geographical Situation | : | AV. INGENIERO HUMET 27-ENSENADA-PCIA DE BUENOS AIRES |
| Main Products | : | PRODUCTION OF AROMATICS, OXYALCOHOLS AND OLEFIN PRODUCTS |
| Personnel | : | 1.094 employees. |
| Assets | : | 732,542 (in thousands of US$) |
| Sales | : | 151,693 (in thousands of US$) |

(2) PETROQUIMICA BAHIA BLANCA S.A.I. Y C.

| | | |
|---|---|---|
| Geographical Situation | : | POLO PETROQUIMICO BAHIA BLANCA-BAHIA BLANCA-PCIA.DE BUENOS AIRES |
| Main Products | : | ETHYLENE |
| Personnel | : | 374 employees. |
| Assets | : | 302, 020 (in thousands of US$) |

(3) D.G.F.M. TOLUENO SINTETICO

| | | |
|---|---|---|
| Geographical Situation | : | CAMPANA-PCIA. DE BUENOS AIRES |
| Main Products | : | ALIPHATIC COMPOUNDS, AROMATIC, FAMITOL, FAMIGOA, INDUSTRIAL SOLVENTS. |
| Personnel | : | 135 employees. |
| Assets | : | 10,151 (in thousands of US$) |
| Sales | : | 15,969 (in thousands of US$) |

(4) D.G.F.M. ACIDO SULFURICO

| | | |
|---|---|---|
| Geographical Situation | : | BERISSO-PCIA.DE BUENOS AIRES |
| Main Products | : | SULPHURIC ACID |
| Personnel | : | 70 employees. |
| Assets | : | 12,302 (in thousands of US$) |
| Sales | : | 7,655 (in thousands of US$) |

(5) D.G.F.M. FABRICA MILITAR RIO TERCERO

| | | |
|---|---|---|
| Geographical Situation | : | RIO TERCERO-PCIA.DE CORDOBA |
| Main Products | : | CANNONS, MORTARS, AMMUNITIONS FOR CANNONS AND MORTARS, TURRET CASES FOR TAM AND VCTP, TURRET BASKET FOR TAM AND VCTP, TURRETS FOR MACHINE GUNS, RAILWAY CARS REPAIRING, TAM FRONT PANEL, AMMONIA, NITRIC ACID, AMMONIA NITRATE, SULPHURIC ACID AND TOLUENE. |
| Personnel | : | 1,168 employees. |
| Assets | : | 120,998 (in thousands of US$) |
| Sales | : | 21,177 (in thousands of US$) |

**C SHIPYARDS AND DOCKYARDS**

(1) ASTILLEROS Y FABRICAS NAVALES DEL ESTADO S.A.

| | | |
|---|---|---|
| Geographical Situation | : | ENSENADA-PCIA DE BUENOS AIRES |
| Main Products | : | SURFACE MILITARY VESSELS, MERCHANT SHIPS, NAVAL ENGINES, LOCOMOTIVES, BOGIES, RAILWAY CROSSING CORES, COMPONENTS FOR POWERHOUSES AND HEAVY BOILERS. |
| Personnel | : | 2,246 employees. |
| Assets | : | 445,224 (in thousands of US$) |
| Sales | : | 3,694 (in thousands of US$) |

(2) ASTILLEROS MANUEL DORREGO GARCIA S.A.

| | | |
|---|---|---|
| Geographical Situation | : | AV.CORRIENTES 672, CAPITAL FEDERAL |
| Main Products | : | SUBMARINES BUILDING AND REPAIRING |
| Personnel | : | 446 employees. |
| Assets | : | (in thousands of US$) |
| Sales | : | 893 (in thousands of US$) |

## D CIVIL PRODUCTION

(1) FABRICA MILITAR DE AVIONES (AREA MATERIAL CORDOBA)

| | | |
|---|---|---|
| Geographical Situation | : | AV. FUERZA AEREA KM.5 1/2 - PCIA.DE CORDOBA |
| Main Products | : | MANUFACTURE OF AIRCRAFT COMPONENTS, AIRCRAFT MAINTENANCE, ENGINES, AVIONICS AND FITTING FOR THE AIR FORCE AND THIRD PARTIES. |
| Personnel | : | 2,950 employees. |
| Assets | : | No balance sheet submitted since it is incorporated to the National Administration Budget. |
| Sales | : | 17,631 (in thousands of US$) |

(2) D.G.F.M. FABRICA DE VAINAS Y CONDUCTORES ELECTRICOS "ECA:

| | | |
|---|---|---|
| Geographical Situation | : | AVELLANEDA-LAVALLOL, PCIA. DE BUENOS AIRES. |
| Main Products | : | METAL DRAWING AND LAMINATION, COPPER AND ALUMINIUM CABLES. |
| Personnel | : | 135 employees. |
| Assets | : | 160,482 (in thousands of US$) |
| Sales | : | 64,807 (in thousands of US$) |

(3) D.G.F.M. FABIRCA MILITAR SAN FRANCISCO

| | | |
|---|---|---|
| Geographical Situation | : | SAN FRANCISCO-PCIA.DE CORDOBA |
| Main Products | : | PLOUGHING DISKS, ARMORED VEHICLE COMPONENTS, RAILROAD BOGIES, BLADES FOR BULLDOZERS, CASES FOR COVERED RAILWAY CARS, MAGAZINES FOR RIFLES, METALLIC BOXES FOR AMMUNITION. |
| Personnel | : | 245 employees. |
| Assets | : | 17,557 (in thousands of US$) |
| Sales | : | 4,196 (in thousands of US$) |

(4) D.G.F.M. FABRICA MILITAR GENERAL SAN MARTIN

| | | |
|---|---|---|
| Geographical Situation | : | MIGUELETES-PCIA. DE BUENOS AIRES. |

| | | |
|---|---|---|
| Main Products | : | ARMORED VEHICLES CHASSIS, REBUILDING OF RAILROAD ELECTRICAL CARS, COMMUNICATIONS EQUIPMENT, TRAILERS FOR VEHICLE TRANSPORTATION, BOGIES, DRY LOAD CONTAINERS, MULTIPURPOSE RAILWAY CARS. |
| Personnel | : | 498 employees. |
| Assets | : | 52,507 (in thousands of US$) |
| Sales | : | 5,761 (in thousands of US$) |

**E MILITARY PRODUCTION**

(1) D.G.F.M. FABRICA DE ARMAS PORTATILES DOMINGO MATHEU

| | | |
|---|---|---|
| Geographical Situation | : | ROSARIO-PCIA. DE SANTA FE |
| Main Products | : | 9MM. PISTOLS, FAL/FAP RIFLEGUNS, MACHINEGUNS FMK3 AND MAG, BAYONETS, HUNTER'S CUTLASS, S1/20 SHOTGUNS, ADJUSTMENTS FOR LITTON SIGHTS, FAP WITH TWO CHARGES. |
| Personnel | : | 850 employees. |
| Assets | : | 61,461 (in thousands of US$) |
| Sales | : | 12,377 (in thousands of US$) |

(2) D.G.F.M. FABRICA DE POLVORAS Y EXPLOSIVOS AZUL

| | | |
|---|---|---|
| Geographical Situation | : | AZUL-PCIA. DE BUENOS AIRES |
| Main Products | : | AMMONIA AND SEISMOGRAPHIC EXPLOSIVES, BLASTING AGENTS, NITROCELLULOSE FOR LACQUER - SPORT AND MILITARY POWDER. |
| Personnel | : | 307 employees. |
| Assets | : | 18,305 (in thousands of US$) |
| Sales | : | (in thousands of US$) |

(3) D.G.F.M. FABRICA FRAY LUIS BELTRAN

| | | |
|---|---|---|
| Geographical Situation | : | SAN LORENZO-PCIA. DE SANTA FE |

| | | |
|---|---|---|
| Main Products | : | AMMUNITIONS, SHELLS, EIRVO FUSES, PAMPERO ROCKETS, ROCKET LAUNCHERS |
| Personnel | : | 850 employees. |
| Assets | : | 51,470 (in thousands of US$) |
| Sales | : | 3,585 (in thousands of US$) |

(4) D.G.F.M. FABRICA PILAR

| | | |
|---|---|---|
| Geographical Situation | : | PILAR-PCIA.DE BUENOS AIRES |
| Main Products | : | PYROTECHNICAL MATERIAL |
| Personnel | : | 118 employees. |
| Assets | : | 7,248 (in thousands of US$) |
| Sales | : | 2,361 (in thousands of US$) |

(5) D.G.F.M. FABRICA DE POLVORAS Y EXPLOSIVOS VILLA MARIA

| | | |
|---|---|---|
| Geographical Situation | : | VILLA MARIA-PCIA. DE CORBOBA |
| Main Products | : | NITROCELLULOSE, TROTYL AND GEONYL, BLASTING AGENTS, DYNAMITE, PROPULSION LOADS AND POWDER. |
| Personnel | : | 562 employees. |
| Assets | : | 23,610 (in thousands of US$) |
| Sales | : | 6,197 (in thousands of US$) |

(6) TANQUE ARGENTINO MEDIANO SOCIEDAD DEL ESTADO (TAMSE)

| | | |
|---|---|---|
| Geographical Situation | : | BOULOGNE SUR MER-PCIA.DE BUENOS AIRES |
| Main Products | : | MANUFACTURE OF COMBAT ARMORED VEHICLES |
| Personnel | : | 259 employees. |
| Assets | : | 766,853 (in thousands of US$) |
| Sales | : | 4,695 (in thousands of US$) |

# EGYPT

# CONVERSION AS CONCEPTION CONFUSION

**Dr.Moustafa Ahmed Moustafa**
*Institute of National Planning, Egypt*

Conversion faces a very challenging triangle. (This triangle symbolised in its three pillars: Technology, Environment and Security).

There are interrelationships between every two pillars of this critical triangle:

eg. T - E : expressing techno-economic relationship

T - S : expressing techno-strategic relationship

S - E : expressing strategic-economic relationship.

This techno-economic-strategy is active and expresses that the "Conversion process" is in theory at the cross roads and still in the grey area and has no norms/measures with which it can be handled very certainly.

The strategic corner here expresses:
- political
- cultural
- social
- organizational
- ecological dimensions.

So, we have to understand and perceive (from my own point of view) the concept of "conversion" as processes which depend on: continuity, adjustment, feedback and retrieving interlinked processes.

The conversion process qualitative matrix consists of:
- confidence
- consistency
- continuity
- concentration
- co-operation

I wish to draw your attention to the fact that the model of thinking covers the following levels:

- international
- regional
- national
- sectoral
- enterprises.

Really, it means establishing **Value** Systems which have their impact on Development as **A PROCESS** as well.

This development process consists of an integrated set of:

- democracy
- dynamism
- dialogue
- deliberations
- demilitarization

and this expresses another or second qualitative matrix of the conversion concept.

The development process is trying to express itself sustainably. It has to guarantee and warrantee the perpetuated process of environmental dimensions and mechanisms, or let us put it as **Third Matrix** at the Matrix of E because we have the Matrix of C "First" and D "Second".

This Third Matrix or E Matrix consists of a very serious set of variables:

- legislations
- management
- finance
- re-allocation -- physical resources
- re-mobilization -- related and concerned with:

+ people's participation
+ NGO's role
+ trade unions' responses
+ syndicate's alternatives
+ scientific force mission

So, the dilemma of "Conversion" is still and will still be for a considerable period and range of time a "conception of confusion".

# FRANCE

# BETWEEN DISARMAMENT AND SOVEREIGNTY? THE ISSUE OF THE CONVERSION OF ARMS INDUSTRIES/THE FRENCH EXAMPLE

**Jean - Paul Hébert**
*CIRPES, France*

The French arms industry has already known some periods of conversion in the fifties or after the end of the Algerian war. The "direction technique des armaments terrestress" (state's ordnance, become recently "GIAT-INDUSTRIES") closed seventeen bases from 1950 to 1971 (and a third of the number of employment opportunities has disappeared).

Of the whole of the "delegation generale pour l'armement", from 1952 to 1972, forty establishments were closed or converted. But since that time, during the seventies and the major part of the eighties, the success of the French arms export has, in a way, "doped"[1] this sector and prevented "conversion" from being implemented. It was all the more difficult since the increase of unemployment made the problem of employment still more tricky (employment was implicitly considered as threatened by the hypothesis of conversion). So it's only recently that the question of conversion appears up to date in France.

However, we can note, since the mid-eighties, "de facto" conversion, but it is a "creeping" conversion, a non-deliberate conversion. It is very distinct depending on the kind of weapons. One of the most important explanations for this process is the significant European restructurations in the arms industries. Lastly, this conversion is more an adaptation to the new market situation than an actual reconversion, as it is for military bases which have been closed in Germany and France, since last year.

Actually, in France, the wish for sovereignity leads us to examine the possibility of conversion in arms industries with special care.

## In France, a "De facto" Conversion

Although there are no public decisions nor substantial comments on this phenomenon, it is true that we can notice a contraction of the arms sector in France. According to "delegation generale pour l'armement" data, the defence's employment was 310,000 staff in 1982 and, after regular decreasing, only 251,000 employees in 1991, that is to say, a twenty per cent decrease (more substantial than the decreasing of whole industrial employment, in the same period).[2] This evolution has been conversely of official provision : in 1983, during the 1984-1988 military programmation discussion, the report to the "Assemblee nationale" wrote that the voluminous expenses previous to procurement will increase from 35,000 to 310,000 workers of French defence industries.[3] That increase has not taken place. On the contrary, a decrease has taken place.

This trend is still more striking if we study the data of the most sizeable firms of the sector.

| FIRMES | EFFECTIS | EFFECTIS 1982 |
|---|---|---|
| 1991 | | |
| Thomson-CSF | 40148 | 22890 |
| Aerospatiale | 36450 | 33800 |
| Dassault Aviation | 15782 | 12390 |
| Matra | 5933 | 3300(1) |
| Snecma | 12595 | 13750 |

| | | |
|---|---|---|
| Dassault Electronique | 3238 | 4331 |
| S.N.P.E. | 6843 | 5741 |
| Turbomeca | 4351 | 3934 |
| Sagem | 7787 | 6392 |
| Giat-Industries | 17000 | 14000 |

The phenomenon is all the more clear since, in the last months of 1991, several French companies have announced different programs of employment removal : SEXTANT AVIONIQUE plans to remove 1160 employments out of 6570, for mid 1993. DASSAULT-AVIATION announces a decreasing of 700 to 800 jobs for 1995. In GIAT-INDUSTRIES, they estimate the departure of 800, in one year. In the TOULOUSE region, AEROSPATIALE will remove 400 jobs and S.E.P. about 250. THOMSON-CSF has prepared a plan for 4000 jobs' removal (of which 1600 will be in 1992). Even SNECMA (however comforted by the CFM 56's success (civil engine) foresee a small decrease of 300 persons in the next year.

To sum up, global estimations evaluate 20,000 to 50,000 the numbers of threatened employments in the French defence industry (including electronics). So, we can say there is really, from the mid eighties, a substantial "de facto" conversion of an important part of the French defence industries.

**A Creeping Conversion**

However, if we admit that this decreasing of employment in the French defence industry can be looked upon as a conversion, we must say that it is a creeping conversion, i.e. a non-deliberate conversion. Not long ago, the word "conversion" was missing in thevocabulary of persons in charge of arms policy or arms industries. It was a word only for "non-violent" or "pacifist" persons. But, nobody said "conversion" outside this club. The more audacious person said "diversification", but he did not say "conversion" or "reconversion".

What has happened in the French arms industry results from the conjunction of several causes. Some causes are specially French, some not. First, there was a certain hypertrophy of the arms industry. More than 40 per cent of the production went for export (and so 40 per cent of the employment). Hypertrophy was all the more striking as the debt and the financial burden were particularly heavy for the companies active in this sector.[6] So it was foreseeable that a volte-face of the mundial market would have very bad consequences for the French arms industry. It was all the more foreseeable because these exports were concentrated on a small number of products (mainly aeronautical products) and on a small number of countries (essentially in the Gulf). These difficulties were made worse by the not enough concurrencial conditions of productions of workshops as the "arsenaux terrestres", before they will be reorganized and consolidated in one company (GIAT-INDUSTRIES).

The other causes are well known : the fall of oil price which cut the means of traditionally buying countries, decreasing of armament expense of a lot of countries in the "third world" (except Far East and South Asia), fast evolution of the international situation since the Washington Treaty (December of 1987) on elimination of euromissiles, the fall of Berlin wall and the end of the United States/USSR face to face in Europe. Weakness or decreasing of the military budgets in the world and more and more heavy costs of the weapons' programs (weight increased by the frequent "derive ofprices" for these programs) oblige one to imagine another's conditions of production, specifications, and choices for set up.

Therefore, all these causes explain the noticeable increasing of arms employment in France. This process went along with a very strong wish of the persons in charge of companies to make a diversification of their productions. They did not want to be wholly dependant on military sales.

But this "conversion" is not preliminarily decided conversion, it is not a conversion chosen for political and strategical reasons. In the French case, it is rather the result of foreign causes : the general evolution of the international relations lead to decreasing of resources, therefore decreasing of demand from foreign customers but also from the French State; in another part, the market induces the industrialists to consider that it is wise to limit their supply. Here, it is the international market and mundial situations' pressure which are the fundamental causes. It is not the endogene choice of the French executive in charge.

So this conversion is not final. A new volte face of the mundial market should carry along probably a resumption of the French arms activity (as we can foresee it due to the intensive commercial prospection in the Far East and in South Asia, where there are still demanding markets. Moreover, conversion is a tactical adaptation but not a strategical orientation. This conversion has appeared through urgency, under the pressure of economy trends, not as a well prepared evolution, and the result is socially detrimental, in terms of loss of jobs.

Ultimately, this conversion has developed so far under restrictions on exports, which are actually not basically a result of the French strategy. What seems essential is not altered: it is the French overall capacityconcentrated on major armaments programs, because this capacity is considered as a sovereign basis. It is true that conversion has not the same meaning in France as in the USA or in the ex-USSR : the size of production lines is completely different (as is the ratio of defence budget to GNP). French employment generated by arms production is about 250,000. According to the above related trends, it may be estimated to be 210,000 in the mid-nineties. It is probably a fair estimation of the minimum required for industrial independence. By this term, the French government means the political sovereignity which is so far, the basic concept of French defence aims[7]. The above developments also explain why this conversion is to be "differential".

**A Differential Conversion**

It is a noticeable fact indeed, that this conversion does not involve the various armament industries (in the same way or to the same extent).

The nuclear armament sector, for obvious strategical reasons, is a distinct sector. It has so far been practically unreduced in terms of employment. Most of the staff are employed by the "delegation general pour l'armement" (DGA) and by the Nuclear energy board (CEA). Their employment decisions are based more on general policy than on industry or trade factors. One can only so far notice a decrease of about 300 staff in one AEROSPATIALE workshop due to the cancellation of S 45 missile (for "plateau d'Albion" site) production, and further of HADES missile. It is true that some more employment decrease may be expected from now in view of the considerations in France regarding "sufficient" level achieved by nuclear equipment, unless nuclear weapons dismantling develops enough to compensate,particularly for neutralization or destruction of the kind within the ex-USSR republics.

Navy shipyards in France are mainly under the control of the "direction des constructions navales" (DCN); their staff is around 31,000 since the seventies. Unlike the "arsenaux terrestres" (which have been converted in a public company), the navy shipyards remain state administrations. This is relevant to the fact that they are in charge of nuclear navy ships such as missile launchers, sub-marines, attach nuclear sub-marines, and nuclear aircraft carriers. They are priorities in the French armaments programs. Furthermore, these ships need important and sophisticated maintenance and repairs at short notice, even if the necessary sites and equipments are not fully used during intermediate periods. Last but not least, the DCN uses civil shipyards for part of its production, which leaves the employment of its own shipyards unaffected. In case of too much unemployment, part of the civil shipyards will bear the greatest share of unemployment. But in the future, even the DCN might have to be reduced by further reduction of the major nuclear programs : for instance, the attack nuclear sub-marine No:8 has been cancelled and the building of the new generation missile launching sub-marine (SNLE) "Le triomphant" is slowed down. The building program of the nuclear aircraft carrier "Charles de Gaulle" is delayed by several months, and the second one which was supposed to be ordered in 1990 has not actually been ordered so far.

The Air Force situation is still more complex, because of mutual intricacies within each production company, if not within each production site between the civil and defence air ministries. The French aircraft industry has slightly increased its employment during the last ten years - an addition of 5,000 to reach 120,000. But in the meantime, a remarkable change of balance has occurred:from about 30%, the ratio of civil aircraft production has jumped to nearly double: about 60%. This is due to the success of AIRBUS, of short range aircrafts, and of the CFM 56 engine. The companies involved are subject to differential evolutions : DASSAULT-AVIATION, largely devoted to the military, has suffered much more than AEROSPATIALE. But it is noticeable that, in 1992, each company aims to develop civil production. From 1980 to 1990 even DASSAULT-AVIATION has increased its ratio from 12% to 28% and aims at 45% in 1995. Similarly, AEROSPATIALE, who sold 2/3 to the military in 1985 is aiming at lowering it down to 40% in 1995. The same is true for SNECMA who sold 80% to the military in 1980 and only 36% in 1990. It is undoubtedly the aeronautic industry which has been more deeply reshaped : each company has initiated diversification, civil production development, and, in some cases, recentered their production, agreeing to cancel the non-profitable ones in order to develop R&D more aggressively.

It may be added at this stage that the successive uncertainties governing programs of European fighting aircrafts, or worldwide situation in this sector, confront the companies in their realistic programs for the future.

As far as defence electronics is concerned, it is increasing its share in defence programs. There, the situation is largely different. In this industry, it is widely known that THOMSON-CSF, during the ten last years, has become the market leader in France. Contrary to the aeronautic industry trend, THOMSON-CSF has mainly concentrated on military electronics--the so called "professional electronics". It is certainly a hazardous choice, if the general decrease becomes also the trend for electronics. But by accepting this challenge, THOMSON-CSF (a state-owned company) confirms that it isthe "national manufacturer", which the government will bc compelled to support. THOMSON-CSF staff has been subjected to significant ups and downs within the general decrease trend. But a precise analysis in terms of conversion is hardly possible, because of too many reshufflings of the corporation's companies.

Various large or small size companies in THOMSON-CSF conditions are meeting serious production program problems (SEXTANT-A VIONIQUE, SFIM, SAGEM). They are compelled to enter into significant civil diversifications (DASSAULT ELECTRONIQUE, SAT, etc).

It is for ground forces arms that conversion is the most urgent necessity. This is due to general export weakness, as well as decrease on national programs size, often as a result of price increases, and it is also due to the joint action of old and new aggressive competitors, and to technical evolution towards less conventional mechanics in modern arms systems. Several manufacturers have been closed, or merged. GIAT-INDUSTRIES, as LUCHAIRE and MUNURHIN-GIAT-INDUSTRIES, having already decreased their staff, manages now a twin strategy to become a European industry leader (buying out of EN-HERSTAL and PRB trademark) and to diversify towards civil production. So far it has been very small, but is aiming at 2,000 employment against a total of 12,000 in the mid-nineties.

The future of the French defence industry looks therefore very distinctly contrasted from one sector to another. But the general employment reduction trend is a key factor for each of them. And it is the reason why the Defence Minister recently commissioned a director for "restruction".

**These French Evolutions are Steamed by European Reshuffling**

It is not necessary here to detail the reshuffling now in progress in Europe's defence industry[8], as its large range is well known for Germany and the United Kingdom, and also for Italy, Spain, Belgium or Netherlands.

**Military Bases Policy**

Before ending this paper, some consideration should be given to foreign military bases which are in the hardcore of French reconversion. The reunification of Germany has led to the French decision aimed at a significant reduction of French military bases in Germany. Furthermore, a reshaping of the French Army known as the "plan army 2000" (plan Armee 2000) includes a drastic reduction of army bases in France : 35 military units will be reduced within 2 years in the French forces in Germany, and, in 1992 alone, the ground forces will lose 19,300 staff. These changes

This general trend is linked with the French arms industry trend. It is a strong incentive for French manufacturers to seek foreign co-operation and to enter into joint-ventures. It also favours merging of small French producers. It encourages French executives to consider Europe as a frame for their future. But this strongly involves a basic change in the relation between the State and the defence industry companies. This relation, based on the State strength as it was, is now moving towards a more conventional industrial basis. But the arms manufacturers are on the other hand looking after civil diversification to decrease their dependance on the military sector. Until recently, a good political understanding with the Administration was a safe insurance cover against future odds. But when the relations with the State become more industrial, as they are in civil production, each company has to pay attention to avoiding exclusive dependance on the military.

This extra incentive for conversion is strengthened by the European market conditions : in order to be in a good position on this bigger market, larger size is a must for each domestic manufacturer. And it could well be very unrealistic to hope for arriving at this minimum size exclusively in military production.

will bring important consequences for local environment : drastic fall in buying in small towns and, suddenly empty buildings ... The town authority sometimes try to find alternative tenants, as in EVREUX (Normandy) a university base for instance. But in other towns, as in BARCELONNETTE (in the Alps) the authority simply acts to obtain the cancellation of the withdrawal of military units. For the time being, it looks more like case per case local answers rather than a general political strategy.

## Conclusion

The large size of the reconversion of military industries in France must be stressed and it has been so far largely overlooked. This is true, irrespective of the fact that this trend is more a result of the market trend, both in France and worldwide, than a deliberate strategy. The worldwide situation is indeed the key factor.

It is also important to understand that this reduction movement is more significant on the production side than on R&D. It is necessary to take into consideration that it involves a weaker State control on arms production, where executives are now more market minded than in the past--just as otherkind of products manufacturers have been for a long time.

Considering this, the arms production conversion is not in itself the final answer to all the problems of this industry. Particularly new is the question now raised as follows: if it is true that the arms manufacturers, now partly diversified, are less dependant on the government, how will the authorities -- whether national or European -- be able in the future to accurately manage the arms production policy?

**Endnotes**

1. See J-P HEBERT "Les ventes d'armes" (SYROS-Alternatives, Coll.Alternatives Economiques. Paris 1988).

2. See J-P HEBERT "Strategic Francaise et Industrie D'armement" (F.E.D.N./La Documentation Francaise. Paris 1991). Chapitre 7.

3. See L.TINSEAU "Rapport sur le Project de Programmation Militaire 1984-1988" (Assemblee Nationale 11 Mai 1983. Doc.No.1485) page 73.

4. NB effectifs des societes non-consolidees.1) donnees non strictement comparables etant donne la filialisation des activites defense. Pour d'autres points de comparaison, voir J-P HEBERT "France : Ajuster le Poids des Armes" in "Economie et Humanisme" No.316 Janvier-Mars 1991.

5. nb : employment of society : non consolidated data. 1) data are not similar because Matra has put his defence activities in a subsidiary company. For other data See J-P HEBERT "France : Ajuster le Poids des Armes" in "Economie et Humanisme" No.316 Janvier-Mars 1991.
6. See in J.FONTANEL et J.ABEN (dir) "Economie de la defense" (ARES Vol.XII.1990-4. Decembre 1991) J-P HEBERT "L'effort Militaire Francais et ses Retombees sur L'economie".

7. In 1983, nobody foresaw this evolution but an armament engineer wrote : "a trois cent mille, on tourne tant bien que mal, a deux cent mille, il faut songer a fermer boutique" F.VARENNE "Ventes d'armes: le juridisme et l'incantation" (PROJET No.177).

8. See in B.ADAM (dir) "Memento Defense-desarmement 1989" (GRIP. Bruxelles 1989): J-P Hebert "1993: Le Grand Marche del L'armement. Redistributions en cours et bouleversements a venir" et B.ADAM (dir) "Memento difense-desarmement 1990" (GRIP Bruxelles 1990) : J-P Hebert "Reorganisations dans le secteur de l'armement en 1989".

# GERMANY

# SUSTAINABLE DEVELOPMENT - A CHALLENGE FOR THE WORLD ECONOMY

**Torsten Amelung**
*Ruhrkohle AG, Germany*

## Introduction

The economic history of the so-called industrialized nations reveals a remarkable correlation between economic growth and the use of natural resources. Even before the industrial revolution, irreversible forest exploitation for energetic use and manufacturing had depleted all the natural forests in Europe. Thereafter, in the course of the industrialization process, economic growth took up momentum, leading to an unprecedented increase of fossil fuel consumption. As a result, today's industrialized nations are highly dependent on fossil energy resources. Due to the rising energy prices and increasing public awareness with respect to environmental damages, some industrialized nations are currently attempting to decouple the use of natural resources and economic growth. In economic to decouple the use of natural resources and economic growth. In economic terms. This process involves substitution of natural resources as a factor of production for financial and human capital. Given their relatively rich endowment with these factors of production, industrialized countries can relatively easily adjust their economies.

However, this does not apply to developing and newly industrializing countries. Given their comparatively low developmental stage, these countries are eager to seize the growth potential that can be realized by using fossil fuels and their domestic natural resources. In this respect, these countries are on the way to follow the wake of industrialized nations, though the world has definitely changed in at least two respects:

Firstly, a number of environmental problems reveal a global dimension, while it is questionable, whether the global ecological system can sustain further economic development based on the growing consumption of fossil fuels and other non-renewable natural assets. The reason is that some of these assets like the ozone layer, biodiversity or the stability of the world climate have become extremely scarce on a global level.

Secondly, the use of natural resources has irreversible impacts on the welfare of these developing countries. Thus, even in their own self-interest these countries should refrain from exploiting their natural resources excessively.

As far as the first aspect is concerned, most developing countries tend to reject any responsibility for the so-called global environmental costs. This position can be supported by two arguments. On the one hand, it is undoubted that the prevailing scarcity of natural assets is primarily the consequence of economic development in today's industrialized nations. Hence, it cannot be justified that the developing countries as latecomers should refrain from using these natural resources and thus renounce valuable opportunities for economic development. On the other hand, developing countries are not prepared to give up their sovereignty over their natural resources or accept any limits for the use of these resources, if this has a detrimental impact on economic growth. This is of special importance for those countries that has unique ecosystems like, for instance, natural rain forests. As a consequence, there is an environmental component adding to theexisting tension in North-South relationships calling for multinational institutional arrangements (Simonis, 1991).

Hence, sustainable development and its implementation is not simply a national issue for developing countries but involves questions on the future distribution of economic welfare. This paper will attempt to briefly outline this global aspect of sustainable development and discuss some international mechanisms that can deal with these distributional problems. The next section will survey the main-stream definitions of sustainable development. Thereafter, section three will briefly discuss options for worldwide economic adjustments in the light of global environmental problems. Finally, section four yields a short description of prevailing concepts to solve these problems.

## Sustainable Development: Concepts and Definitions

The term "sustainable development" (SD) has emerged as the latest development catchphrase, since a number of governments and NGOs have embraced it as the new paradigm of development. Most people use the phrase interchangeably with ecologically sustainable or environmentally sound development. The World Conservation Strategy developed by the International Union for Conservation of Nature and Natural Resources in 1980 was probably the first attempt to carry the concept of sustainability beyond simple renewable resource systems (IUCN, 1980). It suggested three ecological principles for ecological sustainability, namely maintenance of essential ecological processes and life-support systems, the preservation of genetic diversity and the sustainable utilization of species and resources (Lele, 1991). Though this definition constitutes a useful starting point, it is clearly recursive in the sense that it lists theconditions for sustainability without really defining it.

A more deductive definition of SD has actually two components: development and sustainability. Even the first component is highly arbitrary because development is a value word implying change that is desirable, while there is no consensus with respect to its meaning. In this paper development is meant to cover a vector of desirable social objectives including:

- positive increases of per capita real income
- improvement in health and nutritional conditions
- educational achievement and improvement in social security
- a fairer distribution of income and access to economic resources.

Since there is a high degree of correlation between these elements, development can be measured by using a single proxy indicator. The measure most widely entertained is real income per capita. Sustainable development is then a situation in which the development vector D or per capita real income does not decrease in the long run. In theory, this implies the adoption of an indefinite time horizon. The aim is to achieve lasting development. Practical decision-making, however, requires some finite horizon. Since the choice of this horizon is arbitrary, sustainability depends on the time period considered. Moreover, there remains the question whether the development vector must be positive for each and every time period (strong sustainability) or whether only the long-term trend of the development vector should be positive (weak sustainability). For the latter, it is equivalent to say that the present value of development benefits is positive (Pearce, Barbier, Markandya, 1990).

Both concepts are subject to social and environmental conditions. The former will be skipped in this paper though the social conditions are one of the main determinants for developing societies. Environmental conditions are, however, equally important, as following the belief of SD-proponents environmental quality and economic development are interdependent and in the long run mutually reinforcing. The preservation of environmental quality can be based on the requirement that the natural capital stock should not decrease over time. Natural capital stock in this context is equivalent to the stock of all environmental and natural resource assets from oil in the ground to the quality of soil and groundwater, from the stock of fish in the oceans to the capacity of the globe to recycle and absorb carbon and other waste materials.

The rationale for a constant natural capital stock is based on a notion of justice in respect of the socially disadvantaged future generations and nature. In most developing countries it is the socially disadvantaged people that reveal a higher dependence on natural resources. In the same vein, a reduction of the natural capital stock neglects the interests of future generations. Finally, some environmentalists attach some existence value to nature itself, which is independent of their actual user value.

The constancy of the capital stock can take on several different meanings. A common interpretation is in terms of constant physical capital stock, e.g. the Brundtland Report or the World Commission for Environment and Development (Lele, 1991). This can be a sound strategy, as long as there is a high degree of uncertainty with respect to the interactions within the environment or between environment, economy and society. It reflects an aversion to risks arising from our ignorance of the functioning of ecosystems and from lowmargins of resilience towards external shocks such as drought and other stress to the ecosystems. There is, therefore, a rationale in terms of uncertainty and irreversibility for conserving the existing stock, at least until we have a clearer understanding of what the optimal stock is and how it might be identified. Though this concept seems to be very appealing for renewable resources, it cannot be applied to exhaustible resources, since by definition, any positive rate of exploitation reduces the natural capital stock irreversibly. An alternative interpretation focuses on the constant economic value of the stock. This allows for declining physical stock with rising real price over time, as long as the rise in prices is sufficient to maintain a constant economic value. However, this concept raises serious problems regarding the determination of prices and economic values for natural resources. Especially for renewable resources, current prices are less likely to reflect future scarcity.

A broader version of the constant value rule requires that the total value of all capital stocks remains constant. This concept allows for substitution between natural capital stocks and man-made capital goods or financial capital. The underlying assumption is that there is some degree of substitutability of these capital stocks both in production and consumption, while complementarities of use are assumed. The basic idea is that it makes no difference to future generations, whether they inherit man-made capital or natural capital, as long as the combined capital stock is not smaller. Under this concept, the depletion of a resource, e.g. oil can be justified, if it is compensated for by other investments generating the same income. This concept has been already used for the environmental adjustment of GNP of resource using countries. The idea is to deduct depreciations for resource use from GNP growth and thus calculate the net growth effect. Although this method looks veryappealing at first sight, it does not reduce the degree of arbitrariness which is implicit in GNP calculations. For instance, a country like Saudi Arabia, which does not reinvest all of her petrodollars for good economic reasons, would reveal two-digit negative rates of per capita GNP growth, which does not go in harmony with the undoubted positive economic development in that country.

For this reason, economists, or more correctly proponents of neoclassical environmental economics, argue in terms of costs and benefits of changing the natural capital stock (Van den Bergh, Nijkamp, 1990). A reduction of the natural capital stock is due to some economic use like, for instance, conversion of forest areas into agricultural land or natural habitats into housing development areas. Thus, each destructive activity has benefits in terms of gains from the use. The environment as a waste sink, a consumption good or a factor of production reduces the overall costs of production and consumption, thus adding to the individual and national economic welfare, even when the cost of environmental destruction is included. Hence, the optimal use of resources can be determined on the basis of such cost-benefit-analysis, which, however, reveals many conceptual shortcomings, i.e. the determination of the discount rate for future costs and benefits as well as the valuation of environmental goods in the light of uncertainty (Bojo, Maler, Unema, 1991). As this concept imposes the weakest restrictions on the preservation of the natural capital stock, such an economic optimization must produce different results when compared to the other concept. Despite this inherent weakness, the concept will be used in the following because it is more suited for the discussion of interlinks between economic mechanisms and the ecological system.

## Implementing Sustainable Development Strategies

Following Repetto (1986), sustainable development calls for a "development strategy that manages all assets, natural resources, and human resources, as well as the financial and physical assets for increasing long-term wealth and well-being. Sustainable development as a goal rejects policies and practices that support current living standards by depleting the productive base including natural resources, and that leaves future generations with poorer prospects and greater risks than our own".

More than that, an implementation of such a development strategy requires institutional adjustments and systematic changes in social values. This section will discuss only a small segment of these requirements, namely the multinational institutional capacity required to handle sustainable development policies.

There are a number of reasons as to why developing nations are not in a position to adhere to sustainable development policies:

There are a number of countries which lack the institutional capacity to assume control of their natural resources or take into account the costs caused by natural resource-based economic activities (O'Connor, Turnham, 1992).

Even given these institutional capacities, it can be in the interest of developing countries to excessively exploit their natural resources. Environmental damages usually cause economic costs in the medium turn or even in the long-run. For a large number of economic agents, these costs are of comparatively less importance, since they have to exploit their natural resources in order to guarantee their existence in theshort-tun. The same does apply to governments in these countries that have to come up with high economic growth in the short run, in order to provide basic needs for the population. Many countries even run the risk of being caught in the so-called environment-poverty trap. This means that the destruction of the environment fails to enhance the economic process, while the structure of production continues to be heavily dependent on the exploitation of natural resources. As the latter become more scarce, the development potential is ultimately lower than it was before (Pearce, Barbier, Markandya, 1991, Chap.2).

Moreover, governments and economic agents in developing countries might have different attitudes, as far as the global scarcity and the value of their natural capital stock is concerned. For instance, countries like Indonesia and Zaire are not likely to perceive their stocks of tropical rain forests as scarce from their national point of view, though tropical deforestation and the related climatic effects are global issues. Though these global externalities ultimately affect all nations, the respective developing countries are not willing to give a high priority to these future costs, given their more short-run economic goals. Nor are these countries prepared to apply resource-saving sustainable development strategies which increase costs of production and consumption, thus reducing the short-run economic welfare and growth potential. For instance, developing countries renouncing the use of their tropical forests would accept an excess burden for the preservation of the world climate (Amelung, Diehl, Chap.6), while a part of the economic benefits due to conservation are due to other countries (positive international externalities). Hence, the implementation of sustainable development policies is hindered by the fact that the respective developing countriescannot realize a moderate share in the economic benefits of this strategy.

As far as the lack of institutional capacities and environment-poverty-traps are prevailing, development aid has to put these countries in a position to realize their long-term national optimum (Runnal, 1991). After a period of transition, during which development aid supports structural adjustments and institution-building, the respective countries will, in their own interest, adhere to a strategy that is less natural -resource-consuming.

However, even if developing countries realize a national optimum, the respective level of natural resource exploitation will not be sustainable from a global point of view. A policy change towards a more resource-saving economic structure would not only involve benefits to the country that implements this strategy, but these benefits accrue to neighbour states in the case of regional environmental problems or to the world community, as far as global environmental issues are at stake. In a sense, these countries are engaged in exports of environmental services, even though these exports do not create any contribution to the country's national income. For this reason, one cannot expect that developing countries will implement sustainable development strategies to an extent that takes account of worldwide scarcities of natural capital. This, however, is a necessary condition, since it is unlikely that today's developing nations can pursue an industrialization strategy that is sustainable with respect to climate stability and biodiversity.

The next section will show that in these cases multilateral institutional mechanisms that create incentives in developing countries are needed to convey policies that are notsimply subject to a national but a worldwide optimum.

**Multinational Agreements and Institutions for Implementation of Sustainable Development Strategies**

There are mainly two environmental issues that seem to play a major role in the North-South relationship, namely the stability of the world climate (greenhouse effect, ozone layer) and the deforestation of tropical rain forests. Though these issues are often treated separately in political discussion, there is a high degree of interrelatedness between them, since 30% of all $CO_2$ emissions in 1989 were due to deforestation in the tropics (Myers, 1989). Moreover, the conservation of tropical forests can be regarded as a major contribution for preserving genetic diversity. This additional environmental service and the simple fact that $CO_2$ emission due to deforestation cannot be controlled as easily as in industry (Fischer, di Primio, Stein, 1991), justify a separation of these problems.

Moreover, there is an additional reason as to why rain forest conservation and greenhouse gas emissions are treated as different issues in the political debate. The world climate, which is likely to be destabilized through greenhouse gas emissions (Schmandt, 1992), can be regarded as common property for all nations in the world. Hence, a multinational agreement that can be negotiated between a number of nations though not necessarily all nations of the world, is likely to have an environmental impact. By contrast, rain forests and biodiversity cannot be treated as common property, since the respective natural resources are subject to the sovereignity of the respective countries, in which these resources are located. As long as these countries are not prepared to participate in a multinational arrangement,there will be hardly any long-term effect. On top of that, a number of countries have become very sensitive regarding international attempts to negotiate proposals for the conservation of these natural resources. For instance in Brazil, there have been complaints against so-called "ecological imperialism", questioning the property rights of tropical countries with respect to their forests (Amelung, 1989). For this reason, the political rationale calls for a separation of the greenhouse gas issue and the problems concerning tropical deforestation.

Nevertheless, both the stability of the world climate and biodiversity are global environmental issues in the sense that a resource-saving policy in one country does not only yield benefits to the respective country but to all other countries in the world. In economic theory, this situation calls for a compensation of countries that export environmental services. Such transfer payments make sense, as long as they lead to a more profitable accumulation of the world's natural resources and factors of production (Amelung, 1989).

Basically, the shift towards a more resource-saving structure of production involves costly processes of substitution in which natural resources as a factor of production are replaced by a combination of human and physical capital. In this respect, the technology in industrialized countries is very advanced compared to the world average. For instance, the energy efficiency of fossil-fuel-fired power plants in Germany ranks among the highest in the world (Scholtholt, 1992). The reason is that developed countries face a relative abundance of human capital and physical capital, while they have already been experiencing a regional shortage of natural resources, for instance in the form of clean air. Even though the industrialized countries are responsible for the bulk of greenhouse gas emissions, their substitution process has already advanced. This means that anadditional unit of physical/human capital will lead to a relatively modest saving of natural resources, while the opposite is true in developing countries. Since in these countries the efficiency in natural resource use is remarkably lower, a transfer of capital from industrialized to developing countries and a subsequent decline of resource wasting production, may help to fight global environmental problems more effectively than a mere substitution process in industrialized countries.

This does not mean, however, that all substitution processes have to take place in developing countries. Depending on national environmental regulations and the availability of new technologies in industrialized countries, the delinking of natural resource use and economic growth can be definitely further fostered in industrialized countries as well.

In this respect, there is a conceptual problem, as there is a need to optimize the efficiency of natural resource use both in developing newly industrializing and industrialized countries, given completely different structures of production in the individual countries. Basically, capital abundant countries have to implement transfer payments to less capital abundant countries, thus achieving the best environmental impact resulting from a substitution of natural resources for physical or human capital.

Such transfer payments put less developed and capital-poor countries in a position to finance a sustainable development program, while the losses in terms of GNP growth are compensated by the transfer payments. In theory, this form of international cooperation gives the less developed countries the opportunity to simply leap over a developmental stage based on excessive resource use. To a certain extent, thesecountries will not be forced to copy the development process in today's industrialized countries, but rather create a new economic structure that is environmentally and technologically more advanced than the respective economic structure in industrialized nations.

As far as the greenhouse-gas emissions are concerned one can think of several examples.

- Renewable energy system, e.g. photovoltaics, wind energy, biological waste, mini-hydro, etc., are relatively expensive in countries with an advanced interconnected system and unfavourable climatic conditions. The opposite holds true in many developing countries. Instead of developing their system of distribution, it may pay for some countries to concentrate on renewable energy sources and decentralized systems.

- Many countries of the third world are still to develop their power generation capacities. Given financial support in the form of transfer payments, many countries can afford the import of highly energy efficient generation capacities. On the global level, this is a more efficient approach than to force all electricity producers in industrialized countries to replace parts of their existing capacities that are not in line with the latest environmental technology but compare relatively well with average energy efficiency in the world.

- In the same vein, it should be more profitable to converse existing sinks for $CO_2$ in developing countries than engaging in costly afforestation programs in more developed and densely populated parts of the world. This economic rationale calls for aconservation of tropical forests through transfer payments.

Both for the problem of climate change and the tropical deforestation issue a number of proposals have been developed. As far as the problem of climate change is concerned, it makes sense to develop a flexible system of emission quotas and compensation payments (Cline, 1992). Mainstream models for the control of climatic gases stipulate that all emission should be divided by the world population. This average should be multiplied with every country's population size in order to yield the maximum emission allowance for each country. This emission quota has to be corrected by different climatic conditions prevailing in different countries, thus affecting the need for heating and respective energy demand. Hence, industrialized countries are given emission allowances that are lower than their actual emission, while the opposite holds true for developing countries. Industrialized countries may decide, whether they should reduce emissions or purchase emission allowances from developing countries. In the same vein, the developing countries have to decide whether they should foster their development based on the emission of climatic gases or sell their emission allowances. The latter solution ultimately calls for the implementation of a sustainable developement strategy. However, the model described above does not lead to a reduction of climate-destabilizing gas emission but rather to a stabilization. A gradual reduction requires that the sum of all national emission allowances is reduced by a prefixed percentage every year.

Yet one can think of three ways of fund-raising for these transfers. Firstly, it is possible to raise a $CO_2$ - levy or energy tax in order to give incentives for more efficient energy use and finance the purchasing of emission quotas from developing countries. Though in theory a $CO_2$ - tax comes moreclose to the polluter-pays-principle, an energy tax should be considered as more reasonable, since it takes account of limits to replacing coal and oil by gas and nuclear energy. The former is not in a position to replace other fuels while nuclear energy bears other environmental risks.

Secondly, one can allow for a free worldwide trade of greenhouse gas emission permits, so that the individual companies have to decide, whether they reduce their emissions, pay the price or provide costless energy saving technologies for developing countries in exchange for additional quotas. Both ways are efficient in a sense that they give incentives to reducing gas emissions both in industrialized and developing countries, thus fostering sustainable development in the third world or newly industrializing countries.

Likewise compensatory finance solutions have already been developed for the tropical deforestation issue. A number of organizations have stipulated a so-called tropical forest fund (Amelung, 1989). While tropical countries contribute protected primary forests to this fund, non-tropical countries have to supply the compensatory finance or resource-saving technologies, which can be used for the conservation of tropical forest areas. The GNP of the individual non-tropical countries may serve as a useful indicator for determining their share in the fund. The GNP consists of two components, as it is the product of per-capita-income and population size. Since every individual participates equally from the benefits of rainforest conservation, size of population should be the one criterion for the financial shares in the fund. Moreover, ability and willingness to pay should also be a criterion. Basically, more rich countries have a more long-term interest in global environmental issues, because the governments and individuals in these countries face less insecurity in the future,while most basic needs are guaranteed for the majority of the population in the long-run.

It has to be noted, however, that a fund relying on the GNP-key of financing has been challenged by some industrialized countries. Moreover, it has to be questioned whether the industrialized nations will agree on a system of emission quotas, which certainly will constitute a substantial financial burden for them. Nevertheless, compensatory finance solutions are the only mechanism in economic theory that can tackle global environmental problems.

The present state of empirical economic research cannot come up with estimates on the probable financial needs to implement compensatory finance solutions. However, there are reasonable indications that the compensation that has to be offered to developing countries will not exceed the ability of the more developed countries to pay.

Following a study by Ruitenbeek (1990), transfer payments of 1060.15 ECU/sq.km/year are necessary to conserve the tropical rain forest in the Korup National Park in Cameroon. This is the result of a cost-benefit-analysis which takes into account the cost of environmental destruction in the respective tropical country and thus the environmental service function for the region. Though the author concedes that this transfer should be considered as relatively high compared to other rain forest areas, let us assume for a moment that this is a reasonable on price for the bulk of the world's rain forest areas. Accordingly, a yearly transfer payment of 6.4 billion ECU or 8.5 billion $ would be sufficient to protect half of the world's tropical rain forest areas which were 12.01 million sq.km. in 1989.

The dimension of such transfer augments certainly does not go beyond the industrialized countries ability to pay, especially if one considers the future opportunities resulting from disarmament, conversion and the subsequent financial flexibility in budget decision. As table 1 in the appendix shows, the defense expenditures of the world's 20 richest nations add up to roughly 480 billion $. It is not too heroic to assume that a cut of approximately 30 percent can be achieved in the next five years. Even if we assume that only a third of this peace dividend is allocated for global environmental problems, while two-thirds are devoted to economic reforms both in Eastern European and industrialized countries, there is a financial potential of 48 billion $ that can be utilized for compensatory finance as well as research & development. By contrast, costs of climate change have been calculated to amount to 6 percent of GNP in the USA in 2010 (Chine, 1992).

Though all these calculations are highly tentative, they point to the fact that both the financial capacity and the requirements for compensatory finance are not impediments for implementing compensatory finance solutions that can manage global environmental issues. Moreover, these calculations lend support to the hypothesis that it is less costly to manage global environmental issues instead of remaining passive.

**Summary and Conclusions**

The switching from current development paths that are heavily dependent on natural resource consumption to so-called sustainable development strategies is a simple process of substituting natural resources as a factor of production for human and physical capital. As most developingcountries are not well-endowed with financial, physical and human capital, they can be hardly expected to sustain a costly substitution process, especially because such a switch in the development strategy may incur a loss in the growth momentum in these economies.

On the global level, there is some evidence that a development strategy based on the same structures of industrial development as had been prevailing in the old industrialized countries can hardly be sustained. The bottle-neck for such a natural resource - consuming strategy is given by the destabilization of the world climate and the reduction of biodiversity. Since these issues are not in the interest of the respective developing countries alone, the former have to assume a responsibility in the sense that capital should be allocated to those countries that are willing to pursue a sustainable development path.

In economic terms, such compensatory finance arrangements are much more effective than simply stipulating far-reaching environmental goals only in industrializing countries or - even worse than that - in some industrialized regions as the EC. The industrial structure of many newly industrializing and developing countries reveals much more flexibility as far as the implementation of modern and resource-saving technologies are concerned. On top of that, these countries have not yet reached a high efficiency in natural resource use, so that the capital allocated to these countries is likely to show substantially more environmental effects compared to the same amount used in industrialized countries.

As the paper shows, there are various proposals how to organize multilateral funds that can be used to tackle global environmental problems such as deforestation in tropical countries as well as climatechange. Although calculations on the costs of such fund solutions or compensatory finance agreements have not been subject to empirical analysis, the future opportunities for disarmament and the financial needs of developing countries seem to indicate that it is not the fiscal problem that is hindering a multilateral solution. As earlier negotiations have shown, there is a free-rider-problem as far as the potential payers, i.e. the industrialized countries are concerned, while most developing countries have been considering international arrangements as an attack on their sovereignity. For this reason, more international coordination is needed to find a consensus on the management of global environmental issues by international arrangements, as these issues cannot be settled by letter of intents or other less binding agreements (Simonis *et al.*, 1990).

**References**


Amelung, T., Diehl, M. (1992), *Deforestation of Tropical Rain Forests : Economic Causes and Impact on Development.* Tübingen : J.C.B. Mohr, forthcoming.

Amelung, T., (1989), "Zur Rettung der tropischen Regenwälder: Eine Kritische Bestandsaufnahme der Wirtschaftlichen Lösungsvorschläge". Die Weltwirtschaft, No.2, 152-165.

Van den Bergh, J.C.J.M., P. Nijkamp (1990), "Ecologically sustainable economic development: Concepts and model implications". Research Memorandum No.1990-2, Free University Amsterdam, Department for Economics and Econometrics. Amsterdam: January 1990.

Bojo, J., K.-G. Maler, L. Unemo (1990), *Environment and Development: An Economic Approach*. Dordrecht: Kluwer.

Chine, W.R. (1992), Global Warming: The long-term stake. In: Institute for International Economics (Ed.), The economics of global warming. Washington D.C.

Fischer, W., J.C. di Primio, G. Stein (1990), "A convention on greenhouse gases: Towards the design of a verification system". KFA Forschungszentrum Jülich GmbH, Report no. Jul-2390, Julich.

IUCN (1980), "World Conversation Strategy: Living Resource Conservation for Sustainable Development". Gland/Switzwerland: International Union for Conversation of Nature and Natural Resources, United Nations Environmental Programme, World Wildlife Fund.

Lele, S.M. (1991), "Sustainable development: A critical review". World Development, Vol.19, No.6, 607-621.

Myers, N. (1989), "Deforestation rates in the tropics and their climate implications", London: Friends of the Earth Report.

O'Connor, D., D. Turnham (1992), "Managing the environment in developing countries". OECD Development Centre, Policy Brief No. 2, Paris.

Pearce, D., E. Barbier, A. Markandya (1990), *Sustainable development: Economics and the environment in the third world.* London: Hants.

Repetto, R. (1986), *World enough and time.* New Haven: Yale University Press.

Runnals, D. (1991), "Environmental management or management for sustainable development". Erocal, D. (Ed.): *Environmental Management in Developing Countries.* OECD Development Centre Seminars. Paris: OECD.

Schmandt, J.A. (1991), "The regions and global warming: impacts and response strategies". *Journal of Environmental Law and Policy*, No. 2, 133-157.

Scholtholt, H. (1992), "Kohlekraftwerke sind längst nicht ausgereizt". Die Welt, 27.02.1992, S.2.

Simonis, U.E. (1991), "Globale Klimakonvention". WZB Papers No. FS II 91-404. Berlin Science Centre for Social Sciences Research. Berlin

Simonis, U.E., E.U. von Weizsäcker *et al.*, (1989), "The crisis of global environment - Demands of global politics". Foundation Development and Peace, Working Paper No. 3/1989. Bonn.

World Bank (1991), World Development Report 1991 - The Challenge of Development. Oxford: Oxford University Press.

**Table 1 : Defense expenditures in the 20 richest nations of the World, 1989**

| Country | Defense/ Gov. exp % | Gov.exp./ GNP % | Defense/ GNP % | GNP/ Cap % | GNP Mill.$ | Defense exp. Mill. $ |
|---|---|---|---|---|---|---|
| Singapore | 21,2 | 23,3 | 4,9 | 10450 | 28215 | 1394 |
| New | 4,8 | 45,9 | 2,2 | 12070 | 39831 | 878 |
| Zealand | 8,9 | 27,0 | 2,4 | 14360 | 241248 | 5797 |
| Australia | 12,5 | 34,6 | 4,3 | 14610 | 835692 | 36144 |
| UK | 3,6 | 47,9 | 1,7 | 15120 | 869400 | 14992 |
| Italy | 5,0 | 54,5 | 2,7 | 15920 | 235616 | 6421 |
| Netherlan | 19,9 | 31,0 | 6,2 | 16150 | 32300 | 1993 |
| ds | 4,7 | 50,7 | 2,4 | 16220 | 162200 | 3865 |
| Kuwait | 2,7 | 39,3 | 1,1 | 17300 | 131480 | 1395 |
| Belgium | 6,1 | 42,6 | 2,6 | 17820 | 1001484 | 26025 |
| Austria | 43,9 | 13,0 | 5,7 | 18430 | 27645 | 1578 |
| France | 7,3 | 23,1 | 1,7 | 19030 | 498586 | 8408 |
| UAR | 8,7 | 29,0 | 2,5 | 20440 | 1267280 | 31973 |
| Canada | 5,4 | 41,8 | 2,3 | 20450 | 104295 | 2354 |
| Germany | 24,6 | 23,0 | 5,7 | 20910 | 5202408 | 294352 |
| Denmark | 6,5 | 40,6 | 2,6 | 21570 | 183345 | 4838 |
| USA | 5,1 | 29,3 | 1,5 | 22120 | 110600 | 1653 |
| Sweden | 7,8 | 42,7 | 3,3 | 22290 | 93618 | 3118 |
| Finland | na | na | 1,0 | 23810 | 2931011 | 29310 |
| Norway | 15,1 | 13,3 | 2,0 | 29880 | 197208 | 3961 |
| Japan | | | | | | |
| Switzerla | | | | | | |
| nd | | | | | | |
| Total | na | na | na | na | 14193462 | 480447 |

Source : World Bank, *World Development Report* 1991, own calculations.

# ARMAMENT CONVERSION

**Dr.Christoph Butterwegge**
*IFK, Germany*

**Armament Conversion : Concept Definition and Defence of an Integral Conversion Concept :**

Certainly, "Armament Conversion" is no longer a foreign word for the well-informed, politically interested German people, but the problems started therewith already with the conceptualisation. The word, "Conversion", is of Latin origin and means "Rearrangement, Change". The term originates from theology, more precisely: the Catholic dogma[1] and means the passage from the "false" to the "right" belief.

This etymology of conversion concerns resources, and along with it a fundamental withdrawal from military security, armament and war preparation. Armament conversion begins even in the head, and is therefore no pure (production) technical event. During the cold war definitions concentrated themselves mostly on the production area (armament conversion), as hardly any one thought of the possibility of a troop reduction, worth mentioning, on the part of the numerically weaker NATO units (Federal army and stationing of troops), so long as two military blocks in central Europe stood facing each other, armed to the teeth. Thus it was reported under the caption "conversion" in yet another small dictionary of terms such as "armament", "disarmament", "peace", that by the beginning of the 80th year there appeared: "Reorganisation of the armament-and - war economy on the manufacture of civil goods, or on peace production"[2].

The view narrowed itself frequently on the armament industrial production areaand with that on a conception, how to render this civil or to demilitarise. The complex, comprising not only the soldiers and the civil employees of the Federal Army and the stationed troops, but also the supply industry, the retail business and the service sector (from the baker "at the corner" to the brothel owner), of a group of people which is existentially concerned with disarmament measures, remained shielded. Klaus Schomacker identifies the conversion event further, namely as "reorganisation of militarily used installations (barracks/sites-administration) and means of production of the armament industry towards civil utilisation and production of civil goods"[3].

One such definition, as Peter Wilke also suggests, certainly includes the problem of the dimension of troop reduction, but allows the rebuilding of military equipment (Example: change of battle tanks into bulldozers of the really existing tank type "Leopard" into a coal vehicle) not precisely considered as the reutilisation of installations of armament research. "With the concept of armament conversion one signifies the incident of a rearrangement of military used establishment and means of production towards civil utilisation and the production of civil goods"[4].

Werner Hünsel defines armament conversion as the "Totality of measures aimed towards the reorganisation of the personnel, material and financial capacities hitherto utilised for military purposes towards the determination of civil purposes", whereby he uncovers several conversion forms.[5] Consequently conversion allowed itself to be understoodmore concisely and significantly as rededication of militarily used resources, capacities and competencies to civil purposes.

The expression "industrial armament conversion" offers itself as a mark of a reorganisation of the production of military goods (weapons ammunition, war apparatus) towards the production of civil goods. To be distinguished therefrom is the product conversion (reutilisation of military goods), the research conversion as also the personnel or troop conversion, and the site conversion, that is the employment of former buildings and/or pieces of land (sites) with military applications towards civil purposes. Gerhard Banse speaks of it as being outside the social conversion:

> That is indeed the most difficult of the tasks before us, truly it is valid from the way of thinking and behaviourwise, to change fundamentally those which have developed in hundreds of years if not in thousands of years. These thought-and-behaviour schemes have thereby impressed that conflicts which can arise from the living together of people, races and nations with means of threat and power, in particular the military power, were given up and are still being given up. But future thought-and-behaviour modes must be marked by reason, humanity and solidarity.[6]

In connection with the troop reduction among the Federal army and the stationed forces one speaks frequently also of regional conversion, in order to differentiate this from industrial armament conversion, overlooking however in doing so that a geographical space (also through the concentration on armament production) can be constrained to cross warp with an alternative regional development.[7] Regional and armament industrial conversion accordingly form no opposition, as experts of matter themselves allege, but depend a lot more on each other. Since without the reorganisation of military production towards civil production a region remains dependent on armament, there is no comprehensive regional conversion, and vice-versa: so long as a region remains characterized by its military infrastructure and excessive troop concentration, a reorganisation on the industrial or undertaking level is still more difficult to accomplish and is of only limited value.

In the Federal Republic of Germany, relatively favourable assumptions field themselves certainly on the basis of their economic efficiency, but in this country the conversion research has been developed till now only to a bare and initial stage.[9] For a long time a conversion concept prevailed, which was not only fixed on the production area, but was also in itself slightly differential, abstract and not specific in its objective view. Typical thereof is Gerda Zellentins' attempt at a definition: "Armament conversion means the reorganisation of weapons production towards general goods production".[10]

The term "Armament Conversion" must, in the scope of a discussion on the future development of society, be rendered precise and must become politically suitable and sharply pointed. It was sufficient after the two world wars to demobilise and reorganise the production apparatus of a war economy towards civil goods production. Today one must care for an ecologically replaceable and socially compatible conversion process. The produced goods may have no military character, but also do not encumber the environment and bring with themselves material offering for the broad strata of society -- exactly so little as the productionprocess from which they result.

To produce civilian goods is an essential, but not an adequate criterion for real conversion.

An integral conversion concept must stretch out beyond armament conversion, but also demands for a social-ecological transformation of production modes which are organised on private capital and are profit-oriented, must be taken into account. One speaks in support of a conversion concept enlarged in such a manner that it is no more a bare matter of the security of peace, but one of the maintenance of the fundamentals of human life, above all, of Nature and of Environment. The demilitarisation/ civilisation of all spheres of life, an entirely new policy in many fields, not only in the key areas of industry, and the participation of the concerned (with a stress on the industrial undertaking and total economic level) are urgently needed. Conversion proceeds, constrained by a review of the political ecology and civic-parliamentary democracy, which must be supplemented by plebiscite elements (Referendum, Civic Participation, Round Table), and calls for, along with this, social emancipation exactly as democratic transformation.[11]

Peace is more than a few weapons less, armament conversion not just a concomitant symptom of disarmament. Without a breakthrough in the area of armament conversion the success of conventional disarmament measured will be kept within limits. Military dependent employees and their family members are seized by fear of losing jobs, and also the open climate influences and politics. Narrow-minded economic interests, disarmament and armament conversion stand opposed.[12] Since job security must be goal-directed and may not at all costs occur, further criteria (social-andenvironmental compatibility, social meaningfulness, irreversibility of conversion processes) of assessment of armament conversion should supervene. The Barn-Formula is to be employed, according to which, "that which is striven for must be scientifically possible, technically-technologically realisable, economically manufacturable, socially desirable and acceptable, humanly and ecologically replaceable as also clearly verifiable".[13]

Out of this results the necessity to place armament conversion in connection with the far-reaching societal changes and to acknowledge it as formation principle for a peaceful world society. "Armament conversion includes (...) the societal reproduction process in its totality".[14] It cannot be confined to division areas and individual segments of political economy, but must include collective trade levels, and has an economic, an ecological, a political, a social, a psychological and a cultural dimension. Peter Wilke points to the fact that conversion as a special case of an aimed economic structural change touches the basis of modern industrial society: "A comprehensive restructuring of social systems is necessary, and exactly this makes all conversion demands and considerations so difficult".[15]

The enlarged conversion concept, in contradiction with heresy, the market economy, would be finished effortlessly with the economic and social consequences of the disarmament process but for the fact that state interventions, taxation measures and subsidies were necessary. Against such interventions Klaus Schraden, Scientific coworker of the Institute for World Economy in Kiel, makes it valid above all that diversification (enlargement of product palette) or the military compensation through civil tasks was a saleable practice of the industry: "Finally it depends on theability of every individual armament undertaking to renounce the amenities of a state contractor, and instead of them to place the inclement climate of competition along civilian production."[16] The state quota should be reduced through disarmament, the resulting "peace dividends" should be related to a reform of the undertakings, taxation should be distributed to the private big shareholders.

On the same lines lay an advice of Ifo Institute of Munich for the federal government. It was certainly not excluded that conversion in the core area of weapons technology proves to be unsuccessful and plants must be shut down. The study, however, believes that industrial or `economic'-structured political aid measures can be waived: "Since neither on the total economic nor on the sectoral level frictions are to be expected, which could demand too much of the adaptation mechanism of the market economy, there exists no need for special State help for undertakings producing weaponry."[17]

**The example of Bremen : Armament Conversion and Development Perspective for the Unterweser Region**

Armament conversion is, in view of the private capitalistic production and parliamentary majority relations in the Federal Republic of Germany, realisable neither on the individual business, or undertaking level, nor on the central-state level. A lot more is offered by a regional and/or commune trade framework as mediation level and crystallisation core for conversion strategies,[18] particularly when it synchronises with the widening land borders, as in the case of the free Hanfeatic town, Bremen.

Bremen is the smallest Federal State with the greatest dependence on armament. By virtue of its traditional concentration of military production in the areas of ship-building, electronics and also air-and space flight, one speaks of an "Armourer of the North". The Bremen Committee for Economic Research, an Institute of the State, produced an inventory in April 1991, according to which exactly 2500 jobs are directly or indirectly affected thereby, which is so much more important, as the entire region is extremely structurally weak.

Since the early summer of 1989 the Bremish Foundation for Armament conversion and Peace Research has been in existence.[20] Established by trade unionists, work councils and scientists of the University of Bremen, it seeks to disseminate the conversion idea, to make political decision-bearers concern themselves with it, and to promote corresponding projects. Regional-political or regional-economic disposition binds it with an institutionalising strategy, on which it is mainly based, so that Bremen by virtue of its political culture offers favourable framework conditions for a continuation, through nationalisation, of former private initiatives towards armament conversion.

At the Congress, "Chances for Armament Conversion", in November 1989 an explanation was released, wherein three main demands were contained, which since then -- further concretised and better formulated -- have been remaining at the centre of the foundation work: the establishment of a Conversion Council, the creation of a regional conversion Fund and the institutionalisation of the Conversion - or peace Research at the State level.[21]

1. The Bremen Senate should appoint a Conversion council, which is composed ofrepresentatives from armament undertakings, employers' associations, chambers, trade unions, work councils (work circle "Alternative Manufacture") and the peace movement as well as scientists of different special fields. The panel would advise the senate and have a conversion plant for Bremen to be developed, which should be laid out as structure-political promotion programme

and integrated with the "Economic-Structural-Political Action Programme for Bremen and Bremenhaven till 1995". The Bremen conversion council must regularly make reports about its work, the State Parliament (Citizens) should debate upon the outcome and draw up possible consequences for legislation and administration in the two-city state.

2. Armament conversion must be promoted through financial incentives for the concerned firms, particularly smaller ones and supply works. The establishment of the "Bremen Conversion Fund" would have signal operation and would push forward not only diversification measures in the warehouse compound (marine engineering, preparation of ship building force), but would provide easy means of armament conversion, aimed at subsidy for research projects and investment aid for the development/test of adapted manufacture methods or products. A programme is thought of which promotes the reorganisation of the military production towards the civil production that is useful to society and environmentally compatible as also to the marketing of the new goods. So far the undertakings themselves are not in a position to do so. For a short period about 10 percent of the "Economic-Structural-Political Action Programme" funds (exactly 30 million German marks annually) must be rededicated in this way.

3. For the creation of a scientific infrastructure for the reorganisation process in the Unterweser region and towards the support of the conversion Council (Building of corresponding research and planning capacity) the establishment of an Interdisciplinary Institute for regional armament conversion and peace politics would be offered, which would have to critically follow the conversion process and prepare further measures.

The Bremish citizens occupied themselves often with questions of armament and site conversion. On 14th December 1990, the State Parliament called upon the Government "to establish a Conversion Fund, which under inclusion of proposed EG-and Federal programme combines the economic-political measures of the senate with this area and develops further. Thereby a concentration of small and middle level undertakings are to be taken up".[22] On 6th June 1991 the citizens concretised their suggestion and demanded the creation of a Conversion Council. Besides the Senate (state administration) representatives from undertakings and their associations, associations of workers, industrial councils, as also representatives from research and science, belong to it. "The Conversion Council has the task of presenting proposals for the programmed shaping of the Conversion Fund on the basis of the reports produced by the Senate".[23]

Now it is a matter of making the Funds and the Council (in the sense of an integral Conversion concept) useful as a democratically participating model for a social-ecological structure politics on a regional level. Armament Conversion is no self-sustainer but remains confused about direction and administration. After the formation of a so-called Ampel coalition within the state administration, formed nolonger only by the Social Democratic Party, but also the `GRÜNEN' and the Free Democratic Party, possible conflicts with regard to the further conduct/shaping of the "Bremen Model" (Burkhardt J. Huck) for Armament Conversion stand out prominently."[24]

## Endnotes

1 Vgl. Marcus Breitachwerdt. Einführung, in : ders. (Hrsg.), Rüstungskonversion. Facetten einer Strukturfrage, Stuttgart 1988, S. 10.

2. Lorenz Knorr, Kleines Lexikon Rüstung, Abrüstung, Frieden, Köln 1981. S.95 (Stichwort : Konversion)

3. Stichwort "Rüstungskonversion", in : Dieter S. Lutz (Hrsg), Lexikon Rüstung, Frieden, Sicherheit, Mit einer Einleitung von Egon Bahr, München 1987, S.275.

4. Stichwort "Rüstungskonversion", in: Ekkehard Lippert/Günther Wachtler (Hrsg.). Frieden. Ein Handwörterbuch, Opladen 1988, S. 348.

5. Siehe Werner Hansel, Möglichkeiten und Methoden der Rüstungskonversion : Die Konversionspraxis in der DDR, in : Lutz Köllner/Burkhardt J. Huck (Hrsg.), Abrüstung und Konversion. Politische Voraussetzungen und wirtscharftliche Folgen in der Bundesrepublik. Frankfurt am Main/New York 1990, S.678.

6. Gerhard Banse, Chancen fur Konversion? oder : Wie aus Schwertern Pflugscharen werden könnten, in : Wolfgang Kohnein u.a. (Hrsg.) Wissenschaft und Verantwortung, Friedliche Wege in eine gemeinsame Welt. Dokumentation des Kongresses "Weiter abrüsten Friedliche Wege in die Zukunft Wissenschaft für eine gemeinsame Welt" veranstaltet von der Naturwissenschaftler - Initiative "Verantwortung für den Frieden", Münster/Hamburg 1991, S.160.

7. Vgl.z.B. Olaf Achilles, Militärische Belastungsanalysen und Regionale Konversion. Kommunale Instrumente für eine abgerüstete Region. Alheim bei Kassel 1990.

8. Christian Wellmann. Regionale Konversion muß auf der Tagesordnung stehen. in : ÖTV Argumente 4/1990, S.11: "Regionale Konversion ist nicht nur vermutlich schwieriger, sie ist auch aktueller als rüstungsindustrielle Konversion.

9. Vgl. dazu : Ulrich Albrecht. Entwicklung und Stand der Konversionsforschung in der Bundesrepublik Deutschland im internationalen Vergleich, in : Lutz Köilner/Burkhardt J. Huck, (Hrsg.), Abrüstung und Konversion, a.a.O., S. 25 ff.

10. Gardem Zellentin. Möglichkeiten und Grenzen der Rüstungskonversion. in: (Tagungssekretariat) Für das Leben produzieren " (Hrsg), Waffenplatz Schweiz, Beiträge zur schweizerischen Rüstungsindustrie und Waffenausfuhr, Bern 1983, S.93.

11. Vgl hierzu : Christoph Butterwegge. Rüstungskonversion als Möglichkeiten alternativer Produktion und einer Entimilitarisierung der Unterweserregion in : antimilitarismus information [ami] 5/1991, S. 15 ff.

12. Vgl. Z.B. Christian Wellmann Arbeitslos durch Abrüstung? Über den Realitätsgehalt von Ängsten vor Arbeitsplatzverlust durch Abrüstungsfolge, in : Wolfgang R. Vogt (Hrsg.), Angst vorm Frieden Über die Schwierigkeiten der Friedensentwicklung für das Jahr 2000. Darmstadt 1989, S. 169 ff.

13. Siehe Gerhard Banse. Chancen für Konversion? - oder: Wie aus Schwertern Pflugscharen werden könnten, a.a.O, S. 162 ff.

14. Siegfried Schbonerr, Abrüstungswillen setzt Konversionsfähigkeit voraus in:

Gerhard Knies u.a (Hrsg.), Betriebsbedingung Frieden Herausforderungen der Hochtechnologie -- Zivilisation für eine nachmilitärische Ära, Berlin (DDR) 1990, S.207.

15. Peter Wilke, Rüstungskonversion - wirtschaftspolitische Konzeptionen und Instrumente, in : Lutz Köllner/Burkhardt J. Huck (Hrsg.), Abrüstung und Konversion, a.a.O., S.311.

16. Klaus Schrader, Warten auf die Friedensdividende, in : FAZ 6.10.1990.

17. Manfred Berger u.a., Produktion von Wehrgütern in der Bundesrepublik Deutschland, München 1991 (ifo-Studien zur Industriewirtschaft 42), S.272.

18. Vgl. dazu : Hermann Biehler u.a., Regionale Konversionsforschung -- Bestandsaufnahme und Handlungsorientierung : ebd., S. 434.

19 Vgl. dazu im einzelnen : Wolfram Elsner/Gerhard Voss, Bericht zu den Abrüstungsfolgen für das Land Bremen und zu den Handlungsmöglichkeiten, erstellt im Bremer Ausschuß für Wirtschaftsforschung, Bremen 1991.

19. Vgl. dazu : Hermann Biehler u.a. Regionale Konversionsforschung - Betandsaufnahme und Handlungsorientierung : ebd. S.434 vgl. dazu im einzelnen : Wolfram Elsner/Gerhard Voss, Bericht zu den Abrüstungsfolgen für das Land Bremen und zu den Handlungsmög lichkeiten, erstellt im Bremer Ausschuß für Wirtschaftsforschung Bremen 1991.

20. Vgl. hierzu und zum folgenden : Christoph Butterwegge, Bremen entrüstet sich : Rüstungskonversion in der Unterweserregion, in : Johannes Esser u.a. (Hrsg.), Friedenswissenschaft und Friedenslehre an Fachhochschulen und Universitäten, Frankfurt am Main 1991, S, 106 ff : ders., Möglichkeiten zur Rüstungskonversion im Frieden gestalten nach dem Kalten Krieg,. Neue Projekte der Friedenswissenchaft Beiträge der Münsteraner Tagung der Informationsstelle Wissenschaft und Frieden und des Arbeitkreises Münsteraner Wissenchaftler/Innen für den Frieden am 16/17 November 1990, Bonn 1991, S. 82 ff.

21. Vgl. Entwurf für eine Bremer Erklärung, in : Bremische Stiftung für Rüstungskonversion und Friedensforschung (Hrsg.), Chancen für Rüstungskonversion. Dokumentation und Materialien des Kongresses "Chancen für Rüstungskonversion" am 18 November 1989 in Bremen 1990. S. 38 : Forderungen der Bremischen Stiftung für Rüstungskonversion und Friedensforschung, in : Bremische Stiftung für Rüstungskonversion und Friedensforschung (Hrsg.), Rüstungskonversion vor Ort., Probleme Alternativen Perspektiven. Bremen 1991, S. 30 ff.

22. Siehe Beschlußprotokoll der Bremischen Bürgerschaft (Landtag) 12. Wahlperiode, 79, Sitzung, 14, 12. 1990, S.3.

23. Beschlußprotokoll der Bremischen Bürgerschaft (Landtag). 12. Wahlperiode, 39. Sitzung, 6.6.1991 Anlage 1. S. 2; vgl. auch : Abrüstung gefährdet 10,000 Arbeitsplätze, Bürgerschaft : Firmen müssen

umdenken, in : Bremer Nachrichten/ Weser-Kurier v. 7.6 1991.

24. Vgl. dazu neuerdings : Christoph Butterwegge, Rüstungskonversion im Rahmen einer regionalpolitischen Alternativkonzeption. Zur Diskussion über Konversionsmaßnahmen für die Unterweserregion. in : ders./Eva Senghaas-Knobloch (Hrsg.), Von der Blockkonfrontation zur Rüstungskonversion? - Die Neuordnung der internationalen Beziehungen, Abrüstung und Regionalentwicklung nach dem Kalten Krieg. Münster/Hamburg 1992 (i.E.).

# TECHNOLOGIES FOR CONTAMINATED MILITARY SITES

**Jürgen Fortmann**
*Ruhrkohle Umwelttechnik GmbH, Germany*

I am glad to present some experiences with technologies for contaminated soils. The Ruhr is still the industrial heart of our state even with all the problems of the past. The structural renewal of our towns, industrial sites and military bases requires various technologies for the decontamination of soil and groundwater.

Several former gas works and coking plants, chemical plants, refineries and air fields have been polluted by process wastes and accidents or the bombs of the last world war.

Previously, these areas had been decontaminated by excavation and disposal of the most heavily polluted soil in a hazardous waste *Land Fill*. Since the number of these contaminated sites is very high and the disposal sites are limited, this is no longer justifiable.

For decontamination there are in principle several methods at our disposal, for example:

- Encapsulation or insulation,
- Off-site removal to special waste deposits,
- Pumping of water or air out of the ground
- Physico-chemical or washing treatments,
- Microbiological treatment and
- Thermal treatment by incineration or pyrolysis.

**Thermal Treatment**

The last named treatments are the most effective ones. They have been developed in the USA and the Netherlands since 1981. They are not aggressive and do not destroythe soil minerals in the temperature range till 600$^{o}$C. The temperatures can be set in accordance with the contamination. Dust and flue gas release are very low and there are less than 1% proportional low rests of waste from the cleaning system.

**Microbiological Treatment**

The microbiological treatment makes use of the natural purification effect, that means the capacity of microorganism to convert organic contaminants into the components, water and carbon dioxide. The microbiological decomposition of organic contaminants has been practiced in Germany for the past 5 years.

For this process the oil contaminated soil is prepared by crushing, sifting and adding of straw, bark, compost and microorganism. After this the mixture is deposited in a tent 1 m thick over the soil prepared ground. The microbiologically activated soil has to be turned every month because he needs oxygen from the air. The microbiological treatment takes, 12 - 24 months depending on the contamination.

**Physico-Chemical or Washing Treatment**

The physico-chemical or washing plants have been developed since 1985 in the Netherlands and have been tested in Germany since 1986. With these systems the soil is divided in to very fine components of minerals and wastes and the greater particles. The treatment works with high pressure water streams or washing-drums, sometimes using chemical additives. The contaminants are washed out of the soil, combined with the very fine mineral particles. These substances have to be deposited on special waste deposits. This treatment ischaracterized by a proportional high rest waste of 10 - 30%, which is to be deposited.

**The thermal treatment** has the widest range of use and is suitable for all organic substances.

**The physico-chemical** treatment is suitable for light oil contaminants and certain heavy metals.

**The microbiological treatment** is characterized by the smallest field of use in cases of light oil contaminations. It takes much place and time, but it enjoys the best acceptance.

If we compare the treatments, we see the differences in their effectivity. But the most effective is not necessarily the most accepted.

Our **thermal pyrolysis concept**, working since 1987, is as follows: After passing through a soil preparation plant, the contaminated soil is fed into an indirectly heated rotary kiln, at a mass flow of 7 t/h, where it is thermically treated.

The soil will be heated in the rotary kiln, which is 25 meters long. The heat is transferred to the soil by the hot oven surrounding it. The rotary kiln plant is heated intensively from the outside by a range of nature gas burners. The produced gases are heated in an incinerator and burnt at temperatures up to 1,000 degrees centigrade and more.

The burned out flue gas is cooled in a flue gas cooler by water injection. The acid parts of the gas are neutralized by the injection of lime.

The removal of dusty particles takes place in an after-switched flue-gas deduster, a bag house. At the end of the rotary kiln,contaminated soil is moistened and discharged by a cooler auger. The soil is clean and can be brought back onto the site.

We use also thermal plants with direct heating of the soil for cleaning such areas in Germany. These plants have capacities of up to 40 t/h. With these kinds of plants, which are not yet approved in Germany, RAG works already in the Netherlands. The first plant will be built in 1993.

The concept of this plant includes the following process steps: the soil is crushed before entering the rotary drum. This soil is treated at temperatures between 450 and 600 degrees centigrade. The capacity depends on the moisture of the soil, on the concentration of contamination and on the artificial temperature. The cold soil goes into the rotary drum. The soil is heated in the first part indirectly by the heat exchange and by the hot gases. The further heating of the soil takes place in the second part of the drum. Here the soil is heated directly by the flame. After leaving the drum the soil is cooled by water and is prepared to be brought back. The flue gases leave the drum through a cyclone precipitator. After this, they are brought to a thermal after-burning chamber, where they are cleaned from organic elements between 950 and 1,200 degrees centigrade. Then the exhaust fumes are further cleaned until they leave the stack.

The cleaning results of these thermal systems are higher than 99%.

I thank you very much for your kind attention.

# UTILIZATION OF HARDWARE - OPTIONS AND CONSTRAINTS THE NVA CASE

**Dr. Hans-Joachim Giessmann**
*IFSH, Germany*

## Introduction

With the unification formally completed, Germany ranks number two -- after the CIS --on the list of disarming European States. Germany alone is challenged to reduce more than 10,000 systems (TLIs) under the settlement of the Treaty on Conventional Force Reductions in Europe (CFE).

The rate of military deployment on German soil is expected to fall by more than 50 per cent over the next 4-year period.

While national forces in all other European States are facing tasks of reductions and of restructuring as well, it is only the unified Germany that came into an inheritance of two armies, which had been completely different with respect not only to their purpose, training and structures but also to procurement of arms and armaments. The 2+4 formula provided the political and legal framework for incorporating the personnel and all property of former GDR's National People's Army (NVA) into the *Bundeswehr* from the first day of unified Germany's existence.

Because of the *Bundeswehr* is firmly integrated into NATO structures and for its generic procurement programs are adjusted to the co-operative needs of the Western Alliance, the overwhelming part of techniques to be utilized will probably originate in Ex-NVA stocks.

## The Scope of NVA Arms and Armaments

The NVA of the GDR was equipped with 3,130 types of arms and armaments including 1,200 different types of technical items. Commodities amounting to an overall value of 25 bn. DM were stored in order to supply the requirements of personnel and equipment.[1]

After the official revision of data related to CFE in February 1991 the Ex-NVA TLI stocks amounted to numbers as following.

| TLI | Original data | | Revised data | |
|---|---|---|---|---|
| | Germany | NVA | Germany | NVA |
| Main Battle Tanks | 7,133 | (2,334) | 7,000 | (2,274) |
| Armoured Vehicles | 9,598 | (5,469) | 8,920 | (5,817) |
| Artillery | 4,644 | (2,216) | 4,602 | (2,140) |
| Aircraft | 1,064 | (408) | 1,018 | (392) |
| Helicopters | 357 | (51) | 258 | (51) |

(Source: idds 3-91, Vienna - Fax, 28.2.1991)

Even though data were to be revised, the NVA's comparatively abundant mass of equipment is obvious.

The huge dimension of NVA-stocks resulted from an extensive operational planning for the former GDR territory including the final assembly of not only six active Category I divisions but also of five additional reserve divisions stockpiling 100 per cent of wartime armament, weaponry and ammunition.

Thus the official structure was exceeded by approximately 1,500 tanks and armoured troop carriers, 2,600 armoured vehicles, 2,000 pieces of artillery, 22,000 wheeledvehicles and more than 25,000 tons of ammunition and explosives.

The NVA arsenals were further enlarged by another 900 armoured carriers and 870,000 submachine guns after the army took over arsenals of non-NVA military and paramilitary institutions of the former GDR.

Another reason for crowded depots was attributed to the limited life-time and the high rate of attrition of most technical systems.

The overwhelming part of major weapon systems was technically ready for use when the NVA came to its end.

**Destruction and Reprocessing of Munitions**

Lasting for decades, the armament factories in East and West Germany were prepared for high rates of production instead of massive scrapping or even conversion of military goods.

That is why at the beginning of the nineties capacities in appropriate numbers for a massive "denaturization" of arms and armaments and for reprocessing of ammunition, were not available.

The biggest challenge of technical conversion and scrapping/disabling/ reprocessing of arms and armaments with regard to Germany was obviously related to the needs for destruction and reprocessing of enormous stocks of NVA ammunition.

Concrete numbers for ammunition stocks of the NVA provided by the former East German Ministry of National Defence[2] in 1990 enlisted 295,430 tons of ammunition split in 532 different types with a share of 50,000 tons of explosives.

Besides ammunition and explosives some additional 350,000 detonators previously stored in NVA ammunition dumps and factories for further processing were to be dismantled. Another 50,000 tons of ammunition were taken over from non-NVA depots. Estimates about waste material as a result of contaminated soil in the former GDR tend to be still vague, within a range of additional 20 to 40 percent of known stocks.

More realistic assessments are probably connected with upper-level estimates since landed property of more than 240,000 sq.km that was used by the West Group of the former Soviet Forces has to be also taken into consideration.

Owing to technical or ecological reservations only about 25,000 tons of NVA ammunition were officially intended to be incorporated into the arsenals of the *Bundeswehr* on a temporary basis. Therefore approximately 300,000 tons of ammunition were subject to utilization. At the beginning it was widely taken for granted that for political and for safety reasons most ammunition would be destroyed. The perspective has changed, however, primarily for economic and fiscal reasons.

Destruction/reprocessing of ammunition, at first, turned out to be complicated because of unproportional relations existing between huge quantities of stored ammunition on the one hand and only a limited number of appropriate capacities for manufacturing on the other.

Calculations of expenses for destruction/ reprocessing efforts of ammunition are within a range of 1,000 DM to 15,000 DM per ton, depending on its specific substances and composition. On an average basis of 5,000 DM per ton the dimension of NVA ammunition theoretically offers a dimension of contracts for industrial manufacturing amounting to an overall value of 1.5 bn. DM, including 750 mn. DM for the disposal of explosives.[3]

Prior to unification the capacity for processing and reprocessing of ammunition both in East and West Germany covered less than 9,000 tons annually. The full rate of manufacturing was achieved through regular maintenance measures.

After the massive investments that have been made since 1990 the capacity meanwhile has increased up to about 34,000 tons annually. A further expansion to a final capacity of 50,000 tons is scheduled to be achieved within a medium-term period.

Experts have estimated an annual market volume for manufacturing of ammunition in the Federal Republic of about 200 mn. DM over the next 10-year period.[4]

Another, second, complicating factor with regard to utilization of ammunition is connected with the selection and preparation of appropriate technological processing.

Due to the quantity of ammunition, the implementation of a particular cost effective but at the same time ecologically compatible processing is needed. Extreme diversification of munitions, as noted above, points to necessities of effectively diversifying reprocessing technologies particularly if a civil use of specific substances is intended.

On the other hand, for safety reasons, clear priorities have to be given to specific orders and sequences of manufacturing. Owing to technical safety risks it is fill ammunition that has to be manufactured first. Particular attention also has to be paid to obsolescent or improperly stored material and to ammunition in overcrowded depots.[5]

A major part of destruction, however, is being pursued still under questionable separate settlement. Explosives, for example, are being combusted under the open sky without exhaust-gas utilization[6] and although harmful consequent reactions for the immediate environment have become obvious already.[7]

Since the manufacturing is subject to authorization, it takes considerable time and complicated preparations for investors to meet technological requirements and safety standards. While the construction of modern low-pressure reactors would take a time of approximately nine months, technically it might well consume almost two or more years to pass necessary procedures of authorization.

If all ammunition to be utilized would be subject to manufacturing and if all criteria that were announced officially as prerequisites for industrial contracts (environmental compatibility, cost efficiency, expenditures of time, personnel and resources, safety, creation of jobs and contract guarantees[8]) would be accepted -- destruction and reprocessing of all NVA stocks and waste ammunition could take theoretically about eight to ten years.

However, the efforts to sell ammunition to foreign consumers have started to confuse the schedule -- and observers -- as well.

But even a reduced time schedule points to a third and final complicating factor: a safe and cost effective temporary storage of obsolete ammunition.

Safety regulations thereby apply not only to ammunition and explosives but also to respective dumps and depots. The NVA-ammunition was dispersed over 800 different sites on the territory of the former GDR. Estimates about additional expenditures onlyfor guarding and supervision of Ex-NVA-Depots amount to 60 to 100 mn. DM annually.

**Utilization of Arms and other Armaments**

Only very few experiences with such a complex challenge of utilization exist. Even less experiences is available with respect to the entire dissolution of an intact and modernly equipped army.

The dissolution is accompanied (and further complicated) by a parallel restructuring of the *Bundeswehr* including adjustments of longstanding procurement programs to an upper-level force of 370,000 military men. And in the course of ongoing transformation in Europe's landscape of security, further cuts in force levels are likely to occur.

The real and expected social costs of transformation (i.e. costs related to reductions in personnel) will increase pressure on the budget and procurement plans[9] of the *Bundeswehr.*

For that reason, on the one hand, certain Ex-NVA material might well be welcomed in order to bridge some gaps expected in technical supply. On the other, the *Bundeswehr* is facing many problems which are to be solved prior to any further usage and technical integration of that material.

The Federal Ministry of Defence (BMVg) decided during late summer 1991 to single out more than 60 per cent of the 1,200 different types of the former NVA's technical systems.

At least 28 per cent of these systems are scheduled for scrapping (or technical conversion), including about 8,000 major weapon systems like main battle tanks, self-propelled artillery and ships not for sale.[10]

About 36 per cent of systems to be singled out will be utilized otherwise (through purchase, military assistance etc.).

The BMVg numbered the lot of Ex-NVA weaponry to be utilized with 13,386 for major weapon systems and 1,076,612 for small arms.[11] Besides these weapons, masses of other armaments and commodities have been singled out for the purpose of utilization, for example, fuel and patrol amounting to 21 mn. DM of value.[12]

Those arms and armaments that have been singled out are facing three different practical options for utilization.

Firstly, arms and armaments can be **supplied/transferred**, to other users, including the option of re-export.

Secondly, one might think of technical **conversion** for non-military purposes.

Finally, systems not covered by the first and second option would have to be **scrapped**.

In limited numbers, some "demilitarized" major weapon systems might also serve for various "passive" purposes, for example, as training simulators, as earthbound targets or exhibits.[13]

Utilization related to conversion is confined only to either a civil use of degraded/disarmed military systems, i.e. after their substantial technical resetting or of material components after complete dismantling of systems.

*a) Supply (Transfer) of Arms and Armaments*

The supply/transfer of arms and armaments to other users through measures of purchase, military and technical assistance or re-export is politically and legally restricted(Foreign Trade Act, Arms Export Control Act, COCOM-Regulations etc.[14]). These restrictions affect also the case of arms and armaments of the Ex-NVA.

During 1990/1991 almost forty states expressed interest in acquiring Ex-NVA material. Among these countries are, for instance, former WTO states like Bulgaria, Poland, Hungary, Czechoslovakia, Romania, states of the former USSR like Lithuania, neutral states like Sweden, Finland, Switzerland, Malta and Austria, and also Third World countries like Algeria, India, Thailand, Indonesia, Madagaskar, Pakistan,

Singapore, Korea and several countries from South America.[15]

For the time being, in many cases, any kind of supply is clearly prohibited. But political transformation in some countries as well as the financial and economic burden imposed by scrapping for federal expenditures have fueled speculations about partial removals of existing transfer restrictions.

The Federal Security Council, the highest federal control authority for arms exports in Germany, has started to discuss possible contracts of sale with Poland and Czechoslovakia.[16]

Financial and capacity reasons and individual decisions of the Federal Security Council may dictate as many exports as possible and other transfers of arms and armaments can also be expected. That includes ammunition of which more than 60,000 tons were already earmarked for purchase by the end of 1991.[17]

A major part of transfers, to be effected free of charge, has been and will be realized within the framework of existing bilateral agreements and of federal directives fortechnical cooperation and assistance with other countries.[18]

During the Gulf War the Federal Government provided military assistance amounting to 760 mn. DM, including former NVA material, to the United States.[19]

Further assistance was provided, for example, to France,[20] Turkey and Egypt.[21] Israel,[22] Great Britain, Spain, the United States[23] and the Netherlands[24] received items declared as measures of technical cooperation.

Assistance with respect to equipment was given, among others, to Greece[25] and Turkey.[26]

However, though end-use control-mechanisms are unprecise or even lacking, it is possible that weapons might be transferred either directly or through third channels to tense regional spots, contradictory to the law and justice of the Federal Republic.[27]

The failed attempt of the Federal Secret Service (BND) to transfer illegally armoured air defence technics to Israel in October 1991[28] proved as evidence for both sensitivity of military transfers and existent deficits in political control.

Further control mechanisms and sanctions which the Parliament, the *Deutsche Bundestag*, voted for in January 1992, can be seen only as initial reactions to obviously existing defects of transfer control. The general sensitivity of arms transfers in terms of international and regional security implies that more attention should be paid to options of conversion.[29]

*b) Technical Conversion of Complete Systems*

Technical conversion -- as alternative option to supply -- i.e. the civil-oriented reconstruction of items formerly used for military purposes is both promising and at the same time difficult. On the one hand, once invested, resources and intellect would eventually not be wasted if existing systems could be used in civil applications and could be converted, therefore, cost effectively. The latter, however, is the most striking difficulty for a successful conversion of *complete* systems.

Original concepts for utilizing NVA arms and armaments tended either to run into expensive projects of reconstruction or to get decoupled from any rational civil demand. A popular example has been provided by several attempts in former WTO States to convert tanks for civil engineering, low cut digging, construction of tracks and roads, fire-fighting and other purposes. As late as earmarked for such civil uses.[30] Respective conversion more or less was confined to the elimination of the tank turrets and of specific military equipment.

Effective conversion, however, will hardly be accomplished simply through changes in primary functions of a system. The technical features and parameters of a tank, for example, are completely different from common civil standards and demand profiles whether the safety of the working platforms, or traffic-engineering and traffic-safety standards, or exhaust-gas emission and level road fuel consumption, or weight-power ratios are considered.[31]

Tanks, and many other systems in the Ex-NVA stock, were optimized purely for wartime usage, regardless of costeffectiveness and of environmental compatibility.

For that reason, even with a technical reconstruction consuming more than 40 or 50 percent of additional production costs, conversion of aircraft, helicopters, vessels, tanks and armoured vehicles in order to use them as complete systems are not likely to prove very effective. Production costs for these are higher compared to their civil counter-parts as are maintenance and costs for overhaul. Expenditures for consumption usually are also unduly excessive. The Soviet army lorry URAL 375-D, for example, which was extensively used also in the NVA, consumed on an average 1 litre of diesel fuel for a road distance of only 1 kilometer. Consumption, however, is not only an issue of financial costs and waste of resources but also of further pollution. Finally, the supply side by far exceeds any conceivable civil demand, so that whatever technical conversion may be considered, it will certainly not meet utilization challenges sufficiently.

*c)Decomposition and Scrapping*

As the scope for `systemic conversion' is limited, the major part of Ex-NVA arms and armaments not for sale will have to be physically eliminated i.e. dismantled, decomposed and scrapped. With respect to quantity and diversification of items, technologies, costs, time frame and environment, however, scrapping is a comprehensive and expensive challenge as well.

First of all it is necessary to implement appropriate mechanisms of organization and coordination.

Since October 3, 1990, utilization has been under the control and responsibility of the VEBEG-GmbH, an Incorporation in charge of utilizing so-called occupationalassets, a subsidiary company of the Industry-Administration Company (IVG). The VEBEG is organized as a trust company and was appointed through the Federal Government to serve as a federal institution for all issues related to utilization of military hardware. Its first task was to count and to concentrate items for evaluation and to prepare scrapping lots for submission.

According to the circumstances, the criteria for scrapping follow different priorities. As far as CFE regarded items (TLIs) are concerned, a verifiable sequence of operations in order to eliminate visible and hidden military capabilities is specified. Other criteria remain comparatively subordinated. Unilateral decisions of dismantling, probably turn primary interests towards economic and fiscal output or impact on environment. That may lead, for example, to decisions for a stepwise process of scrapping, including intermediary storage of components if adequate mechanisms of utilization are lacking.

Scrapping and all the other inherent conversional benefits require comprehensive investments in advance, covering technologies, production facilities and qualification.

Taken these necessities into consideration, the economic and financial output of scrapping tend to be rather low. In the long run, however, benefits increase, particularly resulting from absolutely diminished maintenance costs for systems in service.[32]

Technological skills to extract as many resources as possible for conversion from systems to be scrapped are unfortunately still limited. The CFE related standards of physical scrapping like fragmentation and cutting, blasting and bursting, deformation and dashing are as coarse as wasteful. Inmost cases the loss of resources will clearly outweigh potential economic benefits.

Only a few industrial competitors can engage in scrapping without substantial federal subsidies. Even fewer are capable of gaining unsubsidized profit. On an average the costs of decomposing a main battle tank can be estimated as high as 900 DM per ton. Scrapping costs for Soviet Tanks with an average weight of 42 tons therefore amount to about 38,000 DM. In other areas of industrial decomposition, the chances for making profit are comparatively less because of the lower value of material that was used for construction. Light-weight construction, for instance, that was commonly used in the former GDR for multiple ship production, reduces the residual value of scrap by 110 to 120 DM per ton.

Profit can be realized mainly through recovery treatment and purchase. Main battle tanks consist of about 18.3 tons of specially hardened high-alloy steel, 8.1 tons of low-alloy steel, 4.5 tons of non-ferrous steel and 1.160 lbs. of aluminium. About 290 lbs. of brass and bronze, 160 lbs. of copper and 104 lbs. of lead can be extracted from one single tank. Estimating a rate of 2.200 MBTs to be scrapped the overall extract of copper could amount to 176 tons. High-alloy steel amassing to 40,260 tons could be offered for civilian use, however, only if technological processing in practice would cope with provided factual opportunities.

Economic benefit has to be calculated with expenditures necessary to scrap waste material, particularly used oil (about 300 lbs. per MBT), waste grease (about 60 lbs.), hydrogen mixtures (200 lbs.), asbestos joint packings and various coats of paint.[33]

Former military producers in East Germany, particularly those suffering fromlosses of contracts and markets, perceive opportunities for scrapping as most promising to adjust without severe social damages to new profiles of production and market demands.[34]

With regard to their critical financial and social situation and to their specific capital investment, scrapping provides for the most of them, for the time being, the only source of possible contracts in order to survive.

As the scrapping of CFE related TLIs is scheduled to begin only after the treaty has come into force and because of time-consuming submission procedures imposed by the VEBEG the room of former producers of arms and armaments to manoeuvre (and to survive on its own expenses) is getting smaller. On the other hand, costs of capacities at a standstill and of storing masses of unused material, inevitably go up.

Estimates regarding the fiscal scope of scrapping amount to 3,5 bn. DM, including approximately 40 per cent only for reprocessing/destruction of ammunition and explosives.

Incentives for federal authorities to expand opportunities for supply to other users are connected with the diffuse cost-benefit ratio of scrapping in the light of an exhausted budget.[35] Even gifts -- although questionable in terms of both political morale and security policy -- may well prove more effective in economic and financial terms than scrapping or any other use.[36]

**Conclusions**

1. Scope, complexity, technical diversification and obvious technological difficulties in the course of utilization demand a **comprehensive and coordinated management** assisted by all responsible and affected authorities and institutions whetherthey are based in political administration, in the military and military engineering or in economy.

The coordination that is required between federal, state and local authorities could be managed with the help of a **Federal Inter-agency Commission in charge for technical conversion** (departments of Defence, Economy, Research and Technology, Econology, Foreign Affairs and Finances) under the auspices and the regular supervision of the Parliament.

2. As the NVA case marks only the initial phase of a far-reaching process of arms reduction and restructuring in the wake of the East-West conflict, it is necessary to **allocate adequate resources** in order to make disarmament work effectively in economic and ecological terms.

If all systems of the NVA to be singled out are counted together with those likely to be eliminated in the course of and following the reductions in personnel of the *Bundeswehr* to an upper-level of 370, 000 a scope of utilization for the next **15-Year period** of as many as **25,100 major weapon systems** of the army, the air and naval forces of Germany can be forecast. Within that period, NVA technics will most likely will be completely eliminated from active use.

3. Utilization of arms and armaments is characterized by a stress field of potential political and military implications through purchase or supply, of economic expenditures and benefits with respect to conversion, and of financial costs of scrapping.

Most technical indicators point against options of using reconstructed complete systems for civilian applications effectively. Similar conclusions can be drawn with respect to explosives.

Reprocessing of ammunition substances for a civilian industrial recovery is expensive but -- compared to systems and explosives -- rather promising and in any case useful for ecological reasons, if only the authorization procedures for new installations could be tightened. No acceptable case can be made for further burning of ammunition under the open sky.

Options of **effective** utilization in terms of economic benefit appear to be confined to either **transfers** or to a limited **component conversion** following decomposition. Although fiscal benefit usually turns out to be higher as a result of selling complete and intact systems, a strong case can be made against transfers of weapons for security reasons. **Existent legal restrictions should not fall a victim of short-term economic or even fiscal considerations**.

4. Technical conversion requires available technologics which are cost effective, compatible to environmental requirements and saving of resources.

Research and development for adequate technologies should be stronger subsidized with public and private funds. Incentives (bonuses or allowances) to private business for investments should be considered as well.

Inadequacy of technological reprocessing accompanied by deficits in organisation and time pressures effect irresponsible waste of material and fiscal and intellectual resources.

Further, sophistication, proliferation and application of modern destruction/

reprocessing technologies call for **re-allocating more resources to research and development from military R&D to disarmament R&D**. With a rate of a 5 per cent cut annually in military R&D (chapter 551 of the Federal Budget of Defence, Single Plan 14) conversion research fundsamounting to 750 mn. DM could be provided for the 1992-1996 period. The technical specifics of the NVA equipment make a reasonable case for considering not only enhanced research but also options of **cooperative projects between Germany and the CIS** or other former WTO States.

**Endnotes**

1. Figures taken from Joachim Goldbach, "Bericht über Abrüstung and Konversion", Forschung für den Frieden, Berlin 1/1990, p.121, and from Militärreform, Berlin, 1990 No.16, p.2.

2. Not including additional stocks controlled by other military and para-military organizations like border troops, `Stasi', `voluntary battle groups' etc. and ammunition stored for export purposes.

3. In late 1991 press reported about shifting priorities in favour of purchase of ammunition and even explosives, Rene Heilig, "Altlasten verschwinden einfach so im Nichts...", in: Neues Deutschland, December 10, 1991.

4. Der *SPIEGEL* No.27, July 1, 1991, p.47.

5. Jürgen Erbe, "Entsorgung von Wehrmaterial", *Soldat und Technik*, 1991, No.6, p.414.

6. Vgl. Wolfgang Flume, "Deutschland: Die Munitionsentsorgung läuft an", *Soldat und Technik* 1991 No.11, pp. 795-796.

7. Waterworks in the immediate neighbourhood of the biggest ammunition reprocessing facility near Torgau were to be closed down owing to alarming ground water contamination. *Neues Deutschland* op. cit.

8. *Berliner Zeitung*, July 20/21, 1991.

9. Further stretching of generations and life-time enhancement of technical systems are most likely.

10. Minister of Defence Gerhard Stoltenberg, quoted in: *Frankfurter Allgemeine Zeitung*, July 2, 1991.

11. *Hamburger Abendblatt*, February 1, 1992.

12. The Federal Minister of Defence (Ed.) *Ein Jahr deutsche Einheit. Eine Leistungsbilanz der Bundeswehr*. Bonn, October 3, 1990 p.16.

13. For example the teaching collection armoured battle troops/tank museum in Munster/Westfalia was given two Main Battle Tanks and two armoured troop carriers. see: *Wehrtechnik*, 1991, No.9, p.54.

14. See the information of the Bundesamt für Wirtschaft (Ed.), *Die Ausfuhr von Embargowaren. Ein Leitfaden für die Praxis*, Eschborn 1990 (with amendments of the Foreign Trade Directive until September 1991).

15. The BMVg announced officially requests only from 27 countries, Der SPIEGEL, No.45, November 4, 1991, p.33.

16. *Frankfurter Allgemeine Zeitung*, December 10, 1991.

17. 44,000t of ammunition were exported to Finland.

18. Germany has concluded respective agreements with Belgium, France, the United Kingdom, Italy, Canada, the

Netherlands, Norway, Portugal, Spain and the United States but not withGreece and Turkey! Transfers exceeding official agreements ought to be confirmed through decisions by the Federal Security Council.

19. Including for example 800 transport -and special lorries 5,500 folding-beds, 200 tents and living-containers, 60 mobile shower cabins for decontamination (@-50 LA/A/C).

20. 24 mine detectors (KMT 5,6 M 2 and MLG 60 M).

21. *Ostseezeitung*, October 11, 1992.

22. For example 7 air-surface missiles (CH-25/-29/-58), 1 ship-to-ship-missile (P-12), 2 self-guiding missile warheads (P21/22), 2 torpedoes (Saet-40), 2 ship-to-ship-missiles (P-21/22).

23. One missile launcher complex OKA (NATO-Code SS-23) and one fighter aircraft Mig 29 were delivered for technical assessment. Another 70 main battle tanks (including 59 T-72), 34 armoured troop carriers, mobile multiple missile launchers, 2 fighter helicopters Mi 24 and respective shells and other ammunition were transfered for training purposes.

24. Automatic rifles AK-74 for technical assessment.

25. Including one air defence complex (12 launchers and 408 missiles), 120 mobile four-barrelled anti-aircraft systems (ZSU 23/40, 317 mobile anti-aircraft systems (ZSU 230,150 missile launchers (RM 70), 200 armoured infantry vehicles (BMP-1), 1,000 lorries (URAL 375 D), more than 8.5 mn. charges of ammunition (calibres between 73mm and 122mm) etc.

26. Including 105,000 anti-tank weapons (RPG 7; RPG 18), 5,000 automatic guns and rifles, 20 mine detectors, 300 multiple rocket launchers (BTR-1), 10 surface-to-air missiles, 280 special fuel trucks and more than 183 mn. charges of ammunition (calibres between 7.62 mm to 14.5 mm).

27. The Society for Threatened Peoples reported, for instance, a purchase of 200,000 automatic rifles from stocks of the Ex-NVA to Aserbaidshan via Turkey. The rifles were delivered to Turkey declared as equipment assistance. *Hamburger Abendblatt*, February 2, 1992.

28. This supply consisted of missile launcher Complex SA-6 (NATO-Code Gainful) and two `spare packages' for armoured infantry vehicles (MBP), *Kölner Stadtanzeiger*, October 31, 1991.

29. A completely different case are supplies for humanitarian assistance. About 10,000 different positions were delivered, among others to Albania, Bulgaria, Russia, Kasachstan and the Ukraine.

30. Joachim Goldbach, op. cit., p.120.

31. Tanks of Soviet origin consume on average 12 ltrs. oil per hour or 30 ltrs. for a distance of 100 km. *Wehrtechnik*, 1991 No.6, p.25.

32. Ibid.

33. See Jürgen Erbe, "Entsorgung von Wehrmaterial", op. cit. p.415.

34. 14 enterprises have applied for scrapping contracts, *Hamburger Abendblatt*, December 11, 1991.

35. The ultimate federal expenditures for scrapping are estimated as high as

260 mn. DM, *Frankfurter Allgemeine Zeitung*, December 18, 1991.

36. In many cases costs of scrapping of technical systems did outweigh their actual value. Militärreform, 1990 No.14, p.2.

# DUAL USE TECHNOLOGIES, CONVENTIONAL ARMS CONTROL AND CONVERSION

**Burkhardt J. Huck**
*Stiftung Wissenschaft und Politik, Germany*

During the years of the Cold War the military use of research and development reached a level unparalleled in history. According to the estimates of Colin Norman, a quarter of the global R & D expenditures was feeding the world's military machine in the 1980s. Another similarly inaccurate calculation adds up global outlays for military R&D to some 100 billion dollars or 10% of the worldwide military expenditures.[1] In the highly industrialised countries the greater part of these outlays were spent in industries as aerospace, electronics and components, instruments, electric machinery and advanced material. This led to a great demand for related equipment like machine tools which are frequently cited as an industry that has been harmed by military research and development projects.

The research and development of numerically-controlled machine tools induced by the US Air Force in the 1950s led to the production of machines inefficient for commercial use. "The same numerically-controlled tool technology that the U.S. companies declared inefficient and unprofitable formed the basis for the Japanese machine tool industry's production".[2] This concerns also the German machine tools industry which is similarly linked with large durable goods manufacturers. The products of the machine tools industry can illustrate the problem of control with respect to the conversion of dual-use technologies. Japanese or German machine tools, developed for civil application, can nevertheless be adapted for military production. Military production lines all over the world are already working with suchmachines. But the same machines can be found in civil production lines for durable goods in the industrialised and industrialising countries. Machine tools and computer work-stations are the basis of the industries manufacturing durable goods for the world market.

The export of such equipment to all countries which were not members of the former COMECON was only in some cases regulated by the CoCom. The more or less unrestrained proliferation of machine tools and computers in combination with the diffusion of other high technologies enabled a number of less industrialised countries to develop a strong and competitive industrial basis. Most of these newly industrialised countries have already built up substantial defence industrial capabilities or have the potential to do so. Also in many less industrialised countries the development of the defence industry goes on ahead of the industrialisation of other industrial branches.

There are surely some good reasons to develop and produce fighter aircrafts in a country which is completely dependent on imported road vehicles. But since the Gulf War, the fear of the destructive power of an outlaw nation like Iraq brought the question of weapons-proliferation, which up to then had always been a second-order security concern, to the top of the agenda.

The guidelines for arms transfers to the Middle East, to which the five permanent members of the United Nations Security Council agreed in October 1991, relate exclusively to the transfer of conventional weapons. The five permanent members declared that they will consider carefullywhether proposed transfers will contribute to legitimate self-defence, to an appropriate response to security and military threats and to the capability to participate in regional or other collective arrangements. Furthermore they declared that they would avoid transfers which would probably prolong or aggravate an existing armed conflict, increase tension in a region or contribute to regional instability, introduce destabilizing military capabilities in a region, contravene embargoes or contribute to other illegitimate purposes.

The agreement does not contain guidelines for the export of dual use products or technological know-how for the development or production of conventional weapons. The most-discussed and partially-realised proposal to restrain such exports focusses on the adaptation of the CoCom-regime to countries in the Southern hemisphere also. The question as to how the export controls of CoCom hindered the development of the armaments industry in the former Warsaw Pact is not yet answered. "But there is no doubt that Western export controls have imposed essential, though practically impossible to calculate, costs on the civilian economies of proscribed countries by forcing them to rely on less technically sophisticated, and thus less efficient, technology in many branches of industries and services and by forcing them to significantly delay the realization of certain economic goals".[3]

The export controls not only affected the economies of the countries under the regulations of CoCom but also the economy of countries with related export controls. The direct costs of US export controls borne by US firms totalled upto nearly 10 billion dollars in 1985.[4]

Even after the liberalization of the core list of technologically-advanced products whose exports are controlled by the UnitedStates and other advanced industrial nations associated with the CoCom, an extension of the CoCom-regime to new countries would only raise the costs for both sides. Regarding the ongoing efforts in the industrialised countries to maintain the defence industrial base by strengthening the dual use potential, it can also be expected that such an extended regime must consequently be inefficient.

This concerns especially the spreading of the production of technologically advanced conventional weapons. The above mentioned proliferation and diffusion of dual use technology and capabilities to industrialising countries enables some of them not only to develop and produce medium range missiles or fighter aircrafts in the coming years but also enables them to produce multiple rocket launches and the related ammunition with a range of up to 100 km as is being done by more than ten of them. In most of these countries the armaments industry is the most advanced sector of the economy. Furthermore, through this industry these countries are participating in the worldwide development and diffusion of high technology. Even if spin-off or spill-over effects are difficult to calculate, it can be assumed, that the demand of the defence industry for specialised equipment, high quality raw materials, parts or subsystems and also for skilled workers and technicians contributes essentially to the development of the civilian parts of industry.

The planned controls and restraints of arms deliveries combined with similar regulations for the transfers of dual use technologies and equipment would surely slow down the arms race in these countries. But simultaneously a strict application of such regulations could also strangle the development of the civilian economy. Therefore the question arises as to how to combine the controls and restraints withcompensatory measures to facilitate the structural adjustment process of the concerned economies.

But it would be as difficult as the feasibility of regulations for control and restraints to develop and implement the related measures and means for adjustment. And this is especially true if the industrialised countries continue to develop and produce more and more sophisticated conventional weapon systems. So may be a fundamentalist approach will lead to more practical results. This approach must logically include the suppliers of weapons and the technology provided for controls and restraints. Jozef Goldblat proposed as a first step to start a "production-cum-transfer limitation process with the most advanced as well as the most destructive categories of weapons. This could be the weapons which are particularly effective in offense, or particularly threatening to the civilian populations or particularly suited to provide an incentive for a conventional first strike".[5]

Compared with the problem of controlling and restraining the transfer of armaments or dual use products and technologies with all their negative lateral effects, the problem of implementing and controlling such a production-cum-transfer limitation process seems relatively small. The first attempts to introduce such limitations into the negotiations on CFE seem to be at variance with that assumption. But meanwhile the events in the Gulf War or in Yugoslavia and the possible developments in the successor states of the Soviet Union gave a foretaste of the extent of the forthcoming conflicts without such limitations.

The specific issues, discussed with regard to confidence and security-building measures, dealing with the control of emerging military technology in the framework of the CFE, could be used as a first steptowards such limitations. "They could include the following:

- modernisation of existing weapons systems, for example, modernisation which involves more than minor adjustments;

- development of common understandings about the types of new weapons concepts which could contribute to stability and those which would not;

- project definitions of new weapons systems; for example, in initial stages of development when components of technology are assembled into a prototype, discussions would serve to assess whether a new weapons system might actually have a destabilising offensive capability or if it would be perceived to be destabilising only if deployed in great numbers in a particular manner;

- ways of controlling destabilising developments".[6]

A dual track approach which combines the control of the development and transfer of armaments with security and confidence-building measures could in the short term lead to a freeze of the conventional arms race. After implementing the necessary first steps, e.g. registers not only for the transfers of armaments but also for the holdings and related planning, in a second step the restraints proposed by Goldblat on weapons systems could be implemented.

By following that line, the problem of controlling the proliferation of dual use technologies without damaging the economic development of industrialising countries could be minimalized. Also the problem of conversion in the industrialised as well in the industrialising countries would be alleviatedby the free flow of sensitive products and technologies. The last is not only the central proposition of conversion but also of the environment's protection in this and the coming decade.

## Endnotes

1 Ulrich Albrecht, "The Aborted United Nations Study on the Military Use of Research and Development: An Editorial Essay". In: *Bulletin of Peace Proposals*, Vol.19, No.3/4, 1988, p.252.

2 David Gold, "Military R & D: A Poor Scapegoat for Flagging Economy". In: *The Bulletin of the Atomic Scientists*, January/February, 1991, p.40.

3 Andrzej Rudka, "Western Export Controls: An East European View". In: *Technology Markets and Export Controls in the 1990s*. Davifd M.Kemme (Ed.) -New York/N.Y.:New York University Press, 1991, p.23.

4 Stuart Macdonald, *Technology and the Tyranny of Export Controls: Whisper who Dares.* - Houndsmills: Macmillan 1990, p.96.

5 Cited in Ulrich Albrecht, p.256.

6 Olivia Bosch, "Arms Control after CFE I". In: *JANE's NATO Handbook 1990-91*. -London: JANE'S 1990, p.171.

# STRATEGIC CONCEPTS FOR REMEDIATION OF SOIL CONTAMINATION IN EASTERN GERMANY

**Dr. Dieter Joos**
*IABG, Germany*

The withdrawal of Soviet troops has been agreed upon for the period 1991 to 1994 by the contract between the Federal Republic of Germany and the Union of Socialist Soviet Republics. This contract also regulates the conditions for the definite stay and the modalities of the planned withdrawal of Soviet troops from the territory of the Federal Republic of Germany. The current facts provide for the transfer of about 1,000 properties (total area about 250,000 hectares) of widely varying structure and size to the Federal Property Offices in the five new Federal States. It is expected that a considerable number of residual loads such as contaminants, fuels and ammunition will be found.

In order to understand the potential risks and minimize the potential liabilities associated with the sale or management of contaminated property, the Federal Ministry of Finance, as the future owner of the Soviet properties, supported by the Federal Ministry of Environment, has selected IABG as the prime contractor for the site investigation and preliminary assessment. In view of the great number of objects and the high complexity of the problems involved a project has been created entitled

**Determination of the Areas with Possible Residual Load Contamination on the Properties of the West Group of Soviet Troops (WGT)**

The aim of the project is:

1. To record, describe and document areas with possible residual contamination on the properties of the Soviet troops;
2. To put forward emergency measures to the relevant authorities for direct aversion of risk;
3. To create the bases for later risk estimations;
4. To carry out risk assessments on selected properties (about 20) and, based on these, give rough estimations of the total environmental improvement and remediation effort required for Soviet troop properties.

The whole project is divided into the following interlinked sub-projects:

**Sub-Project A: Wide Area Monitoring and Recording of Data Sources**

The essential point of the wide area monitoring is the taking of aerial photographs. Areas with recognizable residual load contamination are identified and documented. Basic map materials are generated in order to help orientate the property inspection teams and to support the specialists like geologists or biologists in interpreting and evaluating the aerial photographs. Further property-specific information is collected from archives and, for example, from geographic and hydrological services.

**Sub-Project B: Property-Specific Determination of the Areas with Possible Residual Load Contamination**

Sub-Project B supplements the global items of sub-project A by more detailed survey and specific investigations. This is done before and during the inspections of the properties scheduled for transfer. Inparticular cases, to support the statements to be made, chemical analyses and, where necessary, exploration by boring are carried out. In every case colour photographs are taken of the possible residual load contamination areas.

The aim of sub-project B is, on the one hand, to assess the residual load situation of the property with regard to the emergency measures to be taken to avert the immediate risk and, on the other hand, to provide specific, photographic and cartographic documentation of the areas in question. This documentation is used to compile the "Environmental Appendix", which is a part of the official hand-over protocol signed by the Soviet and the German authorities. Furthermore, on the basis of the data obtained, on possible repeat inspections and on a special evaluation system, a detailed "Environmental Report" is drawn up. This "Environmental Report" represents a unified approach to a comparative evaluation of all contaminated properties. It contains a preliminary assessment of the residual loads and is intended to enable the authorities to set priorities for the measures to be undertaken (investigation program, remedial action plan).

**Sub-Project C: Risk Assessment for Selected Properties**

The sub-project C is designed to carry out complete risk assessments on some 20 properties, which are selected from the point of view of being a representative sample of the whole ensemble of all properties. An investigation programme is developed in order to evaluate the risk potential by determining carefully the sources and the extent of the residual loads and of the contaminations. On the basis of these investigations and including the latest state-of-the-art of remediation techniques, also taking into account the information on theplanned use of the individual properties, a rough estimate is made of the remediation effort required for each type of property. Based on that, an estimate of the remediation effort related to all sites is given.

**Project Management**

In agreement with the competent authorities, IABG as the prime contractor has selected about 20 co-contractors in the New Federal States, who have been shown to have at their disposal the expertise and the qualified personnel required. The close technical and organisational co-operation between these firms and IABG is ensured by the IABG subsidiary offices in Berlin and Leipzig.

The project is supervised by a project monitoring committee. Representatives of the Environmental Ministries of the New Federal States, of the Federal Ministry of Finance, of the Federal Environmental Agency, of the German Liaison Group to the Soviet Armed Forces and of IABG are members of the committee. Thus what is ensured is that all participants are continuously informed of the status of the project and its progress and, in particular, the special wishes of the members can be discussed and taken into consideration.

The rough cost estimates for the project amount to 70 million DM. More than 2/3 of this sum will be spent in the New Federal States and thus contribute to the economic rehabilitation of the country.

# Proposal to Establish a Federal Agency for Baltic Sea Marine Environment Monitoring and Protection

**Ulrike Kronfeld-Goharani and Christian Wellmann**
*PFK, Germany*

In public debate the task of "conversion" and the "peace dividend" are often dealt with as totally different, if not contradictory. Conversion is viewed upon as an attempt to safeguard vested interests and given assets (jobs, demand, profit); the peace dividend on the other hand is seen as serving the public interest by being invested into solutions for common or even global problems (environment, development). This dichotomic view is by no means self-evident or given by nature. If and to what extent conversion and the realisation of the peace dividend fit with each other depends on how and to what end conversion programs are shaped. One proposal bringing both into line shall be sketched here : the formation of a Federal Agency for Baltic Sea Marine Environment Monitoring and Protection, called BfO (*Bundesanstalt für Ostseeschutz und -überwachung*).

This agency is conceptualized to be a German contribution to a joint effort of the Baltic Sea littoral states to strengthen the effectiveness of the existing Baltic Marine Environment Protection Commission (Helsinki-Commission) by adding to it a civilian multilateral and multipurpose operative organization located in former naval bases. It will be largely composed of ex-navy personnel, and equipped with converted military planes and vessels. The task's most important aspects would be

(a) to monitor the pollution of the Baltic sea in a more comprehensive and more efficient way than it is actually done,
(b) to provide politics with a better database on the environmental situation of the Baltic Sea and on which protection measure should have first priority,
(c) to give technical and logistic support to oceanographic basic research on the very specifics of the Baltic Sea,
(d) to identify and track ships letting off oil,
(e) to fight ecological hazards caused by major accidents,
(f) to salvage the huge amounts of World War II chemical warfare agents from the bed of the Baltic Sea.

The larger part of the new Agency's employees could be recruited without necessity of requalification out of the military and civilian personnel still on the payroll of the *Bundeswehr* but earmarked to be surplus in the near future. The same is true for a major part of the small vessels and patrol airplanes the Agency is in need of. They could be granted by the *Bundeswehr* by taking them out of the excess capacities existing especially since the *Bundeswehr* took over the Peoples' Army of the former GDR. No major reconstruction would be necessary, only some new sensors would have to be integrated. Such conversion of military hardware would make much more sense than trying to sell it for dumping prices to Third World countries or to a scrap metal dealer.

Last but not the least, the real estate and the infrastructure feasible for operating the BfO-Agency do exist and are already earmarked by the *Bundeswehr* to be in excess of its future demand. The headquarter of the BfO as well as its ships should be placed at Kappeln-Olpentiz, a small harbor town in an economically depressed area and facing the loss of about 20 percent of its present workforce by the decision already taken to withdraw large parts of the existing naval garrison. A Navy Weapon School thatwill be closed in 1994 would well suit the Agency's offices, laboratories and communication facilities. The Navy's harbor at Kappeln-Olpenitz would provide enough free capacity and appropriate surrounding infrastructure to house the fleet of the Agency by 1994 when the number of vessels stationed there at present will be reduced by half. To establish the BfO at Kappeln-Olpenitz would be relatively cheap as most of the physical assets necessary are not only waiting for some kind of alternative utilization but are already owned by the German Federal Government. And in terms of regional economic development the existence of the Agency would effectively contribute to the efforts to reduce the negative economic consequences of arms limitation the town of Kappeln-Olpenitz and its people would face if there were no compensating measures at all.

The patrol aircraft in mission for the BfO-Agency should and could be stationed in Kiel-Holtenau, at present a marine airbase but also functioning as a civilian airport, the only one near Schleswig-Holstein's capital, the city of Kiel. The problem ahead is that on the one hand the military has no more interest in keeping the airfield under operation and to pay for its maintenance, for air control, weather forecast service, and so on. On the other hand, the given civilian airtraffic is far from adequate to allow the airport to be operated in economically acceptable terms. To keep it running, the civilian usage of the airport has to be intensified. The airborne wing of the BfO would contribute to this task in a technically and economically sound way: the Agency would use the same type of planes already stationed at Kiel-Holtenau.

To establish a Federal Agency for Baltic Sea Marine Environment Monitoring and Protection is, of course, not the only possibility to link the task of conversion to the one of fighting environmental hazards or to realize a peace dividend by some otheraction. But the proposal represents the type of activity relevant to meet both tasks.
It is evident that conversion aiming at minimizing the negative economic consequences of disarmament and action aiming to optimize a peace dividend need not be contradictory. The more narrow economic perspective can be integrated into the broader political context, giving conversion an objective other than merely conserving economic assets. By this comprehensive approach, conversion might be developed to a practical explication of what is necessary to meet the given challenges in the spheres of peace, development, and ecology : think globally - act locally!

**Endnote**

1 The proposal is developed in detail in: Ulrike Kronfeld-Goharani/Christian Wellmann; Saubere Ostsee -ein Ziel für Konversion. Vorschlag zur Gründung einer Bundesanstalt für Ostseeschutz und-überwachung (BfO); Projektverbund Friedenswissenschaften Kiel: Kiel, September 1991 (=PFK-texte Nr.9); Address of the authors: Projektverbund Friedenswissenschaften Kiel (PFK), c/o Universität, Preußerstr. 1-9/III, D-2300 Kiel 1, telefon and telefax : +49-431-56 04 75.

# THE USE OF REMOTE-SENSING METHODS FOR ASSESSING MILITARY EXERCISE AREAS FROM THE VIEWPOINT OF ENVIRONMENTAL GEOLOGY

**Dr. Friedrich Kühn, [1] R. Glaser, [2] S. Dech, [3] H.G. Carls, [4] F. Böker,**
*Germany*

## Landscape and Environment in the Colbitz-Letzlingen Heath

Seen as a whole, the landscape of the Colbitz-Letzlingen Heath area is part of the Altmark. Like the Fläming area further eastward, the Colbitz-Letzlingen Heath is mainly a sandy highland area formed during glacial times (see also Wagenbreth & Steiner 1982). Its present appearance is characterized by a succession of three NW-SE trending terminal moraines. These are the Planken terminal moraine in the southern part of the heath, and the Letzlingen and the Neuendorf-Brunkau terminal moraines as well as several outwash fans. Although all glacial forms of this older moraine landscape were considerably modified locally during the post-Saalian Ice Age, the moraines are still clearly visible as ridges. They are cut by numerous wide drainage channels. The Magdeburg Börde, a loess zone, adjoins them in the south. The military exercise area covers the whole series of glacial landscapes.

The glacial sediments were deposited in a basin-like hollow lined by impervious Tertiary Rupelian clays. These sediments consist mainly of gravel and sand interbedded with clay and silt. Owing to their good storage and filtering capacities, this sequence of sediments is of considerable interest for water supply.

The heath area is one of the German regions that are characterized by a dry, rather continental climate. Owing to the fact that during the last few decades this large arealost almost all its vegetation, the local climate shows sharp contrasts between day and night temperatures. Due to the high albedo, prevailing high-pressure during the summer causes strong thermals (updraughts of warm air), which in turn cause strong surface winds and thus increase desiccation and the danger of forest fires.

The Colbitz-Letzlingen Heath was once counted among the most important forests in the state of Prussia. Upto 1945, there were only a few exclusively wooded areas in Germany (Karlsch 1991). The last reforestation of the Colbitz-Letzlingen Heath took place in 1907. With the construction of the artillery range in 1935, the center of the heath area began to change. Already at that time, the villages of Salchau and Schmöggensburg had to be abandoned to make way for a firing range 30 km long and 750 m wide. After 1945, the military exercise grounds were steadily enlarged. Numerous wood fires -- in some years upto 105 fires were counted -- contributed to the present state of the heath.

## Description of the Problem

Now that the Soviet troops are gradually returning to their own country -- the withdrawal will last until 1994 -- it is essential to evaluate the present state of the abandoned areas. Due to the fact that military exercises took place here over several decades, the impact on the environment in and around these areas was often considerable.

maps, air photos and other documents, if available, as well as by the results of interviews with persons directly concerned with these areas. Often, such interviews help to detect and locate contamination stemming from military activities.

There is a general lack, and locally even an absolute lack, of written information about the geology of the military exercise area. Often, the old geological maps and sections are not sufficient to reconstruct the stratigraphic succession in the hundred meters immediately below the surface. During the second phase of the project, a so-called "environment-geology model" is derived from the geological information on the study area and its surroundings (groundwater aquifers, aquicludes, groundwater level, direction and velocity of groundwater flow). On the basis of this model, predictions can be made as to possible contaminations and their dispersion as a function of space and time. Information on possible effects on the groundwater resources in the surroundings of the military exercise area are of particular importance. Reliable and early predictions, for example, may provide a valuable basis on which to initiate target-oriented measures to protect the groundwater, particularly near waterworks that depend on it.

In the current project, relevant remote sensing methods are developed and tested on the basis of a test area at the margin of the military exercise area. This area provides all necessary prerequisites for testing a great variety of sensors and for carrying out field studies. Work in the test area has the objective of preparing a methodical basis well before it is needed, both for possible investigations carried out with the approval of the legal owner during military use, and forpreliminary assessment of the areas after they have been released.

**Remote Sensing Data**

Remote sensing data is particularly suited to obtaining information on areas to which access is restricted. An essential advantage of remote sensing methods is that they permit large tracts of land to be mapped rapidly and synchronously. This mode of acquisition of information has a multitemporal aspect, allowing the evolution of the area to be documented by a series of repeat surveys at suitable intervals. Moreover, the different sensors and systems make it possible to determine spectral properties and assess their changes with time. The whole spectrum of digital processing can be used on the data obtained. The images can also be used as maps in the field, particularly since modern maps are not always available. The essential characteristics of the most important remote sensing systems are shown in Table 1. In the case discussed here, the following types of remote sensing data were acquired:

- Thematic Mapper data recorded by the Landsat 5 satellite,
- Spectrozonal and panchromatic kosmos-KFA-1000 images,
- air photos taken by the German forces and the Allied forces during World War II.

In the summer of 1991, air photos of the test areas were taken using color and color-infrared films (CIR).

Due to their high resolution of 30 x 30 m (Landsat TM) or 7.5 x 7.5 m (KFA-1000), satellite images provide an excellent view of the exercise area, which shows continuous self-contained forest only at the margins. Certain signs of forest exploitation indicate that formerly the whole area was forested. The present state of the area can be clearlyseen on the images: the main tracks, large buildings, and firing ranges, as well as the area itself in relation to its surroundings. There are also distinct traces of maneuver movements, especially traces of movements and firing practice by tank units. To a certain extent, the remains of the vegetation can be assessed, too. Those parts of the area where the water table almost reaches the surface can be distinguished clearly. The test area is located at the south-eastern end of the maneuver area.

The available air photos of the Colbitz-Letzlingen Heath exercise area were taken during World War II and during a flight after the end of the War. They allow an overall assessment of the study area shortly before and shortly after the end of the war. Using them it is easy to identify the site as a military firing and exercise area. A firing range over 14 km long and about 750 m wide covering much of the state-owned forests of Planken and Colbitz can be recognized. Various military installations are located at the head of the firing range. Therefore, it is very probable that the soil and groundwater here are locally contaminated by waste. A valuable feature of these photos is the fact that they were taken at different times. This shows for example that the exercise area were considerably enlarged after World War II.

A detailed account of the results of evaluation of satellite images, air photos taken during the war, as well as papers, records and interviews was published by Dech, Glaser, Kühn and Carls in 1991. While the satellite image shows the general situation of the area and its setting within the natural environment of the heath, the large-scale air photo provides a detailed documentation of its present state. Large tracts of destruction and degradation of the vegetation are conspicuous over the entire area. Complete destruction of the topsoil due to the impact of numerous vehiclesencourages modifications of the area by soil erosion. As has been demonstrated by spectroradiometric measurements, which will be described later on, the darker color of the soil along the main tracks is caused by oil, possibly lubricants from tank tracks, oil from leaky gaskets or exhaust gases. The entire area shows traces of excavations (field positions, shelters and dugouts, underground pipelines and cables, etc.). All excavations are suspect : hazardous substances or live munitions may have been left behind or disposed of and may have contaminated the subsoil. Strong suspicion is particularly justified if anomalous discoloration of the soil is found or indications that hollows in the ground have been filled up. Stereoscopic evaluation of conventional air photos yields a very precise idea of possible problem areas in the field; each individual case, however, needs to be "ground-checked" separately.

Contamination can be detected in wooded areas not only by multispectral recording systems, but also by the evaluation of CIR air photos. They show healthy vegetation in strong red colors. Vegetation whose growth is adversely affected, for example by contamination in the root zone, display shades of pink to whitish instead of a strong red color. These color changes are due to the fact that the leaves reflect the light in a different way. When the site was checked, we found indications of large-scale contamination of the soil. This was confirmed by trenching and the analysis of soil samples, which yielded traces of mineral oil. In wooded areas, vegetation which is obviously "under stress" must in principle be regarded as an indicator of contamination. However, the possibility of the damage being caused by "acid rain" or other atmospheric pollution must be checked and if necessary excluded. Air photos taken during seasons when the trees have no leaves are very helpful, since they sometimes allow an unimpeded view ofthe ground surface, which is otherwise concealed.

The large tracts of land where the soil is exposed, along vehicle tracks for example, can be assessed initially using conventional black-and-white, colored or color-infrared air photos. Normally, air-photo interpretation of the textural characteristic of the surface provides unambiguous information, but assessment of material anomalies is often very uncertain. This is particularly the case with remains of humus, soil-moisture anomalies, fire scars and oil contamination, which appear on all the above types of photos as dark patches and are practically indistinguishable. Except for soil-moisture anomalies, which with a stereoscopic viewer can generally be correlated with depression in the terrain, the possible interpretations of these dark patches can only be narrowed down if multispectral remote sensing systems are used. For this purpose, extensive preliminary studies using various spectroradiometers have been carried out. These investigations focused on particularities of the diffuse reflection spectrum; knowledge of these particularities is a prerequisite for the thematic programming of multispectral remote sensors. The studies are included in the program of field investigations in the test area.

**Field Investigations Ground Checks**

As has been mentioned before, ground checks are carried out for the verification of statements made on the basis of air photos and satellite imagery. The ground checks carried out in the test area were unusually comprehensive in order to obtain information required by the methods used. In comparison, routine ground checks tend to be relatively rare or their scope has to be restricted to a minimum due to the large size of the study areas and lack of time.

Ground checks, which need not be described in detail here, allow identification of the natural surface features, the type and condition of the vegetation, visible evidence of modification or contamination of the ground, as well as the comparison with the same features as seen on air photographs and satellite images.

**Spectroradiometer Measurements**

Spectroradiometers are used to determine the spectral signatures of the surface of certain objects. Spectral signatures are a measure of the amount of the incident light reflected from a given surface as a function of the wave length. The studies include series of measurements in the laboratory as well as in the field along traverses across typical natural surface features in an unaffected condition and across ground contaminated by oil and other hazardous substances.

Most of the measurements were carried out with a Mark IV spectroradiometer using 842 individual channels within the range of 400 to 2500 nm. Initially, a Mark II hand-held radiometer was used; however, the values were not sufficiently reproducible. Measurements on the soils not covered by vegetation were generally carried out using a tripod at a range of about 1 m (field of view about 10 x 5 cm). To determine the spectral signature of the vegetation, a larger field of view was obtained by using a crane arm mounted on the vehicle.

All objects were surveyed under various conditions (position of the sun, horizontal visibility, inclination of the object). The spectral signature of each measuring point was obtained by averaging five successive spectroradiometer measurements of slightly different fields of view.

The object studied is a patch of soil contaminated with oil, across which a profile was laid consisting of several points where spectroradiometer measurements were made. The profiles were normally arranged to include the transition from the uncontaminated surroundings to the center of the contaminated part.

The spectral signatures of selected objects from the test area are shown in Fig.2. Curve No.111 represents a pale sandy soil devoid of vegetation and humus. The other curves represent the same sandy soil; in this case, the soil (113) is moist, 119 is blackened due to a fire, and 107 is contaminated by comparatively clean motor oil. The results, which were checked by several repeat measurements, show wavelength-specific changes in the energy reflected by the object in the spectral range from 400 to about 1000 nm, for which the airphoto films are sensitive. The diagnostic features, however, are extremely indistinct. Even the use of multispectral systems does not guarantee that these features are transformed into distinct shades of grey which can be interpreted reliably. The spectral range from 2000 to 2500 nm is quite different. The spectral signatures of objects compared using this range show quite distinct characteristic features, which could be used for the direct identification of oil-contaminated soils by suitable programming of multispectral airborne imaging systems, e.g. an airborne imaging spectrometer (AIS) (Fig.3).

Further studies should show whether analog conclusions can be drawn about other types of soil contamination.

**Soil Samples**

All spectroradiometer measurements in the field and checks of modifications to the vegetation and of other suspicious tracts of land were accompanied by collection of soiland water samples. As a result, most of the suspect tracts of land that had been investigated in detail were identified as being contaminated (Teschner 1991).

The primary objective of collecting and analyzing soil samples at the points where spectrometer measurements were made was to correlate the intensity of typical signature characteristics with the degree of contamination.

**Conclusions**

The data, results and interpretations in this paper must be regarded as a part of ongoing studies and are therefore preliminary in character.

In principle, it can be said that remote sensing methods can contribute a good deal towards the identification and assessment of the present state of military exercise areas, which are invariably subjected to highly unusual conditions. It is possible to delineate the potential dispersion trends of contaminants in the subsoil by using the results obtained from the above methods in combination with models showing the aquicludes and aquifers underground and the direction and velocity of groundwater flow. Potential environmental risks to the land surrounding the exercise areas then become predictable.

In general, three types of information are available upon which an environmental assessment of a given military exercise area can be based :

- Historical air photos, ordnance survey maps, thematic maps, files, publications, old geological mapping reports, interviews with people who know the area, etc.;
- Satellite data for assessing the general present state of the area and its setting in the surrounding landscape;
- Airborne photographic data for detailed identification and assessment both of the present state of the area and of indications of potential contamination of the soil and groundwater.

Future studies will focus on the spectral reflection behaviour of soils contaminated by inorganic contaminants as well as on the testing of narrow-band multispectral systems within the spectral range of 2000 to 2500 m.

In general, routine use of remote sensing methods necessarily involves some restrictions. In addition, the time and money that can be spent on data acquisition, evaluation and interpretation are generally limited to the absolutely necessary. The value of a methodical approach to these problems is primarily to be seen in the prospect that, as more knowledge and experience are acquired, routine use of these methods may become more target-oriented, more problem-oriented and less expensive.

**Final Remarks**

The authors' thanks are due to the staff of the Remote Sensing Section, and the Head of the Section, Dr.D.Bannert, and the Head of the Organic Geochemistry, Hydrocarbons Research Subdivision, Prof.Dr.Hollerbach, of the Federal Institute for Geosciences and Natural Resources (BGR), as well as to the Head of the Colbitz Waterworks, Mr.Hellmann, for their assistance, support, and advice.

All photographic flights were carried out by Mr.F.Böker (BGR, Hannover).

**Endnotes**

1 Federal Institute for Geosciences and Natural Resources (BGR), Berlin Branch.
2 Geographic Institute, Würzburg University.
3 German Aerospace Research Establishment, Oberpfaffenhofen.
4 Aerial Photo Data Bank, Würzburg.
5 Federal Institute for Geosciences and Natural Resources (BGR), Hannover.

**References**

Dech, S., Glaser, R., Kühn, F., & Carls, H.G.: Ökologische Probleme durch Rüstungsaltlasten in der Colbitz-Letzlinger Heide, DLR-Nachrichten, Heft 64 (August 1991), p.24-29.

Karlsch, G. : Streiflichter aus der Forstgeschichte der Colbitz-Letzlinger Heide, Forst und Holz, 46 (1991) 5, p.118-120.

Teschner, M. : Untersuchung von Boden- und Wasserproben des Truppenübungs platzes Letzlinger Heide (Sachsen - Anhalt) auf den Gehalt von Mineralölkohlenwasserstoffen und Verunreinigungen mit halogen organischen Verbindungen, unveröffentl. Auswertebericht, BGR, Hannover (1991), 8, p.

Wagenbreth, O. & Steiner, W. : Geologische Streifzüge -- Landschaft und Erdgeschichte zwischen Kap Arkona und Fichtelberg, Deutscher Verlag für Grundstoffindustrie, Leipzig (1982), p.204

## List of Figures

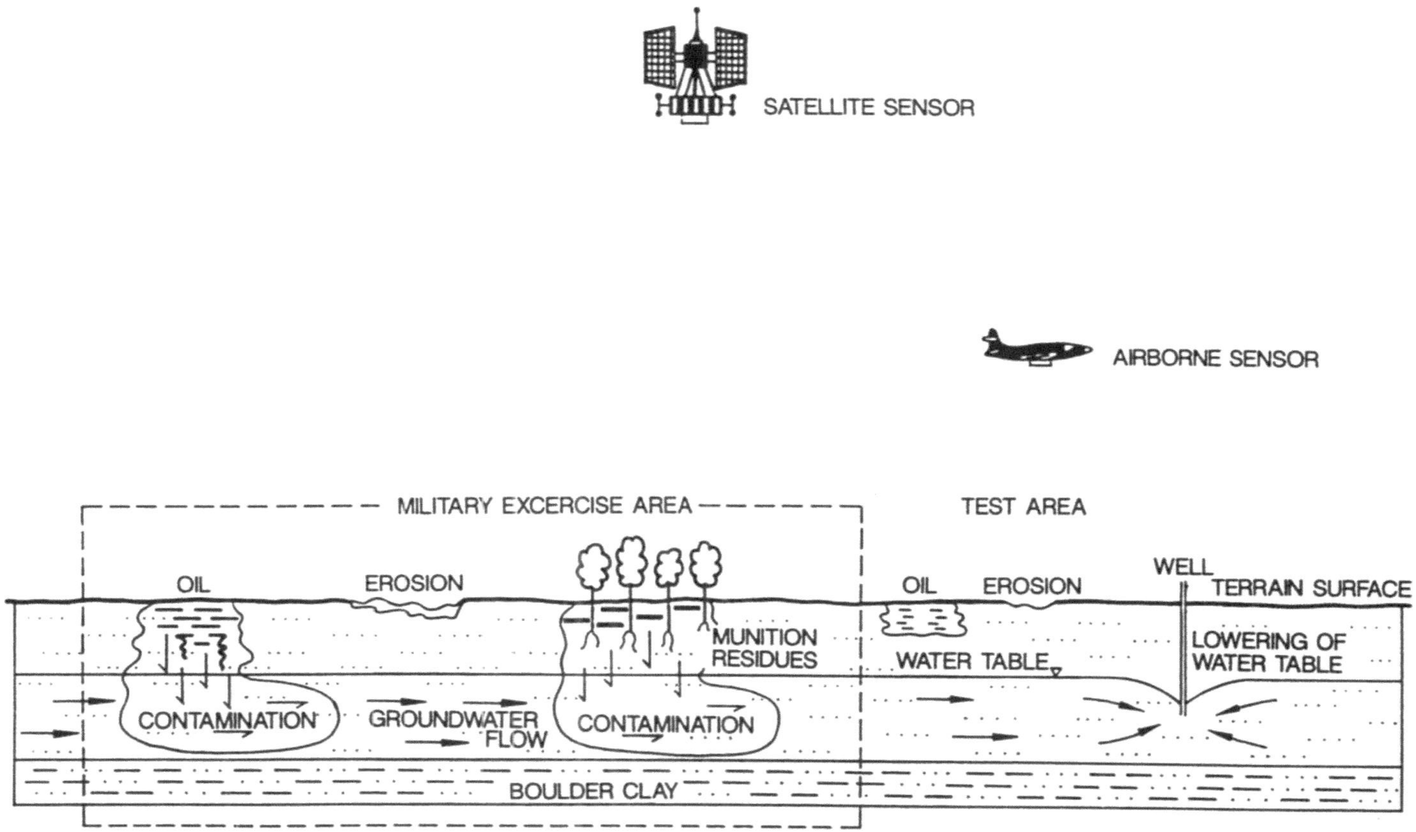
SATELLITE SENSOR
AIRBORNE SENSOR
MILITARY EXCERCISE AREA
TEST AREA
OIL
EROSION
MUNITION RESIDUES
OIL
EROSION
WELL
TERRAIN SURFACE
WATER TABLE
LOWERING OF WATER TABLE
CONTAMINATION
GROUNDWATER FLOW
CONTAMINATION
BOULDER CLAY

Figure 1

# spectral signatures

sandy soils

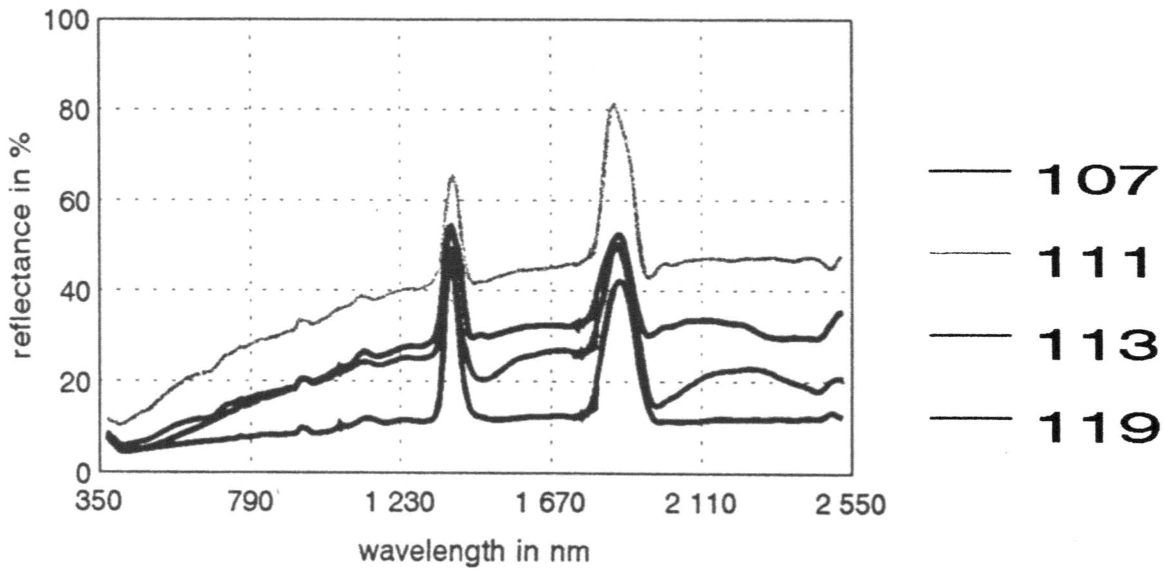


BGR (F. Kühn et al.)

Figure 2

# spectral signatures

sandy soils

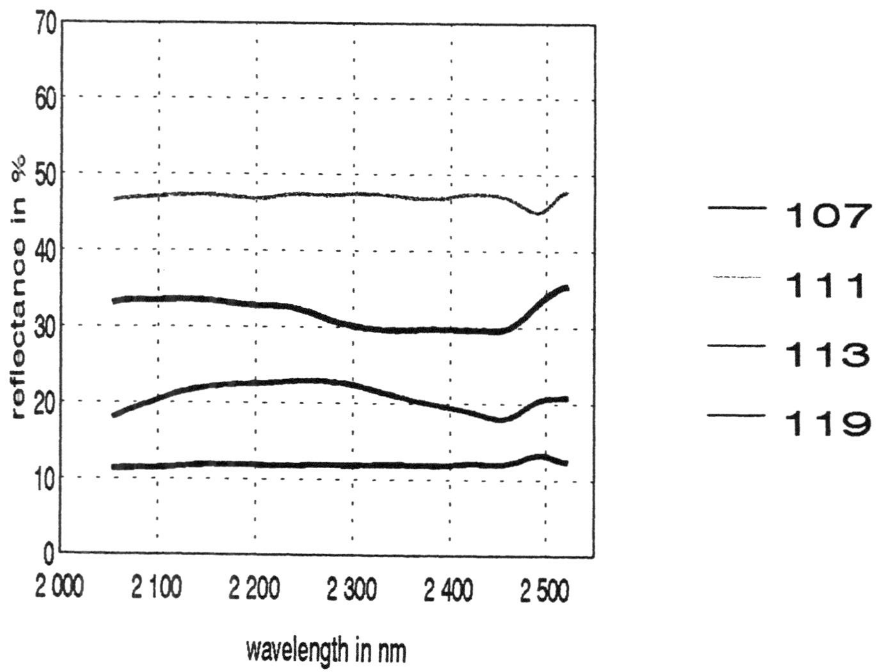


BGR (Kühn et al.)

Figure 3

# THE ECONOMIC COSTS OF PEACE - AN ASSESSMENT OF THE BURDEN TO OVERCOME THE BURDEN OF THE MILITARY-BUREAUCRATIC-INDUSTRIAL COMPLEX

**Dr. Peter Lock**
*Berghof Institute for Peace and Conflict Research, Germany*

For more than twenty years the intricate relationship between armaments and development ranked high on the agenda of the United Nations. Reports commissioned by the Secretary General[1] and dozens of publications produced by many agencies of the UN-system, particularly during the last decade, give ample evidence of the perceived conceptual linkage of the arms race with the manifest incapacity of the existing world order to cope with the imperative development needs of the South.

International non-governmental organisations were even more outspoken in accusing the arms race of having detrimental effects on the fate of the South. The rhetoric of the debate on development yielded the catching concept "peace dividend". Initially the envisioned peace dividend was labelled for overcoming the crying needs of the Third World.

But with the courageous initiative of Gorbachev to end the absurdity of the strategic arms race unilaterally, the logic was shifted to the sheer survival of the Soviet Union. The peace dividend, though still cloaked in declarations of global responsibility, was to play the role of an economic asset that Gorbachev had hoped for in order to advance the reform of the stifling Soviet system. Abstaining from strategic competition with the United States of America would allow the freeing of the enormous resources wasted (virtually unrestrained) in an economy that long since had become dominated by the logic of a permanent war economy.[2] The military-industrial complex had entrenched itself sodeeply in the economic system that it had turned into a self-serving dominant power within the system, not even serving adequately the interests of the Red Army any longer.

The Soviet MIC had turned static and was incapable of generating advanced technology which the military deemed indispensable to meet the target set by the Supreme Soviet, namely strategic parity with the other Superpower as well as an increasing capability to project military power. The Breshnev doctrine had reached an industrial deadlock. The astonishing consequences of this systemic deadlock were sectors within the Red Army militating for radical economic reforms as a precondition to overcome the industrial impediments hindering the Red Army[3] to fulfill its role professionally.

Though military tensions between East and West have diminished considerably over the last five years, cuts in military budgets came about rather slowly and are apparently gaining momentum, if at all, for reasons quite unrelated to the profound changes in the perception of threat.

In the former Soviet Union it was the desperate attempt to mobilize the civilian economy that imposed reductions on the military budget, however, without producing the desired results. Now, the United States is cutting its military expenditures in order to reduce the budget which, due to record levels of debts, has lost its ability to orient and potentially revigorate the dangerously paralysed economy as well as to halt thedecay of the infrastructure. The heralded peace dividend did not materialize in either case, let alone the transfer of resources to the South. Given the wishful thinking ingrained in the philosophy of the peace movement, this has come as a surprise. What was conceptually wrong with the notion of the peace dividend?

**Towards a Realistic Assessment**

As it was politically motivated, it is not likely to deliver the expected results. Worse still, the conversion of the militarized environment into a sustainable condition or, putting it more bluntly, the decontamination of our habitat from the dire legacy of the arms race between East and West, will incur vast additional costs hitherto not taken into account, for they are but beginning to surface.

Responses at different levels will help to explain why the expectation of a material peace dividend was at best an innocent self-deception, possibly a pragmatic argument as a mirror image of the rearmament logic of the Reagan era, and in the worst case, a cynical excuse for the failures of the Soviet system infused into the unwillingly fellow travelling peace movement of the West and expanded into a welfare utopia there.[4] This is, of course, not to say that the peace dividend is not real in the sense that abandoning the strategic arms race constitutes a necessary though far from sufficient condition for global survival that seems, finally, to be met.

Thus, the absence of a peace dividend, permitting the distribution of additional welfare in the terms traditionally envisioned by a wide spectrum of social groups, requires a renewed analytical endeavour in order to comprehend fully the accumulated structural impact of the military sector on the development potential of a less militarized future. The legacy of the arms race has beensystematically underestimated. It will haunt the political agenda, often unexpectedly, for quite some time to come. Conversion, the scheme that was to produce part of the prospected peace dividend, turns out to be far more difficult as well as costly to implement than expected. In some cases conversion is not within reach because it would require exorbitant levels of subsidy and investment, which will make an allocation of the required resources an irrational choice.

Firstly, with an arms reduction on the horizon, the major industrial countries began to assess the strategic value of their military high-tech industries which was willingly taken for granted as long as the arms race provided the military sector with an unquestioned legitimacy and hence an easy access to government contracts. The office of Technology Assessment in the United States and several expert groups in Great Britain were commissioned to assess the contribution of military R & D to enhance national industrial competitiveness.

The results were sobering. In spite of a few industrial highlights related to military R & D, the underlying equation that military R & D lends support to the development of strategic innovations providing the respective industries with a leading edge in the permanent struggle for shares in the world market had to be abandoned. The OTA study observed that military-bureaucratic procedures and their secretiveness inhibit a technological spin-off, and that even within individual companies the military and civilian branches are hermetically separated. British studies found that the partially excellent position of British firms in certain military high-tech gadgets bears little relevance to the competitiveness of British civilian industries and does not help to halt the on-going-de-industrialization.

The restructuring of the German aerospace industry within the multi-domesticDaimler concern clearly indicates the difficulties of shifting military production facilities and R & D laboratories towards civilian activities. The Daimler takeover of MBB clearly brings to bear the multi-domestic strategic interests of the concern upon the restructuring of the highly subsidized and intertwined military and civilian aerospace production. The complete transfer of helicopter R & D to Aerospatiale (France) is indicative of the incompatibility of production structures dependent on protected domestic military demand and of the imperative logic underlying the decisions of a multi-domestic giant trying to keep its oligopolistic position in the world market. Productivity levels are often clearly too low for an easy conversion into civilian markets, even in cases where the civilian-military commonality of the product itself is high, but military modes of organisation of the production came to dominate.

France is bound to reassess its industrial strategy in the near future as economic indicators put pressure on the government to act. The unique blend of an "economic mix" emphasizing the compatibility of a relative technological autonomy and international competitiveness has clearly come to an end. The French arms industry was submerged in the complex network of the "economic mix" and has, with the crisis of its traditional export markets, turned into a burden, sapping scarce government resources. But since the French military doctrine "d'une defense tous azimuts" has retarded the necessary integration of the French arms producers into multinational production structures where civilian and military technologies increasingly develop commonalities, the ongoing crisis is bound to lead to profound changes in the industry and eventually of military doctrine as well.

**The Spearhead Function of the End of a Fiction**

In the former Soviet Union it was assumed that the military sector would have the best chances to take the lead in the transformation process because it commanded the best equipment and monopolized the qualified labour force. What little evidence is available now indicates that the privileged military high-tech manufacturers in the former Soviet Union performed badly and were not able to launch viable civilian production lines beyond the ones that were under their command.

It seems safe to conclude that the logic of military production, whatever the economic system may be, is distinct from the requirements of viable civilian production. Hence conversion towards civilian production is far more demanding than a shift of production lines from one product to another, a shift regularly imposed by the permanent flows of innovation and market conditions on virtually every civilian enterprise. As a result, it may be more profitable to invest in an entirely new plant rather than take the disadvantage of converting a former military manufacturing site.

Military high-tech and years of technological lead, as the stealth technology generated in great secrecy with huge investments in the United States clearly demonstrates, do not easily convert into an improved competitive stance in the embattled international markets. Empirical evidence indicates instead that what national governments subsidize lavishly and perceive as strategic technologies, played a negligible role in determining the relative competitiveness of the respective national industrial sectors.[5]

As long as the fiction of a military superpower competition carried the daymilitary hightech was an ideal partner of any interest group trying to gain access to government resources. Competing military hightech conglomerates in turn portrayed their projects as indispensable to bring about certain breakthroughs with ample industrial applications. Such promises influence the propensity of the legislators to grant funds, but empirical evidence of the claimed linkages is never explored, because all parties involved are inherently disinterested in an evaluation. Success stories are self-promoting and the rest is covered with silence. The parliamentary majority prefers not to reveal the misallocation of huge R & D subsidies; industry and the interested scientific community prefer to discuss "future brilliant breakthroughs" rather than admitting that earlier promises were not kept.

But what is perceived as the Japanese challenge both in the United States and Western Europe casts serious doubts on the prospective spin-off in presumably strategic sectors spurred by profuse military R & D contracts. The military hardware itself shows a tendency to absorb an increasing share of components generated in the civilian economy, an evolution observers have labelled the civilizing of military technology. The electronic sector, more than any other industrial branch, demonstrates the lead of civilian advances over the technologies generated by the tightly closed laboratories of the military-industrial complex.[6] The oligopolistic competition of a few multi-domestic corporations which have access to worldwide sourcing speeds the marketing of innovations so effectively that, in spite of the billions granted by myopic parliaments, the military laboratories shrouded in secrecy and attuned to a bureaucratic logic of lobbyistic reproduction cannot keep the pace and are forced successively to assimilate civilian technologies into the designs of weapon systems.

### The Protracted Fall of the VPK

In the former Soviet Union the absence of multi-domestic corporations has provided for a virtual dominance of the military-industrial complex (VPK) over all other factors. This has led to catastrophic inefficiencies, uncontested duplications and an insufficient division of labour. The intricacies of central planning provoked centrifugal strategies, particularly illicit hoarding of resources at every stage of production, making for substantial cumulative diseconomies. Since the VPK could not draw on innovation generated by the civilian sector, as is increasingly the case in the military production in Western countries, the VPK was desperately trying to circumvent Cocom and selectively acquire western production technology. Raw material exports, particularly the increased oil price levels in the late seventies, provided the necessary hard currency. But the insufficient division of labour, combined with the secretiveness of military production, provided for a systematic underutilization of the imported production technology.

Since the pace of required innovation of military technology was set by the achievements of the other side, the desperate struggle to catch up, or rather not to lose out completely, imposed an often cynical neglect of health protection, environmental hazards and waste of non-renewable resources, the trademark of a war economy where all activities are determined by a single goal and an imperative time horizon.

### The Accumulated Military Debt: Still Disguised

The true social costs of a war economy can only be measured with hindsight. In the case of the former Soviet Union, the accounting process is just beginning, but all evidence so far available indicates that thepeoples of the former Soviet Union have forcibly inherited a social debt of the Cold War economy that they are not likely to be able to shoulder by themselves. More importantly, a considerable portion of the heritage, particularly the still incalculable ecological debt, and also the formidable stock of weaponry and military equipment as well as the productive resources both physical and human, are likely to spread globally. Hence, it will be but rational self-interest if the world community and western industrial nations, in particular, subscribe to a considerable part of the debt that the military confrontation between East and West has left in the East.

The envisioned peace dividend turns out to have been a cruel illusion. The political courage to accept jointly the incurred debt in its full extension is imperative, in order to convert the disarmament process constructively into a condition of common survival on this planet. One does not have to follow the alarmistic headlines of the daily press in order to comprehend that we have failed for long periods to decode the cumulative ecological disaster the present social order continued to discharge blindfoldedly onto future generations.

Somewhat schematically one can classify most of the present-day industrial output as a credit drawn on future generations. Measured with the criteria of a sound ecodevelopment[7] or of a sustainable global environment -- the present generation in the industrial nations lives on credit nobody, least of all the earth with its limited resources, can guarantee. Without qualitative changes lethal disaster is clearly projected.

This general condition of the "industrial culture" has been taken to extremes in the military sector. According to preliminary estimates the budgetary costs of acquiring an arsenal of chemical weapons amounts to less than one tenth of the comprehensive[8] "lifecycle costs", if one makes the positive assumption that this hilarious invention was not put to its conceptual use, namely serving as a weapon of mass destruction. To put it in simple arithmetic, before even a fifty percent reduction of military activities can lead to any peace dividend i.e. at a minimum a tenfold sum of what was cumulatively spent on chemical weapons has to be reserved from a hypothetically stable military budget. However, a systematic account of the present military debt to the environment remains to be established in order to develop an idea of the effort necessary to restore the global life sphere into a sustainable condition as far as the damages owed to military activities are concerned. Even if other investive military expenditures are likely to have lower ratios between acquisition and life-cycle costs, including the recycling into conditions of a sustainable environment, a realistic assessment would absorb stable military budgets for many years to come. At the same time it should be recognized, however, that restoring the environment involves a substantial potential for employment. This employment potential distributes well over a wide range of qualifications including basic research, since for the disposal of many substances applied to military technology there does not yet exist an industrially viable processing.

**Readjustment: A Costly Process**

The burden of the military heritage has many dimensions. The privileged, but isolated R & D establishments that cannot easily be reorganized in order to successfully pursue civilian ends at acceptable cost levels have absorbed the most precious elements of the human capital that the respective societies have generated. In addition military R & D is a major pull factor causing the brain drain from the South as well as migration leading to proliferation. The task of releasing the military R & D personnel from perfecting weapons ofmass destruction and the like turns out to be extremely difficult. What are the concrete civilian research projects they are capable of performing and that society is willing to finance? It may well turn out that their productivity is not sufficient to convince the sovereign i.e. the tax payer to maintain the research laboratories or that the civilian products they propose do not correspond to the priorities the market sets. Whereas in Western countries the sanctions of the market are straightforward, it will apparently take the VPK some more time to adjust to the dire economic realities.

The following quotation mirrors the self-perception of the VPK and demonstrates the isolation of the VPK from the concrete challenges that the former Soviet societies are to face. It is well demonstrated by the conversion aspirations of Energiya scientific-production association (NPO) that built the first Soviet satellites as well as the Vostok and Soyuz space vehicles. In a report in Vechernyaya Moskva a certain Legostayev, who is portrayed as the main lobbyist for conversion in the NPO, is quoted with the following statement which highlights the aloofness of the VPK with respect to the primary needs of the post-Soviet societies. "The most important thing as far as we are concerned, is that our protheses be reliable and comfortable and last a long time. As far as materials are concerned, we are making extensive use of titanium and high-grade steels. This is something that by no means just any country or company in the world can do. And we also use a lot of carbon-fiber reinforced plastic, particularly when it comes to leg prostheses for athletes. That's right, you heard me right the first time -- we are starting to make special prosthetic devices for people who want to continue their sports activities even after they lose a leg. And what's more, I firmly believe that sometime in the future, even in orbit in space, a handicapped cosmonaut fitted with one of ourprostheses will be able to perform as a fully functional member of his crew".[9] This impressive assertion of competence is, of course, totally out of tune with the demand structure of post-Soviet societies.

What was perceived as an asset to develop the productive resources of the former Soviet states turns out to be a highly problematic combination of talents and investments which is difficult or probably impossible to set in motion towards civilian ends. The former prestige and privileges are not the least among the hindrances. The irony of the matter is that the same industrial sectors that for forty years had absorbed virtually all resources and investments as well as the imported hightech machinery leaving behind a dilapidated civilian industry require huge additional investments in order to respond to civilian demands. Preliminary estimates indicate that every rouble of cancelled procurement orders requires at least two roubles of investment before any civilian products can be produced instead. Of course, such estimates reflect only the indispensable technical changes of production and do not account for the unknown costs of reorganizing the distribution or rather the marketing of the newly produced civilian merchandise.

It is evident that a transformation of this dimension leaves the state without a tax base, while many additional functions fall back onto the state because the industrial conglomerates of the VPK have to close down their comprehensive social services like kindergardens and medical care. Thus, the uniquely demanding transformation towards a market oriented economy will inevitably take place within the context of a profound social crisis.

**A Common Problem**

While the NATO-countries are likely to have greater capacities to absorb the changes associated with the disarmament process than the former WTO-countries, since they are not faced with the task of transforming their economic system at the same time, the changes will bring about new challenges to the established market economies as well.

During the last forty years military budgets served indirectly to support less developed regions. The structural deficits of such regions were not overcome by placing military orders or putting up military infrastructures. On the contrary, continued subsidies outside the market economy helped to cover up the lag. As the military induced demand shrinks, the structural deficits will reappear and require comprehensive readjustments, in order to homogenize the regions within the national economies or the larger emerging European market. These readjustments are likely to require explicit regional economic policies at high costs and levels of government intervention that run counter to the presently dominating market philosophy which pretends to reduce the role of the state as much as possible.

In many other fields as well, the burden of our overmilitarized past will continue to present bills of possibly prohibitive dimensions.

The disarmament process will devalue human capital and many trades. In many cases retraining is possible and makes economic sense. In other cases, however, early retirement from military jobs is without alternative. The social costs will be high and not entirely new, since the armed forces have a tradition of solving their manpower planning by pushing officers into early, though costly, retirement.

Largely unmeasured, the ecological burden of the arms race emerges slowly, including an additional heritage of World War II. It may well be that recycling of military hardware and the treatment of ecological damages provides a task where military talent set free can be employed usefully. But the disarmament process must lead to a network of civilian enterprises taking over the clean-up, where former military personnel might find a job. It is important to avoid the perpetuation of the military-bureaucratic structures that were relatively inefficient in producing military input. There is no reason to assume that they will be particularly efficient in the reverse process.

Finally, the maintenance of a minimum defense will be proportionately more expensive than many observers now believe. Smaller numbers will bear a relatively higher level of constant costs. Thus, the sums saved by lowering the military postures will not be proportionate to the reduction. The cost efficiency of the military apparatus will be decreasing in the immediate future. This fact highlights the urgent need to recognize that harvesting the results of the ending of the military confrontation between West and East requires considerable additional investment. This reality has to be faced straightforwardly because the results to be gained in terms of a truly sustainable future are worth the effort.

**Endnotes**

1. The 1977 report of the Secretary General of the United Nations on the "Economic and Social consequences of the arms race and its extremely harmful effects on peace and security in the world" (UNGA A/32/88, 12 August 1977), in particular, established strong causal linkages.

2. This concept has been elaborated by Jacques Sapir, L'economie mobilisee, Paris 1990.

3. Ksenia Gonchar argues convincingly along these lines. She reveals that what was perceived in the West as consecutive generations of tank designs reflected in reality the fierce rivalry of three design bureaus, each powerful enough to have their design produced, instead of having the army select the most appropriate design. Ksenia Gonchar, "Military Reform, Market and Conversion",unpublished manuscript,p.8.

4. Paradigmatic for this trend in the literature : Jorg Huffschmid, Rustungs - oder Sozialstaat? Koln 1981.

5. For a recent assessment see : OECD, "Strategic industries in a global economy - policy issues for the 1990s",

Paris 1991 (OECD International Futures Programme).

6. The paranoia haunting the American public debate because the most sophisticated American weapon systems rely on Japanese electronic components is a bizarre manifestation of the failure of the lavishly financed military R & D sector to render adequate results.

7. For a convincing argument to use this term see : Ignacy Sachs, "Comment Concilier Ecologie et Prosperite", in : *Le Monde Diplomatique,* December 1991, p.18f.

8. As a rule the accounting concept of "life cycle costs" developed for weapon systems encompasses R & D production costs, maintenance, refurbishing, fuel etc. until the system is discharged. It stops short of accounting for the debris left, future generations will be burdenedwith. The United States General Accounting Office criticized already in 1989 that all estimates of savings due to the closure of military bases did not account for costs for hazardous waste cleanup, calculated to require at minimum twice the predicted annual savings. The GAO-report insinuates, however, that environmental studies and tests may lead to a much higher burden. GAO, "Military bases, an analysis of the Commission's realignment and closure recommendations", GAO/NSIAD-90-42, Washington D.C., Nov.1989.

9. Quoted from : R. Kuznetsova, "Space Company Works for the Handicapped", in: Vechernyaya Moskva, Jan.21, 1991. Source : Soviet Press, Selected translations, Vol.1 No.2, p.f.

# POLITICAL, ECONOMIC AND LEGAL CONDITIONS FOR THE PROCESS OF CONVERSION OF INTERNATIONAL SECURITY SYSTEMS - EMERGING PERCEPTIONS FOR GLOBAL SUSTAINABILITY

**Peter Menke-Glückert**
*Germany*

Almost 50 years have passed since the signing of the UN Charter launching the vision of a new peaceful international order. The name United Nations was used during the Second World War to denote all the nations allied against Germany, Japan and Italy. It was later adopted as the name of the postwar world organization. It was the threat of fascist dictatorship which cast different regions and cultures of the world into one international mould.

It was past experience which gave impetus to this vision -- not careful planning for the future of this one earth. It was the classic fabric of international policies -- not a design of a new order "as if people matter".

But inspite of this 19th century BALANCE OF POWER CONCEPT, new issues such as debt and human development, environment crises, the PROBLEM OF THE ABSOLUTE POOR (the Fourth World) found their way into the United Nations Plenary. For The Fourth World, conditions of life are so degraded as to deny its victims basic human rights and bare necessities. Horst Rittel/Berkeley defined these violations of human dignity as W-PROBLEMS (wicked problems) or real-life problems to be distinguished from S-PROBLEMS or scientific, bureaucratic problems of Western elites.

Nine tenths of the 20th century was needed to face the W-problems of our one earth, to learn by catastrophe, civil war, economic or ecological collapse, uprising of the people, especially in Eastern Europe, that lasting peace everywhere in the world canonly come about in societies which really respect human rights, due-process of law and self-determination. Real lasting peace can only be secured when human rights are guaranteed. This message of the European Conference on Security and Co-operation has had a tremendous impact on Eastern European nations and regions.

United Nations and world leaders must act now with courage, designing and building a completely new system for securing basic human rights, and international co-operation to prevent crises of world order and world ecological systems. A mechanism of international sanctions for violating basic principles of international law must be established.

Contrary to the present system, a new system of world order must meet the interests of all nations, strong and weak alike. This would be impossible if military cultures are cut off from democratic processes and international control. From the very beginning, the 1945 World Order has been designed as a two-class-system : veto super powers -- the Big Four -- and normal powers. Membership to the upper class was determined by the size of military potential, not by economic strength, population figures, cultural achievements or share in world trade. In 1945 this BALANCE-OF-POWER-APPROACH was perhaps a realistic one, but to-day military strength does not guarantee the status of world power with respective influence in the international community or in world trade or world culture.

The vision of a new international order has reached a turning point for the very reason that Cold War and the EAST-WEST-CONFLICT have faded away by a strange process of implosion of the Communist power structure. New hopes and new impetus are given to strict arms control, destruction of at least half of existing atomic weapons and gradual disarmament of all other conventional forces, thus redirecting human talents and resources from military to peaceful uses. However, the climate of detente should not underestimate the scope and dimension of threats facing the international community.

**Population Explosion**

In 40 years from now more than 9 billion people will need shelter, food, jobs, health care, education.

**Refugee Streams**

The figure of almost 15 million registered refugees will more than triple in the next 4 to 5 years, if massive aid programmes are launched, offering chances for survival and social well being in Eastern Europe or in the republics of the former Soviet Union.

**Internationally Organized Crime**

From drug syndicates to international terrorism there will be more and better organized crimes with easy access to modern weapon technology.

**Degradation and Destruction of the Biosphere**

Nature is fragile and has been finely balanced for more than 4.3 billion years. These ecobalances are endangered by increased resource use which will lead to dramatic climatic changes if pollution of carbon dioxide (by burning of fossil fuels) is not limited at least by 25% in the next 15 years. Each year 6 million hectares are degraded by desert-like conditions. The Gulf War was a devastating war against the environment of the region, with lasting effects on the biosphere.

**Atomic, Biological, Chemical Warfare and Arms Trade**

There still exists an enormous arsenal of atomic weapons. As the Gulf War has proven, there are many nations with easy access to the atomic weapons of the poor: chemical weapons and the biological weapons.

Concerted action of the United Nations is essential against all these threats, especially that which is backed by the rich, industrialized nations. The complexity of all of these problem areas asks for a world-wide sophisticated logistics of assessment of the respective threat and permanent verification of counter-measures.

Only military logistics are capable of meeting these demands and specifications. One of the objectives of strategic conversion should be the reuse or dual use of large parts of military logistics to assist the international community in fighting these new threats. Gradually national and regional military cultures must be integrated into a new system of internationalization and domestication in the use of military force, thus reformulating the aims and objectives of military forces.

As a first step towards this objective, a whole series of agreement on disarmament and confidence-building measure were signed. Regional pact systems and international conventions or treaties are also proof of the reduction of military power. The next steps must include a much broader scope of internationally designed and controlled social and military service to meet threats (which stem from severe failures in economic, social and cultural development, international crime, environmental catastrophes and lack of progress) and secure basic human rights and democracy.

# CONTAMINATED SITES IN THE OWNERSHIP OF THE FEDERAL DEFENSE FORCES

**Prof. Dr. R. Mull and Wolfgang Schröder**
*Department of Finance of Lower Saxony, Germany*

## Introduction

The Federal Army of Germany, along with the other military forces, is in possession of land holdings encompassing approximately 453,000 ha. The land holdings of the former National People's Army (NVA) in the eastern part of Germany encompass approximately 240,000 ha. Those of the Soviet army are around 540,000 ha. Due to the integration of the army of the former GDR into the Federal Army of Germany, the withdrawal of foreign forces from German territory and the reduction of the total number of troops, consideration must be given to the future utility of these sites. Criteria which must be considered for further utility are the existence of sources from which dangerous contaminants may intrude into air, soil and water and the extent to which these media are already contaminated. Prior to releasing the sites for other uses, it must be made clear whether any clean up is required.

The German Ministry of Defense has started a pilot project in the Eastern part of Germany (New States). 165 land holdings of the former army of the GDR are being investigated initially to determine the extent of contamination, the existence of sources of dangerous contamination and applicable methods of remedial action, if necessary. This report will present the management of the project and preliminary results.

## Management

The Head Departments of Finance (OFDs) in every state are responsible for all construction on sites owned by the Federal Army. For the pilot project, the OFD in Hannover, Lower Saxony, was appointed bythe Federal Ministry of Defense to act as the supervising office. A task force unit was formed, consisting of members of the OFD and experts from the consulting firm Prof. Mull and Partner, which has many years of experience in this field. Contractors were selected in the new states to execute the work on the sites. They got their contracts from the local building authorities via the OFDs of the different states. The results of the first reconnaissance studies were sent to the OFD in Hannover, where the task force unit reviewed and summarized the results. This summary was submitted to the Federal Ministry of Defense as first information about the situation. The organisation chart is shown in Table 1.

The whole operation is split into three phases as shown in Table 2. The first step required a reconnaissance study for each site. This first screening procedure accomplished two things. First, the number of sites with sources of contamination were found. Second, the extent of groundwater and soil contamination was estimated and then verified.

The task force unit in Hannover assessed the results from the various teams on the sites. If no indications of contaminants were found, no further investigation was recommended. Identified contaminated sites require a more intensive investigation to determine whether remedial actions are necessary for clean up.

On some locations, it was obvious that emergency steps towards remediation must be initiated. A special group within the task force unit is involved in these operations. At one site for instance, a hydraulic measure isenvisaged to stop further emission of contaminated groundwater from an area of around 7 km. At this site, ammunition was produced and stored prior to and during the World War II. After the war most of the buildings and the stores were destroyed. However, the dangerous substances from the production of ammunition remained and are polluting ground water and small surface water courses. Located downstream from this investigation site, there are some important well fields for a water work. In some of the wells the groundwater is already polluted. In this case, remedial actions must start soon. The plan is to pump the polluted ground water to the surface, treat it and infiltrate it into the aquifer and surface water courses. This measure will prevent a further migration of the pollutants to other wells. This will also provide the time necessary for a more intensive investigation, prior to a final clean up.

In the second phase of the program, more detailed investigations will be performed to determine the migration of the contaminants in the soil and aquifers. The objects which are endangered by the pollutants have to be identified. The results of these investigations must be reviewed again. This is a prerequisite for any recommendations concerning further investigations aimed at remediation or clean up. Technical and economical aspects must also be considered before a decision is made. Remediation means that the further migration of contaminants in the air, soil or water will be stopped. A clean up involves a removal of the sources of contamination from the site.

The alternative site selected will depend upon the results of the remedial oriented investigation conducted in the third phase.

**Preliminary Results**

At the moment the program has entered the second phase. The reconnaissance study has been completed. The sites where emergency remedial actions will start soon have been identified. The most important sources of contamination are listed in table 3. The percentage indicates the frequency of occurrence. Table 4 shows the contaminants found at the various places. Organic liquids, inorganic acids and heavy metals are the most important pollutants which have been found by the end of 1991. The most dangerous substances seem to be chemical weapons from World War II. The results have been reported to the Federal Ministry of Defense. The second phase of more detailed investigations has started on 136 land holdings.

**Conclusions**

The preliminary results of a reconnaissance study show 1500 locations on 165 land holdings of the former army of the GDR where contaminants have spread predominantly into the soil and also into the groundwater. The further utility of these sites depends upon the extent and the intensity of contamination and risk for the biosphere. This risk must be assessed in more detail. A ranking, based upon information received by the task force unit from the various investigations, must be established. In accordance with this ranking, remedial actions will be recommended to prevent further emission of contaminants into air, soil and water.

Table 1

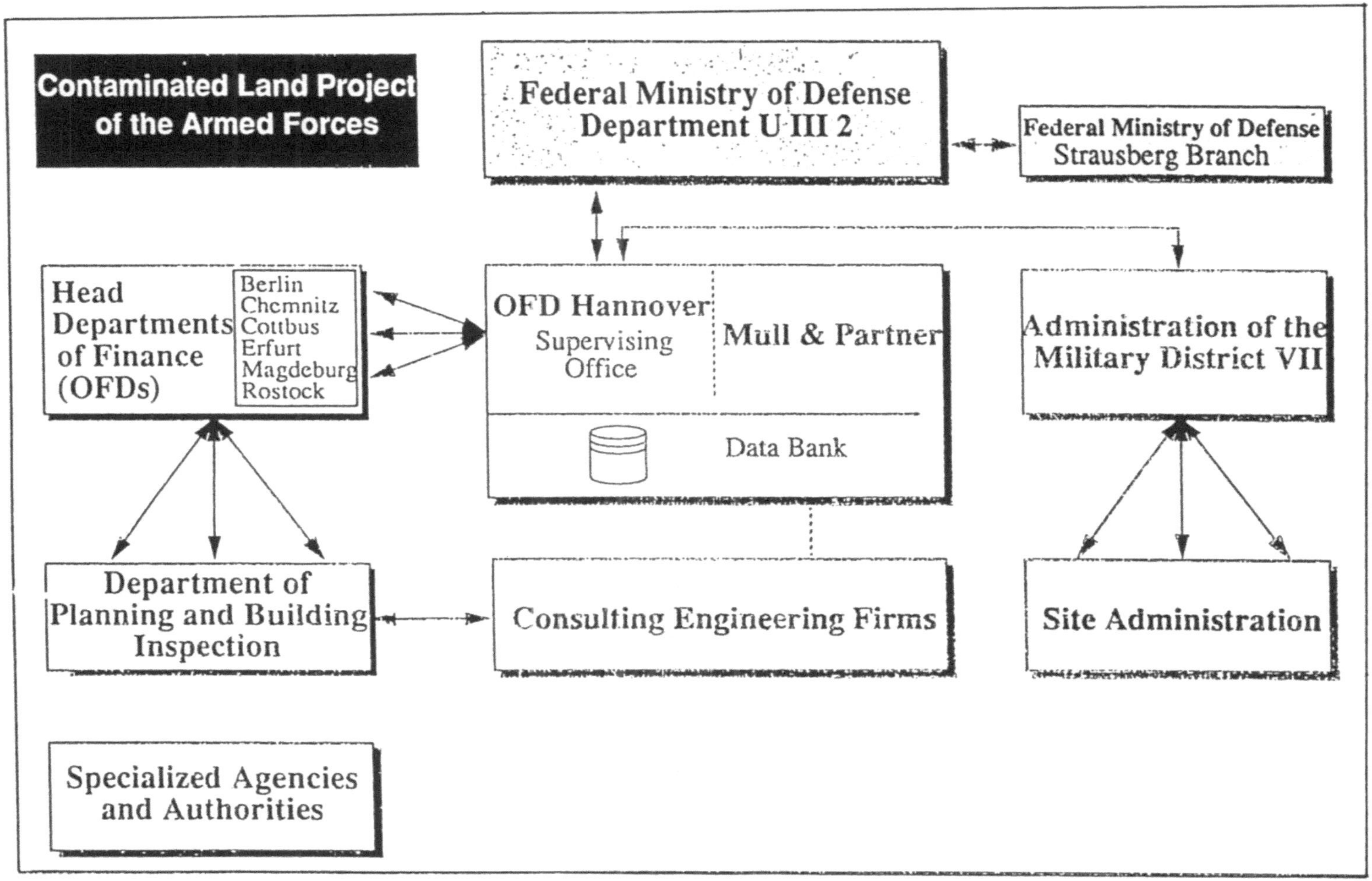

Contaminated Land Project of the Armed Forces
Federal Ministry of Defense Department U III 2
Federal Ministry of Defense Strausberg Branch
Head Departments of Finance (OFDs)
Berlin
Chemnitz
Cottbus
Erfurt
Magdeburg
Rostock
OFD Hannover Supervising Office
Mull & Partner
Data Bank
Administration of the Military District VII
Department of Planning and Building Inspection
Consulting Engineering Firms
Site Administration
Specialized Agencies and Authorities

**Table 2**

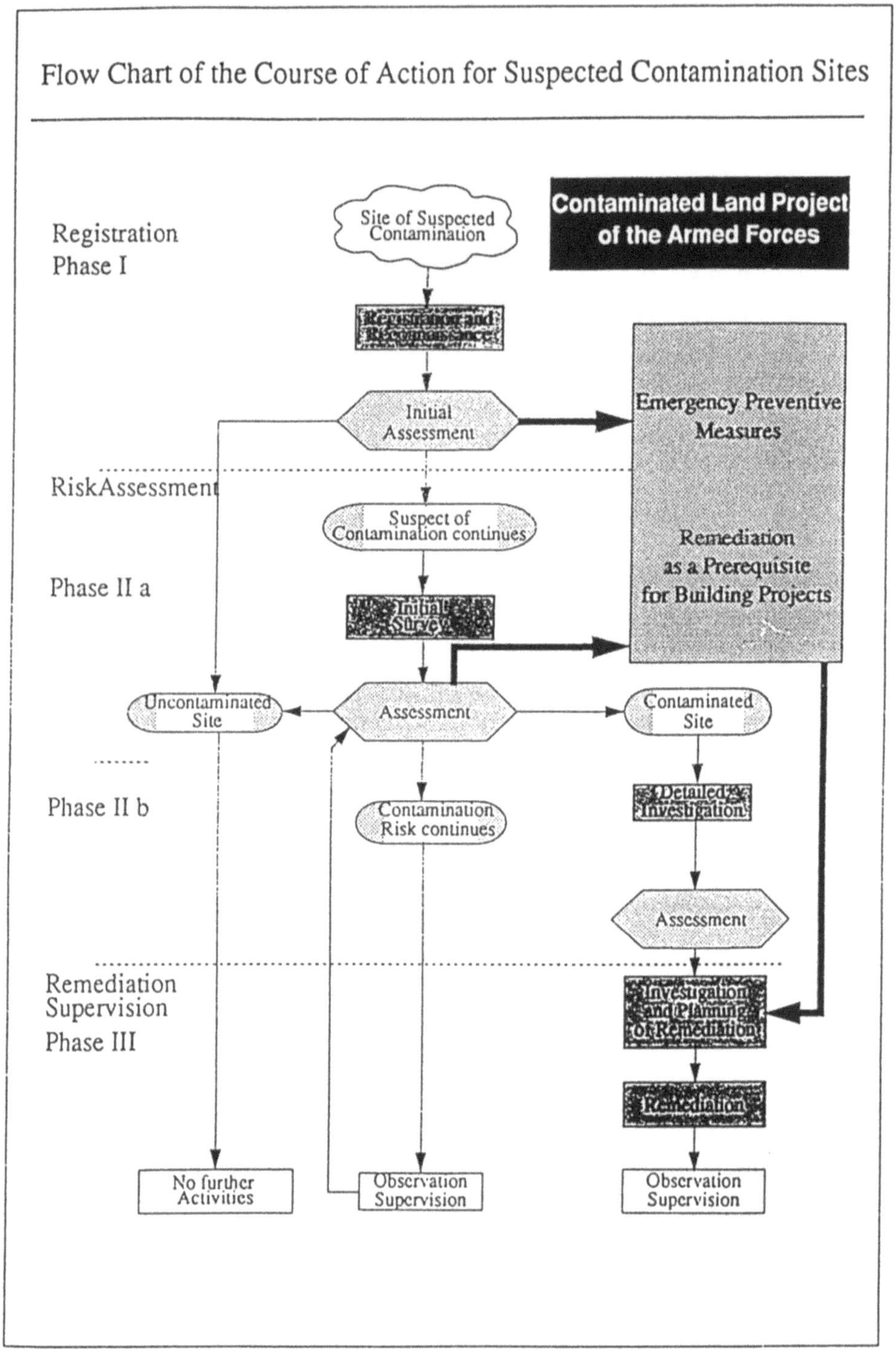

Flow Chart of the Course of Action for Suspected Contamination Sites
Contaminated Land Project of the Armed Forces
Registration
Phase I
Site of Suspected Contamination
Initial Assessment
Emergency Preventive Measures
RiskAssessment
Suspect of Contamination continues
Remediation as a Prerequisite for Building Projects
Phase II a
Initial Survey
Uncontaminated Site
Assessment
Contaminated Site
Phase II b
Contamination Risk continues
Detailed Investigation
Assessment
Remediation
Supervision
Phase III
Investigation and Planning of Remediation
Remediation
No further Activities
Observation Supervision
Observation Supervision

Table 3
**Types of Suspected Contaminated Land**

| **Suspected Contaminated Site** (SCS) | Percentage of Total Amount (%) |
|---|---|
| **Storage sites for** Chemicals, Fuel, Oil, Paint, Varnish, etc. | 21 |
| **Motor Vehicle Service** Car Wash, Maintenance, Workshop, Garage, etc. | 17 |
| **Parking Lot/Garages** | 13 |
| **Gas Stations** | 9 |
| **Deposite Sites, Sanitary Landfills,** Unorganized Deposites, Ashes, Hazardous Waste, etc. | 8 |
| **Separators** Immiscible Liquids, etc. | 7 |
| **Waste Water Treatment Plants** Sewage Treatment Plants, Sewage Purification Pits, Drainage Basins, Feces Pits | 4 |
| **Generation Stations** | 1 |
| **Loading and Transfer Stations** | 1 |
| **Transformers/Emergency Power Generators** | 1 |
| **Neutralization Plant** | 1 |
| **Other Suspected Contaminated Sites (not from armaments)** | 5 |
| **Contaminated Sites due to Armaments** | 12 |

**Table 4**
**Relevant Substances in Contaminated Sites and their Suspected Occurence**

| Substance | Remarks | Suspected Occurence (% SCS) |
|---|---|---|
| Pesticides | Pesticides, Herbicides, etc. | 2 |
| Hydrocarbons | Diesel Fuel, Gasoline, Oil, etc. | 70 |
| OX | Organic Halogen | 4 |
| OX - PCB | OX-Polychlorinated Biphenyles | 10 |
| OX - CH | OX-Chlorinated Hydrocarbons | 6 |
| OX - PCN | OX-Polychlorinated Naphtalenes | 1 |
| PAM | Polycyclic Aromatic Hydrocarbons | 6 |
| BTX | Benzene, Toluene, Xylene | 11 |
| Cyanides | | 2 |
| Phenols | | 4 |
| Heavy Metals | Pb, Cd, As, Hg, Zn, Sn etc. | 33 |
| General Solvents | | 19 |
| Asbestos | | 1 |
| General Explosives | | 3 |
| TNT | Explosives, Trinitrotoluene | 3 |
| Hex | Explosives | 3 |
| Chemical Weapons | | 2 |
| Rocket Fuel | | <1 |
| General Weapons | Standard Weapons, Munitions etc. | 2 |
| Radioactive Substances | | 2 |
| Detergents | Detergents, Cleaning Agents | 9 |
| Dioxine, Furan | | <1 |
| General Chemicals | Other Chemicals, eg. Acids, etc. | 28 |
| Feces | | 4 |
| Other Substances | eg. Building-Rubble, Bulk Refuse etc. | 20 |

# CONVERSION AND DISPOSAL OF EXPLOSIVES AND PROPELLANTS

**Bernd Niemeyer and Theodor Rosendorfer**
*MBB GmbH, Germany*

## Summary

Conversion and disposal of explosives and propellants are important tasks for future activities. Modern technologies have to be employed to meet the given environmental standards and at the same time to be economically competitive. The applicability of recycling as well as chemical conversion as disposal technologies are heavily restricted. Although for both technologies products could be figured out, investigations are underway in order to develop processes and to improve product quality. Experimental studies showed the biotechnological degradability of a series of explosives. A pilot plant is being set up to obtain technical data for a large-scale process. Studies were carried out to optimize the throughput in an incinerator on a minimum level of hazardous components in the flue gas produced. It was found that the oxygen supply and the transfer, as well as the reaction kinetics dependant on the temperature, are main influence parameters on the flue gas composition. With the data obtained, the up scale will be realized. At present the incineration of the explosives and propellants with an additional flue gas treatment is the only fully developed large-scale technique readily available.

## Introduction

New developments in the relations of eastern and western industrialized countries (with a reduced level of ammunition) enable the possibility of a peaceful co-operation. Additionally, weapons produced decades ago have reached their age-limits. For both reasons a significant amount of explosives and propellants have to be converted and disposed.[1]

Conventional techniques like burning or explosion are mainly employed at present. They result in an uncontrolled release of emissions into the environment.[2] Consequently other environmentally conserving disposal technologies have to be developed and used.

## Stages of the Disposal Process

The disposal of ammunition goes through different stages that can be seen from figure 1. The materials to be disposed of have to be sorted out. Modern data techniques are helpful and efficient tools. Subsequently the state of the material including the outward conditions, chemical composition and safety considerations have to be analyzed. After the transport to the disposal facility, the safe delaboration has to be realized. Then the separation into the metals (from the ammunition shelters) and the explosives and propellants are the next steps. The processing of the separated materials may comprise of, for instance, recycling, incineration of the explosives and propellants with an additional flue gas treatment, or the storage of inexplosible waste materials.

In order to gain access to the explosives, the ammunition casings have to be opened by :

* sectioning
* punching or
* cutting.

Studies showed that water jet cutting, employing a high pressurized water jet as well as sawing under special assumptions are safe methods applicable in processing facilities. Figure 2 proves the disintegration bywater jet cutting of a series of 120 and 82 mm projectiles.

## Experiment

After the opening of the ammunition, the explosives can be separated by flushing. Again water jetting has been shown as a safe and efficient method. The processing of the explosives and propellants for disposal can be realized by the following technologies :

* recycling
* chemical conversion
* biotechnological degradation or
* incineration with an additional flue gas treatment.

## Recycling

In general the environmental balance is best if produced substances are reused even in a reprocessed or modified state. Recycling products might result in explosives for civil use in mining industry or for hunting ammunition.

However there are some presumptions for an economical and ecological justification to employ a recycling process. The new product has to meet standard qualification requirements and a customer has to be interested in it. The whole process must be profitable and environmentally safe.

In order to minimize waste from a production plant, a recycling facility for explosives with a capacity of 50 tons per year, delivering a high quality product, was set up.

## Chemical Conversion

The principal aim of processing explosives is to create valuable goods. Our studies evaluated a process in which a base product for chemical industry is obtainable.[3]

Besides this the recovered explosibles can be converted in an inertisation process to inexplosible substances with significantly lower legal provisions imposing the subsequent processing. In this case the products are to be handled as toxic wastes. Today there are hardly any results from experimental studies published. Basic research has to be done to employ this technique in a pilot or even large-scale plant.

## Biotechnological Degradation

The biotechnological degradation in principle is an excellent method because explosives and propellants as feed substances should be decomposed totally in water, carbon dioxide and short chain organic components. Recent studies proved the disintegration of trinitrotoluene (TNT) by biotechnological means. The microbes used were screened from special natural sources. No accumulation of highly toxic intermediates was detected. The problem of the low solubility of TNT in aqueous solutions as in most other explosives could be diminished by the addition of `solubilizers' such as decaline and others. Thus the results of the space-time yield could be enhanced many times to 1.5 tons TNT degraded per year and $m^3$ reaction volume in a laboratory scale setup. Figure 3 demonstrates the flow sheet of the designed three stage pilot plant that is being installed. The dissolved explosive in the feed stream and the substrates like sugar components, molasses and other additivesare fed into the first fixed bed reactor inoculated with TNT degrading microbial species. In order to increase the degree of decomposition and to maintain a high conversion rate if required, the substrate composition is added to the feed stream of the second fixed bed reactor.

Optionally there is a final stirred tank reactor to guide the spectrum of the products. With the scaled up facility it is expected that with optimization programme to be carried out, in future the rate of decomposition of the explosive can be increased further and data for the upscale to an industrial plant will be available.

**Incineration with Flue Gas Treatment**

With the incineration, explosives and propellants are disintegrated by oxidation at high temperatures. Studies on the main influence parameters were carried out on a pilot plant in order to optimize the throughput -- that means, to maximize the oxidation at a minimum production of hazardous components in the flue gas.

Experimental results with the burning of propellants showed that with the decrease of the outer surface by the factor of 10 the specific weight of carbon dioxide ($CO_2$) produced per gram propellant burned decreased by 80-90%. That means that the subsequent flue gas treatment plant would have to be designed in larger dimensions.

The results can be summarized as that the oxygen supply and the transfer as well as the reaction kinetics affected by the temperature influence the flue gas composition most. On the basis of these experimental data, an optimized incineration and flue gas treatment plant is being designed and being set up.

**Endnotes**


1 F. Volk; "Detonations - und Abbrandverhalten von Spreng - und Treibstoffen", 14, *Sprengstoffgesprach, Bad Reichenhall*, 1990.

2 F. Steinseifer; "Art und Umfang des zu entsorgenden und zu verwertenden Wehrmaterials", *CPM-Symposium*, Bonn, 14-15, February 1991.

3 E.Kleinschmidt, Th. Rosendorfer; "Entsorgung von Explosivstoffen", 15, *Sprengstoffgespräch, Eckernförde*, 1991.

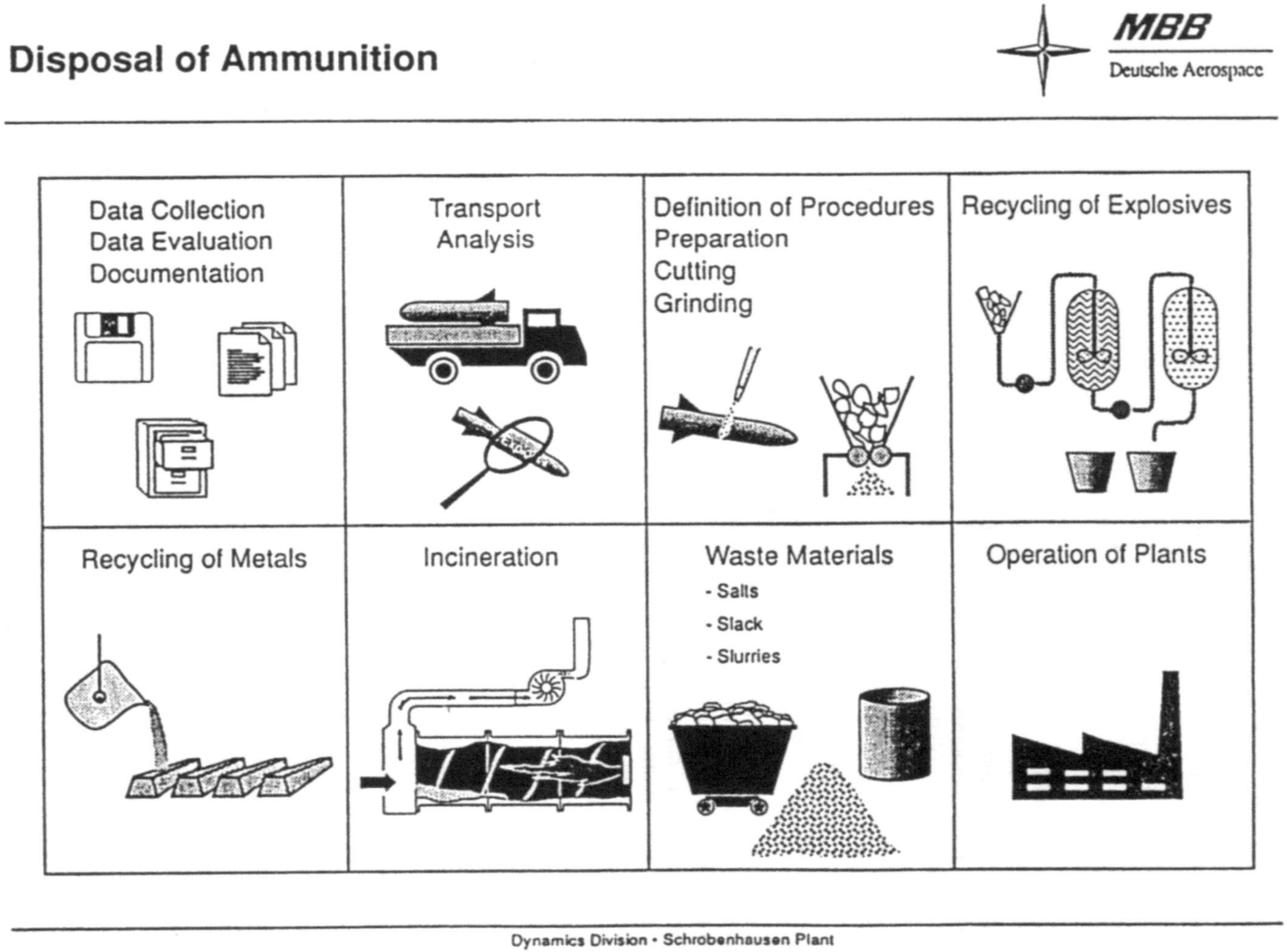


Figure 1: Stages of the disposal process

## Disposal of Ammunition
### Water Jet Cutting

MBB
Deutsche Aerospace

Dynamics Division ■ Schrobenhausen Plant

Figure 2: Desintegrated projectiles applying water jet cutting

## Disposal of Ammunition

**Biotechnological Processes**

MBB Deutsche Aerospace

### Biotechnological Degradation of Explosives and Propellants

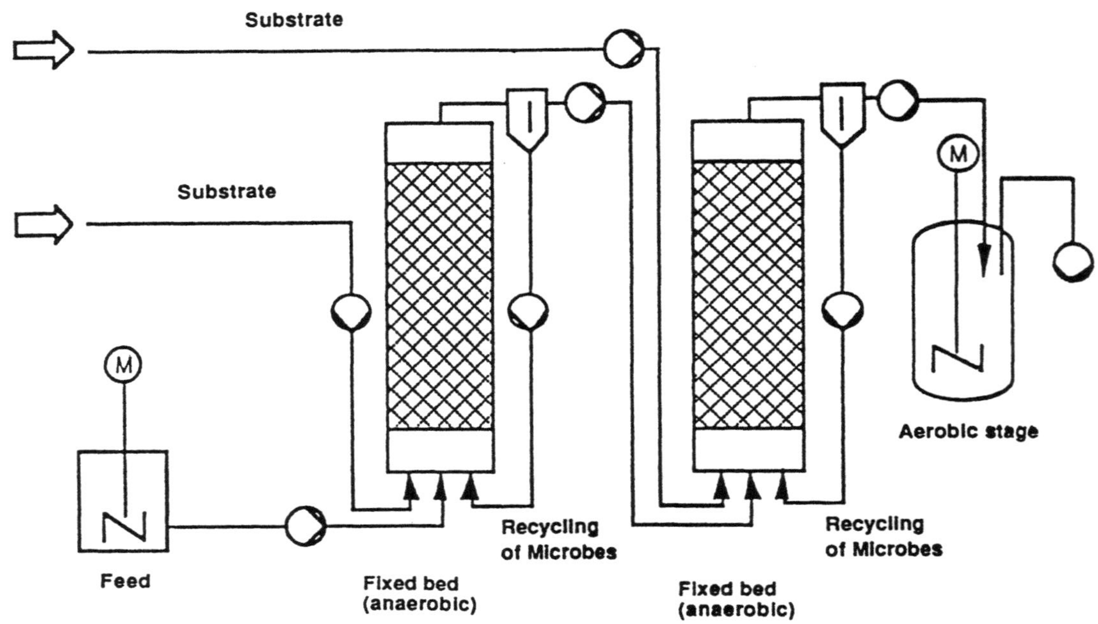


Dynamics Division    Schrobenhausen Plant

Figure 3: Chart of the biotechnological degradation of explosives and propellants

# DISARMAMENT ECONOMY - A NEW INDUSTRIAL ACTIVITY?

**Dr.Petra Opitz**
*Berghof Institut für Friedens- und Konfliktforschung, Germany*

The debate on disarmament, conversion and peace dividends has begun to cause irritation, a development dating roughly from the first attempts to apply these concepts in practice. Once a magic word, "conversion" is rapidly becoming a threatening notion to those affected by it. At least this is the case in those countries that have so far significantly reduced military expenditure -- above all the East European states, including the republics of the former USSR. The causes are manifold; they are, of course, strongly influenced by the specific situation of these countries, which find themselves in a fundamental process of systematic transformation. Despite the specifics of this situation, however, the lack of expected and fundamental consequences of conversion, in the form of the civil use of resources freed by disarmament -- financial and human capital and industrial capacity -- call for a reappraisal of current thinking on conversion.

Expectations with respect to the economic and social effects of disarmament were based on an erroneous calculation: sums amounting to thousands of millions were omitted. Estimates of the costs of the arms race were far too low. Just as the environmental destruction caused by industrial and agricultural production has been left out of account, so too the misuse of apparently cost-free resources has never been included in calculations of military production and running costs. Thus the real costs have been deferred to a later date. The price of the public good "external security" was higher than it appeared.

Today, most people are increasingly aware that the use of FCCH in aerosol sprays, and in cooling and cleaning agentscosts far more than the prices we pay for these goods; the bill for the damage is presented later. The recycling-penny indicates that there is at least a growing awareness of the problem. That the same applies to disarmament, however, is a fact that is still far from being generally recognized. This is shown by the topsy-turvy logic that calls such costs the costs of disarmament, although they are in fact the costs of the arms race.

The necessary inclusion of these deferred costs in national economy accounting reflects the recognition that our living and economic conditions must necessarily be capable of reproducing themselves; that without a sustainable economy the natural conditions of existence of mankind will be destroyed. Wrongly-allocated resources and the still unpaid costs of armaments may thus be seen, like the still unpaid ecological costs of economic growth (also called defensive costs) as a reduction of the substance that is vital for human life.[1]

Disarmament economy is the economics of reducing existing burdens imposed by armaments; it is the economics of reducing damage, and re-creating non-threatening conditions of life.

The forms and the extent of the damage are manifold. Only now, thanks to the possibilities of civilian use of military bases and production sites that have arisen through disarmament, is this damage becoming apparent. The forms include: contamination of ground water and soil by kerosene, oil, detergents and other damaging substances in connection with the use and maintenance of military technology;[2] soil damage, the destruction of forests and wooded areas andthe spoliation of the countryside; noise pollution, potential risks through unguarded military technology; and the deformation of economic structures and the monostructural economic dependence of regions and local communities.

It is hardly possible to measure the real extent of the damage. As arms production and armed forces are gradually reduced, a continuation of the damage will also gradually cease. measuring its extent, and the financial means necessary to repair it, is only taking place step by step. It has begun with attempts to use resources that hitherto were employed for military purposes, either by making civilian use of them, where this appears possible, or, where they represent a possible source of danger, by removing this danger. This means that the deferred costs only become visible wherever disarmament has in fact begun, either as a consequence of international treaties or on the basis of unilateral disarmament measures.

For example, estimates are currently available on the costs of destroying chemical weapons in the Commonwealth of Independent States (CIS). Such estimates vary considerably; depending on the methods used, the process of destruction itself will cost 640 million to 2.5 thousand million roubles (at 1991 price levels) -- without taking into account the indirect costs of the necessary environmental protection measures.[3] 2.5 thousand million dollars are calculated for the costs of safe storage on Russian territory of the nuclear warheads of the CIS that are to be destroyed.[4] Estimates are not yet available on the costs of decontaminating former military-used areas on the territory of the former GDR so as to make them suitable for civilian use. But first estimates suggest that, for the federal state of Brandenburg alone, several thousand million marks will be necessary.[5]

About 250,000 tons of munitions formerly held by the GDR's National People's Army (NPA) must be destroyed. In addition, the agreements reached at the Vienna Conference on Security in Europe call for the destruction of some 9000 armoured vehicles (including 2,800 tanks and 1,900 artillery systems). In this connection, the CIS is confronted by the task of either rendering harmless, or to a small extent converting to civilian use, some 19,300 units of weaponry and military technology (including 1,300 fighter aircraft, 7,600 tanks, 9,600 armoured vehicles, and 760 artillery systems).[6] The cost of dismantling a tank in the Federal Republic of Germany amounts to some 20-30 thousand DM, a sum that cannot be compensated by selling the scrap metal or using components in civilian applications. This list could be continued; the examples presented here are simply intended to illustrate the scope of the overall problem. To these costs must be added others: for re-training measures aimed at reintegrating the military and civilian personnel of the armed forces into civilian life and for the professional training of employees of the arms industries whose jobs are becoming superfluous. Adjustment costs for the re-training of former employees of the arms industry must be included in the calculation of deferred costs whenever arms production is judged a mis-allocation, that is, from that moment when external security cannot be achieved through the military means produced by the arms industry.

The consequences of mis-allocation must be remedied, and this calls for the expenditure of enormous financial means. The introduction of items in the budget to cover these costs does not represent, as is sometimes claimed,[7] a decision of order policy in the sense that the state unacceptably increases the scope of its activities by transferring resources no longer directly required by the military budget toother sections of the budget. Rather, the problem consists in the fact that the state's duty to provide defence -- creating the public good "external security" -- is not limited to producing guns that fire, and for which responsibility must no longer be assumed as soon as they have become military superfluous. The price of this good includes transaction costs that until now have not been included in the calculation. In connection with the responsibility for the development of these transaction costs, the state, i.e. the public purse, must in accordance with the principle of responsibility (in the sense of causality) also bear these costs.

Thus, it is certainly true that given the present possibilities of establishing changed military security structures at a lower level of armaments, the price of the good "external security" is sinking.[8] Since, however, a significant part of the full costs has not yet been included in the price, *de facto* we must reckon with a temporary increase in expenditure to cover the costs of the arms race.

For the disarmament economy the question is thus raised: how can damage be repaired in the most effective way? The following considerations will be relevant:

1. It will be necessary to calculate further secondary costs that cumulate from the hitherto neglected damage (e.g. health problems). It will also be necessary to calculate opportunity costs (e.g. through delays or the lack of possibilities to make economic use of the resources employed in military production). Both secondary and opportunity costs must then be measured against the costs of repairing damage.Already today the consequences of the neglect of damages impose immediate action as in the case of discharged ammunition. As a general rule the transport of ammunition stockturns into a severe problem, once the containers begin to corrode, which is likely to be the case after ten years. The tight time-horizon of the imperative repair and recycling puts the respective industries into a favourable negotiation position.

2. Chances of compensating costs through making use of technical components and raw materials that can be won through "damage repair", must be sought and extended. The competitive commercialisation of disarmament economy is a part of this. Contracts for work in connection with damage repair, including the disposal of military technology and munitions, must be publicly advertised. This will prevent the creation of monopolistic, non-market conforming structures like those of the arms industry, that have an inflationary effect on the public expenditure required for such tasks.

3. Development possibilities must be taken into account with respect to employment, technology and economic policy that arise in connection with economic activities developing from the elimination of the consequences of the arms race.

The elimination of NPA munitions alone can amount to contracts worth 1.5 thousand million DM. The scrapping of armoured vehicles of the ex-NPA will be worth some two to three thousand million DM.[9] Former munitions producers and military enterprises in the ex-GDR previously involved in repair and maintenance work mainly compete for these contracts. Demand for their previous production has ceased to exist. They must either go into liquidation or attempt to become active in civilian spheres, something that will be extremely difficult in the short time available and under the economic conditions currently existing in East Germany. Grantingcontracts to these firms, which were usually the biggest employers and economic factors in their respective regions, would enable them to continue to offer employment to at least a part of their former employees. In addition, this will help to win time to develop civilian production strategies for the period when these recycling contracts come to an end. The

Buck-INPAR-Pinnow facility, situated in Brandenburg makes a good example. Recently a pilot-plant for recycling of various types of ammunition, said to fulfill high standards of environmental safety, was installed. It is intended in the long run to handle dangerous civilian waste as well. This may, however, require additional adaptation of the technology, but would allow it to continue operation after the military waste has been elaborated.

Given the production line of these firms until now, they are experienced in dealing with the military technology and munitions now to be disposed of. Some also have recycling know-how. At the same time, the present level of technology is not adequate for a destruction of munitions that is both truly environmentally-friendly and salvages useful resources as far as possible. Improved technologies must be developed -- at the international level, too. It appears realistic that such research and development work as well as the equivalent practical co-operation could be internationally financed; in certain fields (destruction of chemical weapons) this would appear indispensable. The companies already involved in such processes could play a pioneering role in this regard.

The disarmament economy also has employment effects. But like the economic aspects they are limited, both quantitatively and in time; they cannot directly compensate the employment cuts arising from disarmament. If disarmament economy is not to become a permanent, costly recycling and damage repair as a long-term economicactivity, these will be more-or-less unique processes. Depending on the disarmament measures taken in a particular period of time, they could last for some 5-10 years.

The economic effects of the civilian use of resources of military technology may on the whole be expected to remain within fairly narrow limits. In the CIS, savings of raw and other materials from civilian use are expected to amount to 4.5 - 5 thousand million roubles in the period 1991-95.[10] Unfortunately, the basis for this estimate remains unclear; thus, it is not possible to assess how far it might be relevant to market-economy conditions. Both for economic reasons (high once-off and running costs) and for ecological reasons, a conversion of military technology to civilian use only appears to make sense in a very few cases.

The possibilities of commercializing the recycling of ammunition etc., are generally overestimated. Though it is feasible to elaborate tanks ammunition etc., so that the end product consists basically of the basic substances, the marketing of the substances is hardly economically viable. In order to meet the conditions of natural purity the costs increase again considerably and competitive prices with natural raw materials remain out of reach. At the same time the quantities produced are insufficient to satisfy potential customers who normally absorb much larger quantities. Finally, the output is not steady. This poses additional difficulties for marketing.

The logic of the disarmament economy presented here assumes that the national economy is still functioning, and that the deferred costs can in economic terms be met (although this may be debatable in view of the overall ecological situation). It cannot be transferred to conditions of economic catastrophe such as those that threaten the CIS. There, the re-allocation of resources is taking place on a scale that affects the entireeconomy, going beyond the mere consequences of the arms race. At the same time, the society and its institutions, as the frame of reference for and basis for economic activity, are in a process of dissolution. The framework for economic reconstruction must be constructed anew. Under the conditions, the costs of damage repair cannot be borne from the CIS's own resources.

The alternative is an internationalisation of efforts to counter the negative consequences of the arms race, in other words, the development of a shared disarmament economy. Initiatives such as Western offers to help finance the process in the CIS are not adequate;[11] to the extent that they exploit the desolate situation of the CIS to impose unilateral disarmament, they are politically damaging.[12] Practical co-operation would not only be a form of confidence-building; it could also offer practicable alternatives for those employed in research. If international co-operation does not take place, the damage of the arms race will not be repaired or the necessary financial means may be won through arms exports; however, arms exports themselves will of course not automatically lead to damage repair, since other priorities may be set. But intended practical international co-operation must be controlled and there must be established criteria for it. Projects caused only by political prestige and ignoring their possible negative environmental consequences must be avoided.

## Endnotes

1. Cf.Ch.Leipert, Die heimlichen Kosten des Fortschritts, Frankfurt am Main, 1989, pp.29-30.

2. In the federal state (land) of Brandenburg alone, investigations of areas formerly used for military purposes (by the former National People's Army ofthe GDR and by Soviet armed forces) have revealed 96 different types of contaminataion. Cf.Militarstandorte als okologische Zeitbomben (report in the Berliner Zeitung, 19 and 20 February 1992, p.21).

3. Cf.e.g. A.Kirejew, Cena "mirnovo dividenda", in: pp.16-17.

4. Cf. Dollary na jadernoje razorushenije istsut primenenia v.gosudarstvach SNG, report in Izvestia, 24 January 1992, p.5. G.Attali has suggested that less than 1% of the annual military expenditure of the G7 states be used to destroy all Soviet nuclear warheads. Cf.Rasplatimsja bojegolovkami?, Izvestia, 22 January 1992, p.3.

5. Cf. Militärstandorte als ökologische Zeitbombe, loc. cit.

6. Cf.A.Kirejew, Cena "mirnova dividenda", loc. cit. p.17.

7. Cf.M.Kly,K.Lobbe,"Verteidigungsausgaben der Bundesrepublik und volkswirtschaftliche Auswirkungen verminderter Rüstungsausgaben", RWI-Mitteilungen, Vol.41, (1990), nr.4, p.334.

8. Cf.Wohin geht die Friedensdividende? a project of the Deutsche Bank (project director: O.Storf), Frankfurt am Main, 1991, p.7.

9. Cf. "Explosiv, aber noch lukrativ", Die Zeit, 30 August 1991.

10. Cf. V.S.Gorochov, A.S.Kuznecov, "Konversia voennovo proizvodstvai promlemy racionalnovo ispolzovanija predpriatii miniterstva oborony SSSR" in: Voprosy ekonomikii konversii, vyp.3, 1991, p.95.

11. Cf.Dollary na jadernoje razorushenije istsut primenenia v gosudarstvach SNG loc. cit.

12. Cf. the proposal of the President of the Bank for Reconstruction and Development, G.Attali, to cancel the debts of the CIS in return for the destruction of nuclear warheads that are not currently subject to bilateral or unilateral disarmament measures. Reported in Izvestia, 22 January 1992, p.3.2

# FEDERAL ENVIRONMENTAL AGENCY

**Reinhard Peglan**
*Germany*

## Preamble

The military and political dissolution of the Warsaw Pact, the agreement over the conventional troops in Europe (KSE-1 Agreement), the executive power of the German unity and the reduction of allied troops in Germany linked with it, have considerable effects on the armament industry, the military service sector and the regional garrisons for troops in the Federal Republic of Germany. It is to be observed here that, in connection with the developing sites or regionally related conversion projects, the interest of environmental protection in diverse ways can be touched upon. On the one hand, environmental risks brought about through armament conversion, for example in association with the concern for ammunition and explosives, should, as far as possible, be considered unimportant. On the other hand, the environment related conversion projects can just open up chances for development and environment.

## General State of Knowledge

The relaxation in Europe which became effective at the end of the eighties, the revision of the militarily strategic demonstrations, as well as the modernisation of troops, have resulted in the beginning of the disarmament of the military apparatus. In middle Europe, the region of Germany (formerly Federal Republic of Germany and German Democratic Republic) has been most strongly affected thereby. With the executive power of the German unity, the military apparatus of the Federal Republic of Germany has been united with that of the former Democratic Republic of Germany and has been gradually reduced. Likewise theallied forces reduced their troops in Germany on a long term basis. Disarmament and troop reduction have considerable effects on the armament industry, the military service sector and the regional garrisons for troops.

The strategies for the transformation of military production and service for civil interest are to be developed towards the economic and social mastery over the consequences of disarmament. This has already begun to happen in a few fields and sites. It is shown thereby that these conversion projects with regard to their objective in view for the civil sphere are quite manifold. A stronger orientation towards the importance of environmental protection, the reorganisation of old encumbrances, as also the general environmental care seem to notify the existence of a recovery need. The dynamics of military potential is closely linked with the advances in disarmament negotiations. Thereby the effects on the quality and quantity of the military sectors have often been quite distinctly stamped, that is, on the one hand real disarmament measures such as the reduction of equipment, fixation of upper limits, retreat of troop strength and bases have been agreed upon. On the other hand, these changes in the military structure lead in principle to a new armament base. Then disarmament presents itself above all as change of armament. For this reason, the quantitative statements on the consequences of disarmament agreements abound with considerable uncertainties.

The following developments are to be reckoned with from the year 1990.

## Personnel Development

Around 1.3 million German and foreign soldiers stationed on German soil upto 39% and the employees occupied directly with the troops (civil employees) of 323,000 upto 58% will be reduced by the year 2000. It must be borne in mind therefore, that reductions in regional or individual sites can be omitted. Precise statements on the concerned sites are, at the moment, not sufficiently available.

## Quarters

Among almost all quarters of the west group of the Soviet troops, the exception of a few living areas with grave environmental damages must be taken into account. At the present moment, the Federation Ministry of Environment, allows a stock taking of old encumbrances to be accomplished in the quarters announced for release. The order was given to the Bayerish undertaking 1 ABC with residence in Ottabrunn (the volume of the order : 75 million German marks). The State of Brandenburg, of all the Federal States, has been most strongly affected through old military encumbrances. The troops of the west group of the Soviet forces stationed there occupied almost 50% of the claimed territory in Germany.

## Military Equipment

The quantitative reduction of troops and the planned modernisations, produce a powerful material transformation on military equipment. The forecast over the required capacity for de-experimentation with ammunition and the disintegration of rearmament of military equipment, turn out to be difficult. As this complex almost exclusively refers to the German armies (Federal army, the former National People's Army), the supply of accurate planning data should be feasible.

## Armament Industry

In 1990, in the old Federal States about 190,000 workers were occupied with the production of military equipment. The chief buyer was the Federal army. A smaller portion, not exactly specified, was allotted for foreign troops (armament export).

## Service and Maintenance of the Army

In 1990 around 290,000 employees in the old Federal States, were occupied with the maintenance of the Federal Army and the allied stationed troops. The maintenance and service were related to : building works, supply of energy, stationery and office supplies, items of equipment, transport, care, clothing (uniforms), nourishment etc.

On the whole, from the year 1990, about 1.3 million native employees (workers in armament industries, service undertakings, as also civil employees and soldiers), are to be considered by way of business volume connected with disarmament disposables (see illustration). This corresponds to about 5.2% of all employees in the old Federal States. The retreat of the troops stationed on German soil which will be more than 60% in the next ten years, will have strong repercussions on the (regional) labour market, which must therefore be met with united measures.

## Conversion and Environmental Protection

The transformation of military units, equipment and structures in civil utilisation can be carried out with the aid of conversion projects. These are specific and the location for the present and must be developed with the participation of those concerned, (industries, employees, communists, trade unions and associations). The future mission of the environmental care should therefore be handled with priority. This calls for field-oriented advice from places and organisations entrusted with the development and execution of conversion projects.

The Federal Environmental Agency is an establishment, adapted to support such specific advices and actively collaborate moreover with an environment-related conversion model through a supply of research means. The collaboration of the Federal Environmental Agency appears in the following crucial points conceivable and desirable.

**Department-related Projects**

**Environment Engineering instead of Armament Engineering**

- Environment electronics instead of armament electronics (examples : Telemeter - and surveillance systems for environmentally harmful substances instead of Rocket-and Weapon-guiding systems, systems for monitoring environment (long-term changes), Air- and Satellite control of the spread of harmful substances instead of Guiding- and Spying systems for the military effort)

- International optimised transport system, instead of military logistics (example Energy and Risk optimisation of transport of dangerous quality)

- Development of methods for environmentally compatible concern for military equipment

- Regional - or site - related projects

- Reconstruction concepts for armament industries for civil production (professional accompaniment of a creative conversion programme of the Federal Government)
- Concepts for the utilisation of the quarters and site-bound military installations (equipment) being liberated for the regional environment reorganisation (examples : the reorganisation of old encumbrances, restoration of damages to cultural monuments, reconstruction and management of short-range recovery centres, reconstruction and security of nature protection areas)

- Retraining of Federal army units towards catastrophe defence force

- Incorporation of army sites in investigation areas of national arrangements for environmental protection

- Monitoring of environment through retrained Federal army units (example : air control of shoals and land-locked seas)

**Reorganisation of Military Sites**

- Reorganisation concepts for old armament encumbrances present before 1945

- The grasp and map-drawing of all armament and military sites, above all in the new Federal States

- The utilisation of international experiences in the reorganisation of old armament encumbrances

- In December 1991, the Federal Environmental Agency, arranged a working group on the theme "Environmental Protection and Armament Conversion"

This working group has the following tasks :

- to promote the information flow within the Federal Environmental agency in the sphere of "Environmental Protection and Armament Conversion"

- to initiate and to co-ordinate research and development plans
- to evaluate the extraneous need for advice (particularly in respect of communists) and to submit corresponding proposals for solution
- organisation of co-operation with other places and Federal/State working circles towards environment-related questions and armament conversion

Two research and development plans in the following problem areas are already being carried out:

- "Stock taking of old armament encumbrances - suspect areas in the Federal Republic of Germany"
- "Environmentally compatible de-experimentation with ammunition".

# PROJECT-ORIENTED ENVIRONMENTAL MANAGEMENT CONSTRAINTS AND STRATEGIES FOR RESTRUCTURING LARGE-SCALE INDUSTRY

**Joachim Schatz**
*Treuhandanstalt, Germany*

Improper and unchecked treatment of waste and environmentally damaging material in the former GDR has led to the creation of residual deposits and residual pollution of sites with disastrous consequences for the environment.

These sites have to be remedied, in as far as this is tenable, for industrial and commercial re-utilisation. Priority is to be given, in the drawing up of land-use plans and in traffic planning, to site recycling ahead of area closures. The basic concepts draft on ecological remediation and development in the new Federal States submitted by the Federal Minister for Environment, Professor Klaus Töpfer, pointed the way forward.

A keyrole in the solution of the pending re-allocation of former industrial sites is to be played by the Federal States and the communes.

The Environmental Protection/Residual Loads Directorate in the Treuhand provides the companies with advice and support to help overcome these problems.

11 sites in the iron and steel production industry and non-ferrous metal industry and 4 industrial sites from the chemical and energy industries have been given project management support for the removal of residual loads with job creation programme measures since May 1991 by the Directorate of Environmental Protection/Residual Loads.

The project-related support of the companies comprises of over 1.5 billion DM till the end of 1992.

The background to this task is a project management concept developed by the Environmental Protection/Residual Loads Directorate, whereby on-site project management support for the respective company/project management and project controlling unit based in the Environmental Protection/Residual Loads Directorate are ensured.

The following management structures are envisaged for the conduct of the projects:

- setting up a project management team by the respective company on site.
- project monitors are ensured in the Treuhand companies in addition by means of project controlling in terms of basic data such as duration, costs and workforce.
- programme management (inter-disciplinary coordination of the projects in the Treuhand) which
    * advises the Treuhand companies in identifying, evaluating and setting up of relevant environmental projects using the job creation measures.
    * promotes and uses comprehensive synergy efforts between the projects of the individual Treuhand companies.
    * monitors the Treuhand companies' projects with respect to the new employment, measures the scope of the job creation scheme, measures

the effective use of job creationscheme, non-personnel costs in as far as they are guaranteed by the Treuhand.

External project support is provided at the request of the companies to guarantee the efficiency in the technical and economic construct of the project.

The work to date performed by the industrial companies by the Treuhand shows the possibilities for utilising comprehensive synergy effects as part of the area preparation.

The site plans of the 15 sites mentioned previously have been entered into a databank. What this is intended to achieve is that the privatisation teams can build up their sales negotiations on updated and current area utilisation plans.

**Overview of Contents**

1. Guidelines for Action for the Re-zoning of old Sites for New Industrial and Commercial Use.

2. Environmental and Site-Oriented Recycling using the Steel and Rolling Works Riesa AG as an Example.

   - concept for a comparative industrial centre, Riesa.
   - conversion of the Riesa site into modern company premises in close co-operation with the State and private Industry.
   - overview of job creation programme.
   - realisation phases of the job creation programme.
   - Reisa project examples
   - area utilisation plans.

3. Concept for integrated Project-Related Co-operation.

   - aim of programme
   - basic structure of the site recycling project management
   - project management tasks
   - tasks for the on-site project management
   - basic structure of the programme controlling
   - the Treuhand documentation and information system for the job creation programme
   - EDP-assisted, project-related area development plans for the support of privatisation.
   - tools and aids for job creation programme processing.

4. Initial Situation for Integrated Site Development.

Treuhandanstalt

Environmental Protection / Residual Loads Directorate

## Overview of Contents (1)

1. **Guidelines for Action for the Re-Zoning of old Sites for New Industrial and Commercial Use**

2. **Environmental and Site-Oriented Recycling using the Steel and Rolling Works Riesa AG as an Example**

   - **Concept for a competitive industrial centre Riesa**
   - **Conversion of the Riesa Site into modern company premises in close cooperation with the State and private industry**
   - **Overview of Job Creation Programme**
   - **Realisation Phases of the Job Creation Programme**
   - **Riesa Project Examples**
   - **Area Utilisation Plans**

## Overview of Contents (2)

# ENVIRONMENTAL APPLICATIONS OF MILITARY INFORMATION AND COMMUNICATION TECHNOLOGIES

**Jürgen Scheffran**
*Ianus, Germany*

## Dual-Use and the Conversion of Military Science and Technology

Highly sophisticated technologies (high-tech) are essential elements of the national security and economy of advanced countries. Modern semiconductor technology, nuclear technology, laser technology, computer and communication technologies are employed not only in the manufacture of non-military products but also increasingly in the production of weapons.[1] While in the past the military was the pacemaker in many fields of high-tech development, scarce resources, problems of political acceptance and converging demand profiles of civilian and military technologies lead to a strategy of commercial pre-development: a technology is to be developed in the civilian sector first and used for military purposes afterwards (dual-use).[2] To overcome the enormous financial bottlenecks due to the rising costs for research and development (R & D) in the area of short-living electronic components, NATO increasingly buys civilian parts off the shelf and integrates them later into upgraded weapon systems.

At the heart of high-technology warfare are the systems for command, control, communications and intelligence ($C^3I$). Being the brain and nervous system of the military forces, it comprises facilities, personnel, equipment and procedures used to exercise authority and direction over military forces. Funding for $C^3I$ systems consumes an increasing share of the military budget, reaching more than 10% in the USA. After the Cold War this would be a huge resource available for purposes other than warfighting. New tasks for $C^3I$ would include not onlycrisis management and arms control verification, but also civilian applications, in particular, environmental protection.

Information and communication technologies (ICTs) for the gathering, distribution, processing and modelling of information form the backbone of $C^3I$ systems but are also increasingly important for environmental purposes. A reasonable conversion strategy, focusing on the conversion of science and technology, would try to reverse the dual-use concept by finding and promoting useful civilian applications of military-related ICTs and by converting their development process. A recent UN Study gives a first broad assessment of opportunities, possibilities and limitations of applying military-related resources, in particular those from the field of ICTs, for environmental protection and disaster relief.[3] The Study examines the environmental needs and the military resources available to fulfill these needs. After discussing some general thoughts related to this study, examples of ICTs are given in the following.[4]

## General Thoughts

There is a growing need for continued scientific assessment of data on global environmental hazards like ozone depletion, climatic changes and loss of biodiversity; promotion of appropriate technologies for sustainable development; strengthening of quick response capabilities in case of environmental emergencies; development of human resources to deal with environmental damage especially in the developing countries; and mechanisms for international co-operative ventures. Environmentalstrategies require a comprehensive set of measures, including protective measures for damage prevention and control; restorative measures for repair, rehabilitation and environmental compliance; remedial measures for development of ecologically benign and energy-efficient technologies; and promotional measures for raising environmental consciousness through education.[5]

Military-related resources, in particular, manpower, professional skills and technical know-how, equipment, funds, infrastructure, and technological capabilities, including R & D, may help to support environmental strategies. Military establishments possess highly trained managerial skills, military personnel is well-equipped for dealing with catastrophic situations (environmental disasters) and handling and disposal of highly toxic substances. Advanced research and technology, which is necessary for responding effectively to the environmental challenge, is available in the civilian and the military sector, but is not oriented enough towards environmental goals.[6]

On the other hand, substantial differences exist between military-related resources and environmental applications, making conversion more difficult. While military security is based on destructive measures, environmental protection essentially aims at conservation. Military activities, in particular wars, can be very harmful to the environment. Furthermore, military systems are not explicitly constructed for environmental purposes. In many cases, they are overspecified, too complex and too costly for commercial or scientific use. Often, military thinking and hierarchical organizational structures are not appropriate for environmental protection. Technological requirements are generally more demanding for military purposes (e.g. encryption, survivability against countermeasures,secrecy, real-time operations). Applicability in many cases depends on the technical and operational characteristics, but technical specifications of military systems are often unknown due to secrecy which prevents a detailed discussion of environmental applications. The conversion of military systems has to be compared with existing civilian systems in terms of utility-cost ratios, including the costs for modification, elimination, retraining, operation and maintenance. Marketing strategies and political desirability of the civilian product should also be taken into account.

In order to assess the feasibility and effectiveness of conversion, the following four factors have to be considered:

1. A necessary precondition for the conversion from military to environmental applications is a similarity in the basic physical and technical characteristics of the technologies and systems under consideration. To be more specific, sensors should be similar in frequency, covered area, resolution and availability; communication systems in the number of channels, connectivity, data rate/information flow; computers in storage capacity, speed or compatibility; models in topic, size, adequacy and relevance. Above all, the costs of the systems should be comparable. The dual-use encourages the prospects of redirecting military-related technologies into the civilian sector, but if the differences in the characteristics are too large, costly modifications may be necessary. In many cases, conversion is easier at the subcomponent level while highly complex large-scale systems, designed for a specific purpose under specific external conditions, are quite difficult to convert.

2. During the technical development process, which consists of different *phases* (research, development, testing, production, deployment and use), conversion seems to be easier at early

phases. Research generally offers the widest range of potential applications, and a substantial fraction of the military-related capacities could be transformed into the environmental field. On the other end, deployed and used systems are already designed and optimized for a very specific purpose and thus are mostly unflexible in the change of mission. In between, at the development and production phase, the ratio of flexibility and rigidity differs from case to case. Cost-benefit analysis can be substantially different, depending on the phase of conversion. Despite overlaps in the technical characteristics often the costs of modification may exceed its value. Weapons designed to inflict destruction are mostly useless in the environmental field. In such cases, it may be cheaper to close down or dismantle facilities.

3. Different *stages* of conversion can be distinguished :

   - indirect use of military-related resources for civilian purposes (spin-off), without changing the military structure;

   - direct application of existing military-related resources, with or without modifications, under civilian control;

   - structural conversion of military infrastructure for civilian purposes.

   - A reasonable conversion strategy would be a transitory process towards structural conversion, the desirable end, by using theintermediate stages as far as possible but not more than necessary. According to the historical experience, the spin-off from the military sector was limited to selected cases. At the second stage, applications of military-related resources should be subject to utility-cost ratios compared with alternatives (existing civilian systems). Using military systems under military control for environmental purposes could create an unnecessary justification for the military. Civilian control over useful military systems is essential to realize disarmament, demilitarization and a transition to structural conversion.

4. There are numerous institutional, political or economic *conditions and constraints* which facilitate or limit conversion. In the former Soviet Union, for example, about half of the economy depended on the military. There is no alternative to soon and far-reaching structural conversion, which is essential for stability and survival of the newly formed Commonwealth of Independent States (CIS). The application of useful military-related resources under civilian control would avoid a waste of the existing resources during the transition period. Because of lack of capital, external financial support is a precondition for conversion. In the USA, a smaller but still substantial fraction of the society has been militarized at the price of losses in commercial competitiveness. Conversion would be essential to regain competitiveness but there are strong forces from the military complex opposing deep and quick restructuring.

It seems inconsequent to argue against conversion because of overcapacities in comparable civilian sectors (e.g. aircraftindustry) and at the same time to oppose government subsidies for industries which try to convert from the military to the civilian sector. Both military security and environmental protection are public tasks and require government support. Government defense expenditures, the most important resource, could be directly transferred into environmental protection projects, which cannot survive on the commercial market. They would be more useful than military goods which contradict free-market principles, represent no value for the society, can be risky and harmful to the environment. The technological capabilities of military establishments include research and development, laboratories, equipments, personnel and expertise of the scientific community. Because of their inherent dual-use character, information and communication technologies seem easily adaptable for environmental purposes. The $C^3I$ network, regulating the military information flow, contains many sophisticated systems and technologies which may be adaptable for environmental purposes. The major military missions of ICTs include control systems and command centers; intelligence sensors; communications networks and electronic warfare; models, simulation and testing facilities; automated and intelligent weapons systems. Military systems from these categories more or less could find corresponding environmental applications, although it has to be seen case-by-case whether they can compete with civilian systems. In the following sections, examples are given from the civilian and the military sectors to discuss environmental applications of military-related ICTs.

## Sensors for Environmental Monitoring

The collection of environmental data and the detection of man-induced abuse improve understanding of the complex environmental processes and provide early warning as apreventive measure against environmental hazards. Although a variety of civilian surveillance systems is available for long-term environmental monitoring and short-term emergency response, military sensors, using the visible, infrared, or microwave regions of the electromagnetic spectrum, and being deployed on ground air or space-based platforms, may substantially add to this capability. From various intelligence sources, the defence community has collected a variety of valuable data on the environment. Advanced instrumentation from the defence laboratories can be used to extract more detailed information. Even more important than the existing systems would be the use of the infrastructure, including R & D, production, operation and maintenance facilities and personnel.

## Satellite Remote Sensing

Remote sensing is applied to collect information needed for a better understanding of the various components of the environment: the biosphere (life-bearing layer), the atmosphere (gaseous layer), the hydrosphere (water layer, including the ice-bearing cryosphere) and the lithosphere (solid earth crust). Satellites in various Earth orbits are considered ideal platforms for remote sensors. Major military satellites include: photo-reconnaissance satellites; radar-reconnaissance satellites; electronic reconnaissance satellites; early-warning satellites; ocean surveillance satellites; and weather and geodetic satellites.

The US Defense Meteorological Satellite Program (DMSP) collects and disseminates worldwide weather information to military commanders. Besides lasers and electron spectrometers, a number of different sensors are used in the visible, infrared, ultraviolet and microwave channels to measure cloud patterns, windfields, atmospheric temperatures and charged particles.[7] Usingthe DMSP data for weather monitoring in the civilian sector would pose no technical problem and has been done partly after the DMSP was opened for civilian use in 1973. It would be a political decision to put the DMSP under civilian control or use its capacities, including communication links and ground control stations, more for civilian purposes. The data could complement civilian weather monitoring satellites like Nimbus, NOAA or METEOSAT.

The most advanced military photo-reconnaissance satellites like the US "keyhole" satellites (KH-11, KH-12) carry arrays with millions of small pixels (charge-coupled devices, CCDs) and have a reported spatial resolution of 15-30 cm. This would be much better for civilian photo-satellites which are worse in resolution by a factor of 100. Images at night can be made through the use of infrared detectors and photomultipliers which have a substantially lower resolution than the visible-light images. Lacrosse is a cloud-penetrating all-weather radar in space with a number of sensors specially designed for long life. Its resolution is assumed to be in the range of 2-3m. Depending on the frequency used, the Synthetic Aperture Radar (SAR) of Lacrosse is estimated to be capable of penetrating one to several meters of mature green crops.[8]

Reconnaissance satellites are important for crisis monitoring, early-warning and disarmament verification but could principally perform functions for environmental monitoring with very high resolution. The satellites themselves, in particular their sensors, are not optimized for environmental monitoring but may be useful for data gathering in emergencies, where detailed pictures of an area are required in a short time (as in the Chernobyl case). However, the costs would be immense for this purpose (about $ 500 million per satellite) and cheaper alternatives would be available in most cases,e.g. overflights by airplanes or civilian remote sensing satellites like Landsat, SPOT or the radar satellite ERS-1. These systems are not as advanced in terms of data rate or resolution but are more adequate for environmental monitoring in terms of frequency, resolution and costs. Military reconnaissance satellites mostly use black and white imagery, while the civilian systems usually employ multi-spectral sensors from which colour images can be produced.[9]

On the other hand, knowledge in producing the sensors and satellites, the experience in operating the systems and managing the information flow would be more important for conversion and should be opened for the civilian sector, where a lack in data management is obvious. The communication links, computer systems and ground facilities, including the personnel, would be highly suited for managing the huge data streams from environmental monitoring. Related equipment and facilities are still too expensive to be produced for civilian purposes only, thus modification and retraining of military systems would be the cheaper option.

**Airborne Surveillance**

Airborne surveillance platforms which carry a variety of sensors, including cameras and radars, provide the information on enemy air and ground targets during a battle, but in principle could also gather environmental information. For example, the US SR-71 and
TR-1 military reconnaissance aircraft provide high-altitude systems for near-real-time battlefield surveillance. NASA operates a civilian version ER-2 of the TR-1 as an earth resources research aircraft.

The E-3A Airborne Warning and Control System (AWACS) is a large multi-function phased array radar (PAR) to detect and identify airborne targets and to guideinterception missions with a line-of-sight range of 400 km. The Joint Surveillance and Target Attack Radar System (JSTARS) carries a side-looking PAR that can operate in three modes : to detect moving ground targets at ranges up to 200 km, to locate fixed targets with high accuracy (below 1m) and to allocate fighters and missiles to ground targets (weapons guidance mode). Both systems are based on Boeing 707 aircraft equipped with large computers, multi-purpose display screens and fast communication links to manage the enormous stream of information.[10]

AWACS and JSTARS could perform functions of crisis monitoring, crisis management and arms control verification, but environmental applications are more difficult to imagine. One might conceive of using the AWACS or JSTARS radars for environmental monitoring, traffic monitoring or situation assessment in environmental catastrophes (e.g. in a flooded or otherwise devastated area), when the airborne coordination and management of rescue activities may be necessary. In these cases the computing and communications capacities might be useful, although probably would be inadequate in frequencies and oversized in terms of personnel, data rate, storage capacity or electronic survivability. The whole system would be highly expensive ($200 million for AWACS, $300 million for JSTARS) and complex and would have to be scaled down in size and capacity in order to minimize the costs for operation and maintenance. Software modifications may be necessary. In both the civilian and the military sector, smaller, cheaper and less sophisticated radars and airplanes are available. Unmanned remotely piloted vehicles are another option.

### Techniques for the Detection of Environmental Pollution

Since many substances are hazardous or even toxic at relatively low concentrations, there has been an increasing demand for techniques to detect air, water and soil pollution. Although the military offers no special techniques in this field, the existing R & D capacities, including equipment, personnel and know-how, could be partly converted to such problems. Besides classical methods of wet chemical analysis or ordinary methods like gravimetry and volumetry (titration) a variety of techniques have been developed, based on radioactivity, electrochemical or electromagnetic properties, the scattering of particles, mass-to-charge ratios of ions and other physical properties.[11] A variety of spectrometric and spectroscopic methods exists. Advances have been made in the development of correlation spectrometry (COSPEC) for spectroscopic analysis of the atmosphere and the detection and quantification of certain chemicals. Isotopically labeled tracers have been developed to map pathways and measure contaminant migration times, e.g. in the atmosphere, water or in food webs.[12]

Systems for light detection and ranging (LIDAR) are the complement to radar in the visible and infrared regions of the electromagnetic spectrum. The differential absorption LIDAR (DIAL) has been improved for the identification and quantification of gaseous chemicals (e.g. ozone) in the atmosphere. Various LIDAR systems have been tested on balloons and aircraft, and designed for space applications. Some of the applicable laser types (e.g. semiconductor diode lasers or gas lasers) are also used or developed in the military sector or in factories which produce both civilian and military laser systems. Experts and institutions/companies working in military-related laser researchshould be able to convert their know-how and infrastructure to environmental purposes.

The German scout tank "Fox" carries various instruments, which can work under extreme conditions of temperature and humidity, including gamma detectors for the detection of nuclear radiation and mass spectrometers for identifying organic substances like chemical warfare agents.[13] Meteorological sensors on-board can measure temperature, humidity, wind speed and direction and barometric pressure. A military land navigation system is used to determine the exact position. A data-base stores up to 900 substances, and prints-out are provided. Environmental applications in the civilian sector include water and fire analysis, determination of pesticide and pollution measurement after an accident.

## Communications Networks and Environmental Emergencies

Environmental emergencies, such as large volcanic eruptions, earthquakes, hurricanes, floods, accidents in chemical and nuclear facilities or environmental warfare, are expected to become a growing problem in international relations.[14] Programmes for disaster prevention and damage minimization include the assessment or risks, preparedness activities, and (early) warning systems, which rely on science and technology, as well as human resources. Access to these resources can be improved by technology assistance to developing countries[15] and preparedness training of civil servants, health workers, police and firefighters.

### Civilian Systems

In many cases of environmental emergencies the human and material losses would have been considerably lower if the response time was shorter. Adequateresponse to environmental emergencies requires both prompt distribution of information and effective command and control of remedial actions. To improve their effectiveness, disaster relief teams need the latest portable equipment to carry out immediate on-site analyses of damage and risk caused by environmental disasters. For example, compact mobile satellite systems are capable of transmitting audio, data, fax and telex communications signals, which is fed into computers or other communications networks. Within minutes, scientists and disaster specialists who are thousands of kilometers away could respond to the crisis, aided by sophisticated and user-friendly computer models.

To manage the large stream of environmental information from sensors and distribute messages in case of environmental emergencies, an integrated and reliable communications network is required. For example, Landsat 4 relays data in near real-time to ground stations using the geostationary Tracking and Data Relay Satellite (TDRS) and the Domestic Communications Satellite. Data obtained from Nimbus-7 on sea-surface temperature measurements were made available to commercial fishing vessels through daily radio-facsimile broadcasts. For aerial surveillance a communication set in the very high frequency range is required.

In the civilian sector, several networks have been installed or are under development.[16] The international data network for disaster reduction (IDNDR) links a wide range of information networks, including a fixed satellite system like Intelsat, a portable satellite system, a terrestrial mobile cellular system, similar to a car phone, and a terrestrial telecommunications wire system. CLICOM is a computer information system for climate-data management of the World Meteorological Organization (WMO). TheWorld Weather Watch of WMO transmits information about weather and environmental conditions, as monitored by satellites and more than 10,000 stations. WMO also set up the global ozone observing system, which uses technology of satellite-remote sensing. Complementing these systems are networks such as the international network of earthquake engineering centers. The information is often disseminated through the United Nations computerized International Emergency Network (UNIENET).

Computerized emergency response networks, connecting a large number of sensors with central or distributed processing units, can be used to predict the dispersion and consequences of the release of hazardous chemicals. Such systems could be integrated into safety systems to reduce the risk of technical accidents.[17] The German Federal Civil Defence Agency operates the WADIS surveillance and information system which performs continuous monitoring of surface radioactivity and disseminates information on radioactive and chemical dangers to the public. A similar information system for the physical and chemical surveillance of waters is called WAWIS.[18]

**Military Systems**

Military communication systems from the $C^3I$ network are designed to process and transmit large amounts of data from many different sources. There is considerable experience in real-time transmission of large data streams from military reconnaissance satellites to fixed or mobile ground terminals. Sensor data are collected and evaluated in military control and operation centers. Quick-response capabilities, e.g. the ability to set up teams and mobile communications systems, could be applied for emergency relief and damage limitation. Such capabilities, as far as their technical and operational characteristicscoincide, could be demilitarized and integrated into civilian emergency systems.

Small, cheap and mobile communication systems are better suited for conversion than large, expensive and fixed installations. The NATO Integrated Communications System (NICS), for example, comprises different communications technologies such as radio communications in different frequency bands, telephone and telegraph systems, satellite communications, and fiber optics.[19] The whole network is too oversized, static and unflexible for the purposes mentioned above, but some of the electronic components and subcomponents may be useful in environmental crises if their technical characteristics are adequate. Large military communication satellites, like the Satellite Data System (SDS) and the Military Strategic Tactical and Relay (MILSTAR), which transmit vast amounts of data (upto hundreds of millions of bits per second) in extremely high frequency bands in near real-time, could also transmit data on the environment but would be too costly to be developed, produced or launched for that purpose.

Better suited would be mobile components, like masts and antennas, man-packed radios, field telephones and telegraphs, small satellite receiver stations, audio systems (loudspeakers, headsets) and all sorts of electronic components like receivers, switches, analyzers, demodulators.[20] Particularly appropriate are systems for data communications (modems, terminals, multiplexers, line printers, hand-held microcomputers) although more stringent requirements for data security and encryption exist in the military sector. Sophisticated optoelectronic communication systems, based on mobile fiber cables, optical connectors and switches or laser communicators are increasingly used for military data exchange. Almost useless are technologies for electronic warfare (e.g. jamming systems), which arehighly expensive without any obvious utility for environmental purposes. Systems for direction finding (e.g. interferometers) and global positioning (e.g. NAVSTAR) could technically assist in accurate mapping in geophysics and in oceanographic research, during environmental survey work or environmental crisis management. The infrastructure for military communications, including R & D, measurement and simulation, test and production, operation and maintenance, offers good prospects for conversion, although it exceeds the demand in the environmental sector.

**Computers and Modelling in Environmental Impact Assessment and Decision-Making**

For environmental impact assessment (EIA), decision-makers need appropriate information to evaluate the actual state with respect to gains, costs and risks, in order to choose appropriate measures on the basis of models on the environment. Although human, organizational and financial resources are mostly needed, technology can contribute to EIA decision-making for data processing, modelling and simulation, and systems analyses.

**Civilian Systems**

Computers for data reduction, storage and processing are used for a wide spectrum of environmental applications. Their uses include the evaluation of monitoring data, environmental databases and expert systems, prognosis and visualization, modelling and simulation. For example, geographical information systems collect topographical data, which are the basis for databases describing the pollution of soil, air and water, the forest growth or the occurrence of different animal species. Expert systems have already been implemented in monitoring air quality, pesticide levels, ground and surfacewaters and environmental planning.[21] Fast computers, often working in parallel, have been used for climatology predictions and tracking of air pollution on a global scale. Some smaller simulations can be handled with a normal personal computer. A few examples from the civilian field may illustrate the value of computers in environmental modelling.[22]

1. OECOSYS is a computer simulation model for the prediction of the population dose in case of a large-scale radio active contamination (e.g. the Chernobyl fallout).

2. With the aid of particle-based simulation and meteorological models, the behaviour of exhaust emissions in different atmospheric layers can be computed and displayed on graphic work stations.

3. Modelling and simulation software have been proposed for the analysis of the forest degradation. Sub-models describe the growth of the tree, the water flow in the ground, the chemical reactions in the surrounding soil and finally the exchange of minerals.

4. A model has been developed for personal computers to simulate the flow of refuse and their impact on decision-making. Again there are different sub-models or levels such as waste sources, waste treatment sites and the output of these sites.

**Military Systems**

The military sector, in particular the defence laboratories, possesses enormous computing power, including software, which can be used to interpret results from sensor systems. Supercomputers have been used in global climate research, nuclear winterstudies, multimedia compartment models, and emergency response systems. Managerial skills and knowledge in modelling, simulation and testing of complex and vulnerable $C^3I$ systems may be applicable to environmental impact-assessment, decision-making, utility-cost analysis and risk-management.[23]

Environmental databases could benefit from the results of military surveillance which monitors and assesses the atmosphere, oceans, Earth's surface, and human population. Artificial intelligence, expert systems and robotics could have significant roles in the development of remote sensing and analyzing techniques, in decision-making and process control, accident avoidance and remote analysis of hazardous and radioactive materials. In the laboratories of the US Department of Energy, where a substantial fraction of the military and nuclear research is performed, the application of computer simulation has been demonstrated for a long time. Some examples of environmental applications are given below:

1. The laboratories have been involved in research on the global climatic effects of nuclear war for more than 25 years. After 1983, nuclear winter

studies have been examined in co-operative efforts with the Defence Nuclear Agency and the National Climate Program.[24]

2. Multi-media models simulate and visualize the movement and transformation of chemicals as they spread through the air, water, biota, soils, sediments, and ground water. They evaluate the residual products (e.g. solvents or smoke) from various industrial technologies and can serve as a risk-management tool for the absorption of contaminants by people(e.g. of radionuclides, lead and arsenic, dioxin and benzene).[25]

3. The Atmospheric Release Advisory Capability (ARAC) is an emergency response system that provides real-time prediction of the dose and deposition resulting from accidents involving the release of radioactive or toxic materials into the atmosphere.[26] Real-world events have been analysed and graphically visualized, as the nuclear reactor accidents at Chernobyl in the Soviet Union, at Three-Mile-Island in Pennsylvania and the reentries of several Soviet nuclear-powered COSMOS satellites.

## Conclusions

A reasonable conversion strategy, focusing on the conversion of science and technology, would try to reverse the dual-use concept by finding and promoting useful civilian applications of military-related technologies. To assess the feasibility and effectiveness of conversion, the technical characteristics, development phases, stages of conversion, conditions and constraints should be taken into consideration. Government support and civilian control over useful military systems are essential to realize disarmament, demilitarization and a transition towards structural conversion of military technology development.

The military $C^3I$ network contains many sophisticated information and communication technologies which may be adaptable for environmental purposes. Although several civilian remote sensing systems are available for environmental monitoring, military sensors, using the visible, infrared, or microwave regions of the electromagnetic spectrum, and being deployed on ground-air- or space-based platforms, may add to this capability. Militarycommunication systems, which are able to process and transmit large amounts of data from many different sources in real-time, could be used or modified for environmental applications. Quick-response capabilities, existing in the military sector, could be applied for emergency relief and damage limitation. Computer hardware and software in military laboratories, in particular supercomputers, have been used in global climate research, nuclear winter studies, multimedia compartment models, and emergency response systems. Knowledge in modelling and testing of complex military systems may be applicable to environmental impact-assessment, decision-making, utility-cost analysis and risk management.

However, military systems are often overspecified, too complex and too costly for environmental use. Applicability depends on the operational and technical characteristics. It has to be seen case-by-case whether military ICTs can compete with available civilian systems in terms of utility and cost. For inadequate systems, the best use would be to dismantle them and convert the funding directly for environmental purposes. Finally, technology is only one means for environmental protection and should be complemented by economic, social and political measures.

**Endnotes :**

1. A survey of critical technologies can be found in DoD (1991).
2. On the dual-use of technology and the ambivalence of science see Seiler (1991), Domke (1991), Liebert (1992). Sometimes the different terms "dual-purpose", "dual-capable" or "multi-functional" are used.
3. UN (1991). The Study not only discusses the protection of the natural environment against man-made abuse, but from a broaderperspective protection of the human environment against natural or man-made disasters as well.
4. A more detailed description of some of the examples is given in Scheffran (1991a).
5. UN (1991).
6. UN (1991).
7. Gomberg (1985).
8. Kokoski (1990).
9. Jasani (1987), p.52.
10. A detailed description of both systems is given in Grin (1990).
11. Kalinowski (1986).
12. IAEA (1990).
13. NBC (1988).
14. The 1990s were declared the International Decade for Natural Disaster Reduction.
15. The UN Scientific and Technical Committee has been working with some 30 national committees to assess overall gaps in technology; see Chronicle (1991), p.44.
16. See the survey in Chronicle (1991), p.52.
17. Examples from the civilian sector include the German SAFER, the Austrian UMBLDR and UMBL-NET, the Swiss NABEL and the IIASA RAINS; see Pillmann (1990).
18. The projects are described in leaflets of the German company Rheinmetall in Dusseldorf.
19. Scheffran (1991b).
20. Williamson (1990).
21. In the United States, for example, the Man computer Interactive Data Access System (McIDAS) can integrate conventional and satellite data together and permits real-time monitoring of rapidly changing weather situations. See Lo (1986), p.112.
22. These and more examples can be found in Pillmann (1990).
23. In the field of $C^3I$, a variety of mathematical models has been developed or applied; see van Trees (1989).
24. MacCracken (1985).
25. McKone (1987).
26. Belles (1990).

**References**


Belles (1990) R.D. Belles, H. Walker, T.J. Sullivan, "Data Visualization and the ARAC Emergency Response System", *Energy and Technology Review* (ETR), Jan./Feb. 1990, pp. 3-15.

Chronicle (1991) "UN Chronicle", June 1991, special edition on disaster reduction.

DoD (1991) "Critical Technologies Plan", Washington, DC: Department of Defense, 1991.

Domke (1991) M. Domke, "Kostendämpfungsstrategie : Integration ziviler und militärischer Produktion neuer Technologien", *Informationsdienst Wissenschaft und Frieden*, 4/1991, pp.26-31.

Gomberg (1985) L. Gomberg, "Remote Sensing of the Earth with the Defense Meteorological Satellite", in : A. Schnapf (Ed.), *Monitoring Earth's Ocean, Land, and Atmosphere From Space*, New York, 1985, pp.96-128.

Grin (1990) J. Grin, "Military-Technological Choices and Political Implications", *Command and Control in Established NATO*

*Posture and a Non-Provocative Defence*, Amsterdam, VU University Press, 1990.

IAEA (1990) *IAEA Bulletin*, April 1990, Special edition on isotopes in environmental research.

Jasani (1987) B. Jasani, T.Sakata (Eds.), *Satellites for Arms Control and Crisis Monitoring*, Oxford University Press, 1987.

Kalinowski (1986) M. Kalinowski, "Elemental Trace Analysis in Environmental Physics", Working Paper, Aachen, September 1986.

Kokoski (1990) R. Kokoski, S. Koulik (Eds.), *Verification of Conventional Arms Control in Europe*, SIPRI, Stockholm, Westview Press, 1990.

Liebert (1992) W. Liebert, G. Neuneck, "Civil-military ambivalence of science and the problem of qualitative arms control", to be published in : H.G. Brauch, et al (Eds.), *Controlling Military Research and Development and Exports of Dual Use Technologies as a problem of Disarmament and Arms Control Policy in the 1990s, New York*, St.Martins Press, 1992.

MacCracken (1985) M.C. MacCracken, "Global Atmospheric Effects of Nuclear War", *Energy and Technology Review*, May 1985, pp.10-35.

McKone (1987) T.E. McKone, "GEOTOX : Simulating Contaminant Behavior and Human Exposure", *Energy and Technology Review*, May 1987, pp.14-20.

NBC (1988) "NBC version of TM-170 unveiled", *JANE's Defence Weekly*, 13 August 1988, p.278.

Pillmann (1990) W. Pillmann, A. Jaeschke (Eds.), *Informatik für den Umweltschutz*, Springer-Verlag, 1990.

Scheffran (1991a) J.Scheffran, I. Vydra, "The Application of Military-Related Resources to Protect the Environment: The Case of Information Technologies", Working Paper, Darmstadt : IANUS, 3/1991.

Scheffran (1991b) J. Scheffran, "NATO Command and Control between High-Tech Warfare and Disarmament", Peace Research Institute Frankfurt, Report No.22, 1991.

Seiler (1991) Achim Seiler, "The Civilian-Military Ambivalence of High-Technologies and its Significance for Transnational Co-operation in Western Europe", preferred paper, presented at the 41st. Annual Pugwash Conference on Science and World Affairs in Beijing, September 17 to 22, 1991.

UN (1991) "Charting Potential Uses of Resources Allocated to Military Activities for Civilian Endeavours to Protect the Environment", Report of the UN Secretary General, New York, 1991.

Van Trees (1989) H. van Trees, "C3 Systems Research : A Decade of Progress", in : A.H. Levis, S.E. Johnson (Eds.), *Science of Command and Control, Part II*, AFCEA International Press, 1989.

Williamson (1990) J. Williamson (Ed.), *JANE's Military Communications 1990-1991*, 11th Edition, JANE's Informations Group, 1990.

# ITALY

# FROM MILITARY TO CIVILIAN PRODUCTION: THE PRESENT PROBLEMS OF THE MILITARY INDUSTRY IN ITALY

**Prof. Piero Tani and Dr. Giuseppe Catalan**
*Italy*

## Introduction

The statistical data presented in this paper show the deep crisis of the Italian military industry. The reduction in arms exports has already been described in the first part of this issue. In addition to this, a fall in the purchase of arms by Ministry of Defence has been thoroughly documented.

The Coordination of observers of the *Military Industry* has collected the 1989 and 1990 turnover data of the Italian military firms. Where there is no evidence of a clear reduction, our analysis of the numerous situations of crisis significantly reveal demand trends. For this reason the paper concludes with some notes on the possible reorganization policies in our country.

The analysis of the published data and our common experience in the military area, linked with the exhaustion of the growth characterizing this sector during the last ten years, lead us to note that, in 1989 there clearly emerged a situation of crisis. Consequently, it is not possible to postpone anymore the reorganization of this sector.

In conformity with a notorious European situation (Great Britain, Germany, France), it seems that even the current rationalization process -- which affects major industries such as Alenia, Agusta and Fiat and which concerns consolidation and acquisition -- has not prevented unemployment benefits and staff reduction. The risks of dealing on the arms market is increasing; larger and larger areas of civil society are insisting on a systematic policy, asking for theredevelopment of arms industries in crisis into civil productions.

The new unemployment problem seems to be delicate; but the situation can be successfully coped with if only a clear government policy will supply a suitable industrial policy in order to achieve a drastic reduction of the sector. This can be obtained both by the introduction of social benefits in the case of loss of jobs and by the control and the support of redeveloping and reorganizing firms. The possible interventions are several : research activity support, development and reindustralization programs.

On the contrary, the prospects for industrial policy, which are still mistaken for defence policy, are not clear or comprehensible. This policy is characterized by the patronage system, which supports firms through the relaunching of by-now-unexplained military-commercial orders, or, even worse, through a flood of funds that have no clear targets of efficiency or productivity.

The present crisis in the military-industrial sector is the finishing point of an exceptional growth cycle of military-expenditures, but in the future it can be the starting point of a development cycle only if we become aware that there is already a turnabout.

## The Demand for Military Industry
## Arms exports from boom to crisis

The explosion of Italian arms export had its start at the beginning of the 70s. It was strictly connected with the increase of theThird World military market, which was especially led by some oil-producing countries. In fact, in 1973 weapons exports exceeded imports, and since then, the commercial indicators in arms business was positive and increasing. However, the turning point came in the second half of the 70s: according to ACDA, in 1978 exports doubled "in value" compared with the previous year; while according to SIPRI, in 1979 exports to the Third World tripled. In the early 80s the highest amount of delivery was reached: 1,417 million dollars according to ACDA in 1981; 1,350 million dollars according to SIPRI in 1982 (valuations in 1985 prices). In 1981, the specialization index was also at its highest level, with a value of 0.85. Ranking sixth place in world classification (ACDA) in 1981 and 1982, Italy became one of the chief exporting countries, only behind U.S.S.R., U.S.A., France and in competition with Great Britain, West Germany, China and Poland. Italian arms were mostly destined for developing countries (96.2% in the period 1978-82, and 94.5% in 1983-87), to such a point that in the five-year period 1979-1983 Italy became, according to SIPRI, the fourth exporting country of arms-systems to the Third World.

The boom of Italian exports was based on two favorable "prerequisites". On the one hand, Italian production, which was at the top of the "middle-sophisticated" market range was easily able to comply with the requests of the developing countries; on the other hand, the lack of rules and controls based on the considerations of safety and respect of human rights, gave the industries the opportunity to carry out an effective policy of trade penetration. But the development of exports was also supported by the increase of our national military demand. From 1975 to 1977 the Parliament promoted laws for the modernization of the Armed Forces. They were called "promotional", because they were aimed at promoting and developing a nationalmilitary industry. As a result, the combination of national orders -- more reliable and steadier -- with foreign orders, enabled a series of "average cost prices" not to become a burden on the public administration. In this way, the products turned out to be competitive on the foreign market. Beginning from the second half of the 80s Italian exports began to slacken (figure no.1) : in 1988 they reached the values of 1974-75 according to ACDA, and of 1978 according to SIPRI. There was a real slump in the sales to the Third World : data referring to 1989 and 1990 show a return to export levels of the 60s. That was the end of an era.

The first official data, presented by the Government in the "Six-monthly Report to the Parliament on the Informative Policy and Security" ("Relazione semestrale al Parlamento sulla politica informativa e della sicurezza") (Doc.XLVII), confirm the reduction of Italian export of arms in the second half of the decade, as it is shown in figure no.1. In fact, considering 100 as 1984 reference datum, the exports were halved in 1985, and they settled to one third of the initial value in the following years. Moreover, for the first time, in 1989, the value of the granted licences was lower than the total dispatches, as well as than the licences granted in the previous years. With this situation, a further drop in exports is expected in the coming years.

Also the "Report on Authorized Operations Carried out to Control Exports, Imports and Arms Transfers" (Relazione sulle operazioni autorizzate e svolte per il controllo dell'esportazione, dei materiali di armamento) confirms the current negative trend. The Report, which was expounded to the Parliament according to the new law on arms trade (no.185/1990), also includes a record of exports with statistics on firms, products, values, and countries of destination (Doc. CVII). As it can be seen, 1990 grants forauthorizations to exports -- including the temporary ones -- amounted to 1,265, which were divided as follows : no.1,070 in the period from January 1 to July 28 (equivalent to the value of 1,366 billion Lit.) and no. 195 after the law was passed (equivalent to 387 billion Lit.).

The crisis was due to the reduction in the demand by the developing countries, the economic and financial crisis, the new detente, and the end of the cold war between East and West. In addition to that, there had recently been the achievements of traditional competitors such as West Germany and new exporters like Spain, Brazil, and Israel, that raised difficulties to the Italian offer.

Also the passing of a new law dated July 1990, which establishes the general rules regulating the future policy of Italian arms exports, may have significant consequences. According to this law, exports are forbidden when in contrast with the Constitution, with the Italian international duties, and with the safety of the country. Particularly, it is forbidden to export to countries in a state of war, towards whom UNO declares Embargo, countries whose governments are responsible for transgressions of the International Conventions about human rights, and countries that, whenever there are resources exceeding the needs of defence, destinate Italian funds for the co-operation to the development to their own military budgets. After the passing of this law, the remarkable reduction in the export licences is due to two facts: on the one hand, to the lack of export authorizations of certain items (sport and hunting weapons, conventional munitions, fire-arms, etc.), and on the other hand to the increasing of the crisis in this sector and to the more rigorous regulation in force.

**Recent Trends of Italian Military Expenditure**

After a long period of development, world military expenditures began to fall "in real terms"; at first the situation was slightly noticed, then, after 1990, it became relevant. In particular, the late 80s represent a turning point in the trend of the growth of expenditures for every NATO country : starting from the USA in 1987, through Great Britain, Germany and Belgium in 1988, Denmark and Greece in 1989, to France, Spain, The Netherlands, and Italy in 1990, as SIPRI data reveal.

The total military expenditures of the Ministry of Defence began to decrease in real terms in 1990, and especially in the following year. In 1989, funds assigned to armaments had already been cut in real terms, and in the following years in nominal terms. In 1991 there was a further drop in armament expenditures (-9.9%), and the most considerable cut concerned the purchase of new weapon systems (-13.4%). A new trend inversion is revealed in the data relating to the 1992 budget: in fact, compared with the previous year, the military expenditure increased 9%, the armaments 7%, and, above all, the expenditures for new arms systems over 16%. At any rate, although data concerning armaments show a consistent growth, they are below the 1990 level in nominal terms, and, least of all, below the previous year's data. The whole armed forces is subject to cuts in the funds assigned to the purchase of new armaments. However, the Army suffered from this measure in 1990 (-16.4%) and the Air Force in 1991 (-16.6%). In 1992 increase in expenditures budget will benefit the Air Force and the Navy. After the cuts in the budget and before the reform for a new defence model, armed forces have reduced, postponed, and reorganized some important projects. The main cuts have concerned the Army, whose role in the newEuropean scenery has been significantly reappraised. The Army has reduced its programmes, like for example, the purchase of anti-tank systems Folgore and Milan, light bren-carriers Puma, anti-aircraft systems SIDAM, rocket-ramps FIROS 30, bren-carriers Centauro. Other programmes have been postponed (tank Ariete and Helicopter A 129), and some others have been weakened (ammunitions and Skyguard).

The most considerable cut in the Air Force budget -- which has been made official up to this time -- deals with tactical support fighter AM-X. It has been decided for the suspension of the production of lots no.4 and 5 (from 238 to 136 aircrafts). In addition to that, there has been the corresponding reduction in the commercial orders from the Brazilian Air Force, which brings to a total cut in 4 out of 11 lots formerly estimated. Even if it has not been made official, the programmed purchase of the spoiler-fighters, within EFA project, is likely to be reduced from 165 to 130.

As regards the Navy, the projects of new ship buildings have not been reduced, since some units, provided for by the law of 1975, are in an advanced stage of completion. In the current year the remaining corvettes Minerva will be delivered and in 1992-93 the two fighters Animoso, the fourth series of submarines Sauro and the second series of minehunters Lerici. But if one leaves out some renovations (frigates Lupo and submarines Sauro), the building of new big units is not being programmed at present. In fact, a large amount of the funds has been assigned to purchase vertical-take-off aircrafts STOVL and helicopter-carrier Garibaldi, and to the expensive projects of research and development of helicopter EH 101, torpedo A 290.

**The Trends of the Military Industry**

This section presents a synthetic outline of the recent trends of the Italian military industry, and brings up-to-date the data concerning recent years' turnover, which have already been published in LETTERA2 - for a thorough study on the subject, particularly on the methodological remarks, we refer the reader to that publication. Here is a concise scheme.

Only those factories trading in Italy, with a military turnover equivalent to or over 2 billion Lit. in at least two of the examined years, have been taken into account.

**Unit of the analysis** is the firm directly dealing with military productions. This firm can sometimes carry on the role of a leader in a group of companies or be even a financial holding. Firms carrying on only functions of control in military industries (for instance Finmeccanica and FinBreda) have not been taken into account.

**Group of control** refers to societies that, directly or indirectly, control in a "real way" arms-producing factories. **Sectors** which contain the different firms have been dealt with in the first issue of the periodical newsletter of the Co-ordination. **Turnover** refers to the proceeds of the sales and the services. Of course, **military turnover** is the turnover quota concerning the sales of military products and military service.

The data of the **balance sheets (a)** have been considered the primary source of this research, but the lack of them has led the Co-ordination to refer to the **official statements of the factory (b)**. These **assessments** have been considered **reliable sources (c)** because they are based on the acquaintance with the productions and the programs described in the reports on firmbalance-sheets, as well as in the semi-official statements of the factories. Finally, the **unofficial assessments (d)** are collected from sources other than the previous ones and reliable to a large extent (about 10/15%).

In table no.1, we present the top twenty firms with a budget equivalent of over 100 billion Lit. in at least two of the years included in 1987-1990. Just from the beginning of 1989, Italian military-industry entered a new phase. After a period of uninterrupted increase, there was evidence of a shrinkage in the military turnover: from 7,854 to 7,551 billion Lit. In 1990 there was a slight increment of military turnover of the top twenty firms, but in nominal terms and not in real ones. In fact, only larger enterprises still recorded positive balance-sheets (with the exception of OTO Melara), whereas smaller firms revealed a critical situation. A phenomenon of concentration has been taking place, since the top five firms held about 50% of the total military sales in 1990, whereas they covered only 40% in 1987.

The destiny of some sectors seems to be different: the deepest crises concern the metallurgical and mechanical industry (ammunitions, small arms, heavy arms, and means of transportation); the chemical industrial group (explosives) and the naval group are undergoing a difficult time. The industrial electronics sector is "healthier", even though there have already been the first signs of a crisis, while aerospace industry is still expanding. Also the state-owned industries are more and more expanding (Finmeccanica-IRI). On the contrary, the main foreign industries are leaving the Italian market; they have already handed over the control of some Italian military factories.

In the last few years the total data seem to reveal the reduction of military turnover of the top twenty firms (from 50% in 1987 to 38% in 1990), as well as the start of a newprocess of diversification of this sector towards civil conversion. However, the statistics are influenced by the condition of some big industries, whose positive trend of the civil turnover makes up for the crisis of the military turnover. In fact, nine of the top twenty factories are deeply involved with military production, which represent over 90% of the total turnover. A real diversification process is taking place only in Marconi Italiana, Rinaldo Piaggio and Elsag.

## Military Firms in Crisis

This section supplies an outline of the critical situation that military industries are experiencing. In a word, it can be stated that there are two different forms of crisis. The first one, that can be called "evident", manifests itself in the usual way; that is to say, through announcements of exceedings, requests of dismissals, and recourse to unemployment benefits. The second kind of crisis can be called "crawling"; in other words, it doesn't show itself in a clear way, but its effects are as negative as those of the "evident" crisis. State-controlled firms are usually hit by this emergency. In order to adjust the work force to the production-capacity demanded by the market, these firms have decided to stimulate flight of capitals and to block the turnover; in this way, it is possible to have a gradual reduction of the workers.

The most numerous of all the firms in crisis are the private ones; they have been the first ones to resort to unemployment benefits. Fiat, the most important private enterprise operating in this field, is in the list, together with Fiat CIEI, SNIA BPD, Simmel Difesa and Whitehead. Misar merged with Whitehead Division and gave birth to Whitehead Spa (limited company), while Borletti and Marelli Avio merged in Fiat Cei. Despite this huge reorganization, the occupational problem of the group is stillunresolved. Taking into account the scheduled rotation, approximately a third of the work force of the mentioned enterprises is currently getting unemployment benefits.

According to the agreement signed in December 1990, 600 out of 2,700 workers of Aermacchi are getting unemployment benefits. 350 of them are to be considered structural exceedings; in the mentioned agreement, several plans of unfreezing are described, such as professional retraining, relief for new working opportunities, early retirement. The situation of Oerlikon, Contraves, Europa Metalli LMI, Junghans and Nardi Aereonautica is particularly delicate. In these firms, not only is a large number of workers getting unemployment benefits, but also other economic indicators are very worrying, particularly the turnover and the purchase orders. In short, these enterprises are experiencing both a clear occupational problem and an almost total lack of market prospects.

Right from the beginning of this year, Swiss multinational Oerlikon - after 5 years of negative balance-sheets - has decided to auction the main part of its activities in the defence sector. During these months, a separate sale of Oerlikon Italiana to EFIM and of Contraves to Finmeccanica has also been attempted. However, the only certainty is revealed by the huge reorganization that is taking place in Milan and in the letters of dismissal sent to 214 Contraves employees last spring.

In 1975, Junghans employed 1200 persons, while now it has just 189 workers, 80 of whom are getting unemployment benefits. The plan for reorganization projected an investment of 4 billion liras, 2 of which had to be destined for the civil sector. According to the plan, a thorough production recovery had to occur in March 1990; on the contrary, the immovables are still being restored.Consequently, the production is poor both in the military and in the civil field.

Nardi (aeronautic constructions) is undergoing such a deep financial crisis that the enterprise is on the verge of bankruptcy. A positive solution has been found for the factory in Aprilia (LT), thanks to the intervention of Gepi and Alenia, whereas the situation of the central factory in Milan is still critical and apparently without solution.

Another interesting, and, in a sense, new development is beginning to take place. The State-owned enterprises are starting to resort to unemployment benefits, even if they are entitled to State protection in terms of orders and finance. The case of Alenia is symbolic. An impressive process of industrial and financial reorganization gave birth to this enterprise just one year ago; its target was to oppose international competitors. However, in June, Alenia reported the surplus of 3000 employees. A few days ago, approximately 326 workers started to get unemployment benefits; this solution has been adopted in nearly all the plants of the group. In "absolute terms", the plant in Fusaro is the most damaged by the crisis; however, in relation to global occupation, the one in Palermo is actually in a worse condition : 90 out of 300 employees are getting unemployment benefits. In September, the enterprise and the trade union entered into negotiations again, in order to discuss industrial plans and to re-balance the personnel; Elmer and Meteor will be involved in this reorganization as well. Also SMA (EFIM) has announced the excess of approximately one hundred employees; perhaps, right from the beginning of October, a process of firm reorganization will take place; this means unemployment benefits or early retirements for a large number of workers.

All the marketing fields are undergoing this deep crisis. The first ones to suffer fromits devastating effects have been the technologically mature sectors and those firms that have adopted the wrong industrial policies. Moreover, during the last months, even the most advanced fields have started to reveal the first signs of serious depression in trade. The military ship building is going through a long crisis that started many years ago, due to the exhaustion of new constructions promoted by the naval law of

1975. The unconcluded business with Iraqi fleet contributed to the crisis of this sector, as well. Two further causes led to the excess of production capacity and to the consequent request of hundreds of dismissals: the munitioning (Fiocchi, Junghans, Simmel Difesa, Europa Metalli LMI) after the remarkable shrinkage of the export trade at the beginning of the Eighties and a significant drop in overall domestic demand. The aerospace and electronic sectors have recently undergone cuts in military expenditure; consequently, important plans had to be canceled.

This analysis can be concluded with some observations about the "crawling" crisis. OTO Melara is an interesting case; this enterprise reduced its work force from 2,546 to 2,245 - almost 300 employees - between 1983 and 1990. A similar situation is experienced by Fincantieri (Military Constructions Section); in just three years, its workers were reduced from 3,370 to 2,880. Another significant feature is the "indotto" of these enterprises; as far as OTO Melara is concerned, its drastic reduction of the work force from 580 to 120 employees took place in tandem with the final shutdown of many small enterprises.

Cuts in the orders are, again, clear signs of the approaching crisis; Officine Galileo and Iveco Fiat are two interesting examples. The modernization of land transport is currently in debate; these doubts are already causing troubles to the mentioned enterprises andthey will certainly bring about even more problems in the future.

In the light of the analysed cases, it seems possible to state that all the crises have been dealt with according to the economic situation and by means of a relief programme. They have been faced with the traditional "social benefits". No alternative lines of development have been identified; in other words, no one has tried to find the right measures to adjust the work force to the changed market demands; moreover, no programme has been written out in order to give new qualifications to the workers and to employ the existing professionals in alternative civilian fields.

Consequently, the attempt to include military industry in the law 23/7/91 n.223 is useless. Not only does this law reform the unemployment benefits, but it also allows the enterprises in the process of reorganization to retire the excess labour. Fincantieri is a state ship building industry; consequently, it is already on the list of those enterprises entitled to early retirements. On the contrary, Piaggio, Contraves, Aermacchi, and Fin Breda urge that they should get finance in order to cover their increasing expenses. However, the adoption of early retirements will not be sufficient to solve the huge structural problems of this sector. It will be necessary to take up a specific industrial policy, oriented towards conversion.

## Necessary Tools of Conversion Policy

During the last years, conversion of military industry has been brought to the attention of national and regional legislative boards. Since 1987, the Left has particularly committed itself on this subject and has introduced some bills. Actually 6 different bills regarding military industry diversification andconversion have been proposed. Here is the list, in order of introduction:

- 2/7/1987; introduced by Partito Radicale [Radical Party] both to the Chamber of Deputies no.429 (Rutelli and others) and the Senate no.56 (Spadaccia and others) with the same title: "Devices to achieve industrial conversion of enterprises producing military goods and services";

- 19/1/1988; introduced by Sinistra Indipendente [Independent Left Group] to the Chamber of Deputies no.2178 (De Julio and others) : "Institution of a fund for military industry conversion";

- 28/9/1988; introduced by PCI [Italian Communist Party] to the Senate no.1327 (Salvato and others) : "Institution of a fund to support the conversion of military industry into production of civil goods or services";

- 1/12/1988; introduced by PCI [Italian Communist Party] to the Chamber of Deputies no.3417 (Starda and others) : "Institution of a fund for military industry conversion";

- 2/2/1989; introduced by DP [Proletarian Democracy] to the Chamber of Deputies no.3600 (Tamino and others) : "Rules for the conversion of industry producing armament materials and for the modification of the defence model".

From the day of their presentation, nothing happened; Parliament has examined the bills only in November 1991. Honourable Ravaglia was appointed as proposer of the bills and the debate was arranged. In order to accelerate the parliamentary procedure, some deputies have recently suggested the constitution of a group made up of all the different political parties. Its aim is the implementation of a conversion process bymeans of legislative tools, and the control of the achievement of exportation law. This lobby would also facilitate the drawing up of a standardized letter of the law, which should state a more direct involvement of the government parties in conversion matters. Currently, the internal political crisis has made more improbable the difficult co-ordination between the representatives of the different parties; the situation is worsened by the international crisis due to the Gulf war.

Despite the aforesaid delays in the examination of the bills, some legislative improvements have been made; in fact, Parliament has passed two laws : no.808/85 and no.185/90. The first one deals with the aeronautic field and stimulates its diversification; article 1, paragraph 5 provides "as far as national enterprises are concerned, the increase of civil production to the detriment of military one". In July 1990, a law controlling weapons exports has been passed; it also deals with the conversion matter; article 3 states that Government has to "favour the gradual production differentiation and the civil conversion of defence industries". According to article 8 of the same law, one of the duties of the "Office of co-ordination of armaments production" is to "contribute to the discoveries of possible plans of industrial conversion". However, this office is still to be founded.

The Defence Committee of the Chamber of Deputies did a research on the new defence pattern, that was approved on the 20th of March, 1991. As far as a global defence reorganization and rationalization is concerned, many plans are drawn up in the final account of this research. Here it is stated that many firm projects "will be able to provide for conversion and diversification; however, they will certainly have to provide for investments, researches, and development, in order to reach a qualitative and quantitative level adequate to thedefence and security needs in the third millennium".

On a regional level, since 1988, there have been many initiatives to promote conversion; the most important ones are :

- one motion made by the regional councillor of the Verdi [Green Party] on the 28th of October, 1988, and carried the same year by the Regional Council of Lazio; another motion made by DC [Christian Democrat Party], PCI [Italian Communist Party], DP [Proletarian Democracy], and PSI [Italian Socialist Party] on the 17th of May, 1988, and carried by the Regional Council of Tuscany. Both of them promote researches and initiatives in order to foster conversion;

- bill no.112 proposed to the Regional Council of Lazio by the councillor Mastrantoni of the Verdi Arcobaleno [Green-

Rainbow Party] on the 20th of November, 1990. It promotes the "formation of a committee for the research of the condition of the military production structure; this committee has also to foster the civil conversion of enterprises producing military goods and services";

- research on the "conditions of feasibility" of a process of conversion and/or constant integration of military industry in Liguria; proposed and approved by the Regional Council of Liguria in February 1989; intended for those civil industrial fields that allow the application of a know-how consolidated both on the technological and the work force level;

- the allocation of 666 million Lit. Specific fund for researches on the conversion of military production; approved by the Regional Council of Tuscany for 1989 and 1990, in accordance with the aforesaid motion. Due to this fund, in 1990, Sirio Panel in San Casciano (Florence) firm producing electronicequipments -- presented a conversion research project.

The bills proposed on a national level are mainly aimed at the determination of :

1. the right policy of management in order to help the enterprises;

2. tools for directing and controlling the conversion policies in this field;

3. rules for the distribution of the allocated funds;

4. measures for the protection of the workers.

The main aim of the bills is to make the process of conversion financially attractive both on a techno-economic level and a "social" one. For this reason, all the bills provide for the institution of a "National Fund for the Conversion", kept by the Ministry of Industry.

The features common to all the bills are :

1. the recourse to state "centralized structures" for financial intervention;

2. the choice of exclusively financial incentives for the processes of conversion;

3. the determination of funds in order to cover the costs of scientific research, design, and commercial promotion.

There are two main problems that make the process of conversion particularly hard :

1. the relationship between the structure of the internal demand of this field and the possibility of its re-qualifications;

2. the Italian export policy in this field, that has no relation to NATO. In particular, the structural characteristics of the internal demand and the public contractual mechanisms do not seem to have suggested innovative industrial policies to the proposers of the bills. Moreover, even though the internal demand is the moving cause of the military industry, no alternative markets have been opened for the conversion products; besides, what is worse, no researches on this field have been planned.

From an institutional point of view, the Fund interventions can be summarized in two prevailing patterns : the first one proposes a centralized management in the hands of the Ministry of Industry, which has to operate in accordance with CIPI (Interdepartmental Industrial Planning Committee) or IMI. The second pattern propounds a more joint management, taken over by diversification Committees, CIPI and, in certain cases, even by the regions and the provinces.

Neither trade unions nor industries belonging to this or to other sectors are allowed to take part in the central bodies. This limitation is in accordance with patterns of industrial policy already tested by the laws no.675/77 (about industrial conversion) and no.46/82 (about technological innovation).

De Julio's bill proposes a co-ordination between the interventions in the military sector and other measures of industrial policy. It also limits the availability both of IMI Fund for Applied Research (no.1089/68) and of the Rotating Fund for Technological Innovation. If this law comes into force, both the Funds will have to be mainly invested in the conversion of military production.

Regions and local organizations are allowed to play a marginal role. This limitationclashes with the Regions-CEE agreement on employment protection and regional interventions; it states that regions are free to intervene in all the fields that are undergoing a period of crisis. The only bill that provides for the presence of local union officers and of entrepreneurs is Rutelli's one. Besides, no funds for local or regional plans are provided for.

Nearly all the bills propose :

1. a control over the use of the funds;
2. employment protection arranged by the enterprises;
3. annual parliamentary account of the Fund activities.

In case of non-observance of the legislative provisions, obligations and penalties are provided for by the bill (Strada), the contract (Salvato), or the Minister (De Julio).

As for the employment protection interventions in the enterprises to be reconverted, the bills provide for them in the form of "requisites for admission to the funds"; in other words, they are an integral part of the contract. Work protection has a broad meaning and includes all the provisions meant to guarantee :

1. re-employment of the excess personnel;
2. professional re-qualification;
3. protection of the professional objectors working in military firms;
4. presence of union officers in the organs planning conversion.

The bills proposed by the Radical Party provide for a specific fund. Besides, they propound a more detailed plan of interventions, in order to favour the re-employment -- even in other fields -- of the military firms personnel. This accurate plan provides for the intervention of the conversion provincial committees.

The re-qualification of the managers working in military enterprises has been completely ignored by all the bills. Because of the demand structure and the present characteristics of the military enterprises management, this aspect is fundamental within conversion policies. In fact, all the firms that have difficulties in re-employing their excess personnel could 'possibly' be opposed to the conversion policy.

Finally, the procedures for financial benefits. According to the two bills proposed by PR and the one presented by DP, the conversion fund should be financed either by the enterprises or by the State.

On the contrary, according to De Julio's bill, the Fund set up by the law 675/77 should be used to finance the conversion fund. PCI bill states that the finance is to be provided for by the fund at CEE disposal. The other PCI bill does not propose any solution for this problem.

The debate about conversion is conducted in a quite irregular way, in relation to the structural crisis of military industry and to the Gulf crisis. Consequently, periods of intense activity are followed by moments of total stalemate. Nowadays, this debate has deep connections with the process of reorganization of the defence pattern.

The final result of the official inquiry into the new defence pattern is a document where the Parliament, in relation to the new international situation, expresses the future desirable lines of conduct of the Government:

- to meet the future defence orders;
- to invest money particularly in those goods, systems and components that have a dual use;
- to direct the state financed research to the amelioration of the standards;
- to carry out a policy of re-industrialization for all the sectors undergoing the crisis;
- to deal with armaments industries and military arsenals in the same context;
- to foster those merging processes that are linked with the participation in multinational programmes.

Conversion bills are likely to be set in this same context.

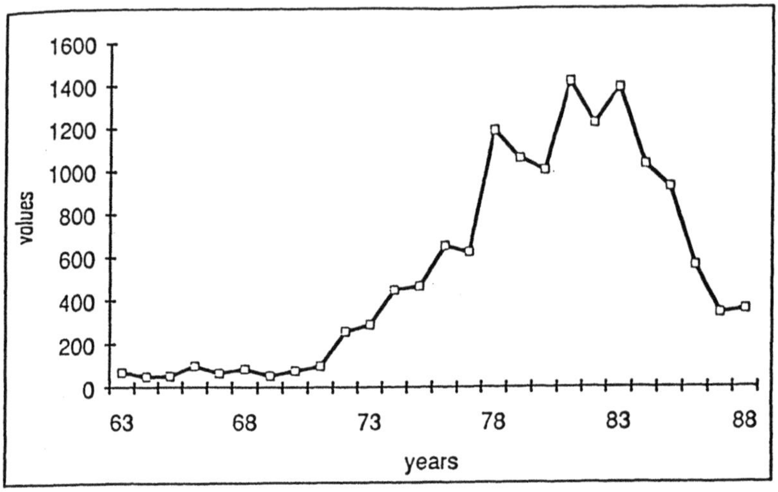


Figure 1 - *Italian arms exports 1963-1988*
(Figure are in million US $ at constant 1985 prices)

*Source*
ACDA, *World Military Expenditures and Arms Transfers.*

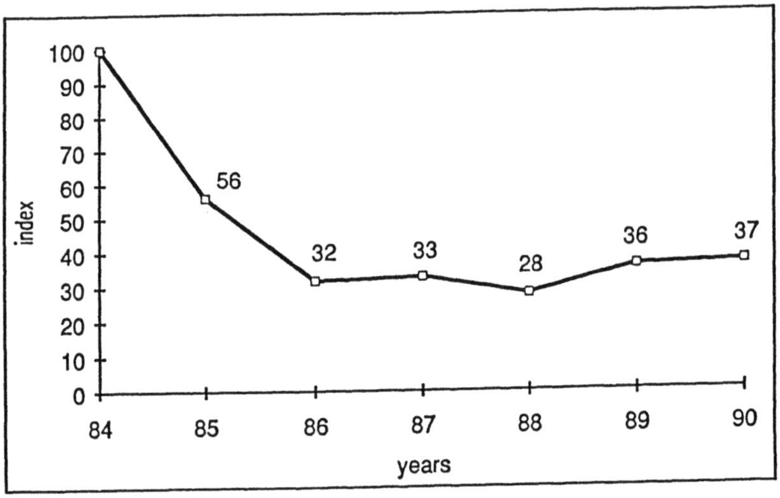


Figure 2 - *Italian arms exports 1984-1990*
(Index: 1984 = 100)

*Source*
Presidente del Consiglio dei Ministri, *Relazione al Parlamento sulla politica informativa e della sicurezza.*

## Table n.1 Italian Military Industry 1987-1990
## The top twenty firms
(billion Lit . at current prices)

| | Control group | Sector | Total Turnover | | | | Military Turnover | | | | Total Workers | | | |
|---|---|---|---|---|---|---|---|---|---|---|---|---|---|---|
| | | | *1987* | *1988* | *1989* | *1990* | *1987* | *1988* | *1989* | *1990* | *1987* | *1988* | *1989* | *1990* |
| Aeritalia Spa[1] | IRI | 51 | 1586 | 1655 | 2052 | 3221 | 1030[c] | 1100[c] | 1130[b] | 1970[c] | 13662 | 14177 | 14983 | 21981 |
| Selenia Spa[1] | IRI | 31 | 652 | 734 | 890 | | 479[a] | 569[a] | 662[a] | | 6850 | 6716 | 6655 | |
| OTO Melara Spa | EFIM | 23 | 660 | 561 | 607 | 570 | 660[a] | 561[a] | 607[a] | 570[a] | 2379 | 2329 | 2294 | 2245 |
| Agusta Spa[2] | EFIM | 51 | 683 | 777 | 952 | 1050 | 495[c] | 563[c] | 580[c] | 630[c] | 7475 | 8866 | 8426 | 8114 |
| FIAT Aviazione Spa | FIAT | 51 | 703 | 1044 | 907 | 980 | 422[a] | 424[a] | 472[b] | 560[b] | 4656 | 4749 | 4651 | 4680 |
| Contraves Spa | Private (CH) | 31 | 206 | 256 | 351 | 390 | 204[c] | 254[c] | 347[c] | 390[a] | 1185 | 1041 | 1020 | 951 |
| Fincantieri Spa | IRI | 41 | 687 | 1061 | 1749 | 2206 | 400[c] | 405[c] | 320[c] | 300[d] | 3370 | 3210 | 3164 | 2880 |
| Aermacchi Spa | Private | 51 | 233 | 261 | 335 | 337 | 210[c] | 255[c] | 318[c] | 320[c] | ·2649 | 2690 | 2715 | 2740 |
| BPD Difesa e Spazio Spa | FIAT | 11 | 335 | 262 | 302 | 374 | 268[d] | 210[d] | 241[d] | 226[a] | 3519 | 2189 | 2120 | 2019 |
| IVECO FIAT Spa | FIAT | 24 | 4148 | 4770 | 5206 | 5364 | 550[b] | 546[b] | 208[d] | 188[d] | 20382 | 21942 | 22969 | 22649 |
| Elettronica Spa | IRI | 31 | 183 | 205 | 196 | 151 | 183[a] | 205[a] | 196[a] | 151[a] | 1319 | 1344 | 1368 | 1363 |
| Elmer Spa | IRI | 32 | 117 | 127 | 158 | 126 | 117[a] | 127[a] | 158[a] | 126[a] | 1024 | 1054 | 1066 | 1060 |
| Breda Mecc. Bresc. Spa | EFIM | 23 | 176 | 166 | 160 | 130 | 173[a] | 160[a] | 155[a] | 124[a] | 758 | 736 | 709 | 692 |
| Alfa Romeo Avio Spa | IRI | 51 | 186 | 208 | 250 | 268 | 93[d] | 104[d] | 125[d] | 135[d] | 1376 | 1509 | 1616 | 1722 |
| Microtecnica Spa | Private (USA) | 51 | 135 | 133 | 139 | 134 | 125[d] | 120[d] | 120[b] | 119[b] | 1109 | 1104 | 1114 | 1076 |
| Marconi Italiana Spa | Private (UK) | 32 | 282 | 359 | 382 | 443 | 144[a] | 159[a] | 117[b] | 142[b] | 2002 | 2058 | 2402 | 2529 |
| Rinaldo Piaggio Spa | Private | 51 | 195 | 192 | 196 | 227 | 117[c] | 115[c] | 115[c] | 114[c] | 1779 | 1934 | 1988 | 1910 |
| Oerlikon Italiana Spa | Private (CH) | 23 | 108 | 139 | 140 | 139 | 81[a] | 107[a] | 103[a] | 104[b] | 882 | 764 | 723 | 300 |
| Elsag Spa | IRI | 31 | 405 | 319 | 342 | 432 | 162[b] | 125[b] | 101[b] | 90[b] | 1872 | 1844 | 1815 | 1780 |
| Vitroselenia Spa | IRI | 32 | 81 | 90 | 103 | 126 | 77[c] | 85[c] | 100[c] | 120[c] | 716 | 759 | 758 | 759 |
| Total | | | 11761 | 13319 | 15417 | 16668 | 5990 | 6194 | 6175 | 6379 | 78964 | 81015 | 82556 | 81450 |

**Notes:**
1) In December 1990 a new company (Alenia Spa) incorpored Aeritalia and Selenia
2) In1989 Agusta incorpored SIAI Marchetti, IAM, Caproni Vizzola, OMI e Breda Nardi.

a) balance sheets
b) official statements of the firm
c) reliable sources
d) unofficial assesments.

# ANNEXURE

## THE CRISIS OF ITALIAN MILITARY INDUSTRY

The Coordinamento Nazionale degli osservatori sull'industria bellica (National Coordinating Unit) was set up in 1989 and is a technical and scientific consultative body among the research institutes, the associations and observatories that study the military industry in various Italian regions. This body came into Government, from the industrial world and the Armed Forces.

The Co-ordinamento receives scientific contributions from prestigious non-governmental research institutes such as the Archivio Disarmo in Rome and the Forum per i Problemi della Pace e della Guerra in Florence, and is supported by public Institutions, private associations and trade-unions. At the present time, the Observatories operate in Lombardia, Liguria, Emilia Romagna, Toscana and South Tyrol. Their work is not limited solely to monitoring and gathering data but they also take part, whenever possible, in concrete projects which deal with conversion of military production into civil production. The work carried out until now has made it possible to bridge the gap which exists between the quality and the quantity of aggregate demand and aggregate supply in this sector.

This paper explains the Italian export trend, form 1975 to 1990. They are analysed with the aid of data coming from ACDA in Washington, SIPRI in Stockholm and data provided by the Italian Government. As far as domestic aggregate demand is concerned, the evolution of the defence budget, dedicating particular attention to arms procurement in 1968 and the one expected in 1992, is carefully examined. On the whole, data at hand show a drop in aggregate demand, both on domestic and international markets.

In the case of the 50 companies whose military turnover is of 10-100 billion lire, data on sales for the years 1987-1990 are provided. The complex analysis of the relation between aggregate demand and aggregate supply must take into account today's most interesting and worrying issues of general crisis in this sector. For this reason, information on those companies that are mainly involved in cases of redundancy and dismissals are given. A synthetic analysis of the most recent balance sheets of the most important industrial groups in this sector is presented.

The Coordinamento, with the intention of finding practical solutions to the issue of converting military production into civil production, provide the results coming from research on possible instruments of industrial policy to be used to this end.

**Ten Ecological Commandments**

**First Commandment**

RESPECT THE LAWS OF NATURE! REALIZE: THERE ARE CLEAR LIMITS FOR THE CITIZENS OF SPACESHIP EARTH SET BY NATURE AND ENFORCED BY NATURE! YOU NEED NATURE AND NATURE NEEDS YOU!

Have respect and respond in humility to the 4.3 billion years old cycles of the universe which guarantee survival of the species HOMO SAPIENS alone by interdependence of billions of very sophisticated ecosystems and cultural systems which are endangered by massive over-exploitation of resources and energy in an over-consumption pattern of carelessness. Therefore, adapt yourself to nature's cycles.

**Second Commandment**

LEARN FROM THE WISDOM OF NATURE! REMEMBER : THE WHOLE IS MORE THAN ITS PARTS!

It is your duty to inform yourself about the rules of ecosystems so that you can act with wisdom and circumspection. Do not curtail ecosystems in their parts without considering what happens to the whole.

**Third Commandment**

DO NOT REDUCE PLURALITY, RICHNESS, ABUNDANCE OF LIVING SPECIES! DO NOT INDUSTRIALIZE NATURE! DO NOT STREAMLINE NATURE!

Every year ten thousands of species disappear because of industrialization, over-fertilization, toxic pollution, careless handling of nature's resources such as the cutting of tropical forests and the sealing-off of larger and larger parts of nature by building roads, industries, cities. Thus the information content of the biosphere, its survivability, diminishes. Every blade of grass, every butterfly, every wild creek or hedge is an important part of the global ecosystem and therefore condition for its survival and our survival as human species.

**Fourth Commandment**

DO NOT POLLUTE!

Avoid pollution of water, air, soil by not overusing your car, by not asking for unnecessary packaging material, by not wasting water or energy. Because of the enormous pollution-load there must be personal zero-level-pollution saving resources and energy -- wherever zero -- alternatives such as walking, bicycling, using renewable resources, are possible.

**Fifth Commandment**

FACE RESPONSIBILITY EVERY DAY FOR OUR CHILDREN AND OUR CHILDRENS' CHILDREN SO THAT THEY MAY LIVE ON AN UNSPOILT EARTH!

Try to leave this small planet earth as a more beautiful place, more livable in, with enough resources for all. Responsibility for earth infers global responsibility taking into consideration causes of nature destruction and economic and social underdevelopment in other parts of the world. Find ways and means of achieving global responsibility by local actions. Earth policy and local policy depend upon each other.

**Sixth Commandment**

FOLLOW THE PRINCIPLE OF SUSTAINABILITY IN ALL ECONOMIC ACTIVITIES!

Do not take more resources out of nature's richness and ecosystems than can be restored and renewed by careful resources planning. Give everything back to nature in substance which can be reconstructed to soil, plants, functioning ecosystems. Start with composting of all green waste in your garden or your neighbourhood. Burning of waste means destruction of matter and off-heat-pollution. Do not allow our earth to lose its soil.

**Seventh Commandment**

ACT AS YOU SPEAK!
EVERY DAY IT IS YOUR VERY OWN PERSONAL RESPONSIBILITY TO EASE ENVIRONMENTAL BURDENS!

Do not lie to yourself by making a distinction between the principal acknowledgement of the environmental cause and your own little sins, which are unimportant in comparisonwith the big sins of others, such as those in the case of industry.

**Eighth Commandment**

PREFER SMALL PROBLEM SOLUTIONS, SMALL TECHNOLOGY TO BIGGER SOLUTIONS AND BIGGER TECHNOLOGIES!
RESPECT CULTURAL AND SOCIAL TRADITIONS DIFFERENT FROM THE WESTERN WASTE-TECHNOCULTURE PATTERN OF LIFE STYLE!

Sustainability of political systems of the future depends on the respect for other cultural and social traditions and identities. Survivability of future political systems and the success of environmental policies depend on many, many small problems solutions which are integrated in existing regional cultures and traditions.

**Ninth Commandment**

INFORMATION ABOUT ENVIRONMENTAL DAMAGE BELONGS TO MANKIND -- NOT TO YOU ALONE! DO NOT MANIPULATE INFORMATION BECAUSE OF COMMON BELIEF AND OF POSSIBLE LOSS OF PRESTIGE OR POSITION!

To protect nature, to overcome environmental damage, needs courage to tell the truth about the effects of certain industrial practices, about the consequences of certain political actions even if it hurts common beliefs, traditional views and interests. One of the great sins of our Media- and Public-Relations-Culture, is the manipulation of information. Therefore resist this distortion of information!

**Tenth Commandment**

LISTEN CAREFULLY TO WHAT YOUR OWN BODY TELLS YOU ABOUT THE IMPACT OF YOUR VERY PERSONAL ENVIRONMENT!DON'T LOSE THIS EARLY WARNING SYSTEM! ALWAYS REMEMBER YOU CAN NEVER STEP OUT OF NATURE'S CYCLES, YOU WILL ALWAYS BE PART OF NATURE!

Environmental awareness and sensibility start with many signals from your own body which link you with the larger ecosystems of our earth, and it is your early warning system that assesses the environmental damage first and foremost. Man and nature are of the same origin: all of us living on earth must obey the same billion-year-old laws of nature!

Some proposals of the State Research Institute of Aviation Systems (SRIAS) for co-operation with companies from Northrhine-Westphalia.

1. Only recently there is the opportunity to receive photomaps in the scale up to 1:5000 using the most advanced Soviet space technology. It is an outstanding result, which might be used in different branches of economics.

2. The problem of Soviet German's autonomy is expected to have been solved. Independently of where the lands would be given (in Russia on the Volga or in Ukraine), SRIAS is ready to carry out topographical and thematical surveys using its flying laboratories with multi-land remote sensing equipment. The results for interpretation would be presented to national or foreign specialists, if there is the permission of the Russian government. If it is necessary, SRIAS could attach scientists in earth biology to this work.

3. SRIAS plans to use airborne multi-land surveillance with the help of its flying laboratories in the interests of ecology and others.

We are ready for international co-operation. We are interested in the initial financing for development. Hansa Luftbild from Munster would become one of the companies which could realize the above mentioned items.

# PAKISTAN

# CONVERSION OPPORTUNITIES FOR DEVELOPMENT AND ENVIRONMENT

**Ather Sultan**
*Deputy Director General, National Housing Authority, Pakistan*

The past two years have witnessed big changes in the structure of the world balance of military and political power. The detente state of relations between the East and West has been transformed into what appears to be an unprecedented, though incipient (as yet), state of entente. An era marked by fear, distrust, tension and uncertainty has ended and it does not seem unrealistic to hope that the present state of transition in the world security system can be guided into an era of co-operation, trust, peace and security. It is a universal feeling that the opportunities offered by the tremendous change should be built upon and that even a partial reversal of the emerging situation would be a tragic setback to human efforts for peace and progress of all mankind.

The transition from detente to an emerging state of entente has two major implications for the world in general and the developing countries in particular. First, diffusion of tension between East and West makes it possible for the developed world to release and divert greater amounts of resources towards the development of less developed countries. Second, it vastly enhances the possibility and effectiveness of concerted political efforts through the U.N. to resolve tensions and conflicts of all types in other parts of the world (i.e. outside Europe and North America). The developments in the East-West relations, therefore, have a direct bearing on the age-old debate between the North and the South because of the possible availability of required political and economic resources to resolve the issues of that debate. It would, however, be simplistic to assume that such a development would bean automatic or necessary consequence of the East-West Entente. It requires conscious decisions, hard choices and determined efforts by the North, matched by equal reciprocity from the South. A failure to apply the lessons of detente and its development into the emerging entente, to the resolution of North-South problems, would be a costly waste of time as it would prolong and might even generate new but avoidable human suffering.

The present state of affairs in the East-West relations has traversed a difficult path. As causes of tension between the two blocks persisted, we saw an exponential rise in military preparedness and a relentless competition in the armament race. The saving grace, however, of detente years with their huge arms build-up, is the prevention of actual military conflict which could have resulted in total annihilation of life and civilization on the globe. From this position of assured mutual destruction, the march to the present state of mutual reconciliation and co-operation between East and West has been made possible by the careful and determined efforts to remove all causes of threat to each other. There can be no disarmament or reduction in armaments until the causes of threat and perceptions of threat have been removed from the scene of possible conflict. In order to extend the spirit of disarmament and peace to all parts of the world, it appears necessary that effective steps be taken to remove the actual and potential threats to peace in different parts of the world. As an immediate first step, the United Nations can undertake a scanning of the world scene and determine the priority areas for its attentionand concerted efforts for resolution of international, regional or ethnic conflicts existing in those areas. It can then give priority to those conflicts where it has an obligation to ensure a just and peaceful settlement of the issues. A delay or neglect in this regard can not only endanger global peace, perpetuate or prolong human suffering but also damage the credibility of the U.N as a cardinal institution for the realization of all that the future, at this benign moment in history, appears to promise. The efforts for the resolution of the conflict and construction of peace may have to be incremental but should be adequate and timely enough to achieve their twin objectives of peace and justice.

The development process of less developed countries is marred by the lack of resources and technology. In addition, it suffers further from a negative flow of capital, an increasing burden of debts, an unfavourable international trading system and a continuous process of brain drain. Mass poverty, mass illiteracy, high birth rates, regional disparities and backwardness, inadequacy of administrative and political infrastructure and processes, cripple their efforts to overcome these problems and at the same time make them extremely vulnerable to manipulations of all kinds. It is in this context that the `conversion' ideology has a relevance for the developing countries because the funds, technology and personnel relieved from the military-industrial sector of the developed world can be utilized for building foundations of peace in the poor and disturbed areas of the globe.

Pakistan is making strenuous efforts to remove poverty, control population growth, increase its agricultural and industrial output, develop its sources of energy, remove illiteracy, provide health care and build infrastructure of road and rail networks and communication systems to meet the needs ofan emerging industrial society. Some of the prominent steps taken by the government include :-

i) liberalization of the economy;
ii) encouragement of the private sector;
iii) disinvestment of state-owned enterprises;
iv) elimination of foreign exchange restrictions;
v) encouragement of foreign investment in the country.

These and other measures are being vigorously pursued to achieve a higher rate of growth to alleviate poverty and build a progressive, democratic and prosperous society. To achieve these objectives, Pakistan needs peace and is strongly committed to the pursuit of that objective. In this connection we have sponsored the idea of establishing a `Nuclear-Weapon-Free Zone' in South Asia and the proposal has been repeatedly endorsed by the UN General Assembly since 1974. Pakistan has also made and supported various other proposals for the promotion of peace and security in the region and the world. The approach of the Pakistan Government to peace in the region is well summarised by the following passage from a recent (June 1991) address by the Prime Minister :

"A regional non-proliferation regime, containing guarantees for non-nuclear-weapon states and agreed to by all the regional countries, is a practical method of resolving the problem in all its dimensions. Such a regime would help usher in a climate of mutual trust and strengthen peace and security in South Asia, enabling the states of the region to concentrate their energies and efforts on accelerating economic development and promoting the well-being of their peoples".

**Environmental Issues**

A recent study by the Government summarises the environmental position in Pakistan as follows :-

"Pakistan, with a per capita gross national product just one-ninth the

World average, is neither a major global polluter nor a large consumer of resources. The average Pakistani contributes only one-seventeenth the world average of carbon dioxide emissions, the greenhouse gas responsible for half the global warming expected in the next few decades. Pakistan produced no chlorofluorocarbons and few sulfur dioxide emissions, and thus makes a negligible contribution to ozone depletion or acid rain. Although global marine fisheries are reaching the limits of harvesting, for example, Pakistan is a small, in fact an under-represented user of the global commons.

Yet Pakistan is a densely settled land, facing rapid population increase and highly dependent on natural resources, with severe physical capital scarcities and limited skill development. As such, it is likely to suffer disproportionately from climate change and other global environmental problems. The implications of the greenhouse effect for Pakistan are difficult to predict, but could potentially be large, affecting patterns of agriculture, fisheries, and forestry. In wet areas, the monsoon is expected to become more vigorous; arid areas could become drier; the receding of mountain glaciers may reduce inflows for irrigation agriculture; and low-lying areas may become salinized by saltwater intrusion owing to sea level rise."

These problems are further aggravated by an inadequate level of energy production, high rate of population growth and an absence of a responsible attitude to toxic waste disposal. To meet these multifarious problems, the Environment & Urban Affairs Division was created in 1974, shortly after the Stockholm Conference on the Human Environment, which in 1991, has been raised to the level of a full-fledged Ministry. In 1983, the Government promulgated the Pakistan Environmental Protection Ordinance which, among other measures, envisaged the establishment of a high-level and powerful Council under the chairmanship of the President of Pakistan. The mechanisms and institutional framework for increased and accelerated activities of the Environment and Urban Affairs Division and the Council to achieve the purposes of the Ordinance and to keep the country abreast of the latest thought and action in the field of environmental protection, are being evolved.

**Conversion for Development & Environment**

Conversion of military-industry and resources for development and environment is primarily an objective of first priority for the developed world, where threats to peace and national security have been largely removed and public and governmental confidence in non-military means of settling disputes and resolving conflicts is well established. Besides, conversion is a highly technical and costly process. The West, however, already has considerable experience in the field, which can be built upon to update and further refine the mechanics and methodologies of the conversion process and the resolution of the problems that inevitably confront or arise out of the process. Meanwhile, efforts of the U.N. can be directed towards the removal of conflicts, especially in the developing world, so that a global environment for peace canbe created by threat-removal and confidence-building measures.

For the development of an environment of peace and security, in addition to settlement of regional conflicts and territorial disputes, some other measures on social and moral planes are also required. First and foremost, there is a need to develop a public opinion ethos that does not eulogise violence and war but regards them as tragic human failures to preserve peace. Secondly, the attitudes and values of those who wield power are of extreme importance for the success of `techniques and practices' of resolving international conflicts. Thirdly, the values and attitudes of the general public and the media, especially in the developed world, who support or criticise the activities of their leaders and governments, are of vital significance for the creation of an atmosphere of global goodwill and understanding. Fourthly, the time has perhaps come for the subordination of the `Power Principle' to the dictates of the `Principle of Truth and Justice'. Only on equitable foundations of universal moral and political principles can universal peace be built.

# RUSSIA/CIS

# CONVERSION AS A CHANCE FOR DEVELOPMENT AND ENVIRONMENT

**Vladimir Dmitriev**
*President, CHETEK International Corporation, Russia*

## Conversion Demonstration Project

"CHETEK" International Corporation is not a State structure (See Annex). The name of our company is derived from the initial letters of three Russian words: Man, Technologies, Capital.

The company is engaged in the implementation of different technologies in the field of electronics, medicine, agriculture, engineering, aircraft industry, space technologies, as well as technologies in other spheres of human knowledge.

The company possesses exclusive rights for commercial use of technologies developed by scientists and not only protects the rights of inventions and authors, but also provides material and social support for the scientists-authors of technologies through financing housing, food, and medical programs for them.

Our company has the right to commercially use the technologies for ecologically safe disposal of chemical weapons and highly toxic industrial wastes by underground nuclear explosion energy. The company has concluded the General Agreement with the Ministry of Atomic Power for financing this program. Rephrasing the well-known saying one could say : WORLD PROPOSES -- CHETEK DISPOSES.

The demonstration experiment, finalizing the technology developed by the scientists of a nuclear research center known as "ARZAMAS-16", is supposed to be carried out in late 1992 or early in 1993 at Novaya Zemlia Northern Test Site. This project is considered by the company as one of theconversion projects.

Upto 25 per cent of the profits received from commercial application of conversion programs shall be used for financing social projects for military personnel and their families, as well as for investments in projects related to re-orientation of military industries towards production of goods for national economy.

It's only natural that the above experiment is considered by our company as a means for obtaining necessary funds to finance social and ecological rehabilitation programs of regions neighboring the Novaya Zemlia Northern Test Site.

The project envisages physical methods to destroy chemical substances, highly toxic industrial non-recyclables, accumulated stocks of pesticides and defoliants prohibited for use, e.t.c. Physical methods mean explosive or vibro-acoustic destruction of substances in closed voids. The most effective example of such technologies is the use of an underground nuclear explosion, which converts the substance to be destroyed into safe neutral plasma and buries it. One can assume that at present all necessary elements of such technologies are available: appropriate and well organized integration of those elements shall contribute to the resolution of the problem of chemical weapons and highly toxic substances disposal. In particular, there are well tested methods for drilling three to six meter diameter holes of necessary depth; there are containers for transporting chemical warheads, which by their status and other reasons rule out any dismantling and must bethus destroyed together with those containers without any risk for personnel. All disposal procedures fulfil the requirements necessary for providing ecological safety.

At present one may speak only about a well-substantiated scientific and technological project, which, however, needs to be validated through a specialized experiment. We believe that it must be carried out under strict national and international monitoring, involving recognized professionals, and within the framework of international agreements. This is of special importance, taking into account the prejudice of the world community against nuclear technologies. However, in this particular case their use is fully justified, because the risk of radionuclides escaping can be completely eliminated, and the cost-effectiveness of that method applied to non-recyclables appears to be sufficiently convincing.

The "CHETEK" company and those who developed the above mentioned technologies are now involved in resolving various technical, economic, organizational, financial, and information problems. Very briefly, I'd like to draw your attention to the features of our project, which reflect its soundness. The project has technological scope, productivity scope, and application scope.

**Technological Scope**

Within a wide range of substances subject to destruction, there are, of course, those beyond the destructive capabilities of other technologies. Those are the substances which we propose to deal with specifically through our project.

**Productivity Scope**

All other technologies, taken together, cannot provide the destruction of all kinds of chemical substances. Without attempting tocompete with other technologies in terms of resources required for their implementation, our technologies shall provide the means to considerably increase overall destructive potentials.

Moreover, the use of that technology does not require other energy sources, whose power could be canalized to satisfy other humanitarian needs of the world nations.

**Application Scope**

The proposed technologies can be applied to destroy not only chemical substances, but also many other substances, including nuclear power plants wastes and exhausted nuclear reactors, at least up to 200 thousand KWt class.

These technologies are also applicable for ecologically safe destruction of retired warheads. These technologies can resolve the global problem of destroying all existing nuclear warheads and chemical agents, which is an age long dream of all nations of our planet.

It is not my purpose to describe in detail all advantages and disadvantages of the proposed technology, for experts could do it much better than me. I would only like to emphasize that even in my country the commercial use of such technologies is possible. The state can invite new national economic structures to participate in those useful activities. Finally, there is an opportunity to initiate international co-operation, both at the level of national or international non-government structures and at the level of joint activities of these structures with government and international intergovernmental organizations, especially environmental, humanitarian, information, etc.

As far as our company is concerned, I can say that we have already reached a goodlevel of mutual understanding with the

Commonwealth of Independent Nations Armed Forces High Command and with the Ministry of Atomic Power of Russia.

We are prepared, both independently and in co-operation with other CIN states and interested organizations, to open a dialogue on the ways and forms of international co-operation to free mankind from stockpiles of dangerous substances accumulated over the years. Further delay in resolving this problem is unacceptable, and this is evident to all who are aware of the real situation.

Even now I could propose to begin negotiations on the commercial use of our unique technology with all organizations, companies, and businessmen, which, no doubt, would speed up the commencement of actual work. This proposal is probably to be addressed, first of all, to the organizations of these nations that conduct underground nuclear explosions for different purposes. As of today, the use of this technology would not cause any further damage to the environment. Later on, one could devise the framework of international co-operation with non-nuclear states, whose population is also concerned with the massive presence of ecologically harmful substances in their territories.

The harm from the dissemination of highly toxic industrial wastes recognizes no state boundaries. Unfortunately, not all the nations, having large amounts of such wastes in their territories, possess enough money and know-how for their ecologically safe disposal. I believe it is exactly for this reason that it would be very beneficial to establish an international fund which could undertake the financing and co-ordinating of the colossal work to develop the technologies for destroying the stockpiles of dangerous wastes. This is the most urgent need for mankind today.

We are prepared to consider other proposals in order to accelerate the integration of the newest technology into that colossal work of ecologically safe disposal of chemical, nuclear, and other dangerous wastes and substances.

The "CHETEK" company, the Russian Ministry for Atomic Power, and nuclear scientists propose a comprehensive pre-and-post-experiment monitoring program involving scientists and representatives of public organizations from different countries.

The very energy of the atom could help the dream of mankind to come true, to eliminate not only its own creation -- atomic weapons, but also chemical weapons, highly toxic industrial and radioactive wastes, particularly those that cannot be destroyed without damaging nature.

If it is true that all nations on our planet share the same home, it should also be true that all achievements of the human mind must serve all those living in that home.

# ANNEXURE

# THE INTERNATIONAL CHETEK CORPORATION
# X-RAY STERILIZER
# RS-20

Within the limits of **Russian Military Conversion** programs a certain class of research accelerators previously used for military objectives changes an area of application to medicine and to radiation technologies providing ecological safety. Accelerators used Plasma Opening Switch seem to be most adapted for frequency operation. On a base of these accelerators pulse frequency X-ray generator have been designed and constructed.

RS-20 is a pulse frequency X-ray generator.

RS-20 could be applied for different purposes:

1. **in medicine** - for disinfection of medical equipment, instruments, preparatory, transplanters.

2. **in food industry** - for foodstuffs sterilizing.

3. for waste products of **agricultural industry** with providing of ecological safety.

4. **in biotechnology** - for genetic correction of microorganisms.

5. **for selection of plants and cultures**.

6. **in production of forage** - for decreasing of its bacteriology population.

Parameters :

election beam energy - 3 Mev

average dose power - 0.5 kGr/h/1 m from the anode/

peak dose power - 1 MGr/h

target sizes - 0.5 x 0.5 x 0.5 m3

electron beam average power - 20 kWt

pulse frequency - 1 - 10 Gz

RS-20 is a stationary device placed in the room with radiation safety. A scheme of RS-20 is shown at figure. Main units : capacity storage system with power amplifying device based on a plasma opening switch, electron diode. Targets should be placed in radiation area automatically.

Generators based on principles as RS-20 does are simple in construction and reliable in operation. They do not require high qualifying of controlling stuff.

Ecology safety is provided after electrical switching off.

With request of a customer it is possible to construct different models of sterilizer: stationary or mobile for dose intensity which is required.

Electron accelerator used in RS-20 has no physical limitations with increase of parameters upto 5-10 Mev of electron beam energy, 1 TWt peak power, 100 KWt average power. Creation of these generators will allow to widen areas to their application and resolve some tasks in radiation chemistry, high resolution medical radiography, geological survey etc.

## MICRO ELECTRONICS AND INFRARED EQUIPMENT AUTOMATED PRODUCTION

The efforts of the Company are concentrated on developing automated, ecologically pure and resources-saving facilities (like minifactories) manufacturing the components of the prospective instrumentation. Highly qualified experts have been engaged in this work since 1983.

The main object is to create, within a very short period of time (5-6 years), the cost-saving technologies, which would provide cutting down the infrared and heat-imaging instruments production costs more than 100 times.

As a result, the Company has the technologies in :

- purification of the substances from pure to super pure level ($10^{-7}$ %/mass);
- growth of defect-free large size single crystals/up to 100 mm of substrate/;
- producing structure-perfect Cadmium- and Zinc-Tellurids substrates;
- synthesizing instrument epytaxial structures under super high vacuum;
- manufacturing of multi-elements photoreceivers on the Cd- and Hg-Tellurids basis, as well SLIS (Superlarge Integrated Ciracuis).

In order to speed up the production work the Company proposes to combine efforts and means with the purpose of :

a) improving the equipment and technological know-how for producing superpure substances, single crystals, epitaxial structures, photoreceivers and chips (microprocessors) for specific applications;

b) building manufacture facilities on a "Turn-Key" basis, which would pioneer creation of :

1. Environmentally pure resources-saving technologies based on the methods of:
   - epitaxial and analytical computerized control under superhigh vacuum;
   - low-temperature microionic and photochemical processes under vacuum.

2. Full-cycle environmentally pure and resources - saving production of :
   - matrix photoreceivers using Cd-Hg-Te material, consisting of 128x128 elements with cooling infrared hardware, no less than 100 pieces per year;
   - superlarge integrated circuits no less than 100000 pieces per year.

Equipment under development is to ensure production of :

- matrix photoreceivers made of Cd-Hg-Te material, consisting of no less than 128x128 elements, for high-resolution thermovision (heat-imaging);
- super-high speed, superlarge integrated circuits, including those based on "HEMT" transistors, used in radars, radiodetection, satellite communication.

Mastering these technologies will provide the long-term superiority in production of prospective advanced instruments and the market monopoly.

**The International Chetek Corporation is a sole proprietor of hardware and the technological know-how of :**

1. **Extended Reproduction of environmentally pure Crops**

ensures :
- increasing of the yield crops capacity,
- reduction of seeding norms,
- keeping the best characteristics of the seeds in the first and second generations.
- increasing the drought resistance of crops;
- shortening of the ripening period.

Commercial offers : To interested firms the Company offers mutually beneficial condition for cooperation in :

- seeds procession of grain, vegetables and other crops,
- establishing joint ventures in seeds procession, processing lines.

2. **Protective coatings on AL, Ti and their Alloys by Anode-Cathode Microarc Oxygenation**

ensures :
- coating microhardness upto 2380 kgs/mm,
- increasing of adhesion upto 10-times of the underlying material strength.

Commercial offers : to establish joint ventures.

3. **Re-Processing Aluminium, Copper and Magnesium Alloys CHIPS**

ensures :
- re-processing chips into quality bars and profiles of different configurations without remelting at machine-building plants.

4. **Re-Processing Steel CHIPS**

ensures :
- producing the axisymmetrical details, the kind of wheels, discs, gears, flanges.

Commercial offers : to establish joint ventures.

5. **Micro Infrared Equipment Automated Production,**

ensures :
- the super pure materials production of completed environmental processes under super high vacuum and super large integrated circuits.

Commercial offers : to establish joint ventures for its production.

6. **The Technological Know-How of Electronic Beam Generators,**

ensures :
- sterilization of objects and materials for medical purposes, food and agricultural products,
- disinfection of drink-water, water sediments and sewerage water flows,
- reclamation of solid industrial wastes using the Electronic Beam Generator, RS-20 (PC-20).

Commercial offers : to establish joint ventures.

7. **X-ray Resistant Rubber, REP-17 (P-17) Production**

Commercial offers : to establish joint ventures in Russia or in another country for production of this material, using the mentioned above technology, on a mutually beneficial basis.

**The International Chetek Corporation is the proprietor owner of the technology for creating protective coatings on Al, Ti and their alloys by anode-cathode microarc oxygenation**

The microarc oxygenation method (in the electrolyte medium) produces ceramic coatings of various compositions, with :

- unique physico-chemical characteristics;
- high resistance to wear heat and corrosion;
- high primitivity decorative aspect.

Coatings on Al, and its alloys possess the following properties :

- thickness of coating equals 5-2000 micron;
- coating microhardness is upto 2380 kgs/mm2;
- adhesion is 0, 9 of the underlying material strength;
- electric resistance is over 1 MOhm;
- resistance to wear exceeds that of the WC-Cu-Ni alloy tenfold;
- with the underlayer adequate cooling the coating can stand the following heat loads : temperature of 3000 C, pressure of 200 kg/cm2, gas velocity of 200 mps (meters per second).

Coatings can be put both on external and internal surfaces of details of any configuration and complexity.

The above method is ecologically pure and highly productive :

- one shift coating capacity 17 dm2;
- three shift work;
- three shift coating capacity 50 dm2
- yearly capacity 18113,0 dm2
- efficiency 80%.

Anode-oxide coatings are widely used in aerospace, electronics, shipbuilding and auto industries, in architecture, construction and production of consumer goods :

- for coating caps and pistons of internal combustion engines;
- for tenfold increase in durability of fuel pump parts made of aluminium alloys with corundum coating;
- for making wear resistant aluminium alloys draw plates with corundum coating.
- for producing electroinsulating coatings of radiators to cool semi-conductor devices;
- for producing corrosion and wear resistant poreles coatings;
- for production of friction couples working in the conditions when the working medium is a lubricant;
- for production of friction couples operating in highly abrasive media;
- for strengthening surfaces of light alloys (parts of spinning machines);
- for protection of electrodes in electric-arc furnaces and of press forms for hot pressing;
- for heat protection of carbonic materials by introducing aluminium, circonium, magnesium, chrome ets. into the coating;
- for coating surfaces of kitchen utensils and other things.

To interested firms the Company offers mutually beneficial condition for co-operation in establishing joint ventures both in Russia and abroad, and in introducing the above technology.

**The International Chetek Corporation offers the PII-7 PC X-ray screening rubber** in the form of
**rubber band** (500*500 cm and 1mm thick) for:

- screening of X-ray equipment,
- production of the protection aprons for medical personnel working with X-ray equipment;

- covering walls of the working spaces to protect against gamma-rays;
- manufacturing of different products (soles for X-ray protected footgear, screening shroud and covers);

**pre-packaged raw material (maxture + catalyzer).**

**The PII-7 PC X-ray screening rubber has the following properties**

- density: kilograms per square centimeter - 2900-3100
- relative stretching strength : MPA - not less than 1.96
- specific lengthening at breaking point (%) - not less than 60
- linear gamma-rays attenuation ratio (for americium 241 isotop) per 1 cm-long monoline: not less than 17.2
- operating temperature range - minus $60^{o}$ - plus $300^{o}C$
- dielectric permeability at 10 Hz frequency and at temperatures $15^{o}$ - $35^{o}C$: - not more than 4.0; after 8 hrs exposure at 95 (plus/minus) 3% and temperature $15^{o}$ - $35^{o}C$ - not more than 5.0.

Address of the Corporation :
103012, 15 Varvaka (Razina) Str.,
Moscow, Russian Federation
tel: 206 57 35, 206 20 86, 206 20 56
fax : (095) 206 20 95

**"CHETEK" INTERNATIONAL CORPORATION IS THE SOLE PROPRIETOR OF THE TECHNOLOGY**

**"RECYCLING OF FERROUS AND NON-FERROUS METALS AND ALLOYS SHAVINGS"**

**PURPOSE**

1. Production of solid briquettes from the metal cutting wastes to be used as blanks for further forging.

2. Production of thin-wall axis-symmetrical parts such as wheels, discs, gearwheels, rings, flanges, conical and spherical cups.

**Features**

The technology can be implemented using specialized crank, screw, and hydraulic type pressing machines not radically different in their design from standard pressing machines with linear tool movement.

The design of the pressing machines includes, but is not limited to, multi-layer forging, comprising settling, preliminary and final forging through shear deformations in the material, and flash trimming.

**THE OFFERED TECHNOLOGY PROVIDES**

Two - to five times reduction in technological effort and tool load;

metal saving through precision forging and waste reduction,

extension of technological capabilities of hot-pressing machines and reduction in design metal consumption;

product quality increase through intensive metal treatment by shear deformations;

capability for manufacturing quality parts from non standard material such as metal shavings;

manufacturing parts (from the shavings) whose mechanical properties meet the requirements of rolled products.

**Technical Characteristics**

| | |
|---|---|
| bay capability per | 5000 tons annum |
| standard power | 710 KWt |
| Space required | 500 square meters |

**International Chetek Corporation offers for use the Technology of re-processing aluminium, copper and magnesium alloys chips**

**Purpose**

The technology ensures re-processing chips into quality bars and profiles without remelting at machine-building plants. It allows to resolve the problem of organising waste-free production, of satisfying the needs in deficit profiles of aluminium, copper and magnesium alloys for construction materials and consumer goods, of de-monopolising the production of pressed details in metal-producing industries.

**Description of the Process**

The process envisages the following sequence of operation :

- chips collection and transportation in specialized containers;
- briquetting;
- pressing of the required profile.

Production of profiles directly from briquets (without remelting) is possible due to special technological regimes and pressing procedures, which ensure plastic welding of chips components. The method and machinery for chips pressing are protected by nine patents.

**Alloy types.**

The process has been approbated (tested) with the following materials :

- casting aluminium alloys;
- deformed aluminium alloys;
- brass;
- copper;
- costing magnesium alloys;
- deformed magnesium alloys.

**Products Nomenclature**

- bars dia. 6 through 25 mm;
- non-perforated open profiles with different shape cross-sections of not more than 70 mm total circumference;
- pipes with external diameter 10-30 mm and internal diameter 5-28 mm;
- length of products : 2 to 10 meters.

**Mechanical Qualities.**

Mechanical qualities of tested alloys correspond to State Standards Strength requirements, and exceed State Standards plasticity requirements 5-8 times for casting aluminium alloys and 1. 7-2 times for magnesium alloys.

**Equipment Required.**

a) Standard conventional press, briquetting press, electrical furnace and inductor, magnetic separator.

b) non-standard - washing machine, pressing device, leveling stretching or roller leveling machine, device to remove press leftovers.

**Production Capacity of the System.**

At two-shift operation :

- aluminium alloys chip profiles - 150 tons per year;
- copper alloys chip bars and profiles -400 tons per year;
- magnesium alloys chip profiles - 100 tons per year.

Aggregate electric motor power - 150 kilowatts space required :

| | | |
|---|---|---|
| - for chip degreasing | - | 50 square meters |
| - for profile pressing | - | 210 square meters. |

The Corporation proposes joining efforts and funds with organisations and enterprises to implement this technology.

**The International Chetek Corporation is a sole proprietor of the technological know-how of extended reproduction of crops**

The suggested know-how of extended reproduction of crops is based on the combination of traditional agricultural methods with the newest scientific and experimental works in the field of individual plant development, bioenergetics, fermentology, theory of species changeability,agrocenology, etc. The present know-how includes specific ecologically pure and safe biopreparations and the means of influencing seeds and plants during vegetation period.

The know-how has been applied in different climatic zones.

Long-term experience has shown that the know-how provides the following concrete and globally systematic effects leading, eventually, to the production growth, raising of products quality and/or reducing expenses:

- increase of the yield crops capacity;
- reduction of seeding norms;
- reduction of watering volumes in the areas of irrigation agriculture;
- earlier ripening;
- far better ecological state of plants;
- reduction of agrotechnical techniques using technical equipment.

The main element of the know-how is direct influence on seeds to be seeded in mass, made according to two methods, which depend on the plant-growing technology of the farm:

- seeds steeping in a working solution during 12-14 hours, and then the seeds are dried and seeded during one day;
- seeds procession during 10-15 min by exogenic fields of a specific nature and dry seeding (dry procession); seeding is made 10 days after procession is completed.

The stimulation effect of the processing lasts 3 months.

**The Seeds Processing Conditions**

The Corporation provides the potential Buyer with the components necessary for the steeping and manufacturing, and operating instructions.

It is recommended to utilize the method of seed steeping for the crops with seeding norm upto 5-6 kg per hectare, as well as for processing of small amount of seeds under field (laboratory) conditions in selection and seed growing.

The dry processing of seeds is made in cases when seed steeping cannot be done, or with the seeding rate of 10 kg per hectare or more. The amounts of seeds to be processed, depending on the kind of crop, vary from 0.5 to 15 tons during a processing session. Since the dry processing is made within the period of 9 p.m. - 11 p.m. according to the local time (which is a specific feature of the know-how) and the time of one processing session is 10-15 min, the daily seeds processing capability varies between 4 and 120 tons.

The maximum number of crops to be processed during a day is 8 or less.

The 1991 crops species revealed the following results of the know-how implementation :

- in buckwheat - upto 52% increase of crops;
- in water-melons - upto 70% increase;
- in cucumbers - upto 28% increase;
- in cereals - 10-12 days earlier ripening;
  1-2% larger grains;
  2-3% protein increase in the grain.

## The International Chetek Corporation

**is the sole owner of intellectual property for creating of the Technology of guaranteed and ecologically pure liquidation of highly toxic chemical materials and industrial wastes using the underground nuclear explosion energy**

Its partners in the project are **the Ministry of Atomic Energy of Russian Federation and the All-Union Research Institute of Experimental Physics**. Thedemonstration experiment based on the Employment of the above Technology is planned at Novaya Zemlya test ground in the Summer of 1992. During the experiment it is intended to destroy, on commercial terms upto 1000 tons of highly toxic chemical industrial wastes from foreign suppliers. The price for the waste disposal rates above 300 US dollar per 1 kilo (gross weight) depending on the chemical composition and the toxic concentration of wastes.

The materials to be destroyed should be placed in special safe containers with 2mx2m section and a mass weight of upto 3000 kilos.

The Technology was discussed by the International Conference of Experts in banning and destruction of chemical weapons in Moscow, USSR, by the International Symposium on Nuclear Weapons Test in Arctic in Ottawa, Canada, and by the International Conference on Environmental Consequences of Hazardous Waste Disposal in Stockholm, Sweden, all held in the first half of 1991.

We would appreciate your interest to any further information on this matter.

Address of the Corporation :

103012, 15 Varvarka Str., Moscow, Russia
tel : 206 57 35, 206 20 86, 206 20 56
fax : (095) 206 20 95.

# DEFENSE CONVERSION ON THE DECLINING ECONOMY: CHANCES AND DANGERS

**Dr. Ksenija Gonchar**
*Russian Academy of Sciences, Russia*

## Introduction

Renewed interest in conversion in the former Soviet Union (papers are based mainly on the topics related to Russia, which concentrates 82% of defense industry potential of the USSR) was caused by the entirely new situation that defense industry faced in the dramatically changed economic and political environment. This situation might be briefly described by the following facts.

It is evident that defense conversion, based on the principles of partial retooling of the limited volume of defense industry capacities (mainly induced by the need to restructure and modernize defense economy itself and to get rid of unneeded obsolete facilities in traditional branches) and diversification of high-tech defense enterprises strictly within the borders of military-industrial complex have failed. This method of conversion, headed by military authorities and affected by ill-defined requirements to preserve mobilization reserves untouched, had completely negative efficiency and brought losses comparable with the economic costs of non-stop arms race. The state program of conversion, worked out according to the mentioned concept, was scrapped before it started to ruin the idea of efficient conversion.

Meanwhile, the pressure of financial bankruptcy and economic realities, as well as political challenges, led to severe defense cuts, which opened the way to radical economic demilitarization. This development is valuable both in itself and as a chance to start the real process of the use of comparative defense industry advantages and released resources for the sake of market-and-civilian needs oriented reform.

But the positive contribution of demilitarization to the fate of the reforming process depends heavily upon the general economic policy and the ability of the government to control the declining economy. The transformation of demilitarization into de-industrialization, accompanied by reallocation and the following sale of released raw and fuel resources for the sake of the solution of budget and debt problems should not be allowed. Direct and straightforward conversion of resources might bring short-term financial benefit, but is not helpful in enabling the nation to produce and survive economically. The unstable balance of chances and dangers contained in these developments still raises a lot of questions that are to be carefully examined.

This paper reflects only the views of the author, and not necessarily the institutions she is associated with, and is aimed at analysing the sometimes paradoxical developments accompanying conversion in a declining economy in these times of unprecedented systemic changes taking place in this country.

## Economic Demilitarization : Positive Developments

Notwithstanding the generally sharply strained (if not catastrophic) situation in the country that excuses many hopeless predictions, there are still some positive developments that might promote a turning point in the course of conversion. First of all

it refers to radical demilitarization that definitely started this year.
Though the statistical figures related to military economic issues in Russia are no better than that of the former USSR (though now the motives are explained more by complete chaos than by conceived intentions), the fact of deep demilitarization is difficult to refute. According to official information, defense expenditures of Russia for the first quarter of 1992 (high inflation and uncertainties in military questions related to the CIS do not allow the appropriation of money for the whole year) accounts for 50.4 bn roubles in current prices (total state budget 258 bn roubles)[1].

Though the announced defense budget is not structured according to functions, the public was informed by the President that it was decreased by 10% in constant prices of 1991, and expenditures for arms procurement are cut by 50%. The last figure was hotly discussed in Parliament, including the initial version of draft budget which proposed complete freeze of purchases of weaponry for a certain period. Finally 6.56 bn have been earmarked for this article, and parliament has added 20 bn more - 10 bn directly from the budget and another half as a loan from the Central Bank of Russia[2]. A lot of questions might be raised about the strange figures, but it seems that unpredictable inflation, unknown borders of so-called, "defense strategic space" that is to be added to these expenditures, do not allow the brightening of the prospect.

In any case the trend is clear : defense industrial contracts and production are to be cut radically, thus creating a real and sufficient space for conversion, which will be able to cause positive changes at the consumption market, enrich civilian technologies and promote much-needed structural shifts. It is the first and absolutely needed condition to start reallocating resources and restructuring defense complexes according to civilian and reforminginterests. Though in the previous years (1988-1991) the defense production was already decreased by 29%[3], the cuts were spread between hundreds of enterprises, only a few of which got rid of military obligations completely. So these developments brought losses and did not inaugurate demilitarization.

The second factor promoting real conversion is connected with guaranteed irreversibility of demilitarization trends. The fact that this country has made its choice and broken off with the past is only part of these guarantees. Another reason is the practical inability of Russia to continue to support a defense complex of its previous size and ambitions.

Three main sources of defense industrial prosperity have been exhausted : price reform has killed artificial cheapness of land, fuel and raw materials. Large-scale investment in high-tech defense branches due to enormous oil revenues are over because of the situation in the world and mainly, the national oil market. The near poverty level of living of the majority of the Russian population and the financial bankruptcy of the state do not allow the concentration of national resources for defense purposes. It might be added that the society itself has moved to a more pragmatic and less ideological mentality. Therefore there are not sufficient grounds for unreasonable economic, political and even military expenses.

What does this mean for a manager of a defense industry plant? The disappearance of expectations to renew military orders and to keep his traditional products in the mainstream of arms procurement. This should not prevent him from taking decisive steps towards conversion in all possible forms. The opportunity to make a fresh start without being burdened by previous responsibilitiesand to use some benefits of creative destruction and to consider conversion to civilian production is not a misfortune, but a chance to survive and to use comparative defense industry advantages, accumulated capital and skill at the civilian market.

No less important is the fact that new wholesale prices, loss of access to secure privileged supply, the mainly barter relations between subcontractors as well as uncertainty in stability of defense market have made military production unprofitable for enterprises and deprived them of nearly all the advantages of working for the army. There is therefore an increasing shift to the hitherto deficit market of non-military goods and decreasing resistance to the lowering tendencies in defense procurement.

It might also be added that the removal of the center of gravity of the decision-making process in conversion to the micro level of defense industry establishments had positive consequences. Conversion is turning from political campaign into hard, everyday and mostly hidden work of the managers of defense enterprises, banks and investment companies to provide conversion efforts according to much more rational criteria of efficiency, profitability, and long-term survival.

New legislative rights were received : to create stock companies and joint ventures, to let released capacities on lease, to split defense and civilian production at the mixed enterprises and count costs and profits separately or even change the form of ownership at the civilian section of the plant. In addition, the past few years helped to gain experience of internal conversion of unused capacities at the small and cooperative enterprises functioning at the premises, and engaging part of the staff of defense plants. All these steps have considerably extended the flexibility of former defense management at the plant level and lowered theirdependence upon irrational political plots to reach the top levels of economic control.

To sum up, deep economic demilitarization has without doubt pushed the brakes for defense industrial conversion policy. It is worth examining now whether we have any conversion policy.

## Conversion and Planned Economic Decline

It is impossible to speak on conversion issues in isolation from the general economic situation in the country. The latter can be described as a threatening combination of several crises accompanying the systemic transformation from command to market economy. It is a really unique situation in international economic history.

Some of these crises were predictable, taking into consideration the chosen policy of financial stabilization. Deregulation of prices, combined with high taxes and expensive credits, could not but cause a severe slump in investments and demand - both in private and industrial consumption. So the decline was "planned" with hopes for limited decrease of production before inflation could come under control.

But the classic picture of stabilization is spoiled by post-Soviet realities and the unfortunate heritage of previous times. An unbalanced and inefficient economic structure (determined mainly by the high rank of defense industries and non-economic priorities for investment), high expenses for foreign and internal debt servicing, as well as the monopolistic behavior of the still state-owned enterprises, are making financial stabilization possible only at the expense of deep economic decline, not to speak of partial de- industrialization. And the government, which was brave enough to start reforms and to insist on severe defense cutsand harsh measures of expenditure savings, seems not to be able to control depression efficiently or suppress the destructive activity of regional authorities, monopoly enterprises and false-market commercial structures. It might be added that the ability of the new-born market tools to control sophisticated and conflicting economy were exaggerated.

What about conversion policy under these circumstances?

One view might be that the government has no clear-cut conversion policy and former defence enterprises will become a bottom dog of "shock therapy" and fiscal stabilization together with personal consumption, savings, investment and subsidized plants and agriculture. If that is true, then the government is losing the opportunity to give birth to one of the few potentially dynamic economic branches able to create competition to the monopolistic producers of consumer goods. This approach might be dangerous for the reforming process, not only due to uncontrolled and unlimited decline. By giving a free hand to the defence industry to survive or die, and moving aside macro-regulation of conversion, the government runs the risk of receiving unwanted results of "liberalization".

Not having received proper signals within the general structural policy, conversion is still considered to be a problem of the military-industrial complex itself, and civilian production is again viewed as a peculiar tax on privileged branches. Not speaking of economic losses, this approach is fraught with serious political dangers of the revival of the MIC of the former USSR on Russian soil. Thus, the State Committee for conversion established by the President of Russia (and headed by the military serviceman) has started its activity with the struggle for the widespread rights to control the whole defense industry together with converted enterprises. In the situation of conceptualvacuum from the side of the government, the military themselves have defined the national goals of conversion as a means "to save the complex from destruction"[4]. Shortages of investment are supposed to be compensated by a sharp switch in arms export policy : the revenues from limited arms sales are planned to be invested in modernization of defense industry and conversion projects.

The following may be proposed :

- to define at last what our defense complex is in terms of the volume and quality of production capacities, labor force skill, technological level and the possibility to produce weapons at a reasonable price;

- to define what our defense complex is to be and what enterprises might be closed, or merged, or totally released from defense industry, sold or given an opportunity for efficient conversion to civilian production, with or without state support. The efforts to survive through expansion of arms sales and ambitious collaborative military projects look dangerous and short-sighted;

- criteria for the decision to close former defense plant or to support its civilian retooling must take into account not only considerations of economic efficiency, but also the state of regional economics (in the former USSR there were 87 closed towns busy only with defense production[5]);

- to work out a flexible mechanism of economic preferences in order to reorient investment, including that from commercial sources, towards demands of a peace economy, promoting competition at the market of consumer goods;

- to get rid of the numerous illusions connected with conversion. The conversion policy is to be worked out, keeping in mind lowering of payable demand and inevitable foreign competition at the internal market, relatively small and short-term return from the direct reallocation of raw and fuel reserves of defense sector in order to suppress the inflation wave;

- international cooperation in the field of conversion looks very attractive for enterprises and gives them a chance to implement their skill and technologies, to get market experience and receive access to the international market. But it would have been better if government officials did not serve as mediators in these projects. They could efficiently influence them with the mentioned tools of economic preferences in the interests of macro-economic normalization.

**Endnotes**

1. He3aBHCHMaR ra3eTa, 29.01.1992, N3BeCTNR 28.01.1992

2. These figures were informed by V.Vitebsky, Vice-Chairman of the Russian Federation Supreme Soviet/s committee on Industry and Power Engineering in Moscow News, No.7, 1992.

3. 3KOHOMNKa M KN3Hb, No.52, ReKaOPb 1991

4. Moscow News, No.7, 1992.

5. KOMCOMOJibCKaR IPaBRa, February 4, 1992.

# PROBLEMS AND POSSIBILITIES OF TECHNOLOGY TRANSFER

**Evgueni Gouklin**
*Urals Research Institute, CIS*

A unique laboratory in one of the Urals Research Institutes is developing highly effective technologies of melting steel and alloys with definite properties and a pre-set structure (MP technology). It is used in aircraft and space industry. In aircraft components, the alloys melted according to "MP" technology are used in an engine of the "Suhoy-27" combat aircraft. The engine was successfully demonstrated at the Exhibition "Conversion-90" in Munich. This technology provides operational reliability of engines and increases their "service life" by 50%. 143 components melted from heat-resistant alloys using the "MP-technology operate in the spaceship "Buran". The "MP" -technology increases the corrosive resistance of stainless steel components by 3-4 times. There are many other possibilities for this technology. In all cases, the "MP" technology helps to improve the quality of metals and alloys without changing their chemical composition. Different methods of influence on liquid metal can be used : ultrasonic and electro-hydroimpulse treatment, electric current effect and the time-temperature treatment as the simplest method of controlling liquid metal structure before solidification. However, the transfer of the technology in a formalized way (tables, diagrams, etc.) does not guarantee metal quality. It is a flexible technology based on the investigation of samples selected over many years. The laboratory is ready to render service using the "MP" technology by sending 2-3 experts to a customer's plant. They will acquaint the customer with melting technology, without going into details and breaking the know-how or other commercial secrets, for the physico-chemical investigations in the Urals in order to work out, during 1-2 months, recommendations according to the customer's order. Paymentfor the service can be made on obtaining the final results of the technological tests and implementation of recommendations. But utilisation of this technology abroad is possible only in case of good marketing. We have used again for this purpose our partner, the JTM company, which has also good connections with some companies involved in technology transfer or metal melting. Next month the first customer from Germany is going to visit the Urals and sign a contract. In case of success, some other laboratories of the institute can be involved in this process of conversion.

The next project we are going to carry out with our German marketing partner is the joint production of multi-purpose $CO_2$-laser with persistent operating beam. The laser has been developed over the last 4 years by a military enterprise for military purposes and has already cost 50 million roubles. There are 2 laboratory samples. They prove the possibility of creating an industrial laser unit for civil use. The most promising field of the laser union application are :

- thermo-treatment of metal surfaces;
- cutting of metals, ferro-concrete, stones;
- wood sawing.

Great economic effect is supposed to be reached while using the laser unit, for thermo treatment of the railways working surfaces. It will help to increase the "service life" of rails at 50-60% and considerably decrease time and expenses for railroad repair. The laser may also be used for thermal treatment and extra hardening of rolling mill rolls, without their being dismantled. The main problem of the scientists, who are the authors of the laser project, is the lack of money. Accordingto a business plan, the laboratory needs about 18-20 months for the creation of the industrial laser unit with the output power upto 70 KW. It will cost some 50 million roubles (still no calculation in dollars). It is suggested that a Russian or foreign partner might invest the necessary amount of money to organise the marketing and get a part of the profit.

# SPECIFIC ASPECTS OF THE CONVERSION PROBLEM IN THE EVOLVING RUSSIA

**Alexander Konovalov**
*Russia*

Dramatic events of the last months, which culminated in the Treaty signed in Belorussia by the leaders of three republics which shortly transformed the USSR into the new version of the Commonwealth created by the eleven republics, have made the Soviet Union the fact of history and have seriously changed the military strategic landscape in Europe and the rest of the world. These factors have led to the necessity to reconsider the all-traditional approaches towards the military security problem. It is evident that in the current economic situation inside the Commonwealth, and the external military-political environment, the real amount of defense spending will be substantially lower than in the previous years.

The general defense spending cuts are sharpening due to the dramatic and unavoidable structural change in the defense budget. In the previous Soviet history, defense money was canalized primarily for military purchases. Even in official defense budgets which did not contend the real costs of the arms production, the military purchases consumed more than 40% of their total amount. This was absolutely crazy distribution of defense money and no state had a comparable distortion among the defense budget's items. Now such a correlation of the defense budget items should be radically changed. The main reductions will be done in the arms purchases and consequently in the rates of arms production.

This has raised a **problem potentially threatening the international security : drastic reduction or complete cancellation of the defense orders in many areas and the necessity to quickly convert the lions share of the Soviet defense industry.** TheSoviet defence industry is objectively a substantial part of the national wealth and it is extremely important to handle it carefully in the crisis situation. The defence industry of the USSR may be the only sector of the Soviet economy which is comparable with the technology levels which exist in developed countries and can serve as a bridge to assist Russia and other republic economies to be integrated in the world structures. Unfortunately, this industry was not prepared by the state for a deep and quick change in economic environment and is now facing the prospect of the complete absence of state orders on many types of weapons and a clear prospect of losing all the privileges the defence sector has enjoyed during the last decades of Soviet history.

All these circumstances make the defence industry possibly even more a socially explosive group of the society than professionally military. There are two opposing trends in this social group now. On the one hand, being well educated, better organized, and more advanced than the others, this social group is better prepared to support the democratic changes and to become not only the proponents but the powerful booster of political and economic development in Russia. On the other hand, the dramatic crackdown in their traditional life style, the uncertainties and a possible decrease in the living standards, could put them on the other side of the barricades.

Let us not forget that among the leaders of the August coup there were two representatives of the defence industry,

Baklanov and Tizjakov. It looks as if at least part of the defence industry leadership has tried to replace on the Soviet political scenethe almost dead communist party. It might be the first precedent in the Soviet history when the industrial leg of the military industrial complex has tried to get the political power which would correspond with its economic weight. That is why the successful conversion of the Soviet defense industry should be put among the other key aspects of international security.

**General economic indicators of the military share in the soviet economy**

Up till now it was hardly possible to get reliable and exact information about the real amount of financial and material resources the defense sector of the Soviet economy consumes. But recent assessments made by the prominent national experts and the foreign analysts, who relied on the indirect economic indicators and who used different methodological approaches but came to the same conclusions leading to judgements which are very far from official data. Well-known American economist I. Birman who analyzed the Soviet defence spending in 1989 came to the conclusion that it should be evaluated as at least 200 billion roubles.

Relying on the assumption about keeping the same defence share of the Soviet GNP during the next two years, which seems likely, and taking into consideration the boosting inflation, the real amount of defence spending in 1991 should be evaluated not lower than 400 billion roubles. Official data showed that in 1991, Soviet defense spending had grown from 71 to 96 billion roubles in current prices. Then this level was corrected by the Planning and Budgeting Committee of the Soviet Parliament which added to the basic sum an extra 36 billion roubles. Finally, by the special decision of the President M. Gorbachev, another 40 billion roubles was added to compensate for the price rise and maintenance costs growth. Thus, the publicly presented, (directly and indirectly), Sovietdefence budget in 1991 was 172 billion roubles, but it only seems to be the tip of the iceberg.

As is known already, the former Prime Minister of the Soviet Union, V. Pavlov, mentioned several times that the Soviet's defence spendings are very close to the American's; in other words they are around 300 billion dollars. Taking into consideration the fact that the purchasing power of the Soviet "defence" rouble was much higher than its "civil" counterpart, the sums of 400 billion roubles and 300 billion dollars are very close values. For instance, the internal price of the well-known main battle tank T-72 was 960 thousand roubles but in the international markets it was sold at 1.8 million dollars, i.e. in this case, the correlation between dollar and rouble in this area was 2:1 in favor of the rouble. Professor I. Birman, mentioned above, considered that the Soviet defence spending consumed 25% of the national GNP during that period.

An even more impressive estimation of the Soviet economy's defence share was given by another analyst named V. Pervyshin who studied all the shipments to the military-industrial enterprises, defense R & D institutions, armed forces, KGB, and MVD, (The Interior Ministry). Together with the military, R & D paid 15 billion roubles from the state budget and military construction funded 10 billion roubles from the same source the value of 30 billion roubles worth of consumer goods produced in the defence sector and so he concluded that the amount of the Soviet defense spending in 1991 was 480 billion roubles which is close to the assessments of I. Birman because his result included some items related to KGB and MVD expenses which are not formally a part of the defence budget. In accordance with Pervyshin' analyses, the military share was 52% in 1991 of the Soviet GNP and 73% of its national income.

The same result, 50% of the GNP, was independently arrived at by the other specialist Dr. B. Lagutenko which shows the high probability of the correct analyses. Let me remind you that during the second world war the military share in the Soviet GNP was around 45%. Dr. Lagutenko considers that there were two main sources to pay such a huge military burden : the enormous taxes, direct (~10% of the income) and indirect (extra ~60-70% via state-controlled prices), and the growing emission of the "empty" paper money. Before 1988, the average annual money emission was around 3-4 billion roubles, in 1988 - 11.5 billion, in 1989 -18, in 1990 - 25 and in 1991 - 250 billion roubles. This specialist estimates that the necessity to pay the military share of the economy led to the artificial extra growth of consumer prices at least by 25 times.

There was some limited information about the minimal defence spending of the year 1992 in the Commonwealth. The Chairman of the budget committee of the Parliament, A. Pochinok, has stated that in the context of the price liberalization, the military spending of all the states forming the new Commonwealth should be around 500-600 billion roubles in 1992. Russia itself is going to pay 300 billion roubles. What is more important, the Chairman has stated that the lion's share of defence money will be used for personnel payments, social programs to support the active-duty professionals and the officers who are to be demobilized, on housing construction for the military, and defence industry conversion programs.

The prominent Russian economist V. Seljunin has calculated that before 1992, upto 80% of the Soviet machinery industry production was used for military purposes, either directly as weapons and military technique, or indirectly as production equipments used by the weapon's manufactures. I. Silaev, working with thePrime Minister of the Russian Federation, has stated that "70% of Russia's industry is represented by the defence industry enterprises". In general, the defence industry has a very high level of regional concentration. Around 80% of the former Soviet Union defence industry is located in Russia. Among the biggest plants and design-research centers, about 20% are in the Ural area, 17% in Central Russia, 13% in Volga-Vjatka region, and 10% in the Saint-Petersburg area. The best engineering talents, the most skilled workers, and the best production equipment in the country are concentrated there.

One of Russia's defence industry officials V. Glukhikh recently stated that the republic has got as its share in the Soviet defence industry heritage 1500 industrial plants and more than 900 defence research centers located in its territory with more than 9 million scientists, engineers, and workers. Of this labor force, 1.7 million are working at the design bureaus and research institutes. There are some industrial centers in Russia which are highly militarized. For instance, in the city of Omsk, (Siberia), with 1.5 million population, 70% of the industrial production is for the military and in Udmurtia, (the autonomous area in Russia), 83% of the industrial production is used for military needs. Besides Russia there are some defence industry centers in other republics, primarily in Ukraine, two big ballistic missiles production plants in Pavlograd, (SS-24) and in Dnepropetrovsk, (SS-18), the biggest Soviet shipyards in Nikolaev, Kherson, and Kerch, and prominent aircraft design bureau named after Antonov. The total number of people engaged in the military R&D and arms production in the former Soviet Union, as V. Pervyshin estimated, is 14.4 million persons, which corresponds to the official information given by V. Glukhikh separately for Russia.

**The challenges and the limits of the conversion in Russia**

The scale and the complexity of the conversion problem in Russia make the situation absolutely unique. Firstly, there are three political and economic factors which impose the depth and impede the speed of the conversion process now. These are : radical change in the military threat assessment and approaches towards the insurance of the military security of state, defence budget constrains, which are explained by the economic difficulties and changes in correlations among its items not in favor of defence purchases and finally the shift towards the market economy which brings the real prices in the defence sector without state subsidies and privileges. The first factor mentioned above should be left aside in this paper as it is a political one. Among the two others the last seems the strongest in the current economic conditions.

Of course the goal to prevent the social explosion inside the armed forces leads to the serious change in the defense budget structure. In the last January statement President B. Yeltsin has announced that besides reducing the defence budget Russia will make it socially oriented. In 1992 it is supposed to reduce military spending by an extra 10% in constant prices but the military purchases are to be reduced by 50%. Some other Russian Government officials spoke privately that the real defence purchases will have to be cut by 8 times. This factor alone already puts the serious limits on the state contracts for the defence industry, but it seems that the marketization of the whole economy, and its defense sector in particular, might put more serious limits. It seems as well that in the new economic environment, the local and republican authorities will insist that the military pay for the land used for their needs which can ruin the defence budget. Now military bases, test fields, and otherdefence installations occupy 42 million hectares or 2% of the former Soviet Union territory. Of course all of them are extremely polluted.

The defence industry conversion in Russia has three main dimensions:

- conversion in the classical sense of the word, i.e. rearrangement of military industry for the production of civil goods;
- liquidation of the weapons already produced, which in the current situation is becoming an extremely difficult technical and economic problem in Russia;
- conversion of the "human factor" i.e. the perceptional change in the public mentality of the highly militarized society and professional and social re-adjustment of the personnel in the defence sector of Russia's economy.

The second dimension of the conversion problem in Russia among these three has the most direct environmental consequences be it in the area of nuclear weapons or in a conventional one.

**Liquidation of nuclear weapons** :
It was declared that the nuclear warheads should be delivered in the production plants associated stores, (located only in Russia), for further dismantling. Let me remind you that this is a principally new problem. Never before had Russia to liquidate the nuclear warheads. There was a process of "refreshment" of the stored warheads because some of their components have a limited life time, for instance tritium (radioactive gas whose life time, is 12.5 years), the others are losing their qualitative parameters during storing and are becoming the sources of radiation, threatening human life. In the previous years there was a routineprocedure to treat such weapons. The nuclear warheads which have exhausted their storing time were taken out, delivered to production plants, and disassembled.

The fissile materials extracted from the warheads, enriched uranium and combat plutonium, were technologically developed and used in the new warheads of the next generation. Now the situation is principally different. There are no prospects for the new orders on nuclear warheads and Russia, together with the developed countries, has to find out a rational and secure solution to the combat fissile materials problem. The problem seems so serious that the US Congress granted Russia 400 million dollars to assist in its solution. There are two main alternatives: it is possible to store these materials till their natural liquidation (some million years), in the safe and secure stores or to redevelop the combat fissile materials into fuel for the civil nuclear reactors. Of course the second alternative is preferable but there are serious technical obstacles to it.

In principle, the combat plutonium could be used as a nuclear fuel but up till now there are no reactors of this type. The existing design projects do not give any hopes for a quick solution to the problem especially taking into consideration that combat plutonium is an extremely active and highly poisonous material. This means that right now, in both cases we have to speak about the stage of the long storing of the combat fissile materials. Recently, the President of the International Bank, J. Attaly has proposed an interesting idea which would link the liquidation of the nuclear warheads of the former USSR with its international debt (around 60 billion dollars). In accordance with his scheme this debt should be gradually paid from the defence budgets of the developed states in exchange for the liquidation of the Soviet nuclear warheads.

The idea looks very attractive but some doubts were expressed in the Russian media about its adequacy as to whether Russia should transfer the fissile materials extracted from the nuclear warheads to its creditors. These doubts were based on elementary calculations. The tactical missile warhead contains 7 kilograms of combat plutonium and its strategic counterpart two or three times more. This means that the liquidation of one thousand warheads corresponds to the extraction of 7-20 tons of combat plutonium. The prospect of the deep reductions in nuclear arsenals leads to the reasonable conclusion that the current soviet nuclear heritage, which is assessed by different analysts from 27 to 33 thousands of nuclear warheads, is to be reduced at least by 20 thousand warheads. Consequently, we have to speak about the necessity to store for a long time an extra 80 tons of highly dangerous combat plutonium. In accordance with the western evaluations the USSR had already in store around 100 tons of combat plutonium and 500 tons of enriched uranium.

It is hardly possible to speak about the market price of these materials because, fortunately, they never were sold. But the expert's assessments demonstrate that the price of enriched uranium is about 100 thousand dollars per kilogram and combat plutonium is 5-10 times more expensive. This means that the price of the fissile materials extracted from only 6,000 tactical nuclear warheads covers the full amount of the former Soviet Union foreign debt. One has to have in mind the limits of such analyses because the previous costs to extract the combat fissile materials do not mean the possibility to rationally use them for civil needs right now. But the specialists have paid attention to the fact that starting this year, Japan has launched the long-term project to stockpile during 30 years 400 tons of plutonium and also its plans to double the Japanese nuclear power stations output bythe year 2010. In any case the probability of future usage of combat fissile materials as a nuclear fuel cannot be excluded completely.

Consequently, regardless of the alternative which can be foreseen for the further use of combat fissile materials, the long-term storing seems the necessary step. In other words, Russia has to plan the construction of secure and safe shelters for these products. Recently, during the international seminar in Kiev, the Deputy Minister of Atomic Industry V. Michailov has proposed the project to invest all the 400 million dollars, granted by the US Congress as an assistance in the liquidation of the former Soviet Union nuclear weapons, in the construction of the giant shelter for the combat fissile materials in the Ural area. It is hardly possible to imagine that one can find the place in Russia's territory where the local authorities and local population would agree with such a project. But even if the problem of such a shelter's location is successfully settled, it would not mean the complete solution. The specialists in Russia consider that we need at least five new stores of this type whose construction would cost about 2.5 billion dollars.

**Liquidation of conventional weapons**

The problem of conventional arms liquidation seems not the less complicated technically and economically and brings no less devastative environmental consequences. Let us review the only one example which is linked with the combat ships which have exhausted their service time. During the last year, the Soviet Navy has taken out of combat duty and put on an anchor in the naval bases areas as many as 265 combat ships, surface and submarines, with the submarines displacement 170 thousand tons. Besides that, there are an extra 242 combat ships partially or completely sunk in the close vicinity of national beachesof low depth. Each year the Soviet Navy is taking out of active duty the combat ships with the total displacement of about 80-100 thousand tons. Corrosion eats each year around 30-35 thousands tons of high quality steel in this ship's graveyard polluting the ocean water. In the world markets the price of such steel is 130-180 dollars per ton. At the same time the metallurgy plants are being stopped across the country because of the shortages in getting metal scrap.

But to prepare the combat ship for sending it into a melting port is a costly and complicated operation. It is necessary to put it into the dry dock, to clean it from corrosion, to disarm the ship, to make it hermetic to be drawn in the harbor where the proper metallurgy plant is located. To make all these operations with a destroyer or a cruiser costs 350-400 thousand roubles. For these operations, the naval authorities have to use the dry docks which are intended to be used for active duty ships service, reducing their ability to perform combat missions. But the current industrial bases allow for secondary use less than 12 thousands tons of the shipping steel. Of course the problem can be solved only by cooperating with foreign manufactures.

The problem of the retired ships illustrates only a tiny fragment of the liquidation problem. Recently President B. Yeltsin has announced that he is going to send the CFE Treaty for ratification in the Parliament. The other European republics of the former Soviet Union have committed themselves to ratify and to fully implement the CFE Treaty as well. But it means that only in the European territory which used to be a part of the USSR should it liquidate 19689 pieces of different armaments including 7575 main battle tanks, 9890 ACV's, 767 artillery systems, and 1461 combat aircraft. Let us not forget as well that during two years, 1988-1989, the Soviety military withdrew fromEurope and redeployed behind the Urals around 70000 units of the same treaty limited equipment whose maintenance does not have military purpose and should be utilized in this or that form, mainly by physical liquidation. Thus it is logical to urgently organize a new branch of industry in Russia which would work out the efficient technologies of weapons-utilization and redevelopment and contribute a lot to clean the environment of the pollutions of the cold war period.

**The "human factor" conversion**

The human dimension of the conversion problem is very complicated as well. The well-known example which of course does not exhaust the whole problem is linked with a possible mass emigration of the nuclear specialists from the former Soviet Union and the proliferation of knowledge in the nuclear arms technologies. Special efforts are to be launched to avoid such a prospect. The Deputy Minister for Atomic Energy, V. Mikchailov, already mentioned in this paper, has reported recently that around 100 thousand people are engaged in the design and production of the Soviet nuclear weapons. Among them around 10-15 thousand are really acquainted with top secret information, and not more than 2-3 thousand are familiar with all the details of the high technologies in the nuclear sphere and their knowledge represents the highest potential threat. The Deputy Minister has estimated the probability of a massive emigration of such specialists as very low but mentioned as well that it can rapidly increase if the Government should ignore their needs.

The other prominent figure, the Deputy Director of Russia's leading research center in the field "Kurchatov Institute"and academician, N. Ponomarev-Stepnoi has shared this assessment. He stressed the specific mentality of nuclear physicists who are ready to afford voluntarily the socialdiscomfort and limits of the life in the secret towns in exchange for the possibility to satisfy their scientific interest. That is why it is crucially important to provide them with challenging work besides more or less modest living standards which would correspond to their intellectual capabilities. Under these circumstances the most efficient way to block the dangerous emigration of the nuclear specialists to the third world countries with totalitarian regimes is to increase the scientific exchange with developed countries and, what seems even more important, to organize the set of joint projects with their western counterparts which would allow the nuclear specialists of the former Soviet Union to be more or less normally paid on the one hand and to continue this work in their own laboratories or plants on the other hand.

Of course, the nuclear specialists engaged in the military R & D and nuclear weapons production represent only a small fraction of the fourteen million defence industry labor force and we cannot forget the rest of the arms manufactures. There are a lot of urban industrial organizations around the big defence enterprises in Russia. The absence of the state defense contracts puts them in an extremely difficult position and threatens the very existence of such cities and areas. It is hardly possible to work out the single universal recommendation of how to settle the problem. The approach should depend on the branch of the industry on the geographical location of the defence industry enterprise and on numerous other factors. It will be much easier to convert the enterprises in the shipbuilding, electronic, or aircraft industries than the highly specialized missile plant.

There were several proposals to provide these defence plants with an average level of salary indicated with inflation and paid from the state budget for a limited period of 1.5 to 3 years. The only condition for such subsidyshould be to stop or to drastically reduce the arms production and to use this time to re-adjust their plant for any civil products. Such an approach does not look like a waste of money because it helps to save a lion's share of the financial and material resources used in weapons production because the personnel payments usually consume 15-18% of the total costs in weapons manufacturing.

## Conclusive remarks

It is evident that the Soviet defense industry conversion is a huge and multi-dimensional problem and its solution will be a painful and time consuming process. To settle it, Russia needs wide-scale international co-operation. As an urgent step it should be necessary to organize the special informative network which should include a number of small, well-equipped bureaus located in the most militarized areas of Russia. Such bureaus should collect the projects and proposals of the local defense enterprises which are searching for the co-operation of the developed countries and quickly pass them for analyses by the western experts and potential partners. Such a network should allow the break through of information barriers which up till now have been the main obstacle in the contacts and collaboration with this sector of Russia's economy. It would serve as a politically confidence building measure as well as help change the previous approach towards military secrecy in the industrial sphere. The secrecy regime which existed in the Soviet defence industry during the past decades primarily damaged the civic sector of its economy. Keeping the military secrecy cost the Soviet tax payers 30-40 billion roubles per year and ruined the consumer goods industry which was cut off from all the innovations.

It is hardly possible to expect that the commonly-acceptable joint projects, especially those which need essential investments, canperform an urgent rescue mission. The other possible option is to widely use the infrastructure, the well trained, qualified and relatively cheap personnel of the defense enterprises as a short-term temporary solution. It is possible to quickly install, without substantial expenses, different types of equipment to make some relatively simple subcontract works, with the developed countries as partners to the common benefits. The information about the industrial capacities and the main specializations of the defence enterprises which lost the state orders and are looking for help and partnership could be accumulated by the same bureaus.

The level of risks, disorders, and uncertainties in the Russian economy is still high enough but opportunities do exist and should be used.

# SOME MILITARY PRODUCTION CONVERSION ASPECTS

**Vladimir Nikitin**
*Central Design Bureau for Machine Building, Russia*

Within the framework of the former Soviet Union, since 1989 the work on military-industrial complex conversion has been undertaken on a large scale. This is a complex and multisystematic process concerning many hundreds of thousands of specialists. Suffice it to say that in Russia alone, conversion involved about 500 enterprises and 200 scientific-research and development organizations. The main deficiency of the programme is the difficulty of putting it into practice. In this case one should take into account that curtailment of production runs with double speeding-up against the plan. Because of this, many enterprises of defence branches are put into a hard financial position. Lack of the conversion fund prevents the organization of effective use of released capacities and scientific-engineering potential for new technologies, to expand the output of civil-purpose products and consumers' goods. Organizational structures being created, such as the Conversion Activities Committee under President of Russia, commissions in St. Petersburg's Council and municipality coordinate this work poorly. More than 20% of all the people working in the region are employed at the defence complex enterprises in St. Petersburg, and they provide an output of marketable products of 30% in volume for all the region and render 50% of services to the population. In practice, all the defence enterprises take part in the output of consumer goods and civil-purpose products, with dynamics being as follows : (in mln. of roubles).

1989 - 1361990 - 1751995 - 301

As an actual defence enterprises' contribution to conversion, one can giveexamples, such as the creation of filling-and-packing equipment, automatic hot-liquid-filling machines, production lines for production and processing of chemical fibres, powder, ceramic composite materials, local fiber-optic networks, truck tractors and fire trucks on the base of the tank T-55, high-speed photographic equipment and others. At the same time the practice of carrying out conversion in the region is indicative of curtailment of the production volume at enterprises of defence branches. The enterprises experience great economic difficulties. Upto now, the general direction of conversion has not been determined, and programs of this process in the region are not available. We think that in order to realize the state conversion program, the required means must be found both in the state budget through saving of military expenses and by a special conversion fund creation. And for the time being, every enterprise being engaged in military-industrial affairs seeks, without assistance, salvation from a critical state, including foreign conversion funds, investments and promised assistance to nuclear scientists.

The present report partially pursues this object. The question is about the Central Design Bureau for Machine Building, which is a member of the Ministry of Nuclear Power Engineering of the Russian Federation. We want to attract attention to our enterprise's possibilities, in connection with conversion, and the creation of new science-consuming technologies and design works which can be performed by our specialists on the basis of our scientific-engineering potential. Our enterprise, organised in 1945, has been working in several thematic directions on the non-standard equipment development fornuclear power engineering. One of the directions is the creation of the equipment for uranium isotope separation by ultracentrifugation. Our domestic industry for enriched uranium production is equipped with these facilities.

On this basis we propose :

- a plant for the production of super pure gases by centrifugal process. Cascade of centrifuges with special fittings and service lines produces gas free from impurities. Transportation and storage of super pure gases is provided in special bottles of capacity 0.5 - 40 l. Plant output is upto 100 kg a month, gas purity is 99.99% - 99.9999%, content of limiting impurities is $10^{-4}$ - $10^{-0}$;

- a gas separator for final cleaning of electron gases (fluorides, chlorides, hydrides and others) from suspended particles of size more than 0.01 um by centrifugal process;

- a plant to measure impurities microconcentrations in super pure gas by means of centrifuges - concentrators, mass spectrometer, control system, measuring system with microprocessor. Limit of measuring impurities concentration is $10^{-6}$ - $10^{-8}$%.

Another direction is small-scale power engineering. We created the unique space power nuclear plant "Topaz" with direct energy conversion. The work is debugged to standard ground prototype. The USA took an interest in these works, and we made a contract for demonstration of this plant. The possibility of subsequent joint work on the creation of higher-capacity plant is not excluded.

On the basis of this work we are ready to propose :

- transportable nuclear power plant TOC-M of electric capacity upto 1 MW with serviceable life of 20 years, reactor weight with coolant 30t. Two-circuit nuclear power plant with lead-cooled reactor using free air as a secondary circuit coolant and working medium in the open gas-turbine cycle;

- space nuclear power plants with reactor-thermionic converter of electric capacity 13, 20, 30, 40 KW which can be used as self-contained power source for space objects of voltage 30-40kW, serviceable life 7-10 years, mass 1.3 - 2.8.t.

We carried out a great complex of works on creation of remote-controlled equipment (manipulators, robots, machines and others) for work with radioactive materials.

On this basis, we propose in the field of handling of spent nuclear fuel development and manufacture of remote-controlled equipment complexes intended for

- transportation from the nuclear power plants to fuel reprocessing plants;
- preparation for fuel reprocessing at fuel reprocessing plants;
- servicing of the spent fuel storage;
- decommissioning of the nuclear power plants.

In the field of weapon stock liquidation development and manufacture of remote-controlled equipment complexes intended for

- handling of ammunition;
- handling of bacteriological and chemical components.

Not long age we organized a Soviet-Dutch Enterprise "Terrablock" for the creation of equipment for manufacture of TERRABLOCK bricks.

We think that we can organize the full-scale production of such equipment e.g. for building houses for service men transferred to the reserve forces.

We hope that our science-engineering potential will arouse interest in foreign companies and will assist them in solving many problems at the difficult time of conversion.

# MAIN DIRECTIONS OF RADIOCHEMICAL TECHNOLOGY CONVERTICE AT A.A. BOCHVAR AUSRI OF INORGANIC MATERIALS

**Anatoli Poljakov**
*AUSRIIM, Russia*

A.A. Bochvar All-Union Scientific Research Institute of Inorganic Materials (AUSRIIM) is a large research centre within the Ministry of Atomic Power and Industry and is a leading one in tackling some important problems of nuclear technology.

The Institute has a chemico-technological department, the laboratories of which developed processes and equipment that were embodied in commercial utilities producing materials for military purposes and nuclear power.

As for their technological and engineering parameters the developed processes are most perfect of all those presently used in traditional "non-radiochemical" branches of the industry (chemical, hydrometallurgical, medical-biological; machine-building and others).

During a 45 year period a highly qualified staff of researchers gathered at the Institute. Their knowledge and experience in a wide range of problems (dissolution of almost unstrippable substances, filtration, extraction, ion-exchange, separation, production of solid metal compounds, evaporation, gas and aerosol clean-up solidification of different activity level wastes, decontamination of equipment) in combination with original instrumentation and control systems for the above purposes, allowed them to work successfully on conversion. In particular, they cover processes of extraction and separation of rare-earth elements from different ores and ore concentration;production of high purity silicon and its compounds; integrated processing of noble metal ores; contraction of noble metals; clean-up of commercial utility effluents and off-gases from aggressive, toxic, ecologically dangerous substances and elements etc.

The Institute is responsible for not only process developments but also for design of production, facilities, hardware, gauges, automated systems of data representation and process control, provision of highly qualified analytical control with express remote and continuous analyses of many components; technical documentation and if needed, distribution of orders for manufacture of hardware; participation in start-up and adjustment and bringing production to a commercial level.

Collaboration with the Institute can be as follows :

- sale of "know-how" of process developments;
- implementation of orders for development of processes, hardware, means of process control ;
- joint developments in different fields.

Based on autoclave, extraction, sorption and membrane processes the Institute has developed an integrated technology for gold -and silver-containing ores with a parallel extraction (different ore types) of valuable components (W, Os, Co, Zn, Se, Te, Br and oth.)

Among the problems resolved by the Institute is production of polycrystal silicon, silicon fluoride, dioxide, nitride and other compounds of semi-conductor purity; purification being accomplished by extraction of both starting materials in the single-crystal silicon production and semi-products in the fiber optics technology. The Institute has established processes for noble metal (Pd, Rh, Ru) extraction from nuclear fuel cycle waste; the extracted noble metals are used in engineering.

In the area of the environment protection the Institute has developed processes for clean-up of effluents and off-gases from galvanic and other commercial production facilities in the direct ions:

- extraction of mechanical impurities from effluents;

- extraction of micro-emulsion and organic solutes (phosphorus-and halogen-containing compounds, phenol-and aniline-derivatives, sulphanol, other surface-active agents, petroleum derivatives);

- extraction and separation of water soluble salts of non-ferrous and heavy metals (Cr, Ni, Co, Hg, Cu, Cd, Zn, Sn, Pb, Bi and others to be used or stored to be disposed of);

- additional extraction of gold and silver from effluents;

- coarse and fine clean-up from aerosols and vapours (acid mists, salts of organic compounds);

- filtration of off-gases to remove dust and solid aerosols;

- clean-up of gases from halogens and halogen containingcompounds, ammonia, nitrogen oxides and others.

At the Institute, developments are under way to obtain compositions of high-efficiency surface active substances to prevent consolidation of inorganic compounds (fertilizers, cosmetics and other flowing materials).

Recently high-molecular demulsifiers have been developed to break aqueous organic emulsions.

Recently processes have been developed to produce granulated flowing and readily compactible powders from different metal compounds (individual and as mixed) by autoclave denitration and precipitation methods with surface active agents to manufacture nuclear reactor fuel compositions; to produce granulated materials for different purposes in non-nuclear industries; to localize radioactive and inactive wastes from different branches of industry. Processes have been developed to clean up off-gases from radioactive aerosols, iodine and inert gases as well as from nitrogen oxides. The technology and equipment have been worked out to solidify liquid low and intermediate activity level wastes by bituminization and cementation as well as liquid high and intermediate activity level wastes by vitrification.

A large scientific stock has been accumulated by the Institute on the integrated technology of purification and concentration of salt and low salt-content liquid radioactive wastes using methods of filtration, ion exchange, evaporation, electrodialysis, bio-decomposition, ultrafiltration and others as well as the technology of solid radioactive waste reprocessing by compaction and burning.

At the Institute the work is under way to scientifically back up and practically decontaminate contaminated radioactive surfaces of hardware as well as to decontaminate soil, engineered structures and large territories as applied to the conditions of major accidents at nuclear installations. Work is also being carried out on radioactive waste disposal.

The Institute has finished the development and mastered the full-scale production of ß-sources for medical purposes to treat oncology diseases in ophthalmology, otholaringology and others.

High-intensity _-sources for technical purposes have been produced and are widely used to irradiate microcircuits, X-ray-radiometric, neutron-activation and other methods of analysis of elements in different solid, liquid and gaseous materials.

The chemical technology department of the Institute has developed and employs the nuclear physics methods for studying irradiated and un-irradiated materials.

Among them is the use of Mössbauer spectrometry to investigate atomic and magnetic structures (ordering phenomena, phase transformations) for the design of new high-quality materials and the latest technologies, including

- phase analysis of alloys;
- studies of phase transformation mechanisms during heating and cooling;
- studies of irradiation, mechanical and other effects on the surface of iron-carbon materials and oth.

Among them are :

At the Institute the method of positron annihilation is widely used to studythe actual structure of substances at the atomic and microscopic levels to gain understanding of the microproperties of materials. Investigations are under way to study the processes of defect nucleation in the crystal structure of a solid body as well as defect behaviour under the conditions of strain, irradiation and temperature.

The method is indispensable in assessment of technical characteristics :

- structural materials (metals, ceramics etc);
- conductors, superconductors;
- non-destructive control of product quality (individual samples);
- detection of causes of material quality degradation, fatigue control etc.

The Institute has developed the method of control based on the content of "pure" ß-nuclides in environmental objects. Samples are investigated by liquid scintillation that allows determination of a content of a particular nuclide (e.g., tritium, carbon-14, strontium) when mixed with others. The method of ecologic control has been worked out for the contents of _-and - active contamination of the environment, food etc.

In addition to the above, the Institute has carried out a large amount of work to develop non-standard technical means of control of different process parameters of chemical production for especially rigid (or specific) conditions of sensor operation (aggressive media, high temperatures, suspended matters, crystallization in a medium etc) under which conventional instruments cannot be employed.

1. Thermal flow meter PTB-3 designed for remote contactless

control of flow rate of both electroconducting and non-conducting process solutions.

2. Flow meter with electromagnetically suspended float designed for remote control of flow rate of electroconducting and non-electroconducting liquids containing suspended matter.

3. Signal-measuring device (CNY) "Sloi" designed to determine the coordinates of phase separation boundaries (levels) of a multiphase medium and to signal the accumulation of a layer of light (heavy) liquid above (under) the level of heavy (light) liquid.

4. Device to measure the vibration swing and shape of liquid level ("Uroven"). It is designed to measure the parameters of quickly changing liquid level.

5. Pressure pulsation analyzer (Pul's") designed for measuring amplitude, shape and frequency of phneumopulses ("Puls-EM").

6. Valve control units ("5YK") designed to control solenoid valve windings in air pulse generation system.

7. Level gauge,extra equipment HCY-H-1 designed to control the level of hydrogen containing environments of any extent of aggressiveness, toxicity, foaming, crystallizing, radioactive, conducting, non-conducting. It is recommended for hydrometallurgical chemical, petroleum and other industries.

8. Loss of tightness alarm (e.g., membrane break-through) (CHM).

9. Hydrostatic device for measuring pressure difference deviation (HOPD) designed for continuous remote measurements.

Thus, the scientific potentialities and developments (methods and devices) accomplished at the Institute can be employed in different civil industries at minimum expenses for transmission of experience.

# ECOLOGICALLY SOUND CONSTRUCTION TECHNOLOGIES FOR THE RUSSIAN MILITARY

**Alexander Rackitsky**
*Terrablock-Conversion, Russia*

The joint project of Terra Block Soviet Union and Building for Peace Foundation of constructing housing for the Russian military, is an effort to derive benefit from the assets of the Terrablock construction systems for the socially unstable Russian military.

The pilot project of 50 family houses to be built in the Moscow suburbs has been started by the President of the Building for Peace Foundation, Ms. ter Borg, with the cooperation of the Russian military authorities.

The houses to be built must meet the following requirements :

- good quality
- low-cost
- short term construction (out of big cities)

We believe that the Terrablock construction technique meets them all.

Presently the project is at an initial stage, but the first houses can be expected to accept their dwellers in autumn this year and the whole project will be finished within one year.

Now I would like to make a short introduction of the Terrablock construction technology :

Today, earth is being rediscovered as an ideal building material though the building process called earthwall construction has been with us since ancient times, and no wonder. The earth buildings are beautiful, functional, energy efficient and extremely durable.

Nevertheless, until quite recently, earth technology had changed very little down the years and earthwall construction remained a time-consuming, labour-intensive process.

The introduction of the Terrablock system is a recent technological breakthrough that transforms earthwall construction into the simplest, most efficient, and cost-effective construction technique in the world.

Historically the image of earth blocks has been that of a mud brick produced by hand in a crude wooden mould. Terrablocks change all that.

They are highly compact, quickly produced building blocks of superb quality, dimensionally uniformed to the finest tolerances.

The wide range of soil types that can be utilized make it universally practical. The Terrablock System literally transforms the earth right beneath your feet into a valuable construction commodity.

Terrablock walls are chemically stable, soundproof, insulating, fireproof and non-toxic.

A 12 inch thick Terrablock wall is easy and cheap to build. It has an extremely long life and once protected from erosion it is virtually indestructible.

The walls are treated with a chemical sealant that is quickly and easily applied and are now completely prepared for a finishing coat of stucco which is applied directly to the blocks. Protected in this way, physicalcharacteristics of Terrablock are identical to that of adobe but at a mere fraction of the cost. In many parts of the world no sealant is ever required. An adobe house can be designed with proper drainage away from the house and good roof overhanging, that would never need sealant of any kind.

The blocks can be dry-stacked into the wall, a technique that does not require a traditional mortar. This time-and-money-saving method is possible because of the accurate molding of the Terrablock with plain flat bedding surfaces.

Dry-stacking requires no masonry experience or skill, provided the layers are properly leveled and set on a suitable footing. The blocks are easily handled by gripping tongs and stable enough to build a structure with amazing speed and efficiency. Best of all, the blocks are produced right on the job site and can often go directly into a wall without any curing or drying time whatsoever.

The Terrablocks meet and exceed the building codes of Russia, France, Italy, the Netherlands, Georgia, the USA etc. This means that from an engineering point of view, there is virtually no restriction on the use of Terrablocks even in the places which have the strictest building codes due to continuous earthquake activity.

Building up to two stories high can be realized, providing a concrete ring beam with two reinforcement bars which will be inserted between the first and the second story, each story being 120 inches high. With these reinforcements you may go upto six stories in most seismically safe places.

The Terrablock System begins with the most basic, plentiful and inexpensive raw material on the face of the earth : common soil. Soil with a natural moisture content of four to fourteen percent is ideal for theTerrablock System. The moisture is only needed to act as a lubricating agent between the soil particles while they are compressed.

Finished Terrablocks formed by hydraulic pressure, upto 280 ato, are produced continuously at both ends of the machine at a rate of six to ten blocks per minute.

The Terrablock Duplex Machine is designed for simplicity of operation and maintenance. All components are treated for severe industrial usage and designed to operate efficiently and continuously in extreme dust and tropical heat.

Once on the site, the Terrablock Duplex can be set up and made ready for operation in less than 30 minutes. It is self-contained and completely mobile and it is also compact enough to fit into a standard shipping crate.

# PULSED EQUIPMENT - PROBLEMS OF CONVERSION

**Leonid Raginsky and E.I. Zaharov .**
*Research Institute for Non-organic Material, Russia.*

The scale of modern chemical industry places progressively more stringent requirements upon the apparatus which is used. The high efficiency and reliability of equipment that provides for the specified production in a single line must be also associated with high effectiveness. The latter circumstance requires a turbulent motion to be established in a reactor which permits the removal of an external diffusion barrier between reagents and conversion of the process to the sphere of chemical kinetics. It is a pity, but it is only in a liquid-gas system that it is possible to establish conditions of this type; in liquid-liquid, liquid-solid and even liquid-solid-gas systems the rates of reagents do not permit the above regime and additional power must be supplied to the reactor. For this purpose mechanical energy is used. In most cases an agitator is introduced into a reactor and pneumatic mixing is also sometimes employed. The principal disadvantage of these methods is a nonuniform distribution of power across the section which limits the volumes and efficiency of the reactor.

In our country, vibrational and pulsed extraction columns appeared in 1960 for the atomic industry. However, later pulsation was found, realized and is being developed now for the intensification of a great variety of processes in different apparatuses 1.7%. The spheres where the developed apparatuses are presently commercially operated are shown in Fig.1.

As a rule, pneumatic pulsators with a control-valve mechanism are used as drive for pulsation apparatus (Fig.1.)

The diagram of connecting up the apparatus 1 to the pneumatic pulsator 2 is similar for all pulsators. Pressure pulses agitate the liquid in the apparatus and as a result the process is intensified.

The pulsator is an integral part of any pulsation apparatus - a reactor, a column, a filter, a pump, a mixer-settler. As a rule, the pneumatic pulsator in all the above mentioned apparatus is used with a drive gear.

The pneumatic pulsator is never subjected to direct contact with reagents and the pulsing system is an ecologically clear system for supplying energy in clean chemical processes. Its construction is simple, and its size is considerably smaller than that of a piston pulsator.

Since the apparatus is connected to the pulsator only by means of an air main, the pulsator can be doubled if necessary, which is impossible with mechanical pulsators. Only the pulsator has moving parts. The apparatus does not have any other moving parts.

One of the important results of the developments of pulsed equipment was production of multipurpose column reactors which can serve as the basis of further developments of many new chemical industries and intensification of those in operation. In columns KRIMZ packings are employed (1.2). This is a packing with perforated plates having numerous rectangular nozzles of a relatively large section (from 10x20 to 100x200 mm depending on the diameter and purpose of the column) (Fig.2).

Each flow from a single nozzle sets in motion the continuous phase producing numerous microvortices. As a result, the total volume above and under the plate is mixed. The interaction between the streams sets the liquid on a plate in rotary motion producing a homogeneous flow structure --uniform radial mixing that provides the uniform distribution of reagents across the section and relatively low longitudinal mixing.

In all pulsed columns, independent of their purpose, packings of KRIMZ type are employed, which allow work with two-phase and three-phase systems. For each system, the optimum hole sizes have been found. In processes which need reagent breaking, e.g. for extraction in a liquid-liquid system, they are usually not more than 40x80 mm and in a system containing a solid phase they can be increased to 100x200 mm.

The fact that it is possible in a single column, under continuous conditions, to perform processes formerly conducted in batch tank-type reactors or in their cascade to ensure continuity, is a great advantage of pulsed technology. As can be seen from the table (fig.3) columns can be used for the full dissolution of solids to prepare reagents for the extraction of the required component from a solid leaching, washing valuables from solid and liquid impurities, precipitation from solutions, their homogenization and so on.

The column flow hydrodynamics ordered through KRIMZ packing and the developed turbulence ensure the absence of stagnant zones, effective reagent contact, the minimum time of the reagent contact close to that theoretically required and as a consequence, minimum expenditures for their manufacture and operation.

Washing, i.e., separation of the main liquid or solid product from impurities, that was more often conducted in several tank-type reactors operating in series, is successfully accomplished under continuous counter-current conditions in a single column; in this case washing sewage is reduced by a factor of 5-10.

Columns operating with a solid containing system have found a wide use. In particular counter-current columns are successfully used for leaching (extraction from a solid for medicine; e.g. in cyanidation of gold-containing ores, one column replaces 3-4 reactors with pneumatic mixing that have by a factor of 5-6 higher volume. In that case we eliminate gas wastes. In this case the extraction of gold increases by 2.5-3%. Sorption counter-current columns have been developed and are widely used at hydrometallurgical plants in several versions, in processes with a quick interaction between a sorbent and reagents, including suspensions; the PCK - a column having a common continuous pulsation columns 2.4 m dia operates throughout up to 200 m/h.

In hydrometallurgy slurries are classified and sludges are separated from sands in pulsed columns upto 1.5 m dia, that replaced cumbersome classifiers of different types (3).

For sewage treatment, sorption columns 3.4m dia are employed, having the flow rate of 300 m/h of solutions containing low concentrations of a component to be extracted. In chemical industry at some plants the clearing from mercury and chlorophos was managed with the help of an extraction column 1.2 m dia which resulted in significant ecologic effects (7).

In the well-known mixer settlers the reagent contact and transport are effected by pulses devices viz. centrifugal and valve pumps (discussed below) ejectors etc. that replaced mechanical mixers. Regulating the intensity of pulsation makes it possible to produce "reduced" mixing.

This kind of mixing provides an optimized drop size and as a consequence an increased rate of separation and hence, a reduction of the settler area. Besides, this eliminates the need for the maintenance of a great number of devices inside an extractor, as they are replaced by built-in ones, the operation of which is effected by a single device-pulser that services 10-20 sections (fig.4), so we have fully ecology -safety equipment.

The pulsation effect on the filtering surface made it possible to realize the old idea of permanently cleaning filters, in which pulsation is employed for cleaning the filtering surface from solid particles that stick to it (fig.5). In this case the regeneration of a filtering material that is repeated at the frequency of pulsation permits a continuous operation of filtering facilities that were formerly often stopped for cleaning or replacement of a filtering surface.

Pulse filters of the area 30-50 m provide for the continuous filtration of sugar syrup to produce the required condition 2.

To intensify the operation of the tank-type reactors a series of pulsed mixing devices have been developed. They form a part of the reactor structure where the energy of pneumatic pulses arriving from the pulser is transmitted to streams that turbulize the reaction space and ensure the high level of suspension and uniformity (homogeneity) of mixing 2 (Fig.6).

Pulsed mixing can be effectively used in tank-type reactors containing developed heat-exchange surfaces, where it is difficult and sometimes impossible to place a mechanical agitator, as well as in processes employing highly aggressive media requiring nonmetallic materials (ceramics, plastics etc).

Pulsed mixing devices in high volume (500-3000 m) tank reactors can be effectively used in a petroleum product storage to prevent some fractions and separation and to eliminate precipitation and accumulation of deposits in a NPP waste storage. In this case the use can be of low frequency pulsations and air from high pressure blowers (instead of compressors) which provides low operational expenditures for mixing.

One more sphere where pulsed mixing can lead to a significant intensification is electrochemical processes (oxidation-reduction and electroplating process etc) where, due to the fact that the reaction space is occupied with electrodes, mixing is as a rule absent, although it is known that it appreciably intensifies these processes and improves the product quality.

For processes conducted at an excessive pressure, the usual scheme of pulsation proves to be not only ecologically unacceptable but also energetically disadvantageous. In this case, the most economic efficiency is achieved in a closed symmetric scheme of pulsation for two paired devices. The oscillation of liquid takes place in an antiphase.

This system of pulsed mixing investigated for column devices, viz. autoclaves at a pressure to 2.5 MPa, is presently used in cellulose production (7).

Depending on the process character air, inert gases or water steam can serve a working minimum in pulsers. Steam employed to transfer pulse energy from a pulser to reagents proved to be very convenient for processes conducted at elevated temperatures and slightly excessive pressure (to 0.3 MPa).

The application of pulsed devices to technological lines of chemical processes was aided by pulsed valve pumps, the establishment of which became possible due to the usage of a special form of oscillation of pneumatic pulses in a pulser and a soft fit of a valve to a seat (2). Both these provisions promote a shockless fit of a valve, its long-term (more than 8 years) failure-free operation and the high reliability of a pump as a whole. Pumps developed on this basis, having different efficiencies (0.5 - 50m) are capable of transporting a great variety of reagents at the lead up to 30-50 m (fig.7). Nonmetallic materials used for the main units of the pump enable transporting any acids (e.g. hydrofluoric one), different suspensions and solutions contaminated with abrasive suspensions. The high reliability and gas lightness of the pumps, no moving parts (with the exception of a valve), a remote pneumatic drive from a pulser and low energy expenditures (3-5 m air per 1 m of liquid pumped per one hour) promote their introduction into a great variety of chemico-technological and hydrometallurgical processes.

Summary : Pulsing equipment, which was developed for atomic industry, may be used for different civil branches of industry very successfully (pharm industry, medicine, purification of pig-farm wastes, water, dangerous gas wastes, decontamination of soil, food industry, to produce fertilizers, etc.).

## REFERENCES

1. Karpacheva S.M. Zachrov E.I. Pulsing Column Reactors Basis Theory, 1980, Moscow.

2. Karpacheva S. Raginsky L.S., Muratov V.M., Pulsing Horizontal Reactors Basis Theory, 1981, Moscow.

3. Pulsing Apparatus in Industry Part I, Moscow, 1979, p.180.

4. Pulsing Apparatus in Industry Part II, Moscow, 1979, p.172.

5. Pulsing Apparatus in Industry, Part III, Moscow, 1979.

6. Pulsing Equipment, 1983, Moscow, p.153-163.

7. Karpacheva S.M. Raginsky L.S. Horhorina L.P. J.Pr.Chem. 1986, N.9, p. 1953-1962.

**Commercially used pulsed apparatus**

| Apparatus | Purpose | Characteristics | |
|---|---|---|---|
| | | Max. dia(m) | Min. eff. (m/h) |
| Columns for liquid-liquid system | Extraction, stripping, washing out impurities, sewage treatment | 2.0 | 80 |
| Columns for liquid-solid system | Sorption from solution<br>Sorption from suspension<br>Desorption & washing sorbent<br>Washing solid from impurities<br>Leaching<br>Sorption treatment of sewage<br>Extraction from structures | 3.4<br>1.5<br>2.0<br>3.1<br>2.4<br>3.4 | 300<br>20<br>30<br>15-25<br>50<br>300 |
| Continuous column reactors for different processes | Synthesis and Polymerization<br>Dissociation of solid<br>Reagent mixing<br>Powder and suspension classification | 0.2<br>0.6<br>0.2<br>0.9 | -<br>1.0 t/h<br>4.0<br>16 t/h |
| Pulsing and vibro columns for three-phase systems liquid-solid-gas | Chemical reactions (e.g., cyanidation of gold-containing suspension) | 2.4 | 50 |
| Column autoclaves for pressurized processes up to 2 MPa | Leading from ores production of cellulose from chips | 2.4 | 40 |
| Mixer-settlers | Extraction, washing, stripping | - | 50 |
| Filters with pulsed and vibro regeneration | Thickening suspension to S:I=1:3 | Filtrn. area upto 40m | 0.2-2m/ (m/h) |
| Tank reactors of different types including tubular, of special configuration (pulsing and vibro) | Decontamination<br>Mixing heavy suspensions<br>Mixing emulsions<br>Mixing solution | 1.0 m<br>1.0 m<br>20 m | - |
| Pumps | Supply and metering suspensions to s:I-1.5 (including corrosion media) | - | to 60 |
| Electrolyzers | Change of element valency and increased efficiency of current; use of cascades of electrolyzers in continuous process<br>Development of pulsed conditions in technological units | 0.3 m<br>- | -<br>- |

Volume of apparatus

# THE FOREIGN POLICY ASSOCIATION

## THE CENTER FOR ARMS CONTROL AND STRATEGIC STABILITY

The Center is set up as a non-government, non-profit research organization, acting as a department of the Foreign Policy Association, headed by E.A. Shevardnadze.

**Composition of the Center**. The Center is formed on the basis of the military-political and disarmament departments of two institutes of the Academy of Sciences of the USSR - the Institute of World Economy and International relations and the Institute of the United States and Canada.

President - Dr.S.M.Rogov, Director - Dr.A.G. Arbatov.

Advisers - Dr.J.A.Ivanov, Dr.A.A.Konovalov, Dr.G.K.Lednev, Dr.V.A.Masing, Dr.S.A.Oznobishev, Dr.A.S.Kozlova, Dr.A.N.Kaljadin, Dr.O.A.Amirov, Dr.A.V.Surikov and Dr.J.E.Pinchukov.

**Aims and tasks of the Center**. With radical shifts in the military-political situation in the world and profound changes in the country, the center sees its task in elaboration of fresh approaches towards security and strategic stability, on a global level and towards further reduction of and structural changes in the military potentials of states. Bearing in mind the enormous importance to international security of the internal processes taking place in the former Soviet Union, the Center intends to carry out research projects designed to assess the political and strategic consequences of the disintegration of the Union. Such research should, on the one hand, help the world community to obtain an adequate picture of what is going on inthe country so that it can elaborate the best ways to assist in the processes of demilitarization and democratization in the USSR and, on the other, facilitate the conjugation of the republican concepts of security and the organization of their own armed formations with the obligations of the USSR related to arms limitation and the principles of international security.

In view of the vital importance of internal stability in the USSR for international security, the Center plans to assist the Russian Union authorities in taking practical steps towards setting up an effective mechanism of civilian control over policy-making in defence matters, the defence budget and military programs, social provisions for military personnel and demobilized officers and conversion of the defence industries. High priority of the center will be to work out the fundamental concepts of the strategic security and the foreign and military policy of Russia.

**The activities of the center**. The main activities of the center will consist of carrying out research projects on a contractual basis on behalf of interested persons or organizations, both at home and abroad. A number of papers will be prepared on current security problems in the light of internal developments in the former USSR, as well as a series of other publications. In view of the wide range of problems to be studied, the center intends to form, in addition to its permanent staff, working groups of highly-qualified experts for specific projects of foreign organizations interested in research on given subjects.

The center also plans to organize international seminars and republics of theformer union will be invited in order to participate in a wide and free exchange of views guaranteed by the independent and non-governmental status of the center. Whenever required the center will organize contacts with Soviet and foreign partners who are interested in making practical use of its recommendations in the field of politics and economics.

# THERMIONIC ENERGY CONVERSION, SPACE TECHNOLOGY FOR ENERGY CONSERVATION

**Dr. Lodwijk Reiner Wolff and Dr. ValeryIvanovit Yarigin**

*Energy Conversion Systems B.V., The Netherlands*
*The Institute of Physics and Power Engineering, Russia.*

## Introduction

Only a few years ago, Obninsk could not be found on any USSR map. Still it is a city of 150,000 inhabitants who are employed by about ten different institutes. Most of these institutes are related to nuclear science. There is for e.g. a hospital which specializes in radiation therapy and treatment of radiation diseases. There is a meteorological institute, but one of the most important Obninsk Institutes is without doubt the Institute of Physics and Power Engineering (IPPE). The IPPE was founded in 1946. In 1951 it was charged with the design and construction of the world's first nuclear power plant. On June 27, 1954 this nuclear power plant came into operation.

Since then the IPPE, which presently employs about 6,000 scientists and technicians, has achieved a lot in nuclear physics and (fast breeder) reactor design. In the sixties, the IPPE started developing thermionic nuclear reactors. A Thermionic Energy Converter (TEC) is a devise, which directly, that is without moving parts, converts heat into electricity. This makes a TEC very attractive for nuclear power applications in space. The TEC principles are explained in Fig.1.

In 1970 the first thermionic fuel element of the TOPAZ type passed its in-core power test at the IPPE. Since then, the TOPAZ thermionic nuclear reactor has been designed and tested at the IPPE. Several flight models of the TOPAZ reactor werebuilt, two of which were launched and successfully flight-tested in "Kosmos" satellites. Fig.2 shows a flight model of the TOPAZ thermionic reactor.

About 300 scientists and technicians at the IPPE are presently involved in thermionic research. This makes it by far the largest thermionic group in the world. A few years ago, this thermionic group started a separate line of research on combustion heated thermionics. The idea was that, although there might still be an opportunity to apply the nuclear thermionic technology to civil application in space, it would be a good idea to explore the potential of terrestrial application of combustion heated TEC's.

In 1989, Dr. Yarigin, head of the Thermionic Laboratory of the IPPE, approached Dr. Wolff of Eindhoven University of Technology at the Intersociety Energy Conversion Engineering Conference in Denver USA and asked him to co-operate with the IPPE in developing terrestrial thermionics. Dr. Wolff and his team had been working in this field since 1978 and had successfully developed a composite material for the "hot shell" of a combustion heated TEC. The hot shell is the case of the TEC which protects it against the combustion atmosphere and at the same time acts as the emitter electrode of the TEC. Fig.3 shows a preliminary design of a water cooled TEC.

Dr. Wolff accepted the proposition and invited Dr. Yarigin to participate in a Thermionic specialist meeting at Eindhoven University. During the specialist meeting Dr. Wolff showed Dr. Yarigin the thermionic research facilities at the University and explained the technology involved. Only recently Dr. Wolff learned that Dr. Yarigin was very surprised at the time to be allowed to see everything, ask questions and even get a response to them!

Dr. Yarigin in turn invited Wolf and his co-workers Gubbels and Veltkamp to visit the IPPE. A seminar was organized at which a number of presentations were given by IPPE specialists. The three Dutch scientists also gave a presentation each. Especially Wolff's presentation on the cost aspects of mass produced TEC's, part of a feasibility study that he had recently completed on terrestrial thermionics, got a lot of attention. Finally a letter of intent was signed to start a formal co-operation in the field of terrestrial thermionics. Since the aim was not merely to join research efforts, but also to commercialize thermionics, Wolff decided to set up his own company to this end : ADENCO -- advanced energy conversion.

In 1990, representatives of the IPPE thermionic group came to Eindhoven and signed a co-operation agreement with ADENCO. At the same time a protocol was signed stating that ADENCO and the IPPE would establish a joint venture for the commercialization of terrestrial thermionics. It was decided that it would be best to set up this joint venture according to Dutch law. The reasons being, that the laws of the USSR were changing day by day. Furthermore, most of the financing of the joint venture would have to come from western countries. It was felt that a Netherlands based joint venture would have a much better chance of obtaining thenecessary funding than a Soviet based one.

A business plan was drafted and an agreement was made to hire IPPE personnel to carry out the necessary R & D and to set up a pilot plant. Finally, an IPPE delegation was invited to visit the Netherlands in August 1991 and formally establish the joint venture Energy Conversion Systems B.V. in Eindhoven. These were very hectic times, considering the fact that a coup took place in the USSR only a week prior to the IPPE visit. Everything finally turned out just fine, since the coup failed. But for the Minister of Atomic Energy and Industry, Dr. Konovalov and his Deputy Minister Dr. Kulikov, the complete USSR delegation could be present in Eindhoven. The official documents were signed in the Eindhoven Town Hall by Dr. Kuzin on behalf of the IPPE and by Dr. Wolff on behalf of ADENCO in August 1991 (See. Fig.4).

Energy Conversion Systems B.V. is owned by the IPPE (49.75% of the shares), ADENCO (49.75% of the shares) and Dr. Wolff (.5% of the shares). This way IPPE and ADENCO each own about half the company while Wolff as General Director of Energy Conversion Systems can cast a decisive vote in case the votes are equally divided.

**Research and Development concerning TEC's**

The R & D planning includes the development of combustion heated TEC prototypes to be tested and demonstrated. Most of this R & D work will be carried out at the IPPE laboratories. In all, 135 scientists and technicians (50 full time equivalent) will be involved in this work. It will include setting up a pilot plant facility, capable of manufacturing upto 100,000TEC's a year. By the end of 1993, over a thousand TEC's are to be manufactured. They will be subjected to life and performance tests and they will be used in prototype TEC-systems.

A demonstration type combustion heated TEC was already completed in 1991 (see fig.5).

The present TEC design is based on a design by Eindhoven University of an air cooled heat pipe collector TEC (See fig.6).

Energy Conversion Systems will develop the mass production technology for combustion heated TEC's in close cooperation with an industrial partner with a vast experience in the field of mass production technology. The TEC production line will be highly automated. The first facility which will be operational in 1995 should have a capacity of 500,000 TEC's a year. The planning calls for a steep rise in production. In the year 2000, the ultimate production level of 16 million TEC's/yr should be reached.

Energy Conversion Systems in Eindhoven will also be responsible for most of the R & D concerning TEC-systems. Again this activity will be carried out in close cooperation with industrial partners.

As the industrial partners of Energy Conversion Systems already control a substantial part of the future market, this will have the advantage of a smooth introduction in the market, using their sales, distribution and service networks.

**Research and Development concerning TEC-systems**

Several TEC-systems will be developed in the period till 1995 when the production will start. These systems are :

1. - The TECTEM
2. - The GTB TECTEM
3. - The TECBURNER
4. - The SOLARTEC

The TECTEM[(a)] is a central heating facility, which will also produce some electricity. It is a small cogeneration unit which typically produces about 20 kW of heat for central heating and tap water and about 1.5 kW of electricity. The TECTEM[(a)] consists of boiler separately heated by an array of 10 TEC's. Each TEC in turn is heated by its own radiative burner/recuperator. The TECTEM[(a)] is equipped with a solid state invertor which converts the low voltage DC electricity, generated by the TEC's, into the normal AC electricity before feeding it into the grid.

Apart from the small TECTEM[(a)] to be used in houses, there will other ones that store electricity in batteries. These can be used in many places independent of the grid e.g. in ships. Larger TEC's will be used in hospitals, department stores, homes for the elderly and public swimming pools. The TECTEM's will function as cogeneration systems in the range below 150 kW which is unattractive for conventional systems because of maintenance costs. The TECTEM's have the distinct advantage that since they have no moving parts, they have inherently low maintenance costs and they produce very little noise.

Design and development of the TECTEM[(a)] will be carried out in close co-operation with industrial partners, one of which is a manufacturer of central heating facilities. The other is a manufacturer of DC/AC invertors.

The GTB TECTEM[(a)] is a thermionic cogeneration system for greenhouses. It basically provides hot water for heating greenhouses and steam for disinfecting thegreenhouse soil. It also provides $CO_2$ for fertilization. Finally it provides electricity which can be used for artificial light in order to prolong the day, or when it cannot be used by the farmer it can be fed into the grid. Many of the components to be used in the GTB TECTEM[a] are identical to the ones being used in the TECTEM[a] such as the TEC's and the radiative burner/recuperators. The Inventer/power management system will be a larger, more sophisticated one than the ones used in the TECTEM [a]s. The GTB TECTEM[a]. The GTB TECTEM can be used in combination with conventional co-generation systems. The GTB TECTEM will provide additional power in a low power range (<150 kW) where conventional cogeneration systems are unattractive because of the maintenance costs. The fact that the GTB TECTEM[a] flue gas can be used for $CO_2$ fertilization is an additional advantage.

The GTB TECTEM[a] will be developed together with a boiler manufacturer which serves the greenhouse industry and a company that provides all the equipment needed for greenhouses.

The TECBURNER[a] is an industrial burner equipped with an array of TEC's. It produces process heat for low temperature processes, while at the same time producing some electricity.

The TECBURNER is a thermionic cogeneration system which is optimally geared to generate the process heat required. Thus a conventional industrial burner can simply be replaced by a TECBURNER[a]. As a result, the installation time is attractively short compared to conventional cogeneration systems. This in turn has a very positive effect on the return on investments.

The TECBURNER[a] will be developed in cooperation with a major manufacturer of industrial burners.

The SOLARTEC[a] is a TEC system which is powered by solar heat. In fact two systems will be developed : a small system and a large system. In the small SOLARTEC[a] system, 6 TEC's will be heated by the sun via a facetted parabolic mirror. This system may provide electricity for communication purposes in remote tropical or subtropical places. The larger SOLARTEC[a] will serve as a topping cycle for solar tower plants. In a solar tower plant, a field of flat, sun tracking mirrors reflects the solar radiation to a large facetted parabolic mirror. This mirror in turn will focus the radiation on a solar furnace system like a steam or a Brayton cycle. Thermionics can be used as a topping system in order to raise the overall efficiency. The same approach is possible in the second type of solar tower system in which a field of sun-tracking parabolic mirrors focusses the solar radiation directly on the furnace on top of the solar tower.

Development of these solar systems is carried out primarily by ECS Russia. At this stage it is still unclear whether these SOLARTEC[a] applications are economically viable.

**Market Survey**

A market survey concerning the Dutch market for thermionic systems was carried out by PWK, an institute specialising in cogeneration. The results show that in the Netherlands ultimately 4910 $MW_g$ in thermionic systems could be installed. This estimate is based on TECTEM[a]'s for homes (3000 $MW_g$), Larger TECTEM[a]s for offices, hotels, homes for the elderly, hospitals, swimming pools, ... (720 $MW_g$), GTB TECTEM's (1015 $MW_g$) andTECBURNER'S (175 $MVV_g$). When this is extrapolated to the world market, it means that the market will be limited primarily by the production capacity of Energy Conversion Systems.

In the homes, TECTEM[a]'s could provide about 60% of the electricity used. In the greenhouses, thermionics could provide 2.3 times as much electricity as is needed by the greenhouses themselves.

**Discussion**

Energy Conversion Systems is a fine example of co-operation between a major Russian Institute that was until recently involved in defence oriented R & D and a Dutch company which knows how to set up a new business.

This co-operation has led to a joint venture which will develop, mass produce and sell an innovative product, which will save energy and reduce, $CO_2$ - $NO_x$ - emission.

The market for the products is almost unlimited and if the penetration of this market is successful, this company will not only provide employment for a number of excellent Russian scientists and technicians formerly working for defence, but also set a good example of commercial enterprise in the former USSR.

# ENERGY CONVERSION SYSTEMS B.V.

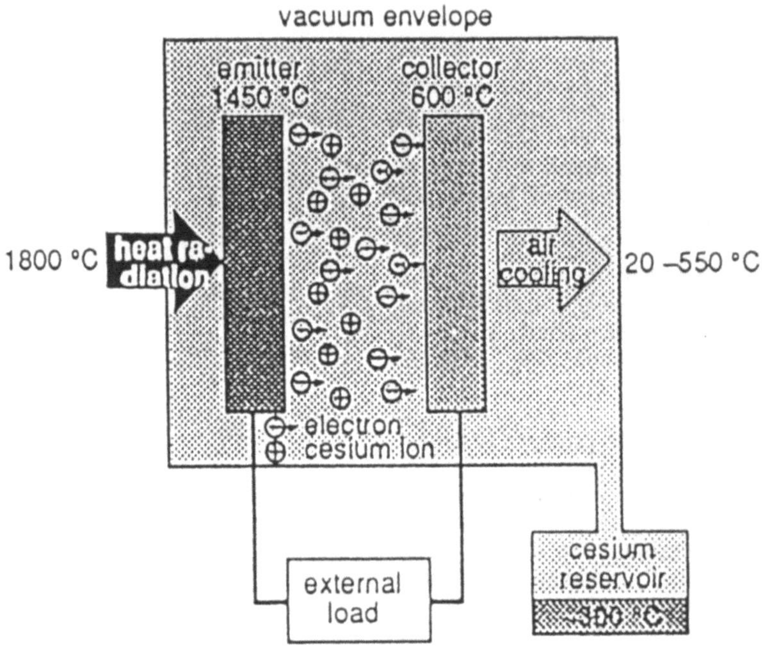


*Fig. 1. Principles of a thermionic energy converter (TEC).*

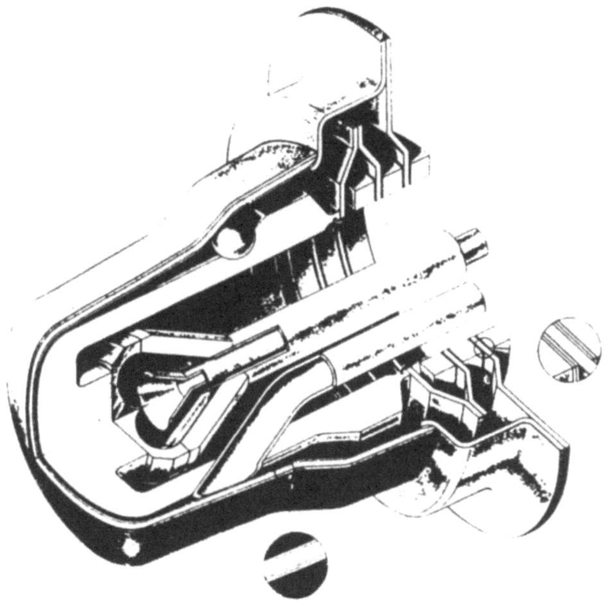

*Fig. 3. Conceptual design of a water cooled TEC of Eindhoven University of Technology, Eindhoven, the Netherlands.*

# ENERGY CONVERSION SYSTEMS B.V.

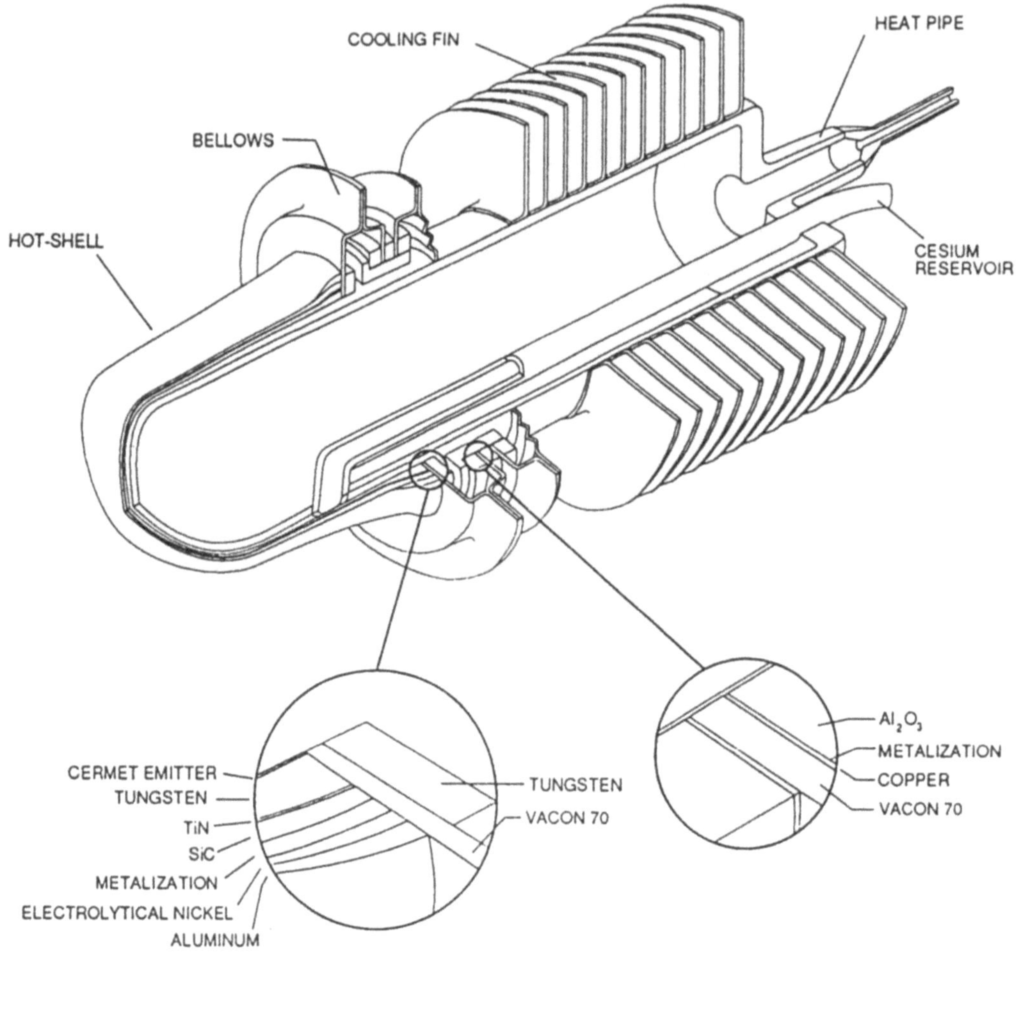


*Fig. 6. Air cooled heat pipe collector TEC, designed by Eindhoven University of Technology.*

# UNITED KINGDOM

# NATIONAL CASE STUDIES IN CONVERSION: THE UNITED KINGDOM

**Paul Dunne and Sue Willett**
*University of Leeds and Centre for Defence Studies*
*Kings College, United Kingdom*

## Introduction

As a country with a historically high defence burden and a history of relative economic decline, the changing strategic environment has considerable implications for the UK. The end of the cold war has led to defence cuts and a restructuring of the defence industry, and promises of more cuts. The pressures upon the defence budget have a longer history, however, with the UK seeking a decline in military spending since the mid 1980s. In addition, there was the application of an industrial policy based upon "Thatcherism", which was intended to open the defence industry to competitive forces. These factors have already led to a considerable restructuring of the defence industrial base and together with responses to the Single European Market initiative, a restructuring of the European arms industry.

The changes in government policy towards the defence industrial base in the UK represents an attack on the cost relationship between the state and defence companies that characterize most economies. This does not mean that the process is by any means complete. The UK still has a high burden of military expenditure relative to its competitors and further cuts are planned. What it does mean is that the UK's experience provides a considerable amount of information on the economic impacts of cuts in military expenditure and the defence industrial restructuring.

There are clear benefits to this process, but care needs to be exercised to manage the structural alignment of the economy in the wake of large scale defenceexpenditure cuts. Increasingly, studies have shown that while a "peace dividend" is available in principle, it can only be achieved by careful planning of the required structural adjustments, see Barker, Dunne and Smith (1991). So far the UK has not done this and would appear to be missing an important economic opportunity.

This paper presents an analysis of the UK experience, considers what can be learned, and critically evaluates policy proposals. The next section looks at the trends in military spending and the changes in defence policy in the UK. Section 3 then considers the changes in UK's industrial policy, followed by an analysis of the effect of these policies on the UK defence industrial base in section 4. Section 5 reviews the possible policies for transition to lower military spending and section 6 presents some conclusions.

## Defence Policies

There has been a considerable degree of continuity between Labour and Conservative government's defence policies in the UK. The Conservative government of 1979, led by Margaret Thatcher, apart from privatizing the arms firms, broadly followed the trends of the previous Labour government's defence policies. After 1985, however, defence spending was cut sharply and competition introduced into procurement. This change in direction was an important contributory factor in making military production less profitable and prompting a massive restructuring and internationalisation of the UK arms industry, a process which is still continuing at both a UK and at a European level.

The broad trends in spending are evident from Figure 1, which shows the shares of total military spending and spending on military equipment as a percentage of GDP. The break around 1985 is very obvious. Till then the Conservative Government had honoured the commitment to expand real defence spending by 3% a year given to NATO by the previous Labour Government. Figure 2 shows that the other main powers in NATO did no follow this policy, except for the US. It also illustrates that the decline in defence burdens was common to all the countries in the late 1980s. There were other elements of continuity -- the commitment to Trident and the four main strategic commitments of independent deterrent, European ground and air forces, North Atlantic maritime forces, and defence of the home base. Baylis (1989) and Roper (1985) provide a more detailed discussion of British defence policy.

In 1984-5 the policy of reducing public expenditure was finally applied to defence. The defence budget fell in real terms by 10% between 1984-5 and 1988-9, that is prior to the ending of the Cold War. As Figure 1 shows, military expenditure as a share of GDP peaked at 5.2% and dropped to 3.8%, by 1990. In 1990 the share of national income devoted to defence was the lowest since the 1930s. The 1990 Autumn Statement on Public Expenditure, following the "Options for the Change" Defence Review, anticipated the decline continuing by a further 6% in real terms by 1993-4, leaving it at less than 3.5% of GDP. The Gulf War has done little to disturb the downward trend, despite the fact that the lobbying for the defence interests has had some impact in scaling down proposed cuts, in particular in the reorganisation of the army.

As Table 1 illustrates, along with the decline in military expenditure in real terms in the 1980s, there has been a marked declinein the share of expenditure on equipment, with a slight increase in that going to personnel. Thus industrial demand from defence is showing a clear declining trend. This trend is reflected in the figures for employment generated by defence spending, both direct and indirect, shown in Table 2. Direct employment related to equipment expenditure has declined, particularly in the late 1980's, with non-equipment expenditure showing a less marked decline. See Dunne (1991) for a discussion of the basis of these estimates.

There has also been a clear variation in the trends across regions, as indicated in Table 3. Defence expenditure has been heavily concentrated in the wealthier regions the SE and the SW. While the SE has seen a marked decline in its share of expenditure and employment, the SW has been surprisingly resilient. The North has held its ground, leading to increased shares, as have Scotland and for rather different reasons N.Ireland. Within these regions there has, however, been considerable variation in the impact on towns and communities of cuts and the general restructuring of the defence industry.

**Industrial Policy**

The period before 1985 saw one major shift in policy towards the defence industry: the privatisation of arms producers. This was not a distinct defence choice but a consequence of a general economic policy, however it had major consequences for military production. In 1979--80, the arms industry was dominated by state-owned (British Aerospace, Rolls Royce, British Shipbuilder's and Royal Ordnance) and smaller amounts were paid to other state owned firms. They are now all private. Rolls Royce and BAe were floated as independent firms, and the remaining government stake in BP sold. BAe acquired BL and the bulk of Royal Ordnance; theBritish Shipbuilders warship yards, which were profitable, were sold off individually; Short Brothers was bought by Bombardier of Canada. In addition private contractors were brought in to operate Government owned facilities such as the Royal Dockyards. With the partial exception of Frigate production, the privatisation left all the monopolies intact.

Initially, the Conservatives continued the Labour policy of using cost-plus contracts and preferential purchasing to maintain a British Defence Industrial Base (DIB). In 1984, however, a new procurement policy, introducing market forces and competition into defence procurement was implemented. There was competition for contracts, fixed-prices rather than cost-plus, payments were tied to performance targets, and risk was shifted onto the contractor. It was initially anticipated that this competition would only be notional, since most major weapons systems were produced by monopolies. to make competition effective required allowing foreign firms to compete and this was perceived as politically unacceptable and, despite strong political opposition, from MPs of both parties, trade-unions and domestic monopolies, the share of imports did rise significantly. Even in cases where British equipment was eventually purchased, having real competition from abroad appears to have driven down prices.

One must be careful not to overstate the changes that have taken place as their effects do vary across industries. As Hilditch (1990) argues the shipbuilding industry, for example, had its moves to competitive procurement compromised by a concern for employment policy in the regions, though the effectiveness of procurement as a means of maintaining employment was also limited.

Influential studies such as Taylor and Hayward (1988) have argued the need for the support of the defence industrial base,suggesting that successful companies should not be endangered by heedless regard for a narrowly defined concept of "value for money". In addition, there have been various failures and lobbying for defence contracts has continued on a large scale: it was probably a significant factor in the 1988 and 1991 decisions in favour of the British Challenger II tank over its foreign competitors.

The competitive procurement policy, was promoted within Europe through pressure on the Independent European Programme Group (IEPG), the organisation designed to coordinate European arms procurement. This and company responds to the coming Single European Market led to a restructuring of the European arms industry. The UK has not only been fairly relaxed about acquisition of UK defence firms by foreign companies, like Siemens, but it has opposed extending the EC Common External Tariff to weapons and EC policies which would restrict US ownership of European arms firms. Given the size of UK arms exports to the US, the dependence of the UK on imports from the US for its nuclear systems, and the number of UK defence subsidiaries in the US, this position is understandable.

**The UK Defence Industrial Base**

Studying the development of the defence industrial base is not a straightforward exercise for a number of reasons. There are problems of finding relevant data and knowing whether to believe it or not. The companies involved in military production are some of the largest in the UK and do not necessarily specialise in military production, making it difficult to find information on the relative importance of military production. There is likely to have been considerable cross subsidisation of product lines from government contracts andcompanies are very sensitive about such information. In addition, there has been considerable industrial restructuring taking place in the UK and it is not all connected with military spending. The changes that have taken place in the international situation, the cuts in defence spending, actual and expected, the changes in procurement practices, privatisation, the single European market, and the increasing competition from second string producers, mean it can be very difficult to disentangle the effects which are specific to any individual factor.

Nevertheless, undertaking such an analysis, even with caveats, is important as it is only by studying past adjustments of the defence industrial base to procurement changes that we can get information on the likely impact of future reductions in military spending.

One can characterise the defence industries in most advanced economies as being highly concentrated, with companies having close links to the state. This has led governments to protect the companies on strategic grounds and has shaped the nature of the companies. The symbiosis inhibited internationalisation of capital in the defence industry and cross country division of labour. True multinational characteristics of most other industries were conspicuously absent in the arms industry, despite large economies of scale.

Fixed costs, particularly R&D, account for about a third of total costs in military projects and there is a steep learning curve, thus the cost of a weapon is largely a function of the length of the production run. The consequence of national preference was short production runs, general excess capacity and higher prices for military equipment, Smith (1990). There was also a particular set of relations of reproduction which led to complex, expensive andimpractical weapons, a baroque arsenal, Kaldor (1983). The rising R&D costs and falling defence budgets have made this position unsustainable, and provoked the massive restructuring of the European industry, Lovering (1990), Walker and Gummet (1989).

To provide some information we created a sample from the Datastream company data bank. The sample used is given in Table 4 which is from ACOST (1989) where, unusually, information on the share of turnover which was defence related is reported. In what follows, the composition of the total sample might differ slightly because of missing data.

The more competitive procedures introduced from 1985 implied higher risks and lower returns for the UK arms producers, and this was reflected in their stock market valuations where their share prices were heavily discounted. As Figure 3 shows, however, the story, looking at the monthly share prices for all the companies except Rolls Royce, is not so simple. The decline in defence share prices relative to the all FT all share index has been apparent since around 1982. Nevertheless, a sharp drop is evident around 1985. The individual companies did, however, have rather different experiences.

As regards competition, evidence suggests that concentration increased substantially with the nationalisation of BAe and British Shipbuilders by the previous Labour government, remained roughly constant from 1980 until 1986-7 when it fell sharply, remaining lower for the next two years. See Dunne and Smith (1992).

It is also difficult to measure profitability in defence production, given creative accounting and the multiproduct nature of many of the companies, but one would expect that it was probably similar tothe industrial average around 1985 and then feel somewhat relative to the average. Using the share of profits (published pretax) in total sales for our sample of companies does indeed provide such a pattern, as shown in figure 4.

Some firms responded to this increased competition and reduced profitability by diversifying out of defence, but diversification by defence companies is difficult and dangerous, the skills required by the military and the commercial markets are quite distinct. This led firms to manocurve into positions of monopoly, either through domestic or international mergers and competition. See Dunne and Smith (1992) for more detail.

It is clear that the defence industry faces major problems because of falling real defence budgets, falling demand for arms exports, rising development costs and increased competition. But, as this section illustrates, the problems are nothing new. A considerable amount of restructuring has already taken place and there will inevitably be more in the future. While the attack on the military industrial complex has been beneficial from a procurement perspective, it is also clear that the "hands off" approach being adopted by the UK government is creating considerable adjustment problems for the defence industry and defence dependent communities. See Dunne and Smith (1992). It is, therefore, important to ask what the likely future impacts are to be and how the transition can be managed better. The risk is that the government will either end up bailing out the companies and recreating the past inefficiencies, or the UK will simply lose those companies with high tech potential. The transition to lower levels of military expenditure will need to be managed within a more general programme of economic restructuring and regeneration.

Given the present government's commitment to *laissez faire* policies it is unlikely to use the peace dividend to aid defence industrial restructuring. A much more likely use of defence savings will be to redress the budget deficit, to provide tax cuts, or as recently illustrated, to increase other items of public sector spending for short term electoral gains. However, a change of government, following a general election in the first half of 1992, could bring about a change of policy towards the DIB.[1]

The Labour Party is committed to setting up a Defence Diversification Agency (DDA) located within the MoD, to support firms in their adjustment to civil production. The DDA, however, is not designed to be a proactive agency on conversion, but rather as an enabling agency designed to support defence companies facing difficulties with market adjustments. Its tasks would involve, researching best practice in diversification, administering a "substantial fund" within the MoD budget to aid companies to diversify, encourage new commercial investments, redirect R&D from military to civil innovation, identify market opportunities, retrain service personnel, management and defence workers and encourage management and unions to co-operate in diversification attempts. The DDA would target mainly small and medium sized companies and would work closely with local and regional authorities. In addition it would work with European partners to establish an international diversification strategy.

These policies are not without their circuits. See Lovering (1991). The insistence on locating the DDA within the Ministry of Defence, is viewed with scepticism, as the MoD has little experience in administering industrial and technology policies outside of its defence remit. A more appropriate location would be within the Department of Trade and Industry, where defencediversification would form part of an overall national industrial strategy.

### Policies for Adjusting to Lower Military Spending

Defence cuts although desirable, representing as they do an improved security environment, are not without their problems. Between 1979-89 125,000 jobs were lost in the defence industry. Since the Options for Change announcement in July 1990 the pace of job loss has escalated, a further 35,000 workers have lost their jobs between June 1990 and July 1991. In certain defence dependent localities the direct and indirect impact of defence plant closures has caused considerable economic hardship. The prospect of further job loss and plant closures has increased pressure on the government to adopt adjustment policies to ameliorate the worst excesses of defence expenditure cuts. The plethora of conversion studies gives testimony to this trend.

Conversion is understood as the transformation of military resources into civil activities and production. It is by no means a new phenomena-during World War Two conversion from the civil to the military and then vice versa was a widespread occurrence - however, the conditions which favoured conversion in the past no longer exist. Dumas (1989) has pointed out that military and civil production have widely diverged since World War II and while in the past military production generated a number of significant spin-offs into the civil economy, the present norm is the reverse. In addition, there are now a whole generation of scientists and engineers who have worked solely on military products, isolated from the competitive discipline of market forces, vital to civil innovation.

Cronberg (1992) argues that the contemporary problem inherent in militaryconversion irrespective of economic systems relates to the military industrial product concept-producing for one customer based on a "performance at any cost" principle - and the specific product development culture created by the closed nature of military secrecy. This combination, of barriers to exit from the defence market explain the fact that despite attempts at conversion, no country in the world has succeeded in converting military industries to civil production. This is in spite of the fact that military and civilian production more usually than not coexist within the same firms.

Conversion as a rather vague concept has come to encompass a multiplicity of meanings. Willett (1990), Southwood (1990) and Cronberg (1992) have attempted to classify the different approaches to conversion. Four major standards can be identified:

**Macro-economic** or socio-economic conversion focuses on the evidence of the negative relation between military spending and economic development. Melman (1974), Smith (1977) are pioneering examples. While there is still some debate, as discussed in Dunne (1991), it is generally accepted that military spending represents an economic burden and that reducing it can lead to improved economic performance. This school of thought argue that equivalent levels of investment in the civil sector creates more employment than in the military sector. Therefore the transfer of defence savings to investment in the civil economy would not only create jobs for those displaced by military expenditure cuts, but also generate more employment opportunities than previously. See Anderson, Bischak and Oden (1990). At the core of this approach is the focus on macro-economic adjustment policies.

**Plant based or technological conversion** is concerned with the transformation of resources which have already been accumulated by the military - namely defence production facilities, skills and scientific and technical knowledge and in some instances military products and components. The essential element here is the re-use of existing resources for civil purposes.

**Local community** conversion accentuates the role of local authorities and activist groups like trade unions and peace groups working together to develop strategies to reduce the vulnerability of local economies on defence expenditure. Local authority and local planning mechanisms are the key components of this approach.

**Political conversion**, an all encompassing concept, corresponds with demilitarization and implies demobilization, reduced defence expenditure and reductions in weapons production. In this context conversion means a general shift in national priorities, presupposing the transformation of resources from the military to civilian economy. The key factors in this approach are disarmament and the political decision making process.

To varying degrees all of these approaches have proponents in the UK.

Barker, Dunne and Smith (1991) provide a recent example of the macro-economic approach. The Cambridge Growth Project inter-industry model of the UK economy is used to investigate the impact of cutting military spending by one half by the end of the century. Chalmers (1991) provides a detailed discussion of the force structure consistent with such a reduction. As Table 5 shows, this implies an 8-9% per annum cut in military expenditures, which is rather more than the 6% per annum beingconsidered at present. The result of such a cut without reallocating the savings is a reduction of about 200,000 defence jobs and almost a 4% reduction in GDP. When this expenditure is reallocated to the other categories of expenditure shown in Table 5, in proportion to their share of the total before the change, 600,000 jobs are generated, giving a net increase of 400,000 jobs, as shown in Table 7, and a net increase of almost 2% of GDP, as shown in Table 6.

This evidence suggests that at a macroeconomic level there should be no problems in adjusting to lower levels of military expenditure and that with sensible problems of adjustment there can be improved economic performance. In other words there would be a tangible "peace dividend". The study also discusses the context of these changes and some of the likely problems of adjustment. In particular it presents estimates of the likely industrial impact, as shown in Table 8 and suggests where there may be problems. Underlying the simulations, however, is an implicit assumption that the defence contractors are able to adjust without government assistance and that the defence workers with their particular skills will be able to gain employment in the jobs created by the increase in expenditure in other parts of the public sector, and those created in the private sector by multiplier effects. This may in practice be problematical. In addition, the costs of troop relocation, retraining, rehousing, and the restructuring of remaining troops could be costly. Unless the policy is part of a well orchestrated plan for restructuring, the UK could risk losing what high technology industry it has and be left with considerable social disruption through regional variations. It is therefore important to consider the other approaches to conversion, within the context of the general macroeconomic picture.

**Company Based Approach**

The terms conversion and diversification are often used interchangeably but are actually quite different. Conversion means the transformation of military production facilities into the production of civilian goods and implies that a defence firm stops making military products altogether. Diversification, on the other hand, involves a company shifting into other areas of production which does not necessarily replace military work. This has the advantage of making a defence contractor vulnerable to a downturn in the defence market.

At the level of the firm diversification is the norm which can either occur in the form of mergers and acquisitions or by strengthening the civil operations of the enterprise. The main characteristic of this approach is that the decision to diversify is made at the enterprise level. Several major defence contractors have declared policies of diversification but so far they have met with limited success. British Aerospace has reduced its defence market dependency, from 60% to 30%, through the acquisition of the Rover Group. But this strategy has not been without its problems. Recent rumours about a possible takeover bid by GEC confirm that BAe, far from overcoming the problems of declining defence profits, has weakened itself through overextension.

Efforts to convert a company level through new product development, based on a conscious effort to transform military technology for civil use, are less in evidence. One such example is Racal which moved into the cellular radio market in the early 1980s, building on their technical expertise in tactical military radios. Racal's commercial success in this civil market has transformed the company from being defence market dependent to civil market oriented. Othercompanies such as Thorn-EMI and Philip's have attempted to divest themselves of their defence divisions and build upon their core commercial businesses, but so far both companies have had difficulties in finding takers for their defence business.

Apart from these isolated examples, defence contractors have shown a reluctance or inability to diversify in the present economic climate. The norm has been for defence firms to seek to increase their monopolies in defence market niches, through mergers and acquisitions, whilst shedding those defence divisions that are no longer lucrative. GEC is a classic case in point. Throughout the 1980s its merger and acquisition activity was heavily biased towards increasing its defence production capabilities.

The general failure of defence companies to adjust to the changing market environment arises from the considerable barriers to exit, which is essentially due to the orientation of defence contractors culture and management style. See Alexander (1990). These include inexperience in commercial marketing, emphasis on product rather than process innovations, a hierarchy of skills and a technological orientation structured by the weapon systems acquisition process, a management organization isolated from commercial practices, a production structure defined by large scale systems integration, protected markets through government procurement and risk aversion through government subsidies in R&D and capital investment.

**Plant-based Approach**

This is the traditional swords into ploughshares approach which aims to replace defence production capacity. It seeks technical solutions to existing production facilities and skills and tends to rely upongovernment or local government markets to guarantee survival. The most well known of such proposals was the Lucas Corporate Plan which was produced in 1975-76. Central to the Lucas Plan was the notion of reshaping technological priorities through the development of "socially useful products" which could be made utilising the existing resources and skills of defence plants. Although never implemented, the Lucas Plan provided a model for many such workers' plans and work place conversion committees. More recent successors include the Barrow Alternative Employment Committee, "Alterplan" at the Llangenerch Naval Stores Depot in South Wales, the Better Future for Defence Jobs Committee based at BAe in Bristol and plans alternative jobs at a number of Royal Ordnance factories and stores. The national trade unions with large membership in the defence sector-IPMS, T&G and MSF-actively support such initiatives.

At the centre of the plan-based approach to conversion is the **alternative use committee**. Ideally they would include management, labour and community representatives who would join forces to draw up an alternative business plan. While the interest in alternative use committees has gained widespread acceptance amongst trade unions, local authorities and peace groups, so far there are few examples of successful alternative use planning to be found. Alternative use committees have found a lack of cooperation from management, but there are other significant barriers to plant-based conversion. The social groups which make up alternative use committees do not always seek the same objectives. Defence workers and the trade unions priority is to retain jobs and this often means continuing to lobby for defence contracts. Peace activists in their moral commitment to achieving disarmament are often insensitive to defence workers' plight and vulnerabilities. Local authorities mayactively promote alternative production at local defence plants without the resources to back up workers initiatives.

In general alternative use committees operate under unrealistic assumptions about how business and markets operate. The lack of technical expertise and knowledge of the innovatory process, may lead alternative use committees to make technical proposals with little realistic consideration of costs of production, market opportunities, or required investment capital. Plans are often drawn up with little acknowledgement of the problems of raising investment funds as few of the participants have entrepreneurial skills.

Unless new markets are identified entry into existing markets is highly risky because competing firms are well up the learning curve and already have the advantage of scale economies. In this context the provision of government markets would represent an opportunity cost as they would not be purchasing from the least cost source of supply. In addition the reliance on government markets may be ruled out after 1992 by EC legislation on preferential purchasing unless a special case is made for the DIB.

The capital/labour endowments of defence facilities are rarely appropriate for volume commercial production, and because governments have guaranteed markets and profits the production and labour processes within defence plants are often archaic. Alternative use committees rarely have experience in commercial marketing, they tend to lay emphasis on product rather than process technologies, they seek to preserve the existing hierarchy of skills and technology culture which is isolated from commercial disciplines. The civil counterparts to the particular combination of capital, skills and technology culture are to be found in civil aerospace, satellites, nuclear power, powergeneration, space technologies and off-shore ring platforms. Most of the major defence contractors are already involved in these types of civil programmes and if anything overcapacity exists in these markets. In addition, such programmes are likely to prove very expensive and involve considerable opportunity costs by diverting resources from other areas of public expenditure or the economy.

Kaldor (1990) has argued that plant-based conversions which seek product solutions to retaining skills and facilities are likely to replace the military industrial complex with programmes that will reproduce the characteristics of defence production by social-industry complex or eco-industrial complex -- in other words grandiose alternative energy or socially useful produce programmes. This would reinforce the present industrial and technological patterns of production rather than transform them, representing a missed opportunity in terms of the potential for economic restructuring.

The emphasis on product solution to employment problems, reflects a poor undertaking of the process of innovation and technological change. Conversion policies tended to promote product rather than process innovation while completely ignoring the innovatory implications of the diffusion of existing process technologies. The great technological successes of the Japanese resulted not so much from the evolution of radical new products, but from the application of existing techniques and technologies to a wide range of existing products is digitalizing television, miniaturising products through the application of microchips, improving the productivity of traditional industries through the use of flexible manufacturing systems and the application of information technologies which has given rise to "just in time production" all contributing to the highly competitive nature of Japanese products.

Lower down the defence production hierarchy those contractors who produce generic technologies, mainly subcontractors, which have dual use applications are unlikely to need to convert their production processes or products. For this group of contractors the crucial issue is for alternative markets for their inputs, which is dependent on healthy demand for civil industries. In this context a demand stimulus is far more important than supply side intervention.

**Community Based Approach**

The emphasis in this strategy is for local communities and local authorities to take the initiative in supporting workers at local plants or bases to convert their facilities or to diversify their local economies away from defence dependency. This approach has largely found its inspiration from the USA where local state initiatives supported by the Federal Economic Adjustment Committee were successfully implemented following the downturn in military expenditures at the end of the Vietnam war.

In the UK a number of local authorities have set up conversion committees such as the Greater London Conversion Committee 1982-86, the Local Government Arms Conversion Working Group, the Local Authority Diversification Network, in addition to Bristol City Council, Lancashire Enterprises, Barrow Local Authority, Sheffield City Council, the Welsh Development Agency, Glasgow City Council Hertfordshire Country Council and the Stevenage and Hatfield Councils to name but a few, actively supported local conversion initiatives.

The majority of local authority initiatives are engaged in promoting plant-based conversion programmes, which are subject to all the limitations outlined above. However, the transformation of localeconomies through small innovative firm led developments, modelled on the experience of the north and central Italy, the Basque country and Southern France offer an alternative strategy for local economic regeneration. However, for such local initiatives to work there needs to be a high level of local autonomy in politics, finance and legislative power.

A central feature of Thatcherism in the 1980s was the increasing withdrawal of central government support for regional economic policy and the systematic undermining of local authority autonomy. Spending on regional policy is now at its lowest level since 1972. These changes are part of a general trend towards the centralisation of government functions and reflect the government's ideological rejection of "regional and local planning" as a process too closely associated with public sector intervention. See Miller (1990). This context limits the ability of local authorities to emulate the US experience or indeed the experience of regional regeneration on the European continent.

**Political Conversion**

The approach emphasises the way in which military imperatives dominate national technological and industrial objectives of high military spenders and define the trajectory of economic development. Its proponents argue for the transformation of resources tied up in defence production as part of a broader social and economic agenda of industrial and technological reconstruction rather than being an end in it itself. See Willett (1990), Renner (1988), Yudkin and Black (1990), Cronberg (1992). Central to this approach is the recognition of "market failure" in advanced capitalist societies in meeting basic human and environmental needs, and the urgent requirement of shifting industrial, technological and scientific resources awayfrom military defined objectives and instead targeting "national needs" like industrial renewal, environmental restoration, sustainable agriculture and renewable energies. Such a national needs policy, although initiated by the government should operate in partnership with industry, finance and local and regional authorities, workers and consumers.

An industrial policy driven by national needs would encompass a broad initiative aimed at improving the full range of manufacturing industries including those that produce goods needed for dealing with the pressing social and environmental problems confronting contemporary society. In this way a number of fledgling industries like solar cells or recyclable plastics would be targeted. But so should selected basic manufacturing industries such as the extractive and service industries, most of which receive little government R&D support but which are important to a balanced economy. If basic industries such as steel and textiles, for example, were targeted for R&D investment to develop new products and processes they could once again become competitive in world markets. The economic successes of the Third Italy, and the Basque regions have given much credence to strategies of regional regeneration through the application of process innovations to mature industries. See Cooke (1990).

At the same time industries are being revived, several "strategic" high-tech industries-microelectronics, computers, telecommunications, advanced materials, robotics and numerical controls - should also be targeted as they are crucial to the performance of other industrial sectors. These industries would no longer be dominated by military objectives, R&D policies directed towards these sectors would instead reflect commercial as well as social objectives.

The mechanisms available to encourage such developments include direct subsidies as well as incentives, for industries to conduct basic research, improve technology transfer, support job training and education, environmental regulation. If the government were to use the peace dividend to meet national needs the economic returns would offset the short term dislocations caused by defence spending cuts. However, bringing a country's industrial and technology policies in line with its national needs requires more than just drawing up a shopping list of social needs and then shifting funds from defence towards other public expenditure programmes. More fundamentally the country requires a new framework for guiding R&D policies, scientific and technological priorities and relationship between the private and public sectors. Such a framework would provide a filter for selecting critical industries and technologies that would receive R&D and other forms of government assistance.

R&D programmes targeted at high-tech performance in computing, advanced materials and optoelectronics for example could be designed to complement technology initiatives in steel processing, public transport, renewable energy systems, environmental monitoring and pollution control. At the same time superconductors and ceramics which are at present being developed largely under military programmes have potential application in fuel efficient engines, cutting tools, power turbines and many consumer items.

Mechanisms such as retraining, start-up incentives, tax breaks, civil R&D programmes and investment programmes should be designed to restructure capital and labour away from the defence industrial base and into more dynamic and innovative forms of production and labour processes, creating a radical break with the pervasive culture of the defence industrial base. The use of suchmechanisms need not be prohibitively expensive and offer more positive benefits to the whole of society in the long run.

Critical to all of this is increasing worker participation in those innovation programmes associated with industrial and process innovation much along the lines of the Swedish, german and Japanese examples. Support for community based technology projects, technology extension services and the establishment of technology advisory centres which can assist local business with the application of new process technologies, product design etc. much on the lines of the highly successful model developed in the Third Italy is central to the revival of mature industries such as ceramics, furniture, clothing and knitwear.

The ability to generate technological innovations and to incorporate them quickly into industrial processes and commercial products--in other words the process of technological transfer--is the key factor in improving productivity and competitiveness. But, a high rate of technological innovation and diffusion requires a strong science and technology base. A strong S&T base also acts as a basic seed corn for the education and training of the next generation of scientists and engineers upon whom future technological and industrial activity will depend. Ideally a nation's science and technology policies should be guided by coherent goals and policies which both promote an expanding science and technology base and that facilitate the transfer of innovations from the laboratory to factory floor to the market place.

Adopting such a major programme requires appropriate institutional mechanisms to oversee industrial restructuring. Clearly an empowered DTI would be the obvious agency to implement an industrial and innovation policy. But significant changes need to bemade to its orientation and decision making process if it is to target its policies towards social and national needs. It should be responsible for ensuring that local and regional authorities and the trade unions are involved at every level of the decision making process as its agenda needs a broad social mandate if its policies are to succeed. A forum needs to be created in which public interest groups such as health workers, housing groups, ecology groups as well as industrial workers and local community groups can define their needs and identify corresponding technical and industrial solutions with the aid of scientists and technologists, thereby setting the criteria for a new science and technology policy.

Another institutional step would be to empower and broaden the remit of the National Economic Development Office, and to increase the R&D advisory process within the parliamentary system. Regional and local interests should have far greater representation on such advisory groups and panels. In addition local and regional planning authorities would need to be empowered to aid regional development programmes which address the totality of a local economy and not just the declining defence sector.

## Conclusions

The government's approach to defence industrial adjustment is clearly seen to have failed. The exposure to market forces has resulted in a general decline in the performance of defence contractors and has led the survivors to adopt defensive strategies based on increased concentration and monopolisation of the defence market, rather than diversification into civil markets.

While support for the use of the peace dividend to convert the defence industrial base to civil production has grown,the evidence of successful conversion in the UK is limited, largely due to the formidable barriers to conversion within defence dependent companies. The alternative of restructuring production and economic priorities towards national and environmental needs implies the creation of entirely different industries and occupational structures. In this context political conversion represents much more than a strategy to save defence workers' jobs or industrial capacity; it workers' provides an important lever for the establishment of a different organisation of production in which social and environmental values have a central place.

Yudkin and Black (1990 p.283) argue that a national needs policy would help:

> "to demystify, demilitarize and return science and technology--which for so long in world history have been the handmaidens of powerful aggressive forces--once again to a human level. In this way, science and technology can play the vital and positive role in human evolution that they should as the principal tools for enhancing the quality of life on our planet".

In this context the peace dividend and conversion can be viewed as economic levers, but ultimately they represent a political choice and are only likely to be carried out in the context of a major shift in political and economic thinking. Without the political will to commit national resources to clearly stated national objectives, it is unlikely that there will be a conversion process let alone a tangible peace dividend.

## Endnote

1 See T.Voss, **Policies For the Defence Industry**, Oxford Research Group, 1991, for a more detailed review of party policies towards the UK DIB.

## References


Alexander A. (1990) "National Experiences in the Field of Conversion. A comparative Analysis" A paper presented at the UN Conference on Conversion, Moscow, 13-17, August.

Anthony Ian, Aggnes Corades Allbeck and Herbert Wulf. (1990) "West European Arms Production" *SIPRI*.

Barker T, Dunne P. and Smith R. (1991) "Measuring the Peace Dividend in the United Kingdom", *Journal of Peace Research*, Vol.28, No.4, November, p.345-358.

Baylis J. (1989) *"British Defence Policy: Striking the Right Balance"*, Macmillan.

Chalmers M. (1990) *"United Kingdom Defence Requirements: 1990-2000"* Safer World Foundation, Bristol.

Cooke P. (1990) "Manufacturing Miracles: The Changing Nature of the Local Economy" in *Local Economic Policy*, Campbell M (ed.) Cassell Educational Ltd.

Cronberg T. (1992) "The Social Reconstruction of Military Industries and Technologies: Seen from a Danish Perspective". Paper presented at the International Institute of Peace Conference on "Conversion" Vienna, February.

Dumas L. and Thee M. (1989) *Making Peace Possible*, Pergamon Press, Oxford.

Dunne P. (1990) "The Political Economy of Military Expenditure: An Introduction" *Cambridge Journal of Economics*, Vol.14, No.4, Dec, pp.395-404.

Dunne P. and Smith R. (1992) "Thatcherism and the UK Defence Industry", Chapter 5 in Michie J (1992) *"1979-92 The Economic Legacy"*, Academic Press.

Fontanel J. & Smith R.P. (1991) "A European Defence Union?" *Economic Policy.*

Freedman, Lawrence (1987) "The case of Westland and the bias to Europe", *International Affairs.*

Hilditch P. (1990) "Defence Procurement and Employment: The Case of UK Shipbuilding", *Cambridge Journal of Economics*, Vol.14, No.4, Dec. pp.453-468.

Kaldor M. (1991) "Problems of Adjustment to Lower Levels of Military Spending in Developed and Developing Countries" paper prepared for the World Bank's Annual Conference on Development Economics Washington D.C. April 25-26.

Lovering J. (1990) "Military Expenditure and the Restructuring of Capitalism: The Military Industry in Britain" *Cambridge Journal of Economics*, Vol.14, No.4, Dec. pp.453-468.

Lovering J. (1991) "The Defence Industry After the Cold War" *CND.*

Melman S. (1987) *Profits Without Production*, University of Pennsylvania Press.

Melman S. (1985) *The Permanent War Economy*, Simon and Schuster.

Melman S. (1981) "From Military to Civilian Economy: Issues and Options" *Occassional Paper Series* No.8, Centre for the Study of Armament and Disarmament, California State University.

Miller D. (1990) "The Future of Local Economic Policy: A Public and Private Sector Function" in Campell C. op. cit. pp.196-211.
Oden M.D. (1988) "A Military Dollar Really Is Different: The Economic Impacts of Military Spending Reconsidered" *Employment Research Associates*, Lansing, Michigan.

Roper J. (ed.) (1985) *The Future of British Defence Policy*, Gower.

Schofield S. (1991) "Arms Conversion in the 1990s" unpublished mimeograph, Dept. of Science and Technology Policy, University of Manchester.

Smith R.P. (1976) "Military Expenditure and Capitalism" *Cambridge Journal of Economics.*

Smith R.P. (1990) Defence Procurement and Industrial Structure in the UK, *International Journal of Industrial Organisation* 8, 1990, 185-205.

Southwood P. (1991) *Disarming Military Industries*, Macmillan.

Taylor T. and Hayward T. (1989) *The UK Defence Industrial Base: Development and Future Policy Options*, Brasseys Defence Publishers.

Voss T. (1991) "Policies for the Defence Industry', Oxford Research group.

Walker W. & Gummett P. (1989) "Britain and the European Armaments Market", *International Affairs.*

Willett S. (1990) "Conversion Policy in the UK", *Cambridge Journal of Economics*, Vol.14, No.4, Dec.pp.469-482.

Yudkin, J. and Black M. (1990) "Targeting National Needs: A New Policy for Science and Technology" *World Policy Journal*, Spring.

**Figure 1.**

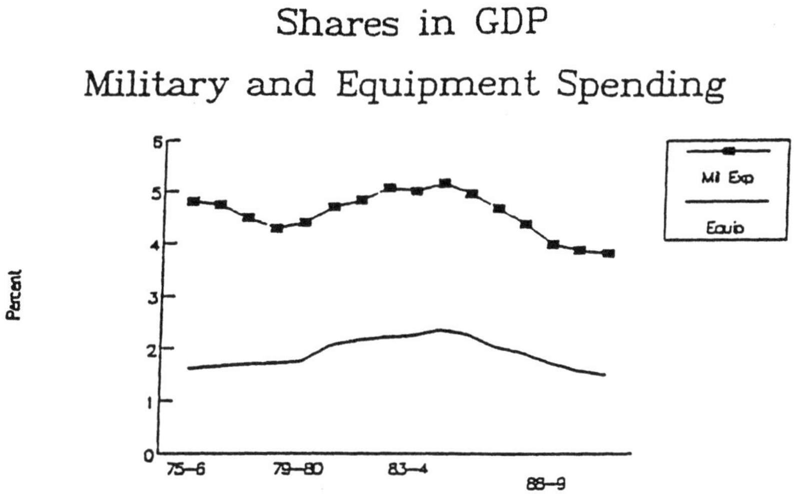


**Figure 2.**

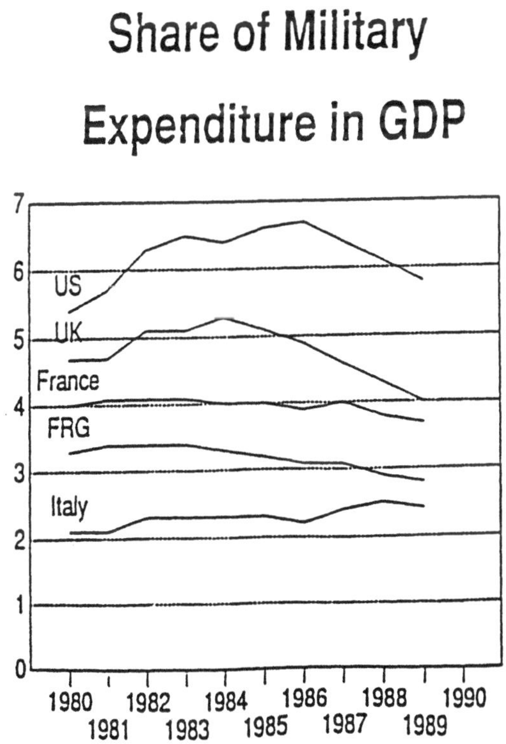

**Figure 3.**

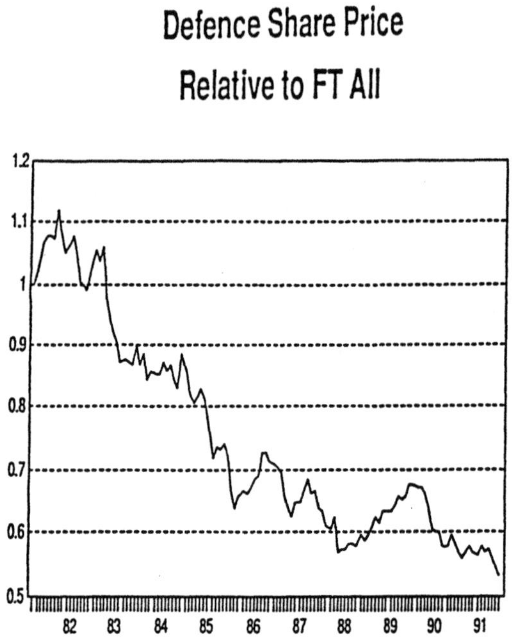


**Figure 4.**

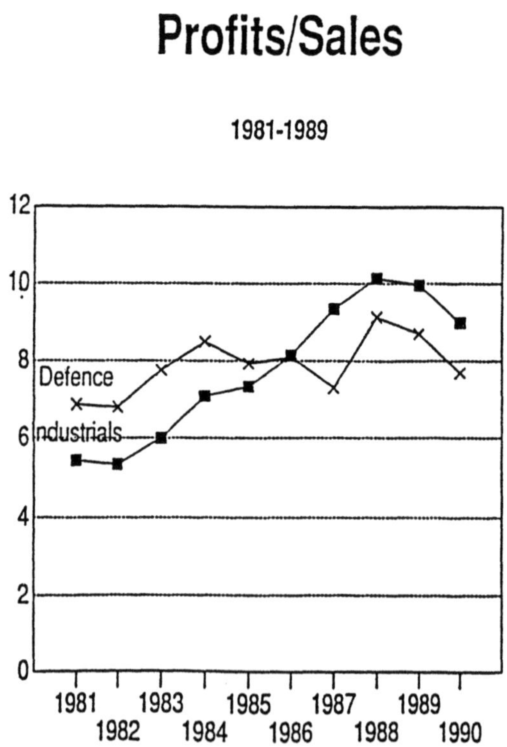

| Table 1. UK Defence Expenditure | | | | |
|---|---|---|---|---|
| | 1983-4 | 1986-7 | 1989-90 | 1991-2 Estimate |
| Total (£m) | 15487 | 18163 | 20755 | 24027 |
| % of total | | | | |
| Personnel | 37.0 | 37.9 | 39.0 | 41.3 |
| Equipment | 44.8 | 43.4 | 41.1 | 38.6 |
| Other | 18.2 | 18.6 | 19.2 | 20.1 |
| Total('90 £m) | - | 21450 | 20755 | 20896 |
| Forces (£m) | 3076 | 3787 | 4539 | 5453 |
| Civilian (£m) | 1873 | 2123 | 2363 | 2849 |

Source: SDE

| Table 2. UK Employment Dependent on Defence Expenditure | | | | | |
|---|---|---|---|---|---|
| | | Thousands | | | |
| | | 1978-9 | 1983-4 | 1986-7 | 1989-90 Estimate |
| Equipment | -direct | 220 | 215 | 170 | 150 |
| | -indirect | 180 | 160 | 140 | 130 |
| Non-equip | -direct | 100 | 85 | 85 | 85 |
| | -indirect | 80 | 70 | 70 | 70 |
| Total MoD | - direct | 320 | 300 | 255 | 235 |
| | - indirect | 260 | 230 | 210 | 200 |
| Exports | - direct | 70 | 70 | 65 | 80 |
| | - indirect | 60 | 60 | 55 | 70 |
| Total | - direct | 390 | 370 | 310 | 315 |
| | - indirect | 320 | 290 | 255 | 270 |
| | - all | 710 | 660 | 565 | 585 |

1. MoD is Ministry of Defence
2. Employment figures do not include MoD employees (service or civilian)
Source: SDE (1991)

Table 3 UK Defence Expenditures and Employment by Region

| Year | Expenditure 85-6 | 87-8 Est | 89-90 | Employment (th) 85-6 | 87-8 Est | 89-90 |
|---|---|---|---|---|---|---|
| UK | 6850 | 7200 | 7300 | 120 | 175 | 150 |
| % Share | | | | | | |
| North | 6.0 | 9.0 | 11.5 | 7.5 | 9.0 | 14.0 |
| Y & H[1] | 3.0 | 2.0 | 3.0 | 2.0 | 2.0 | 2.5 |
| East Midlands | 4.5 | 3.5 | 3.5 | 4.0 | 5.0 | 3.5 |
| East Anglia | 1.5 | 0.5 | 1.5 | 2.0 | 2.0 | 1.5 |
| South East | 47.5 | 42.5 | 39.5 | 43.0 | 40.0 | 37.5 |
| South West | 14.5 | 15.5 | 15.0 | 11.0 | 14.5 | 14.5 |
| West Midlands | 3.5 | 5.5 | 6.0 | 3.5 | 5.0 | 6.0 |
| North West | 10.0 | 12.0 | 9.0 | 11.5 | 10.5 | 8.0 |
| Scotland | 6.0 | 7.5 | 7.5 | 9.0 | 9.5 | 9.5 |
| Wales | 2.0 | 0.5 | 0.5 | 1.5 | 0.5 | 0.5 |
| N.Ireland | 1.5 | 1.5 | 3.0 | 4.0 | 1.5 | 2.5 |

1. Y & H is Yorkshire and Humberside
2. Figures for the UK are in £ million for the expenditures and in thousands of employees for the employment figures.

Source: SDE

Table 4 . UK Defence Contractors' Turnover in 1986 GB £m

| | Total | % Civil | % Defence | % MOD |
|---|---|---|---|---|
| British Aerospace | 3137 | 24 | 76 | 32 |
| Chloride | 310 | 92 | 8 | 8 |
| Dowty | 520 | 58 | 42 | 19 |
| Ferranti | 485 | 36 | 64 | 48 |
| GEC | 5969 | 72 | 28 | 22 |
| GKN | 2200 | 95 | 5 | 5 |
| Hunting Inds | 233 | 48 | 52 | 52 |
| Lucas Aerospace | 285 | 41 | 59 | 15 |
| Pilkington El. Opt. | 120 | 10 | 90 | 42 |
| Plessey | 1461 | 59 | 41 | 26 |
| Racal | 1270 | 68 | 32 | 10 |
| Rolls Royce | 1600 | 44 | 56 | 37 |
| Short Brothers | 200 | 75 | 25 | 20 |
| Smiths Industries | 160 | 45 | 55 | 33 |
| Thorn EMI | 2735 | 88 | 12 | 9 |
| Vickers | 600 | 85 | 15 | 3 |
| Westland | 344 | 29 | 71 | 58 |
| Total | 21629 | | | |
| Mean level | 1272.3 | 811.1 | 461.2 | 271.1 |
| % total | 100 | 57.0 | 43.0 | 25.7 |

Source: ACOST (1989).

Table 5 . Government Expenditure In Disarmament Scenarios, 1985 £m

| | 1992 | 1993 | 1996 | 2000 |
|---|---|---|---|---|
| Defence | 17176 | 15751 | 12145 | 8588 |
| NHS (Central) | 19036 | 19729 | 21662 | 23947 |
| Other central | 13236 | 13785 | 15361 | 17318 |
| Education (local) | 14118 | 14633 | 16224 | 18653 |
| Other local | 16360 | 16956 | 18617 | 20581 |
| Total | 79927 | 80854 | 84009 | 89087 |
| Social capital formation (excluding housing) | 7815 | 8044 | 8035 | 8350 |
| Total | 87742 | 88898 | 92044 | 97437 |

Table 6. Simulation - Percentage Difference From Base

| | Compensated | | | Uncompensated | | |
|---|---|---|---|---|---|---|
| | 1993 | 1996 | 2000 | 1993 | 1996 | 2000 |
| Consumers expend | 0.22 | 1.73 | 5.39 | -0.31 | 1.97 | -4.41 |
| Govt. Consumption | -0.21 | -1.31 | -2.79 | -1.87 | -6.99 | -12.13 |
| Investment | 0.25 | 1.58 | 4.27 | -0.3 | -1.77 | -3.64 |
| Exports (G&S) | -0.02 | -0.25 | -0.79 | -0.01 | 0.06 | 0.69 |
| Imports (ge) | 0.09 | 0.99 | 3.21 | -0.50 | -2.34 | -4.39 |
| GDP factor cost | 0.10 | 0.63 | 1.84 | -0.45 | -2.03 | -3.64 |
| B of P (pp) | -0.04 | -0.40 | -1.36 | 0.16 | 0.93 | 2.17 |
| PSBR (pp) | -0.08 | -0.48 | -1.27 | -0.16 | -0.35 | -0.39 |
| Unemployment (m) | -0.05 | -0.22 | -0.52 | 0.06 | 0.25 | 0.46 |

1. G&S = goods and services
2. pp = percentage point
3. The compensated results are for the simulation where the reduction in military expenditure is reallocated proportionately to other categories of government current and capital expenditure. For the uncompensated simulation there was no reallocation.

Table 7. Public Sector Employment (thousands)

| | Levels 1992 | Difference from Base 1993 | 1996 | 2000 |
|---|---|---|---|---|
| Defence | 419 | -36.5 | -126.4 | -210.7 |
| NHS (Central) | 1413 | 29.7 | 102.4 | 169.8 |
| Other central | 590 | 12.6 | 44.8 | 76.3 |
| Education | 1709 | 36.5 | 131.4 | 233.2 |
| Other local | 1380 | 29.2 | 102.2 | 171.5 |
| Total | 5511 | 71.6 | 254.4 | 440.0 |

1 The employment changes in the Table are computed as being in the same proportion of the changes in expenditure.

Table 8. The Industrial Impact of the Cut in Military Spending with Compensation:

Percentage Difference From Base

| | Output 1993 | Output 1996 | Output 2000 | Employment 1993 | Employment 1996 | Employment 2000 |
|---|---|---|---|---|---|---|
| 1 Agriculture etc. | .18 | 1.04 | 2.62 | 1.0 | 5.6 | 13.5 |
| 2 Coal & Coke | .05 | .38 | 1.06 | .1 | .3 | .7 |
| 3 Mineral Oil & N Gas | .00 | .00 | .00 | .0 | .0 | .0 |
| 4 Petroleum products | -.06 | -.02 | .41 | .0 | .0 | .0 |
| 5 Electricity etc. | .07 | .52 | 1.56 | .1 | .7 | 1.8 |
| 6 Public Gas Supply | .11 | .63 | 1.77 | .1 | .5 | 1.3 |
| 7 Water Supply | .05 | .40 | 1.31 | .0 | .2 | .7 |
| 8 Minerals & Ores nes | -.01 | .07 | .25 | .0 | .0 | .1 |
| 9 Iron & Steel | -.21 | -1.17 | -3.12 | -.2 | -.9 | -2.3 |
| 10 Non-ferrous Metals | -.13 | -.46 | -.83 | -.0 | -.2 | -.3 |
| 11 Non-metal Min. Pr. | .10 | .76 | 2.01 | .2 | 1.5 | 3.9 |
| 12 Chemicals & MM Fibre | .04 | -.30 | -1.69 | .1 | -1.0 | -5.0 |
| 13 Metal Goods nes | -.06 | -.04 | .09 | -.2 | -.1 | .2 |
| 14 Mech. Engineering | -.27 | -.97 | -1.92 | -2.1 | -7.4 | -13.6 |
| 15 Electronics | -.50 | -.65 | 1.96 | -2.1 | -2.9 | 9.0 |
| 16 Elect. Engineering | -.02 | .37 | 2.03 | -.1 | .9 | 4.8 |
| 17 Motor vehicles | -.05 | .21 | .64 | -.1 | .5 | 1.3 |
| 18 Ships & Other Vessels | -2.83 | -9.64 | -15.43 | -1.7 | -5.0 | -6.5 |
| 19 Aerospace Equipment | -3.30 | -10.92 | -17.10 | -4.2 | -12.5 | -17.0 |
| 20 Other vehicles | .05 | .28 | .18 | .0 | .1 | .0 |
| 21 Instr. Engineering | .01 | -.24 | -.98 | .0 | -.2 | -.9 |
| 22 Manufactured food | .14 | .87 | 2.13 | .6 | 3.7 | 8.6 |
| 23 Alcoholic Drinks | .08 | .57 | 1.54 | .1 | .5 | 1.2 |
| 24 Tobacco | .08 | .61 | 1.78 | .0 | .1 | .3 |
| 25 Textiles | .04 | .15 | -.10 | .1 | .4 | -.2 |
| 26 Clothing & Footwear | .02 | .02 | -.80 | .1 | .1 | -2.4 |
| 27 Timber & Furniture | .22 | 1.73 | 5.19 | .7 | 5.5 | 16.3 |
| 28 Paper & Board | .04 | .40 | 1.18 | .1 | .6 | 1.8 |
| 29 Books etc. | .04 | .50 | 1.61 | .2 | 1.7 | 5.4 |
| 30 Rubber & Plastic Pr. | .02 | .48 | 1.72 | .1 | 1.1 | 4.0 |
| 31 Other Manufactures | .09 | .53 | 1.39 | .1 | .5 | 1.3 |
| 32 Construction | .20 | 1.37 | 4.17 | 3.2 | 22.7 | 66.6 |
| 33 Distribution etc. | .14 | 1.25 | 4.04 | 5.5 | 50.1 | 157.3 |
| 34 Hotels & Catering | .31 | 2.50 | 7.93 | 4.4 | 36.9 | 120.0 |
| 35 Rail Transport | -.01 | .73 | 3.05 | .0 | 1.0 | 3.8 |
| 36 Other Land Transport | .10 | .88 | 2.75 | .6 | 5.3 | 16.2 |
| 37 Air Transport | .11 | .99 | 3.09 | .1 | .6 | 1.9 |
| 38 Sea & Other Trans | -.05 | .07 | .70 | -.1 | .2 | 1.8 |
| 39 Communications | .12 | .92 | 2.79 | .6 | 4.6 | 14.8 |
| 40 Banking & Finance | .15 | 1.09 | 3.25 | 1.1 | 7.6 | 22.6 |
| 41 Insurance | .17 | 1.33 | 3.84 | .6 | 4.5 | 13.4 |
| 42 Business Services | .09 | .74 | 2.19 | 2.0 | 16.5 | 51.9 |
| 43 Miscell. Services | .23 | 1.22 | 3.19 | 6.2 | 36.3 | 110.2 |
| 44 Unallocated | -.20 | -1.37 | -4.10 | .0 | .0 | .0 |
| Total | .02 | .52 | 1.96 | 16.9 | 180.4 | 608.5 |

1. The simulation results in this table are for a halving of military expenditure by the year 2000 with the expenditure reallocated to other categories of government current and capital expenditure.

# UNITED NATIONS ORGANISATIONS

# THE MACROECONOMIC AND ENVIRONMENTAL EFFECTS OF MILITARY CONVERSION

**Timothy R. Muzondo**[1]
*IMF*

## Introduction

The events of 1989 in Eastern Europe, the end of the cold war and the subsequent easing of military tensions and conflicts in other parts of the world have created an environment which makes it possible for a number of countries to receive "peace dividends" -- or savings on defence expenditures which can be obtained without sacrificing national "security". The savings arise from the fact that the improvement in the security environment has favourably changed the relationship between amounts of defence inputs that are required to produce the given levels of national "security". The prospect that a number of countries may receive these savings has intensified the debate on military conversion -- or the process of transforming military facilities and retraining human resources formerly employed in military industries to make them suitable for civilian use -- and has rekindled the debate on the economic effects of military spending. There are two broad (opposing) views on the economic effects of military spending.

One view is that military spending creates employment which otherwise would not take place. The alternative to military spending, according to this view, is not a better allocation of given levels of aggregate demand, but lower levels of output and higher rates of unemployment.[2] Military spending is also seen as promoting technological innovation which induces economic growth through spill over effects to civilian use.[3] An alternative view stresses the opportunity cost argument. It points out that economic performance, particularly in the long run, is negatively affected by diversion of resourcesfrom productive investment (including education) to defence.[4] Empirical studies that have sought to establish the relationship between military spending and economic performance have been inconclusive leading one commentator to remark that "despite the volume of work on the subject we do not know whether there is a causal relationship between military expenditure and development, much less what that relationship is"[5] and another to conclude that the relationship between military expenditure and economic growth will "depend on circumstances", and "will not follow some general law applicable to all times and places".[6] A number of other authors have expressed similar views.[7]

The point of departure of this paper is in recognition of the fact that macroeconomic and environmental effects of military spending are likely to vary from country to country and from time to time depending on a number of factors, including, most importantly, the macroeconomic policies adopted.[8] The paper examines the macroeconomic and environmental effects of a cut in military expenditures and shows the various ways in which a number of factors, including the conduct of macroeconomic policies, can affect final outcomes. The paper also examines the impact of conversion on macroeconomic performance and on the environment.[9] Its conclusions are that (a) in the short run, a cut in military spending accompanied by sound demand management policies can reduce inflation, increase output and employment, and improve the balance of payments. In the long run, economic performance can improve if, following a reduction in defence spending, savings and investment were to increase; (b) the environmental effects would depend onthe level and structure of demand that would result from cuts in military spending; and (c) by reducing the costs of adjusting to lower levels of defence spending, military conversion can improve economic performance, but it could also have negative economic effects, especially if it goes hand in hand with economic rigidities and encourages unwarranted and unintended government subsidies.

The rest of this paper is organized as follows: Section II discusses the linkages between macroeconomic policy, the environment and military conversion; the output, employment and environmental effects of reduced military spending are presented in Section III; Sections IV and V discuss the effects of a cut in military spending on inflation and on the balance of payments, respectively; the impact of military conversion on economic performance and on the environment are discussed in Section VI; and Section VII concludes the paper.

## Linkages between Macroeconomic Policy, the Environment and Conversion

The macroeconomic effects of a cut in military expenditures vary from country to country depending on, among other things, the economic and institutional characteristics of a country, the prevailing economic environment and policy responses of governments aimed, in part, at mitigating adverse effects of a reduction of such expenditures. In some countries the policy responses may include various forms of financial, manpower and other forms of government support for military conversion. The environmental effects will vary depending on the effects of military cuts and the effects of policy responses on levels and composition of output. This section examines some of the macroeconomic factors and policy responses (including military conversion) to a cut inmilitary spending that may affect final outcomes of reduced military spending.

A cut in military spending may affect the composition and level of output, the level of employment, the rate of inflation, and the balance of payments depending on (a) the composition of magnitude of the cuts; (b) whether fiscal deficits are allowed to fall or whether they remain unchanged because offsetting increases in other government expenditures are implemented; (c) whether sizes of fiscal deficits are maintained by lowering taxation; (d) whether monetary policy is accommodating or restrictive; (e) whether reductions in fiscal deficits lead to retirement of interest bearing or non interest bearing government debt; and (f) the prevailing economic environment -- and, in particular, whether the economy is operating at full capacity or whether there are unemployed resources. For example, output and employment could fall as a result of a cut in military spending if aggregate demand is not maintained through offsetting increases in other public expenditures or through easing of monetary policy.[10] Similarly, if the economy was initially operating at full capacity and experiencing supply constraints and inflationary pressures, a cut in military spending which is not offset by increases in other government expenditures but is accompanied by a non accommodating monetary policy would ease such pressures while output and employment could fall. Finally, other things remaining the same, a cut in defence imports would improve the balance the payments, while output and employment may not be affected.[11]

Economic activities often require use of environmental resources such as air, water, land and forests and when use of these resources involves environmental externalities, the effects of a cut in military spending could affect amounts of environmental externalities in four ways.[12]First, amounts of environmental externalities could change as the level of output changes in response to a cut in military spending. Second, a change in the composition of output may occur following a cut in military spending which could change amounts of environmental externalities even though output remained the same. Third, the nature of technological developments may be affected by cuts in military spending and this may also have long run effects on amounts of environmental externalities. Finally, a shift from military-related to civilian-related production and expenditures may affect amounts of environmental externalities if, as is often the case in many countries, enforcement of environmental regulations varies between activities undertaken by the military and those undertaken by civilian entities.

As already noted, military conversion is a process that involves transforming military facilities and retraining human resources formerly employed in military industries. It can be viewed as a microeconomic level policy instrument that seeks to increase capital and labor mobility from military to civilian uses by reducing economic friction and costs of adjustment. Military conversion can affect the environment through its effects on the level and composition of output. For example, the presence and nature of military conversion programmes can affect the level of defence expenditure cuts that are considered politically feasible and can, in this way, affect sizes of fiscal deficits, the balance of payments (to the extent that imports and exports are affected) and the level and composition of output (because of the level and composition of cuts).[13]

In the section below we will examine the macroeconomic and environmental effects of a cut in military spending and show how the effects are dependent on macroeconomic policies adopted. We will consider two cases.In the first case, a cut in military spending gives rise to a reduction in the fiscal deficit equal to the cut. Monetary policy is conducted a view to stimulating private investment, and possibly, consumption so as to partially offset the impact of the cut. In the second case, the cut is offset by increases in other public expenditures.[14] The macroeconomic and environmental effects of each of these cases are examined assuming that markets adjust without help from military conversion programs. The impact of such programs is discussed in Section VI.

**Output, Employment and Environmental Effects of Military Spending**

There is a considerable disagreement in literature on the effects of military spending on output and employment and many empirical studies that have sought to establish a relationship have come to different conclusions.[15] As a result, the conclusions from these studies provide little guidance to policy makers interested in the knowledge of the effects of the cuts in military spending on output and employment. Our view is that the output and employment effects of defence expenditure cuts will vary from country to country and, in a given country, from time to time depending on, among other things, macroeconomic policies adopted: that is, whether taxes (and what type of taxes) are reduced, the monetary policy stance that is adopted and the exchange rate policy in effect.[16] This section examines the output, employment and environmental effects of a cut in military spending assuming adoption of specific macroeconomic policy responses to such a cut.

**a. The Effects of an Uncompensated Cut in Military Spending**

There are three reasons why a government may not offset a cut in military spending by increases in other publicexpenditures. First, the economy may be overheated and experiencing inflationary pressures. Second, a government may consider that the size of the public sector is too large. Finally, the opportunity to cut military expenditures may coincide with the need to contain aggregate demand by reducing the fiscal deficit with a view to stabilizing prices and improving the balance of payments. Such a cut will tend to be deflationary and output and employment in the short run may fall. The extent of the decline will vary depending on economic structure and on the extent and nature of the cuts (for example, their import content) and on how the reduced fiscal deficit is financed.[17]

The full impact of a cut in military spending in an open economy may not be felt domestically and can vary enormously from country to country depending on **trade and capital flows** between the military budget and the external sector. In many developing countries direct imports of military hardwares are large. For example, military imports as a percentage of total military budgets have exceeded 20 percent in some developing countries.[18] The total direct and indirect import demand may be considerably higher than indicated by this figure, particularly for countries that have significant import -dependent - manufacturing - sectors. If cuts in military expenditures involve such imports, the extent of the decline in output and rise in unemployment will tend to be less than when the cuts fall on domestic expenditure. In industrialized countries, on the other hand, the import content of military budgets tend to be smaller than those in developing countries. Therefore, other things remaining the same, cuts in military expenditures in these countries may result in greater reductions in output and employment than would be the case in developing countries.

The way in which military expenditures and fiscal deficits are **financed** can also havean impact on output and employment. For example, if military expenditure (whether on domestic goods and services or on imports that are cut were intended to be financed by foreign grants or loans, then domestic financing of the fiscal deficit need not change even though the deficit is lower than before the cut. As a result, financial considerations would not exert a separate influence on output and employment.[19] However, if the cut were intended to be financed from domestic sources then the output and employment effects would vary depending on whether the reduced budget deficit is financial through greater reliance on bank or non-bank sources. For example, following a cut in military spending that was intended to be financed by domestic sources, the monetary authorities may decide to ease credit and increase reliance on bank credit to finance a lower fiscal deficit than previously envisaged. The effect would be to reduce interest rates which in turn would stimulate private investment and, possibly, private consumption. The resulting increase in private demand would partially offset the reduction in aggregate demand caused by the cut in military spending. Alternatively, the monetary authorities may finance the budget in such a way as to keep interest rates constant. The result would be lower output and employment levels than in the case just discussed.

The output and employment effects of a cut in military spending may also be affected by the **exchange rate policy** in place. For example, under a fixed exchange rate regime a cut in military expenditures may reduce the rate of inflation, and so improve international competitiveness compared with what would have been obtained in the absence of the cut. As a result, and other things remaining the same, net exports may increase and partially offset the deflationary impact of the cut. This follows because domestic prices may fall which in turn may have the effect of increasing prices of traded goods relative tothose of non traded goods. This would cause resources to shift to traded goods sectors.

The **environmental effects** would depend on the pollution intensity of output of the cut military spending relative to the pollution intensity of output of the increase of other public spending. Assuming that the pollution intensity of output does not increase, amounts of environmental externalities would tend to decline as output falls but the extent of the decline may be less where the cut involves military imports,[20] or when the adopted policy stance offsets the impact of the cut by easing monetary policy. The extent of the decline in amounts of environmental externalities may be greater in cases where arms production and/or military activities (such as training, use of energy) are, on average, more pollution intensive than all other economic activities. The environmental effects of cut in military spending under the assumptions discussed above would also vary depending on whether the cut is from imports or from domestic production and also depending on the monetary policy stance adopted. For example, a cut in military imports may not affect domestic output and, therefore, amounts of environmental externalities. On the other hand, a cut in domestic military purchases would, other things remaining the same, tend to reduce output and employment and, therefore, amounts of environmental externalities.

### b. The Effects of a Compensated Cut in Military Spending

A number of studies have examined the output and employment effects of a cut in military spending that is matched by increases in other public spending.[21] Matching cuts in military spending with an increase in other spending may represent a policy response of a government that considers that both the size of the public sector and the level of aggregate demand are appropriateand the policy issue it faces is how to distribute the "peace dividend" among other public expenditure priorities. The fact that total government expenditures and the size of the fiscal deficit are maintained does not, of course, mean that output and employment will be the same following a cut in military spending. Output and employment may change in part because the domestic component of military spending that are cut may differ from the domestic component of the increase in other public spending. In addition, the financing of the fiscal deficit with reduced military spending may differ from the financing of the original deficit. The environmental effects would depend on the distribution of the "peace dividend" among other public expenditures and, hence, on the composition of output as well as on the effects of the cut in the level of output.

The **import content** of defence expenditures in many developing countries tends to be higher than that of other public expenditures except, possibly, public investment. Because of this, a cut in military spending that is, for example, concentrated in arms procurement and is matched by equal increase in other public spending (proportionately distributed across all other expenditure items) may increase output and employment.[22] A similar result may obtain for industrialized countries. For example, a recent study in the United Kingdom examined the impact of an 8-9 percent real cut in military expenditures proportionately distributed across all military spending items and matched by an equal increase in other public expenditures (also proportionately distributed among expenditure items). It concluded that such a cut would increase output and employment.[23]

Even though a compensated defence expenditure cut would not change the size of the fiscal deficit, such a cut could alter its **financing** in ways which may affect outputand employment. The purchases of arms in many developing countries is often financed by foreign loans or, in a few cases, by foreign grants.[24] In such countries a cut in purchases of arms may reduce amounts of foreign financing for the budget. Which such a reduction in foreign financing may be welcome because it reduces external debt burden in the future, for the present it means an increase in domestic financing requirements which may be inflationary depending on, among other things, supply constraints. The impact on output and employment will also depend on supply constraints and on the stance of monetary policy. If the monetary authorities choose to accommodate the increase in domestic expenditures they could increase bank credit to meet the increase in the demand for money and if there are no supply constraints, output and employment would increase. However, it credit is not increased, interest rates would rise and this would tend to reduce private investment and, possibly, private consumption. The result may be that output and employment would not increase.

The **long run** effects of a cut in military spending and of accompanying compensatory policies adopted would depend on the impact of such a cut and such policies on savings and investment. For example, economic growth potential could be improved by a cut in military spending which does not adversely affect private savings and investment, and in which the "peace dividend" is invested in, for instance, education and training and physical infrastructure. On the other hand, economic growth potential could worsen if the "peace dividend" were to be allocated to current expenditures, such as wages and salaries. This is because growth enhancing aspects of military spending, such as physical infrastructure that is jointly used for defence and civilian purposes and technological spill over to civilian use of military research and developments, would be lost.

The **environmental effects** of a compensated cut in military spending will vary depending on, among other things, the effects of the cut on output, the distribution of the cuts among different military expenditure items, the pollution intensity of output of military expenditures relative to those of other public expenditures and how the increase in other public expenditures is distributed among other public expenditure items. For example, if the pollution intensity of output of military expenditures is higher than that of other public expenditures, then a cut in military spending (that is proportionately distributed among items) matched by an increase in other public expenditures (also proportionately distributed across other public expenditures items) could reduce amounts of environmental externalities even when output increases. A similar cut that is distributed in a like manner would leave amounts of environmental externalities unchanged if pollution intensity of the output of military expenditures is the same as that of other public expenditures and if output does not change: the amounts of environmental externalities would be uncertain under these assumptions if output were to increase.[25] This means that if, for example, a "peace dividend" is distributed to such areas as education, health care and environmental clean up and protection, then the pollution intensity of output and amounts of environmental externalities may fall even though output increases.[26] The fact that output may increase and rates of unemployment may fall could have positive environmental effects, especially in developing countries where unemployment and poverty are important sources of environmental degradation.

**The Effects on Inflation**

Military expenditures have been viewed as inherently inflationary for three reasons. First, it is said that military expenditures are inflationary because they create an excess demand for goods and services owing to the fact that such expenditures create incomes and purchasing power which is not matched by production of additional goods and services for people to purchase.[27] Second, it is argued that military spending has negative impact on rates of growth in productivity and thus makes the economy prone to inflation.[28] Finally, it is often pointed out that defence industry's pricing and contracting policies and practices, characterized by "cost-pass-along" rather than cost minimization, indicate that the military sector may be a source of cost push inflationary pressures.[29] While these arguments may be valid, they tend to ignore the important role that macroeconomic policies, and in particular monetary policy, can have on inflation and, therefore, provide no guide as to whether in an actual situation a cut in military expenditures will be inflationary.

As already noted, an uncompensated cut in military spending tends to reduce output and employment which in turn may reduce incomes and aggregate demand. This may drive prices down unless monetary policy is not eased to offset the deflationary fiscal impact. For example, following an uncompensated cut in military spending monetary policy could aim at keeping interest rates constant, in which case the deflationary fiscal impact may not be offset by an increase in private investment and consumption. However, it is conceivable that, in an attempt to maintain high levels of aggregate demand, bank financing of the reduced deficit and monetary growth could be increased sufficiently to cause inflationary pressures and thus thwart the (deflationary) fiscal impact.

Similarly, the inflationary impact of a cut in military spending compensated by anincrease in other public spending may depend on, among other things, the monetary policy stance adopted in response to this situation. If the objective of monetary policy is to increase output beyond that exerted by the fiscal impact, bank financing of the deficit and money supply could be increased above levels that would have obtained before the cut. The increased availablility of credit, and the resultant lower interest rates, could stimulate investment and private consumption sufficiently to cause inflationary pressures to mount despite the reduction in military spending.

The existence of a "peace dividend" may provide an opportunity for reallocating resources in favour of productive investment and so promote non-inflationary growth through increases in productivity. However, sound policies would need to be implemented in order to realize this outcome. For example, use of the "peace dividend" to increase expenditures on education and training, productive infrastructure (such as transportation, communication and power) and research and development could significantly increase growth in productivity. On the other hand, using the dividend for public and/or private consumption may have similar or worse effects on productivity growth as military spending.

**The Effects on the Balance of Payments**

The effects of a cut in military spending on the balance of payments would vary depending on, among other things, whether a country is a exporter or importer of military hardware, the dependency of military exports on domestic sales of arms, the extent and nature of the cuts and the short and long term impact on international competitiveness of macroeconomic policies adopted to offset the impact of the cut. These factors vary considerably even among industrial countries.

A cut in military expenditures involving hardware will tend to improve the balance of payments for three reasons. First, the cut will often involve direct military imports as well as indirect imports by industries that supply the military. Second, cuts in domestic sales of arms in countries that export arms may force arms producing firms to stimulate exports. On the other hand, military exports may fall if, as a result of a decline in domestic sales, production costs increased substantially to make arms producing firms uncompetitive in international arms markets. Third, a reduction in domestic arms sales releases resources that can be used for producing non military goods for exports. This factor may be particularly important because capital expenditures on defence related goods tend to be concentrated in export sectors such as machinery, transport and electronics.

Fourth, a cut in military spending may result in increases in rates of productive investment, including expenditure on education, which may increase international competitiveness in the long term. Fifth, a cut in military imports may reduce future external debt burden of countries that borrow to finance military imports. Finally, if a cut in military spending is accompanied by sound macroeconomic policies, including sustainable fiscal deficits and stable prices, this may improve a country's competitiveness in world markets.

While some developing countries are important exporters of arms, as a group they tend to be major net importers of military hardware. For example, between 1984 and 1986 arms imports as a percentage of total imports ranged between 6.1 and 8.3 per cent for developing countries compared with 0.4 and 0.7 per cent for developed countries.[33] Similarly over the same period, arms exports as a percentage of total exports ranged between 0.5 and 1.4 per cent among developing countries compared with 2.1 and 3.1 per cent for developed countries.[33] Thismay mean that a cut in military spending involving arms will tend to improve the balance of payments of developing countries by reducing imports. As for developed countries involved in military exports, the impact of such a cut on the balance of payments would depend largely on how exports of military or non militray goods are affected. As noted above, non military exports could increase are resources are released from military use. Military exports could also increase if the net effect of the cuts is to force arms producing firms to be aggressive on export markets. On the other hand, as domestic sales plummet, production costs could escalate making arms producing firms uncompetitiveness on the world arms market.

An uncompensated cut in military spending will tend to be deflationary and may improve the balance of payments as imports fall in response to declining incomes while exports may increase stimulated by lower domestic costs. On the other hand, a compensated cut in military expenditures tends to increase incomes and, consequently, imports (other than military) while export performance may suffer from higher domestic costs.

**The Impact of Military Conversion**

The impact of conversion on economic performance and, to a lesser extent, on the environment, depends on how conversion is conceived and implemented. There are two broad conceptions of the conversion process. One conceptions defines it, as a process through which production facilities and human resources whose products are no longer in demand are transformed or retrained to make them suitable for production of goods still in demand.[34] Defined in this way, conversion is a process that is intended to improve capital and labor mobility from declining to growing industries during structural adjustment. Whilemost economists welcome efforts to improve resources mobility they differ on how the burden of conversion should be shared between the affected private parties and the public sector.[35] One view argues that a free market places adequate compensation for victims of economic change and that adding compensatory subsidies over and above such compensation can undermine incentives for resources to move quickly and efficiently. An alternative view is that there may be special instances in which compensation to the economically injured represents a necessary price for effecting rapid economic change. In addition, advocates of this view argue, the costs of moving resources from one use to other uses may slow down or even thwart the process of structural adjustment.[36]

Military conversion refers to a process through which *military* production facilities and human resources are transformed and retrained to make them suitable for civilian production. Defined in this way, military conversion is a special case of conversion as defined above. However, military conversion is some times conceived differently. In addition to its role in transforming military production facilities for civilian use, it is also often conceived as a process that stresses job creation and maintenance and, in some cases, as a process that is concerned with production of "socially useful" goods.[37] For example, military conversion has been described as a process that aims at re-employing workers formerly producing military goods into the same industries, and whenever possible, the same factories in which they previously worked.[38] In line with this view, military conversion is influenced by workers and communities who are affected by cuts in military spending or who wish to reduce dependency on military spending. Military conversion, however defined, often calls for comprehensive programs of financial support by government: but the extent and nature of government support and, therefore, the economic and environmental impact may differ depending on how it is conceived.

Military conversion can affect macroeconomic performance in three ways. First, when defence expenditures are reduced, military conversion can improve the efficiency of an economy by reducing economic friction through improving mobility of capital and labor, and by preserving capital which would otherwise become obsolescent. In this way it assists the process of resource allocation in a market economy and helps maintain growth and productivity in periods of structural adjustment.[39] Its impact on output, employment and inflation would tend to be positive because without conversion economic performance could be worse. However, conversion can also have negative impact on economic performance. This may be the case, for example, when conversion focusses on employment creation and maintenance in the same industries and production facilities that have suffered from cuts in defence spending. In addition, the emphasis on production of "socially useful" goods could run counter to market valuation of such production and could result in continuing government subsidies to support entities which may not be commercially viable. The danger, therefore, is that this form of conversion may encourage economic rigidities while discouraging resource mobility to areas where they are most valued.

Second, by reducing social costs of adjustment, military conversion may help to make it politically feasible to effect large cuts in military expenditures and in this way increase the speed of adjustment to the desired levels of military spending.[40] This in turn could reduce fiscal and current account deficits by larger amounts than would be the case without conversion. Finally, if, as is often the case, military conversion is supported by government subsidies, this may reduce the size of savings from defence expenditure cuts and, therefore, the net positive impact on fiscal deficits. The extent of government assistance and, hence, the negative fiscal and economic impact may be greater when conversion stresses job creationand maintenance compared with when it is viewed purely as a policy for improving capital and labor mobility. It may be important in this regard to ensure that government subsidies for conversion are temporary to avoid long term support of entities that may not be commercially viable. Failure to do so may lead to continued state support of inefficient entities with adverse effects on output, employment and inflation. The chances of such a situation developing may be high, especially in countries in which public entities are major producers of arms.

The environmental effects of military conversion are difficult to predict, in part, because they depend on the nature and extent of the conversion process. To the extent that militray conversion places emphasis on production of "socially useful" goods and is sensitive to environmental concerns, it would tend to have positive environmental effects. For example, advocates of military conversion have emphasized shifts in the structure of public spending from the military to education and training, health care, environmental clean up and protection, housing development and civilian physical infrastructure. Except, perhaps, for housing development and building the physical infrastructure, all these expenditures seem to be environmentally as clean or cleaner than militray spending. If military related production and expenditures are more pollution intensive than other forms of production and expenditures, then it could be said that by making it possible to effect cuts in military expenditures, militray conversion has positive effects on the environment. However, the extent of the positive environmental effects may be reduced by the fact that in some cases conversion may also aim at producing products in the same industries in which employees previously worked. If this were to happen, then the pollution intensity of output of the converted facilities may be the same as before the conversion.

Facing sources for military conversion may include "peace dividends" when they exist. When there are no such savings, military conversion could be financed from a variety of sources, including savings from reduction or elimination of government waste in all ares (including defence) and savings from elimination of explicit or implicit subsidies, especially those on environmentally harmful inputs such as fertilizers. Pesticides and fossil fuels. The extent of government financial assistance for conversion programs is likely to vary over the business cycle. For example, when the economy is operating at full capacity and experiencing labor shortages many workers from military related establishments can find alternative employment without any government assistance. On the other hand, during a recession when unemployment rates are high, the number of workers needing government assistance through conversion process would tend to be higher.

## Concluding Remarks

For a number of countries the changed international security environment has meant that less defence inputs may now be required to produce the same level of national "security". Becase of the potential savings on defence outlays resulting from these developments, prospects for real cuts in defence spending have improved dramatically. Such prospects raise three important questions. Would cuts in defencespending improve or harm economic performance? What would be the environmental effects of such cuts? And how would military conversion affect the adjustment process to reduce levels on defence spending? This paper has attempted to provide some answers to these questions by examining the issues at the macroeconomic level.

Its conclusions are that (a) the macroeconomic effects of a reduction in military spending vary from country to country and from time to time depending on a variety of factors, including, crucially, macroeconomic policies adopted to offset and short run possible negative effects of reduced military spending. In general, if sound macroeconomic policies are adopted, output and employment could increase, inflationary pressures could ease and the balance of payments could improve; (b) the environmental effects would depend on whether or not output and employment increased which in turn depends on factors noted in (a) above, and on the distribution of the "peace dividend"; and (c) by reducing costs of adjusting to lower levels of defence spending, military conversion can improve economic performance by improving resource mobility and making the transition to lower levels of military spending rapid and efficient. At the same time, conversion could worsen economic performance if government assistance to affected parties were to introduce economic rigidities in efforts to save or create employment. The impact of military conversion on the environment would depend on the pollution intensities of goods produced by "converted" resources compared with the pollution intensities of output generated by military spending.

## Endnotes

1. I would like to thank Ved Gandhi, Nuri Erbas and Obert Nyawata for valuable comments. The views expressed in this paper are the personal views of the author and do not, in anyway, reflect the views of the International Monetary Fund.

2. This view, which has also been referred to as "military Keynesianism", is based on two tenets of Keynesian economics: that private sector demand may not create sufficient employment, and that government spending can bring full employment. See, for example, Alexander (1989).

3. See, for example, Rosenberg (1976). The importance of such technological spill overs effects need to be compared against alternative uses of resources. In a comparative study of the impact of R and D on the performance of high technology industries DiFilippo (1991) concluded that the major problems that many US and UK high technology industries have faced in losing domestic and foreign markets are strongly related to the fact that military R and D preempts technical talent in these countries.

4. The argument that military spending diverts resources from productive uses dates back to at least Adam Smith. It is a long run argument which assumes that resources are fully employed and that resources diverted from defence are invested rather than consumed. Recently, Aschauer (1989) and Alexander (1991), among others, have provided empirical support for this view. For a discussion of the view that defence spending diverts resource from investment and can be an importantsource of economic decline -- the depletion argument -- see, for example, DeGrasse (1983) and Dumas (1991).

5. See Neuman (1979).

6. See Kennedy (1974).

7. See, for example, Adams *et al.,* (1987) and Smith *et al.,* (1983).

8. Smith *et al.,* (1983) in a survey of OECD economies came to the conclusion that the only single conclusion about the effects of military expenditures that one can come to is that it depends on the nature of expenditures, the prevailing circumstances, and *current government policies*. See also Adams *et al.,* (1987) for a similar view.

9. This paper does not address questions related to appropriate levels of defence spending. For a theoretical discussion of these questions, see for, example, Stiglitz (1988).

10. For studies which incorporate offsetting measures of a cut in military spending, see Adams *et al.,* (1987), Barkert *et al.,* (1991), and Leontif *et al.,* (1965).

11. This conclusion reflects the fact that the full impact of government spending may not be felt domestically. For further discussion on differential impact of budget items on the domestic economy see Chelliah (1973).

12. For a definition and discussion of environmental externalities see Baumol *et al.*, (1988).

13. See Section VI below.

14. In addition to the two cases discussed in the text, one could also examine a casein which a cut if defence spending is matched by a cut in taxes.

15. For a sample of the discussion see, for example, Benoit (1978), Degrasse (1983), Bezdek (1980), Adams *et al.*, Chan (1985), (1987) Dunne (1986) and Barkert *et al.*, (1991).

16. For a similar view see Smith *et al.*, (1983), Adams *et al.*, (1987), and Dunne (1986).

17. For a discussion of fiscal policy in stabilization programs and of budgetary and financial policy, see Tanzi (1987) and Buiter (1990), respectively.

18. See Rivlin (1987). for an analysis of world military expenditures see Hewitt (1991 a and 1991 b).

19. However, over the medium term the fact that amortisation of foreign loans would be lower with than without the cuts means resort to domestic financing would be reduced by a cut in military spending.

20. Because the domestic fiscal impact will be less than when no military imports are involved.

21. See, for example, Barkert *et al.*, (1991) and Leontif *et al.*, (1965).

22. Output would increase because, under the stipulated assumption that the import content of the cut military spending is higher than the import content of other public expenditures, the size of the domestic deficit would increase as a result of the cut and would, therefore, be expansionary.

23. See, for example Barkert *et al.*, (1991). Although a number of studies of OECDcountries have also examined the impact of cut in military expenditures, the policy responses assumed are different from the one considered by Barker *et al.*, For example, studies done for Sweden, Germany, and Norway assumed that the "peace dividend" would be allocated as aid to developing countries. For a survey of studies for Canada, France, Germany, Norway, Sweden, United Kingdom and the United States see Dunne (1986).

24. See Rivlin (1987).

25. This is because if we assume uniform pollution intensity of output for all expenditures, then amounts of environmental externalities would vary directly with output. The impact of changes in the structure of demand that would result from reduced levels of military spending will, in part, depend on the distribution of the increase in other public expenditures. The impact of distributions suggested by some advocates of military conversion such as increased levels of spending on physical infrastructure, education, health, and environmental clean up and protection may have positive effects on the environment. For a discussion on restructuring of public expenditure see Feldman (1991).

26. In order to predict the impact of fiscal policy on the environment it is critical to know or to be able to make reasonable assumptions about the pollution intensities of various outputs.

27. See United Nations Center for Disarmament (1978).

28. For a discussion of the implication of military spending on productivity growth see, for example, Melman (1974) andDumas (1977).

29. See Melman (1974) and Dumas (1977).

30. Allocating the "peace dividend" entirely for consumption could be worse (from a productivity point of view) than reducing military spending because it could result in a loss of investment and technological improvements associated with some military spending and that can be used for civilian purposes.

31. See Rothschild (1973)

32. See U.S. ACDA (1975/86).

33. See U.S. ACDA (1975/86).

34. See U.S. Congress (1980)

35. See U.s. Congress (1980).

36. See U.S. Congress (1980).

37. See for example, Gordon (1984).

38. See, for example, Gordon *et al.,* (1984).

39. Willet (1990) has emphasized the microeconomic and political aspects of conversion pointing at the fact that the observation that a reduction in military spending need not lead to unemployment does not mean that unemployment will not actually take place. It may be because of the large differences in level of skills between defence and other workers.

40. For example, Alexander (1989) had argued that without conversion options, policy makers sometimes support defence expenditures which may not be needed for national security but are vital for maintaining local employment.

**References**

Abell, J.D., "Defence Spending and Unemployment Rates: An Empirical Analysis Disaggregated by Race," *Cambridge Journal of Economics* Vol.14, No.4, (December 1990).

Alexander, J.D., "Manufacturing Productivity and Military Depletion in the Postwar Industrial Economy:," *Towards Peace in the United States: Essays on Military Disarmament and Economic Conversion* ed. Bischak, A., (New York: St. Martin's Press, 1991).

-------, "Peace Without Depression, "National Commission for Economic Conversion and Disarmament (Washington, 1989).

Adams, G. and D.A. Gold, "The Economics of Military Spending: Is the Military Dollar Really Different," in *Peace, Defence and Economic Analysis* ed. by Schmidt, C., and F. Blackaby (New York: St. Martin's Press, 1987).

Aschauer, D., "Is Public Expenditure Productive?" *Journal of Monetary Economics* (1989).

Baumol, W.J., and W.E. Oates, *The Theory of Environmental Policy* (New York: Cambridge University, Second Edition,1988).

Benoit, E., "Growth and Defence in Developing Countries", *Economic Development and Cultural Change* Vol.26, (January, 1978).

Bezdek, R., "The 1980 Economic Impact - Regional and Occupational - of Compensated shifts in Defence Spending", *The Journal of Regional Science* Vol.15, No.2, (1975).

Boulding, K., "The Impact of the Defence Industry on the Structure of the American Economy", in *The Economic Consequences of Reduced Spending* ed. by Udis,

B.,(Lexington, Massachusetts: Lexington Books, 1973).

Buiter, W.H., *Principles of Budgetary and Financial Policy* (New York: Harvester and Wheatsheaf, 1990).

Chelliah, R.J., "Significance of Alternative Concepts of Budget Deficit", *Staff Papers*, International Monetary Fund, (Washington) Vol.20, No.3, (November 1973).

DeGrasse, R. Jr., *Military Expansion and Economic Decline* (New York: Sharpe, 1983).

DiFilippo, A., "Military Spending and Government High-Technology Policy: A Comparative Analysis of the United States, Germany, Japan and Great Britain", In *Towards Peace in the United States: Essays on Military Disarmament and Economic Conversion* ed. Bischak, A., (New York: St. Martin's Press, 1991).

Dumas, L.J., "Economic Conversion, Productive Efficiency and Social Welfare", *Peace Research Review* Vol.7, (1977).

Dumas, L.J., "National Security, Non-Contributive Activity and Macroeconomic Analysis: Theoretical, Empirical and Methodological Issues", In *Towards Peace in the United States: Essays on Military Disarmament and Economic Conversion* ed. Bischak, A., (New York: St. Martin's Press, 1991).

Chan S., "The Impact of Defence Expenditure on Economic Performance: A Survey of Evidence and Problems", *Orbis* (Summer 1985).

Elliot, D., and H. Wainwright, "The Lucas Plan: The Roots of the Movement", in *Economic Conversion* ed. by Gordon, S., and D. McFadden (Cambridge, Massachusetts: Ballinger Publishing Company, 1984).

Feldman, J., "Constituencies and New Markets for Economic Conversion: Restructuring the United States' Physical, Environmental and Social Infrastructure", in *Towards Peace in the United States: Essays on Military Disarmament and Economic Conversion* ed. Bischak, A., (New York: St. Martin's Press, 1991).

Gold, D., "Conversion and Industrial Policy", in *Economic Conversion* ed. by Gordon, S., and D. Mcfadden (Cambridge, Massachusetts: Ballinger Publishing Company, 1984).

Gordon, S., "Economic Conversion Activity in Western Europe", in *Economic Conversion* ed. by Gordon, S., and D. McFadden (Cambridge, Massachusetts: Ballinger Publishing Company, 1984).

------, and D. Mcfadden, "Introduction", in *Economic Conversion* ed. by Grodon, S., and D. McFadden (Cambridge, Massachusetts: Ballinger Publishing Company, 1984).

Harvey, S., W.H. Francis, A.H. Jeffrey, and J.R. Freeman, "The Relationship Between Defence Spending and Inflation", *Journal of Conflict Resolution* Vol.28 (march 1984).

Hewitt, D.P., "Military Expenditure: International Comparison of Trends", IMF Working Paper WP/91/54 (May, 1991a).

------, "Military Expenditure: Econometric Testing of Economic and Political Influences", IMF Working Paper WP/981/53 (May, 1991a).

Kennedy, G. *The Military in the Third World* (New York: Charles Scriber's Sons, 1974).

McKinlay, R.D., *Third World Military Expenditure* (London: Pinter Publishers, 1989).

Melman, S., *The Permanent War Economy* (New York: Simon and Schuster, 1974).

-----, *Profits Without Production* (Boulder, Colorado: Westview Press, 1982).

Neuman, S., "Reply", *Orbis* (Summer, 1979).

Owens, S., "Defence and the Environment: The Impacts of Live Firing in National Parks", *Cambridge Journal of Economics* Vol.14 (1990).

Renner, M.G., "Swords into Plowshares: Converting to a Peaceful Economy", *Wolrdwatch Paper* Vol.96, (June, 1990).

Rosenberg, N., *Perspective on Technology* (New York: Cambridge University Press, 1976).

Rothschild, K.W., "Military Expenditure, Exports and Growth", *Kyklos* Vol.26, (1973).

Sanger, C., "The West and the Rest: Conversion and Third World Development", in *Economic Conversion* ed. by Gordon, S., and D. Mcfadden (Cambridge, Massachusetts: ballinger Publishing Company, 1984).

Smith, R., "Military Expenditures and investment in OECD Countries", *Journal of Comparative Economics* (March, 1980).

-----, and G. Georgiou, "Assessing the Impact of Militray Expenditure on OECD Economies: A Survey", *Arms Control* Vol.4, (May, 1983).

Stiglitz, J. *Economics of the Public Sector* (New York: Norton, 1988).

Tanzi, V., "Fiscal Policy, Growth and the Design of Stabilization Programs", IMF Working Paper WP/87/1 (1987).

United Nations Centre for Disarmament, *Economic and Social Consequences of the Arms Race and Military Expenditure* (New York: United Nations, 1978).

U.S. Arms Control and Disarmament Agency, *World Military Expenditures and Arms Transfers* (1975/1986).

Willet, S., "Conversion Policy in the United Kingdom", *Cambridge Journal of Economics* Vol.14, (1990).

Yudken, J.S., "Conversion in Aerospace Industry: The McDonnell Douglas Project", in *Economic Conversion* ed. by Gordon. S. and McFadden (Cambridge, Massachusetts: Ballinger Publishing Company, 1984).

# MILITARY RESOURCES TO THE ENVIRONMENT

**Maj Britt Theorin**
*Member of Parliament, f. Ambassador for disarmament,*
*Hon. Dr. of Philosophy, Chairman of the UN Study on Nuclear Weapons*
*and the UN study on Military resources to the Environment*

The cold war is over. A totally new situation has developed in Europe. The WP-pact does not exist any longer. Drastic disarmament of both nuclear and conventional weapons now takes place. Common security and cooperation has replaced the former suspicion and rivalry. This new international situation has totally changed the map of Europe both politically and militarily and the risk of military confrontation between the two former opponents has dramatically diminished.

In this new atmosphere of East-West co-operation there are radically improved prospects for significant reductions in weapons and armed forces.

But other threats to the global society are at the same time increasing. The degradation of the environment is a growing threat to global security. The environment does not see any national borders. Pollutions, the effects of the ozone layer, acid rain etc. do not stop at any boundary.

Resources now used on the military side could be used in order to protect, restore and improve the environment. This was the idea behind the Swedish proposal 1990 of a UN-study "Charting potential uses of resources allocated to military activities for civilian endeavours to protect the environment" which should identify resources, currently used for military objectives in technology, know-how, production, the infrastructure etc. which may be reallocated for environmental purposes, both in immediate and a longer term perspective.

There are many areas in the military field which could be used for the environment i.e.

-**industrial and technological** capacity within the areas of transportation, energy and field engineering;

-**organizational** capacity within the military sector to educate and mobilize man power;

-existing **networks** of the military sector, universities and other institutions for the dissemination of **information and data**;

-**satellite technologies**, for instance, in the field of remote sensing;

-**research capacity** in, for instance, laboratories and computer facilities to detect and combat environmental degradation;

-personnel and know-how, military equipment for **disaster relief**;

-military equipment, personnel and know-how for the handling and **disposal of dangerous materials** including the destruction of weapons.

UN decided during the fall of 1990, with a strong majority, to set up a small study group in order to study potential uses of military resources which could be used to protect and restore the environment. Six countries had experts in the UN Group; Brazil, China, Ghana, Soviet, USA and Sweden. I had the honour to chair the Group.

The UN Expert Group worked only during six months in order to introduce their report in time to influence the UN Conference of Environment and Development (UNCED) in Brazil in June 1992.

In July 1991 we finalized our consensus report, which then was introduced to the preparatory committee in August, where it was received very positively, not least due to the fact that it has a lot of concrete proposals.

In the report the expert group describes the background and the environmental challenge, the environmental impact of military activities and some national experiences. It goes on to discuss environmental strategies and needs and how to integrate military-related resources into environmental strategies.

The utilization of military-related resources for environmental purposes would essentially depend on two major steps: an inventory stage, identifying their suitability and a political action plan to ensure their availability. Military-related resources vary greatly from one nation to another in size, composition and technological level. No matter what the specifies are, these resources comprise in principle:

- **Manpower**, including regular armed personnel, conscripts, as well as civilian, technical and support staff;

- **Professional skills** and **technical know-how** of commanders and soldiers, as well as scientists, technicians and other staff in institutions and agencies supporting the military organization proper;

- **Equipment** of all kinds, from hand guns to tanks, ships and aircraft to advanced laboratory instruments;

- **Funds** allocated by the government for salaries to employees in the various branches of defence, for maintenance of existing equipment and for research, development and procurement of new equipment;

- **Infrastructure** covering land use, capital stock, production plants, machinery, factories, design bureaus and buildings in various stages of their useful life;

- **Technological capabilities** including the ongoing R & D endeavours.

In some ways, military establishments are uniquely placed to strengthen the international civilian capabilities for the implementation of environmental strategies. Military personnel are well-equipped for dealing with catastrophic situations, which could be useful in responding to environmental disasters and handling and disposal of highly toxic radioactive or otherwise harmful substances.

The report goes on to describe the technological capabilities of military establishments and ways of making the military sector a stakeholder in environmental protection. It also discusses costs and other considerations.

The defence community has at its disposal a wealth of information, gathered by intelligence sources that can assist in tracking changes in the atmosphere, the oceans and the surface of the earth. Military satellites, aircraft, surface ships and submarines have the ability to collect additional information about climatic changes and the flow and temperature of oceans.

Techniques for military surveillance such as "tagging" could be easily employed for monitoring transportation of pollutants and toxic materials, and for ensuring compliancewith ecologically safe methods of weapons disposal.

In one chapter the report describes the environmental applications of military-related technologies. Of all the military-related resources technologies qualify most for environmental use as most technologies are inherently dual-purpose. Advanced technologies are necessary for responding effectively to the environmental challenge.

Information technology, in particular, is easily adaptable for environmental purposes. Many of the results of military applications of information technology can be utilized for environmental ends without costly modifications of equipment and retraining of personnel.

For the performance of military missions, modern military establishments have developed highly sophisticated techniques, e.g., sensors; platforms, e.g. satellites; equipment, e.g. computers, communication networks, global positioning systems; and exercises, e.g. simulation and modelling.

In some instances, environmental applications are either already undertaken or are underway. In others, the underlying technology can be adapted.

The accelerating impact of human activities on the environment has become a decisive factor in planet Earth's prospects for survival and well-being. Global environment is remarkably resilient and life on earth has survived many onslaughts far more cataclysmic than any human intervention. Yet there are limits to nature's capacity for self-preservation which are not known.

Environmental integrity and sustainability of development are issues of paramount importance. In a fast-changing international climate, preservation of environment mightwell be the next arena for initiatives to break new grounds for multilateralism and the role of the United Nations. The global nature of the environmental challenge makes it imperative that international actions go hand in hand with national actions.

The world of today is very different from the time when environmental issues were first put on the international agenda. The seventies were marked by a series of national initiatives to address environmental problems. The eighties witnessed a global environmental awakening. The nineties, however, are poised for a comprehensive rescue plan for the global environment.

Global environment recognizes no divisions of political blocs, military alliances and economic systems. The opportunities for translating this recognition into practical actions were never more promising than now. Rapid advances in technology have opened up new possibilities for meeting the environmental challenge. Preservation of the environment is one new channel for the vast energies released from the end of the cold war.

Either as a result of negotiated bilateral and multilateral agreements, or as a unilateral action, many governments are reducing the size of their military establishments. The human, material and technological resources devoted to the military sector are either being reduced, released or turned idle. Even where no measures of arms reductions have been undertaken, the existing capabilities of the military establishments could be profitably used as environmental tools.

Protection of the environment requires a comprehensive set of measures: protective measures for damage prevention and control; restorative measures for repair, rehabilitation and environmental compliance; remedial measures for development of ecologicallybenign and energy-efficient technologies; and promotional measures for raising environmental consciousness through education.

There is also an urgency to develop capabilities for responding promptly to environmental disasters and to forestall such occurrences through constant monitoring of the environment.

In many ways the worldwide military establishments have a unique potential for augmenting the civilian endeavours to protect the environment. Military organizations are well-equipped for dealing with catastrophic situations. Suitably qualified military personnel could be employed for responding to environmental disasters and handling and disposal of highly toxic radioactive or otherwise harmful substances.

The defence community has at its disposal a wealth of information gathered by intelligence sources which can assist in tracking changes in the atmosphere, the oceans and the surface on the earth. Military satellites, aircraft, surface ships, and submarines have the ability to collect additional information about climatic changes and the flow and temperature of oceans.

Techniques for military surveillance could be easily employed for monitoring transportation of pollutants and toxic materials and for ensuring compliance with ecologically safe methods of weapons disposal.

Information technology, in particular, is easily adaptable for environmental purposes. Many of the results of military applications of information technology can be used for environmental ends without costly modifications of equipment and retraining of personnel.

Clearly long-term objectives of environmental protection require an understanding of the technical issues underlying natural and man-induced risks, the ability to formulate clear policy decisions and the capability to respond swifty to environmental emergencies. Military establishments can assist in achieving these objectives by contributing their technical expertise, advanced equipment, and communications and surveillance systems.

A considerable portion of the military spending of the technologically advanced countries is devoted to research and development. National decisions to incorporate environmental concerns in the ongoing R&D endeavours could be particularly rewarding in the development of ecologically benign sources of energy, environmentally safe methods of weapons disposal and techniques for treatment of hazardous and toxic substances.

The recommendations in the report are three folded; towards UN, towards national governments and towards UNCED. The Group recommended that UN should create international *environmental relief teams*, what we called "Green Berets" - I recently heard that foreign minister Genscher spoke of "Green Helmets" - making use of personnel, equipment and facilities from the military sector and put at the disposal of the UN by national governments. These relief teams should respond to environmental emergencies on the request of a country.

The Group also recommended that the United Nations should facilitate *global sharing of environmental data*, including that obtained through the use of military-related satellites and other information gathering platforms and act as a *clearing-house* for international exchange of information about actual national experiences in environmental applications of military-related resources.

Since sovereign national cooperation is vital for a global effort in employing military-related resources for environmental purposes, the Group recommended that national governments *make inventories of their environmental needs* and military-related *resources* applicable for environmental purposes, utilize such resources in national environmental *action plans* and report their experiences to the UN.

They should consider what *military-related resources* they could place at the *disposal of the UN* or other international bodies on a temporary, long-term or stand-by basis as instruments of international multilateral co-operation for responding to environmental disasters and emergencies. Governments should also ensure that *military activities conform to environmental norms* and regulations and *correct* effects of past negligence in this respect.

They should develop environmentally sound technologies for weapons disposal and *incorporate environmental concerns in their military R&D programmes* and integrate the aims of environmental preservation and sustainable development in their *concepts of security*.

Let me here add some new facts and thoughts.

In the environmental report STATE OF THE WORLD - 91 from the well known American World Watch Institute states that the military is the single greatest polluter of the world.

In my own country, Sweden, a study from IVL, Institute for research on water and aircare, reports that the military is behind 90% of all collected waste, between 800 to 900 tonnes per year, from transport and handling of airfuel. And the military areexempted from environmental tax for released exhaust. This is of course not reasonable.

Another military environmental problem is the waste from fire practice. Metals like copper and lead are released in great amounts. Studies at the defence committee of Norway showed that one third of all disposal of lead in Norway came from military fire practice.

Soviet nuclear scientists and engineers, now unemployed, could be used to clean up the enormous damage to the environment caused by forty years of nuclear weapons production. Who better to clean up the mess than those who made it ? This could be done in an effective way by the USA investing in ideas and procedures for cleaning up nuclear weapons production sites in Russia.

The United States faces a $ 300 billion clean-up challenge throughout its own weapons production complex. The Department of Energy acknowledges that it still lacks the know-how and technologies to accomplish this gargantuan task. But Russian scientists already have clean-up knowledge from Chernobyl and Kyshtem. A co-operative program could benefit both parties. Employing a qualified Russian scientist costs much less than the price of a comparable American. And this could be financed by a small amount from what was already decided by Congress for warhead dismantlement.

Back to the Study.

Lastly, the Group recommended that the conclusions and recommendations of the report be incorporated into the elements for an "Earth Charter" and "Agenda 21" in the UNCED conference.

The report was, as I said, earlier introduced at the preparatory committee ofthe UNCED in August 1991 and later in the fall, the General Assembly by consensus decided that the Report should be sent to UNCED. The Report as such will by now be on the table of UNCED.

Now it is up to interested individuals and governments to work for its *recommendations* to be adopted.

# TECHNICAL CO-OPERATION TO SUPPORT THE CONVERSION OF MILITARY TO CIVILIAN INDUSTRY

**Dr. Horst Wiesebach**
*Deputy Director-General, UNIDO*

Disarmament is proceeding at an increasing pace in many countries. They include the countries of Central and Eastern Europe in which UNIDO has always had technical co-operation programmes, together with the republics of the former Soviet Union (the CIS states). In the medium term, it is hoped that similar trends will emerge in developing countries as development aid is increasingly being linked to such criteria as the level of military expenditures compared to other key indicators of development.

Two main factors have been instrumental in accelerating the process of disarmament in the recent past. One was the need to reduce international tensions which threatened mankind with global extinction given the destructiveness of modern weapons, and forced governments to undertake ever more complex counter measures against potential military threats. The other was the unbearable cost of modern defence policies which threatened to undermine economic welfare, especially in the poorer parts of the world.

For the first time in several decades the evolution of East-West relations has permitted the reduction of arms expenditures on a much larger scale than seemed imaginable a few years ago. As such reductions will affect both employment and industrial activity by lowering incomes, turnover and profits based on the military industrial complex, they may reduce overall and regional economic activity, depending on the relative size of that sector and the importance of the cuts in expenditures (e.g. as a share of gross national product or of turnover in certain industrial sectors, such as ISIC 38). On theother hand, the savings achieved by such cuts in expenditure (the so-called peace dividend) could be actively used to counter-balance depressive efforts by promoting economic and social development both at home and abroad. Conversion of redeployment thus means shifting resources from the production of goods and service for military purposes to production of a civilian nature. Major shifts of resource use and of production are by no means a new phenomenon. Rather, economic history shows that constant and sometimes brusque changes are quite normal in economic life, due to technological change, discovery of new materials or change in consumer preferences. History also shows that large-scale redeployment of resources caused by massive disarmament can be successfully managed by governments and economic agents, as was the case in the demilitarization of the world economy between 1945 and 1948 which started a postwar boom and led to widespread gains in prosperity.

There are two major types of conversion from military to civilian activities. One of them concerns the military establishment as such. Conversion usually starts with the demobilization of soldiers and civilians directly employed by the army and military administration. This affects the need for military hardware and therefore may lead to rededication of such hardware or to its scrapping. Finally, military sites (barracks, depots, training grounds, etc.) may need to be cleaned up and put to other use. While all of these aspects are important and some of them may affect the national economy, e.g.through the employment problem created by demobilization, I shall not deal with them as they are not a specific problem of the industrial sector. They are taken up else where in this conference by more competent specialists.

The other type of conversion relates to that part of the economy mainly industry and the R&D institutions, that directly or indirectly has the military as a client. Here it is important to differentiate between the major economic systems. While in the Western system, only very few enterprises are completely dependent on military procurement - the large majority having a dependency ratio of 5-15 percent, measured as the share of military sales total turnover- the situation is completely different in the Eastern system, where the ministries responsible for military production have created a huge military-industrial empire, totally devoted (or nearly totally) to military procurement and research. We can, therefore, observe three types of military production systems which will be of relevance for conversion:

a. enterprises of the military-industrial complex fully devoted to military production and research;
b. similar enterprises, devoted partly to military production and partly to civilian "markets" (Soviet authorities claimed that 40 percent of total capacity was devoted to civilian goods and developed in 1988/89, a State Conversion Plan to raise that ratio to 65 percent until 1995);
c. enterprises in the civilian sector that also receive military orders.

It is very difficult to find exact and clearly defined quantitative data, mainly for three reasons:

- military secrecy prevented clear-cut statistics being put up by the official sources of such data.[1]

- methodological differences in national accounting and major structural differences in prices and wages systems make it difficult to compare statistical data;

- drastic changes in the economy of the former USSR are presently affecting both military expenditure and production.

While there is hope for an improvement of the statistical situation in the future as all three reasons will be progressively removed, great efforts will be needed in the near future to establish detailed data in order to give a better basis for decisions on conversion.

In spite of the difficulty of putting together a comprehensive set of data, a few figures will indicate the order of magnitude of military production and thus the size of the conversion problem. It may be safely assumed that military expenditures in 1989 were at a level of US $ 900 billion (in prices of 1988).[2] It can further be assumed that the figure for 1992 will probably be lower by 10 percent. In 1989, military expenditure was furthermore roughly equal between the United States (roughly $ 300 billion current) and the USSR (some $ 270 billion equivalent).[3] The share of military expenditure was about 6 per cent of the US gross national product, while a similar figure of the USSR is hard to establish for the reasons given above. While official figures indicate a share between 8 and 12 percent of the national product, there are also estimates going as high as 20 to 25 percent of the national product, taking into account indirect costs and flawed data.[4] This order of magnitude is realistic as can be inferred from certain other data, e.g. on the share of military production in specific branches of industry. Woprosy Ekonomiki, anoted economic journal, recently gave total direct employment of the Soviet defence industry as 15 million[5] and Igor Kljamkin of the Academy of Sciences of the USSR claimed that the military-industrial complex of the USSR comprised half of the total industrial establishment of the Soviet Union.[6] The former Mayor of Moscow, Lev Zaikov, was quoted by Alexei Izyumov, the well-known Soviet expert on conversion, as saying that one-third of Moscow's workforce and one-half of its research establishment was devoted to military ends. These are indeed awesome figures. They are matched, atleast in absolute terms, by some Western data.

In 1989 the US Defence Department awarded $ 120 billion worth of prime defence countries within the USA, plus another $ 19 billion outside of the USA.[7] SIPRI gives total arms sales of the "SIPRI 100", i.e. the leading one hundred producers of arms in the OECD countries and in developing countries, as $ 169 billion in 1989.[8] However, employment through arms production in the West is much smaller, firms are much less dependent on arms sales than in the East and they have shown remarkable flexibility in the past in adapting themselves to even large swings in the market. Therefore, conversion in the West, even though it may hit individual firms or communities with more impact than others, has a different quality from that in the East.

In the former Eastern block countries the disarmament process coincided with an ambitious attempt, known as Perestroika, to change the guiding principles and the structure of the economic system as a whole. This adds another strain, in addition to the sheer weight of the military-industrial complex, on the process of conversion. It is well-known by now that perestroika has affected economic policy in a very uneven manner, as no clear concept was ever put forward or implemented for dissolving state ownership of the major means of production,for changing the highly centralized planning system, for minimizing public price fixing and intervention, for opening the economy to international co-operation and for liberalizing the economy in general. The uncertainty created hereby has vastly complicated the development of the economy and the conversion process in particular. While earlier on, Soviet economists had assumed that state ownership of the means of production and strong central planning would facilitate the conversion of military industries, it is now increasingly recognized that the magnitude of the task of converting a huge and extremely complex industry to civilian output has been grossly under estimated. The only recent example of a moderately successful conversion is offered by China which managed to raise within ten years (1979 to 1989) the share of civilian production in the military industry from 10 to 66 percent. China, however, always had a less centralized planning system and a much less complex military-industrial establishment than the Soviet Union. It also introduced economic reforms much earlier, namely in the early eighties. Its conversion process received much support from capital and know-how either based in or channelled through Hong Kong. Even so, many plants experienced great difficulties and big losses and there are complaints about low quality production.[9]

Conversion efforts in the Soviet Union, which started in earnest after Mikhail Gorbachev's address to the United Nations General Assembly on 7 December 1988, have run into great problems. Alexei Izyumov gave three basic reasons for the failure to reach the goals of the State Conversion Plan of the former Soviet Union:

- the highly complex and sophisticated nature of the military industry;
- its compartmentalization, due to military secrecy requirements;
- the sheer size of military production and research.[10]

In view of the problems outlined above, one may wonder what will be the effect of the radical disarmament decisions taken by the RSFR at the beginning of 1992. Whatever the exact nature of the cuts, there can be no doubt about the enormous impact they will have at both the national, regional and institutional/enterprise levels. Such changes in policy will need to be synchronized with the restructuring at the micro-economic level in order to provide incentives and guidance and support to enterprise managers who will be faced with unprecedented challenges.

**Action needs to be undertaken at three levels**

**a. Government Level**

Governments must vigorously pursue the transition from a centrally planned to a market economy. It is by no means sufficient to set prices free, indeed complete liberalization may be counter-productive by giving too much leeway to speculators. Rather, governments must act to remove the many structural deficiencies of the economy by reforms in the areas of ownership of the means of production, investment and financial policies, monetary policies (including foreign convertibility), budgetary stability, production incentives via the taxation (and subsidies) system, employment and incomes policy, training of managers and middle level staff, trade policy and other means of international economic co-operation. While these policies will be needed in the short run, they must be supplemented by medium and long-term policies such as education, technology and research, infrastructure, decentralization and regional policies and, last not least, protection of the environment. It goes without saying that this amounts to an enormous task, short of changing the structure of total society. Must one add that such an effort needs similar support as Western Europe received after the end of the last World War?

**b. Institutional Level**

While the attention of policymakers is often directed to the national level, the vital bridge between the macro and micro-level, consisting of thousands of institutional networks, is often overlooked. Modern industry is not only composed of highly efficient enterprises, it could not function without a multitude of institutions and service agencies which create the humus on which individual enterprises prosper. This is the institutional (as opposed to the physical) infrastructure of the economy which covers such functions as training, financial, legal and accountancy services, marketing, design and fashion as well as advertising, R&D, information banks and centres, trade journals, trade and industry associations, consultancy services of all sorts. Such services will be especially needed by enterprises which have just been freed from the shackles of a command economy where many of the above functions were either not needed or assumed by the state's bureaucracy.

**c. Enterprise Level**

In view of the total dependency of individual enterprises on government intervention and their lack of know-how essential to coping with the market process, much of the technical support needed during the process of conversion must be directed to the enterprise level. It should be clear, however, that such effort would be wasted if, at the same time, the government would fail to fashion a proper policy as described in the two preceding paragraphs.

While the restructuring of enterprises in the civilian sector has been pursued already for several years and with some success,especially in Hungary and the CSFR, the situation is quite different in the military-industrial sector. As compared to the civilian sector, military industries offer both advantages and disadvantages in the process of conversion.

- The advantages are their superior technology, modern capital equipment, highly qualified workforce and the constant need for high quality production. This makes them fairly well suited to enter the high tech end of the product gamut, provided they find the proper niches in a highly competitive and dynamic international market.

- The disadvantages are that the firms were highly isolated by their ministries, that their research, their production programmes, their supplies and finance were strictly controlled from above and that they did not have to compete for market shares. Managers, engineers and workers were usually much better paid than their civilian counterparts. They needed to worry only little about the cost of their operations as long as they were able to produce that was requested. Geographic concentration for strategic reasons often placed the firms far away from potential consumer and supplier markets. Their enterprise "culture" tends to give high glamour to military products while civilian goods are considered as secondary, thus creating psychological barriers to conversion.

Clearly, there is a massive case for foreign support in the restructuring process at all three above-mentioned levels if conversion is to overcome its inherent risks. Private funds and initiative are indispensable, yet they will by no means suffice in view of the magnitude and the nature of the task. In some areas such as giving support to the government or creating certain institutions, noprivate funds will be forthcoming. Even at the enterprise level, which is the classic field of private investment, much of the investor interest will only be kindled once it is clear that good opportunities exist to gain market shares and make some profit. As the experience of the Treuhandanstalt, mandated with the restructuring task in East German industry, has shown, equity capital is only forthcoming if the enterprises have first been put on a secure footing. There will also be cases where privatization may not be indicated and where nevertheless successful conversion is needed, in order to prevent the workforce, often highly skilled specialists with sensitive know-how, from leaving their plant which will only be possible if they are offered a viable alternative.

**The Role of UNIDO**

UNIDO, as the organization of the UN system dealing with all issues of industrial development, has in recent years emphasized and expanded its technical co-operation and advice concerning industrial restructuring in numerous countries. These programmes adopt a comprehensive approach in order to integrate the financial, managerial, technological as well as policy dimension. UNIDO could advise both governments and enterprises on the direction of the conversion effort and undertake the needed policy and feasibility studies as well as put experienced teams into those agencies, institutions and enterprises where they are needed. As a side-effect, such advisory work would be used to build up local capacities to undertake similar work (consultancy, training, policy studies, etc.). In addition, UNIDO could use its world-wide network of industrial investment promotion services to draw investor interest to those cases of conversion where privatization and joint ventures are a promising alternative. Through this process it could bring to military industries its rich experience in the restructuring andmodernization of industrial enterprise in economies in transition belonging to Eastern Central Europe.

## Endnotes

1. Even the most recent edition of the SIPRI Yearbook (1991) still has a blank for the USSR in its listing of military expenditures.

2. Computed on the basis of SIPRI figures (1991 Yearbook), assuming a value of $ 270 billion for Soviet military expenditure.

3. cf. SIPRI Yearbook 1991, pp.117 and 145.

4. cf. SIPRI Yearbook 1991, p.145.

5. Quoted by Handelsblatt (German economic journal), Dusseldorf, 20 Dec. 1991.

6. Igor Kljamkin, Der sowjetische Weg zu Markt and Demokratie (The Soviet Approach to the Market and to Democracy), Aus Politik und Zeitgeschichte, Beilage zur Zeitschrift "Das Parliament", No.B 52-53-91, 20. Dec. 1991, p.10.

7. SIPRI Yearbook 1991, p.291.

8. Ibid., p.285.

9. cf. Jin Zhude and Chai Muliang, Conversion: The National Experience of China, and Arthur J. Alexander, National Experiences: a comparative analysis, both in UN Topical Papers on Disarmament, No.5, Vol.II, New York, 1991.

10. Alexei Izyumov, Conversion: The National Experience of the USSR, UN Topical Papers on Disarmament, No.5, Vol.II, New York, 1991.2

# UNITED STATES OF AMERICA

# DEBT REDUCTION IN EXCHANGE FOR ARMS CONVERSION: THE CASE FOR LINKING CONVERSION, DEVELOPMENT AND DEBT REDUCTION

**Gregory Bischak,** Ph.D.
*Executive Director, National Commission for Economic Conversion and Disarmament, U.S.A.*

## Conversion and the Post Cold War Era

The disengagement of the United States and the former Soviet Union from the Cold War brings with it the prospect of deeper reductions in world military spending. Clearly this is a welcome opportunity for these nations to end the nuclear and conventional arms races. Moreover, reducing military spending could free financial resources for these countries to address their respective budget deficits and many pressing domestic needs.

For decades, peace and disarmament advocates have advanced economic conversion as a means to effect an efficient transfer of resources from military production and operations to relevant civilian work. The objective has been to prevent job loss through advance planning and to lower resistance to reducing the military budget. Perhaps more importantly, it has been argued that releasing the capital, labor and scientific resources from military work, will permit new productivity enhancing applications of these resources. Despite this approach, conversion planning has met with an unhappy reception in both the west and the east, not to mention the developing world.

## Conversion in the United States

In the United States, despite the profound lessons from the conversion after the Second World War, the Korean War and after Vietnam, the post-Cold War conversion process is proceeding as an after thought. Little planning or attention has been given to prevent severe economic dislocation arisingfrom current and proposed military spending reductions.

Even limited Pentagon budget cuts passed by Congress to date have produced tremendous economic dislocation. Hundreds of thousands of workers have been affected by these cuts and **the nation has lost an opportunity to retain jobs and develop new business opportunities through a more comprehensive conversion program**. Nearly 300,000 defence industry workers were laid off in 1990 and 1991. Under Defence Secretary Cheney's FY 1992 plan, over the 1991-97 period active duty troops will be reduced by over 500,000, civilian DoD personnel will be cut by 155,000 and defence industry workers will be reduced by over 800,000.[1] Without a conversion program, cutting the military budget will mean more lay-offs and will exacerbate the recessionary pressures both regionally and nationally. Countering these recessionary pressures will require the implementation of a two fold conversion program.

## Structural Dislocation and Conversion

Cuts in DoD spending have caused economic dislocation in defence-dependent industries, occupations and regions. Advance conversion planning, targeted assistance and incentives are necessary to mitigate and prevent further losses in jobs and business. Incentives for contractors to re-tool their plants and restructure their firms for civilian production would help to develop new business opportunities and to preserve jobs. In addition, re-training programs are needed to assist managers, engineers and productionworkers to acquire the knowledge necessary to adapt to the cost and quality control requirements of civilian work. Military-dependent communities already hard-hit by base closures or major contract reductions require economic re-development assistance if they are to develop non-military business to re-employ workers and to salvage the small to medium-sized firms so vital to their regional economies.

**New Budget Priorities and Conversion**

Military spending has imposed trade offs on our budget priorities, leading our nation to make fewer productivity-enhancing public investments. This has been especially true for research and development (R&D) where the majority of federal funding has gone to the military, rather than for basic research and civilian-oriented infrastructure. Moreover, military spending has often crowded our public investments in human resources, particularly education and job training. In the 1980s, military spending increases occurred at the expense of public infrastructural investment.[2]

A re-alignment of federal budget priorities by investing military savings in these critical civilian needs could address fundamental economic and social problems while stimulating key economic sectors to produce jobs. Such public investments should be viewed as wealth-producing assets, rather than as a simple public expense. In the long-run, these expenditures would pay for themselves by generating new economic activity and restoring productivity. Thus it would be short-sighted to exclusively use military savings to reduce the federal budget deficit while ignoring the nation's most pressing economic and social needs.

A compensated shift in federal budget priorities using savings from military cuts to fund critical domestic needs is acomplementary fiscal policy which could cushion the economy from the shocks of the military wind down by stimulating the overall level of economic activity. Such a shift in federal budget priorities could help to maintain the level of employment and output and thereby prevent the current recession from worsening. Perhaps more importantly, public investments in education, job training, infrastructure and civilian research and development could enhance productivity and improve the quality of life.

Numerous economic studies have shown that a shift of federal spending to key infrastructural and human resource investments could generate more jobs and higher levels of economic activities than military spending.[3] Indeed, one study showed that an annual average transfer of 70.5 billion dollars from the military budget to education, infrastructure and other critical needs would generate an annual gain of nearly 4,77,000 more jobs on an average of over a four year period. Moreover, the Gross National Product was shown to grow by an average of $ 17.6 billion annually.[4]

Additional savings from the Pentagon's R&D budget should be transferred to urgent civilian research priorities, such as alternative energy, transportation and environmental research. Such an approach would maintain the overall level of research effort in the United States, and would also create new conversion opportunities for science-based military serving firms, as well as many civilian firms working on these alternatives.[5] The key to conversion in the United States, however, is the claiming of a peace dividend to begin the process of social reconstruction.

For over 45 years, Washington has pursued an industrial policy designed to make this nation militarily second to none. It has succeeded brilliantly in this goal. Today, we need industrial policies geared to restoringour nation's economy. Economic conversion policies should be at the forefront of efforts to rebuild the U.S. economy. If the U.S. is committed to forging a post-Cold War order it must plan for the transition and make the most of the opportunity. Perhaps as important, the U.S. must demonstrate its commitment to peace and economic development by converting its military-oriented economy, rather than merely advocating such a course for our former adversary.

**Conversion in the Former Soviet Union**

In the former Soviet Union, conversion has proceeded as an essential prerequisite of moving to a market-based economy, yet, the experience with centralized conversion planning has been a disaster, with little consideration of the match between the capabilities of the plant and equipment with the nature of the product to be produced.[6] Now under the conditions of so-called "shock treatment," many former military-industrial plans in the Commonwealth republics are supposed to guide their military-industrial output into the uncharted realm of the free market.

The prescribed approach by many western advisors has been to break up enterprises, scrapping of the social overhead, and shutting down the inefficient plants. What is left are presumably the most efficient plants. Meanwhile a fire sale of the most advanced military technology is being conducted to raise hard currency. But how, one might ask, will this help reconstruct the economies of what was once the Soviet Union? The process will be torturously slow and dangerous and does not immediately address some of the most pressing needs. Moreover, how does this help create a market system?

Many western experts and advocates of "sound financial principles," have ignored the obvious: there can be no market when there is no effective supply, i.e. efficiently organised production subject to commercial cost and quality controls; and there is no effective demand, given the collapse of the state supported employment system, and the rapid outfacing of consumer prices from wages.

Conversion in former Soviet Union can help to overcome these problems. The key lies in strategically conceived conversion to improve supply, particularly in food production, and distribution. Economic reconstruction should begin with the building of a new infrastructure for these economies. However, this development can only take place if onerous debt burdens are reduced to permit large-scale and environmentally sustainable infrastructure investment. This is the foundation of sustainable growth. Conversion can be the linchpin to put together the financing and infrastructural development.

To encourage development and conversion in Eastern Europe, the former Soviet Union and the developing world, the advanced industrialized countries should consider entering into agreements to reduce the debt of developing countries in exchange for the conversion of their arms industries to civilian markets.

**Conversion and the Developing World**

For years a debate has raged over relative merits of military spending in fostering development in the developing world. By now the record has become clearer, and the data indicates that militarism is also a drag on development in the South.[7] Despite this record, the developing world has not moved away from militarism by embracing disarmament and conversion. Yet, the ending of the Cold War may open newpossibilities. However, to embark on this path, the northern industrialized countries must lead by example and through innovative solutions.

**Conversion, Development and Debt Reduction**

The basic idea behind such an arrangement is that the external public debts of arms exporting nations that are held by Western countries could be reduced in return for each country's conversion of its military production to civilian use. For example, debt relief could be linked to a reimbursement-in-cash agreement, whereby the debtor country would deposit the savings from reduced investments in military industry into a development fund. Each payment would count as a corresponding reduction in external public debt. The rate of debt reduction would be based on a negotiated formula matching payments to the hard currency value of the debt. The savings-on-debt service could then be channeled back into investments to convert military-industrial facilities to relevant civilian production. The principal in the development fund could be used for a variety of purposes, either to promote other development projects, or to insure against default of the agreement.

An alternative approach would be to negotiate a moratorium on selected external debts with each debtor country, thereby freeing a great deal of hard currency for internal re-investment in the conversion of military-related facilities. Successful compliance with the conversion plan by debtor countries would lead to the commutation of some of the outstanding principal and interest on the debt. Debt service to lenders could be paid for through the issuance of international bonds by international lending institutions or by western governments to cover these charges during the moratorium. Insurance against default onthe obligations to convert could be assured through deposits of the principle in interest bearing accounts. Ultimately, the cost of deferring interest charges will have to be borne by the western governments through the normal channels of public finance. Of course there is the possibility of using some of the so-called peace dividend to fund this debt reduction process.

This approach could work when the debt is held by the leading international institutions like the World Bank, or by the governments of industrialized countries that are interested in promoting conversion of arms industries. Where the debt is held by commercial lending institutions, a *debt reduction in exchange for equity* in the converted facilities might be an attractive arrangement.

Forgiving debt is not a new idea. In contemporary international affairs, however, writing off debt is generally done to "reward" developing countries for complying with certain political agendas. For instance, Washington's decision to forgive Egypt's $ 6.7 billion in military-related debts as a *quid pro quo* for Cairo's active support for "Operation Desert Storm" is a case in point. Perhaps a more constructive example is the recent forgiveness of half of Poland's $ 5.5 billion debt with the German government in order to bolster the economic reform process.[8] This agreement is in keeping with the earlier actions by the so-called Paris Club of sixteen advanced capitalist countries to waive half of Poland's $ 33 billion debt with these governments.

Both the World Bank and the International Monetary Fund (IMF) are beginning to take note of the linkage between disarmament and development. At a recent press conference Lewis T. Preston, the president of the World Bank said that "if we found a situation where defense expenditure was 35 to 40 percent of the [government]budget, we might wonder if it was an appropriate use of [World Bank] funds." The Managing Director of the IMF, Michael Camdessus, made a similar point by noting that "We have to be very outspoken in pressing this point," regarding excessive military expenditures. These remarks made it clear that the two institutions are committed to curbing borrowing for defense spending.[9]

Despite this assessment, reduced international lending and fiscal austerity programs have not significantly affected military spending by the governments of most developing nations, indeed, IMF and

World Bank data indicate that military spending has fallen the least of any component of central government expenditures.[10] By contrast, capital investment, infrastructural project and social programs have borne the brunt of this austerity. Even from the point of view of the IMF, curtailment of capital investment will probably compromise future economic growth.

There are several important reasons why the developed industrialized nations might find debt reduction in exchange for conversion an attractive idea. First, as the World Bank has noted "evidence increasingly points to high military spending as contributing to fiscal and debt crises, complicating stabilization and adjustment and negatively affecting economic growth and development."[11]

This conclusion is amplified by the data presented in Table 1 which shows that several of the leading debtor nations of the developing world carry rather large military burdens as measured by the percentage of GNP and of central government expenditures. Several of these countries are also among the leading arms exporters and importers of the developing world.

Potentially, many of these highly Indebted countries could benefit from a debt reduction for conversion agreement. Other countries not listed in this table, principally eastern Europe and the former Soviet Union, could similarly benefit from such an arrangement. For instance, the Soviet Union owes approximately $ 30 billion to Western governments and had been in virtual default on these debts until a recent debt agreement with the West re-negotiated the terms.[12] Some of these nations may see the debt reimbursement through disarmament and conversion as a reasonable growth-promoting strategy.

A second reason as to why some of the industrialized nations of the North might be interested in debt reimbursement through disarmament is that it might provide an additional means to leverage control over the proliferation of guided missile technologies, chemical weapons production-capabilities, and nuclear weapons technologies. However, the failure of the Northern powers to curb their own appetites for these weapons of mass destruction makes curbs on developing countries seem paternalistic at best, and self-serving at worst. Moreover, the development of a comprehensive international inspectorate, with powers beyond those of the IAEA, could be a more secure basis to ensure that technology transfers will not foment proliferation. But success on this front depends on both supplier nations and recipient nations permitting intrusive inspection.[13]

A third incentive is that debt reimbursement should enhance the political stability of each nation, and perhaps enhance regional stability by decreasing pressures to continually re-arm. Yet, without positive contains on arms export promotion by the industrialized North, such an incentive remains questionable. Clearly, the debt crisis has seriously cut arms imports by thedeveloping countries. More recently, several important regional peace accords will reduce the demand for advanced weaponry and supplies.

Another incentive stems from the very real conversion potential of military-serving industries for civilian use. As several conversion studies have shown, developing countries can avail themselves of the industrial potential of their aircraft and shipbuilding industries to service their agricultural sectors through production of food processing and storage equipment, pumps, irrigation equipment, and dedicated agricultural transportation equipment. Their construction industries could be geared to producing low-cost pre-fabricated housing, and ancillary transportation links to existing agricultural communities, rather than large-scale virgin lands transportation projects. Numerous other industrial conversion possibilities have been outlined in special studies conducted for the U.N. Special Session on Disarmament.[14]

The debt-reduction approach to conversion may overcome one of the main obstacles to conversion, namely the lack of financial means and incentives to undertake the project. In addition, certain reimbursement premiums could be offered for specific types of conversion projects. These premiums could be extended to projects which phase out inherently dangerous technologies to adopt more ecologically benign technologies. For instance, chemical processing and production technologies for the manufacture of pesticides are suspected to have been used by Iraq to manufacture chemical weapons. This threat along with the legacy of Bhopal should make clear the merits of encouraging the development and use of ecological methods of pest control.

**Debt, Conversion and the Commonwealth Republics**

Recent events in the emerging republics in the former Soviet Union raise the urgency of well-conceived conversion initiatives which address the most pressing problems of transition. To date, western interest in the conversion of Soviet military industry has focused on breaking these enterprises into smaller units, focusing on the commercially viable elements of the enterprise. But this piecemeal approach has failed to consider the merits of a bolder conversion initiative focused on productivity-enhancing infrastructure. Indeed, there are many reasons to question the conventional wisdom about converting Soviet military enterprise through simply breaking them up into smaller, more efficient units.

Perhaps the most important conversion project for Soviet military enterprises is the production of food processing, storage and refrigeration equipment and food distribution equipment, especially containerized truck and rail systems. Military industry possesses the necessary capabilities to produce this equipment, including the metal forging, milling and machining equipment, as well as the machinery and refrigeration technologies. Moreover, conversion efforts here will allow dramatic improvements in productivity in the food sectors, thereby easing the pain of lifting price subsidies, which made up over 35% of the central governments budget. Failure to consider the conversion option will overlook a potentially effective means to overcome supply bottlenecks and improve productivity by using existing resources.

It should be stressed that we feel the principal role of joint ventures should be technology transfers of patents and designs for producing such products. Organizing and facilitating information and technologyexchanges on these alternative products could be a valuable role for the West which would not involve large cash investments, but at the same time promote rapid conversion.

These products are highly value-added and would address a critical need in the Soviet Union. The virtue of such an approach is that it allows indigenous producers to supply more of the content of the new industrial infrastructure and thereby minimizes the reliance on imports. Thus, conversion maximizes the return on the nation's existing industrial assets, and furthers its industrial self-sufficiency.

**The Soviet Military Industrial Enterprise**

The unique character of the vertically and horizontally organized enterprise characteristic of Soviet military industry poses a special problem for the transition to civilian, market-oriented production. In particular, these enterprises, in addition to managing the production of a wide variety of industrial products, routinely provide a wide range of social services, thus adding to the costs of running Soviet-style enterprises.

Re-organizing these enterprises will not be easy. The welfare of the work force would not likely be improved by simply scrapping the social overhead in favor of a stripped down enterprise modelled on a version of U.S. firms. It could prove useful to explore some other models of cooperative enterprises, perhaps along the lines of a Mondragon style industrial cooperative, capable of enhancing productivity through cooperative management and ownership schemes, which could provide for social welfare in a self-financing fashion. Failure to consider the alternatives will produce more western proposals calling for integrated production facilities to be broken up.

In addition, there is the need for the elaboration of cost and quality control techniques necessary to reorganize production for commercial, markets-oriented work. This requires that military industry managers, engineers and workers be retrained to implement cost-minimizing criteria.

**Industrial Conversion in the Developing World**

The potential for economic growth provides an incentive for developing countries to explore the conversion option. Developing countries can use the industrial potential of their export-oriented military industries to supply their domestic industrial demands for agricultural, transportation and construction equipment. This strategy would reduce their need for imports, generate industrial employment and improve their balance of trade. A United Nations sponsored study examined the conversion potential of specific industries and product markets in the developing world.[15]

Finally, the adoption of these types of projects should lead to public investments in more balanced, equitable and sustainable strategies of growth and development. A movement away from large-scale public infrastructural projects long favoured by the World Bank and state ministries could only help to redirect resources toward the daily needs of the landed peasantry, the urban poor and the unemployed. A release of these resources may help to build more rapidly the interlinkages between industries and sectors which can serve more domestic and regional markets.

(Numerous other industrial conversion possibilities have been outlined in special studies conducted for the *United Nations Special Session on Disarmament)*

**Conclusion**

Conversion in both developed and developing countries would promote international peace and economic growth. In contrast, the current U.S. Administration's approach to "national security" issues will inspire international conflict and economic decline. It is the failure of past policies and the dangers presented by the Pentagon's current effort that makes consideration of alternatives to militarism essential.

This and other proposals need to be explored if conversion is to advance. However, if the developing world and Eastern European countries are to accept such arrangements they must be assured that the West is likewise pursuing conversion.

## Endnotes

1. Active duty troop reductions from *Annual Report to the President and Congress*, by the Secretary of Defense, January, 1991, Table B-1; Civilian DoD personnel reductions and Defence industry workforce reductions were *Budget Estimates* FY 1992, Comptroller estimated from *National Defense* of the DoD, March 1991, Table 7-7. Out-years for 1994-96 were estimated by using a two-year moving average of the annual percentage change.

2. See Michael Dee Oden, *A Military Dollar Really Is Different* (Lansing MI:Employment Research Associates, 1987), pp.20-35.

3. See Roger Bezdek, "The Economic Impact-Regional and Occupational - of Compensated Shifts in Defense Spending", *Journal of Regional Science*, Vol.15, No.2, 1975. Also see *Converting the American Economy: The Economic Effects of an Alternative Security Policy*,Lansing, Michigan, Employment Research Associates, 1991.

4. Ibid, *Converting the American Economy*, pp.26-28.

5. Total potential savings from military R&D cuts for reinvestment could amount to $ 8.5 billion, according to *Converting the American Economy*, pp.23-24.

6. See "Converting the Soviet Economy, 1989-91, An Interview with Alexi Izyumov" in *The New Economy*, Washington DC, National Commission for Economic Conversion and Disarmament, Fall, 1991, Vol.2, No.5, pp.6-7.

7. See *Military Expenditure in the Third World* by Saadet Deger, Routledge, Keegan and Paul, London, 1986.

8. See "Bonn Cancels Half of Debt Poland Owes" by Ferninand Protzman, *The New York Times*, February 20, 1992, p.D-3.

9. Paul Blustein, "World Bank, IMF to Press Defense Cuts", *The Washington Post*, October 8, 1991, p.B1.

10. Ibid. pp.109-113.

11. See *World Bank Development Report 1988 and 1990*, (New York: Oxford University Press, 1988), pp.106-109.

12. See "Soviet Pact: U.S. Banks Little Affected" by Keith Bradsher, *The New York Times*, November 26, 1991. Also see "Soviets Postpone Some Payments to Foreign Banks" by Keith Bradsher, *The New York Times*, Dec. 5, 1991, p.A-6. Also see "G-7 Nations Consider Aid on Soviet Debt" by Steven Mufson, *Washington Post*, Oct. 22, 1991, p.D-1.

13. See *Economic Conversion and International Inspection: Alternatives to Arms Exports and Militarism* by Gregory Bischak and James Raffel, Washington DC, National Commission for Economic Conversion and Disarmament, Nov. 1991.

14. See United Nations, Department of Political and Security Council Affairs, United Nations Centre for Disarmament, Report of the Secretary-General, *The Relationship between Disarmament and Development*, (New York: The United Nations, 1982), pp.177-180.

15. See United Nations, Department of Political and Security Council Affairs, United Nations Centre for Disarmament, Report of the Secretary-General, *The Relationship between Disarmament and Development*, (New York: The United Nations, 1982), pp.177-180.

**Table 1**

**External Public Debt, Debt Service and Military Burdens for Leading Debtor Nations in the Developing World**

| Country | 1988 Military Spending/GNP | Military Spending/ Central Govt. Expenditures | 1989 Total External Public Debt | Debt Services as % of Exports of Goods & Services |
|---|---|---|---|---|
| | | (Millions $ US) | | |
| Brazil | 0.3 | 2.1 | 84,284 | 31.3 |
| Mexico | 0.6 | 2.2 | 76,257 | 39.6 |
| Argentina | 3.1 | 7.0 | 51,429 | 36.1 |
| India | 3.5 | 15.4 | 54,776 | 26.4 |
| Venezuela | 1.4 | 6.4 | 25,339 | 25.0 |
| Egypt | 7.8 | 21.7 | 39,751 | 20.5 |
| Nigeria | 0.8 | 2.6 | 31,668 | 21.3 |
| Israel | 13.8 | 23.5 | 15,931@ | 18.9@ |
| Chile | 4.0 | 16.3 | 10,850 | 27.5 |
| Morocco | 6.0 | 21.1 | 19,507 | 32.2 |
| Yugoslavia | 3.4 | 67.5 | 14,303 | 14.5 |
| Pakistan | 6.9 | 27.1 | 14,669 | 23.2 |
| Peru | 5.0 | 34.2 | 12,669 | 6.8 |
| Poland | 8.7 | 44.1 | 34,747 | 9.4 |

Sources : World Military Expenditures and Arms Transfers 1989, and World Bank Development Report, 1990, Tables 21 and 24. @ 1986 data.

**Table 2**

**Conversion Prospects for the Developing World**

| Developing Nations' Military-Serving Industries | Alternative Goods for Civilian Markets |
|---|---|
| * Aircraft and shipbuilding industries | Agricultural demand for food processing and storage equipment, pumps and irrigation equipment, agricultural transport, vehicles and equipment. |
| * Construction industries | Low-cost pre-fabricated housing and transport roads to link rural communities. |

# CONVERTING U.S. SCIENCE AND TECHNOLOGY RESOURCES TO CIVILIAN USE: INDUSTRIAL EXIGENCIES AND POLITICAL IMPLEMENTATION

**Antony DiFilippo**
*Lincoln University, U.S.A.*

Nothing new is being said today when academic and government work points to the inability of many American industries to compete effectively with rival nations, most notably Japan.[1] Similarly, nothing new is being said when we hear that U.S. living standards are not markedly improving. However, it is disturbing, though not exactly new, that the most recent data show that in 1990, for the first time since 1982, the U.S. experienced a real decline in its standard of living.[2]

Since the early 1980s when Americans were introduced to the promises of supply-side policy there has been a general sensitivity to U.S. competitive problems and how they ramify on job availability and the standard of living. Rather than deliver the promised industrial renaissance, the "voodoo economics" of the Reagan period only managed to enliven the service sector of the economy. Weakened by foreign competition, manufacturing industries, including many in the hightech sector, never experienced the rosy outcomes predicted by the supply-side pundits. Despite the disappointments of the supply-side policy, the Bush administration has not drifted far from the course set in the early 1980s. In fact, the Bush administration has never really had a well-defined economic policy, other than being committed to the unlimited power of market forces.

So as many U.S. industries continue to show signs of strain, having been sapped by very serious challenges from Japan, policy makers have been remiss in defining and developing substantive economic and technological plans that will notably increasethe prospects for re-industrialization. Considering the inextricable connection of Science & Technology to industrial development in this century,[3] as well as its increasing importance in the next, and because of the significant reduction of superpower discord, the current period provides policy makers with the opportunity to convert U.S. defense spending, and its science and technology (S&T) complex, to civilian use. The rapid transition of America's large S&T complex to the civilian-industrial sector would immediately strengthen manufacturing industries now moving on treadmills and unable to meet the Japanese challenge.

## Recognizing the Problem: Some Do, Some Still Do Not

A big part of the problem is that even after the horrible experiences of the past decade, some leading analysts in the 1990s are still arguing that U.S. manufacturing industries are strong. One policy analyst supporting this position argues that the prospect for the competitive growth in this decade is growing because of U.S. industry's successes in lowering production costs, upgrading product quality, and increasing company funds for research and development (R&D). With regard to R&D, because for the first time during the period 1980-1989 industry funds in real (1982 dollars) exceeded government spending, business is now in the position to take the technological helm. This development supposedly lets the private sector determine the direction of R&D and will lead to improvements in commercialization for industry in both process and product areas.[4]

The empirical argument would make perfectly good sense if U.S. businesses did not have to contend with foreign industry. Like many other arguments that wrongly claim that American corporations are not "hollow," this one fails to consider what is happening on the other side of the fence. A comparison of the average growth rates of company-supported business R&D spending measured in 1985 dollars (at purchasing power parties) for Japan and the U.S. shows another important dimension of the issue. Specifically, between 1981 and 1988 we find that the average annual increase in Japanese business expenditures on R&D was much bigger than it was for the U.S.[5] For U.S. industry just to spend a larger amount than the government on R&D is not enough when its chief competitor's expenditures are increasing more rapidly. Moreover, the trend of increasing U.S. industry, R&D spending may be grinding to a halt : measured in real (1982) dollars, during the three year period 1989-1991 these expenditures have not grown.[6]

While some analysts persist in clinging to an "America still first position," most fully recognize that Japanese industry in particular is a major and increasingly competent manufacturing and technological rival. There is the further general recognition that military preparedness in the U.S. not only has required a massive S&T apparatus to support it, but as the world has changed, these technologies have become less able to be usefully incorporated into the industrial economy. On this issue, nonetheless, there is massive reluctance by most analysts to move too far away from the earlier position which saw civilian industry benefit from government R&D spending through occasional spin-offs. Because of the failure of spin-offs to provide the needed technological sustenance to civilian industry, however, dual-use policy (a plan which sees technology as capable of being used in eitherthe military or civilian sector and supported by either defense or private spending) has become the accepted strategy.[7] Statements made by the Bush administration acknowledging the importance of encouraging S&T notwithstanding, and even though some incentives for cooperative R&D work and tax breaks exist, the focus is largely a dual-use approach for technology diffusion and for industry's technological revitalization. Significantly, like in the past, in this policy military security concerns are still the top priority of government policy makers.

After considering the technology-based problems confronting U.S. industry, their deleterious impact on the nation, and the inability of defense-driven technologies to strengthen civilian industries sufficiently to make them competitive, the conversion of America's S&T resources to commercial use is a policy alternative that deserves immediate and full attention. Changed political relations between America and the Soviet Union even before its dissolution forced policy makers in the U.S. to begin talking about conversion from military to civilian economy.[8] However, there is a disturbing lack of seriousness given to conversion, mainly because political leaders still envision U.S. military solutions to world problems. In the superficial conversion discussions in the U.S. that do occur, policy makers see a definite need for the nation to maintain its lead in military technology.[9] Thus an important alternative intended to help U.S. industrial revitalization is being seen as one that is promising, but does not quite fit with present conditions or those of the near-term future.

**Bad Becomes Worse**

The planned use of most government R&D resources in the U.S. for military and related purposes over four decades strongly suggests that a similar civilian effort wouldhave provided substantial and direct benefits to industries rivalled by foreign producers. That this could have been otherwise is not easy to envision, given the recognized importance of technology to economic and industrial growth. Continued trade deficits with Japan in many manufacturing areas and the erosion of a number of high technology industries are clear signs of fundamental problems facing U.S. producers. This is a principal reason for the nation's scant availability of jobs created by U.S. producers in the manufacturing sector and the depressed standard of living.[10] While the decline of market shares and sizable trade deficits with Japan make evident that major problems exist for American manufacturers, other more recent developments suggest additional ominous outcomes: Japan's formidable S&T policy, which has bolstered its industries' international market competitiveness, is now helping its industries establish technological and production roots within the U.S.

An avoided issue or one given only minimal attention in the debate centering on U.S. industrial problems with Japan is patents. New patents indicate the degree to which a nation is making use of its potential for innovation and in a large part show the extent to which industry is improving processes and products. Whereas the amount of R&D is an input sign of a nation's potential strength, patents are outputs of its S&T efforts; ultimately, inputs and outputs show up in impact indicators such as trade statistics. Although a previous analysis of patent data shows a general decline in U.S. innovation, in relation to Japan, a decrease in patents granted in the area covered by the organization for Economic Co-operation and Development,[11] more recent work indicates a major problem confronting American industry.

As Figure 1 shows, the percentage of patents granted *within* the U.S. to Americanshas almost steadily declined from almost 73 percent in 1970 to about 52 per cent in 1989. This occurred despite the fact that after adjusting for inflation, private sector R&D spending exceeded government expenditures during the 1980s. The Japanese share of patents granted within the U.S. has increased considerably from 4.1 per cent in 1970 to 21.1 percent in 1989. Unlike the second-ranked (West) German share (under 10 per cent) that generally remained unchanged between 1980 and 1989, the proportion held by Japan continued to grow. The U.S. granted 314, 887 foreign patents between 1980 and 1989. Of this amount, the Japanese accounted for almost 40 per cent.

The Japanese have recently experienced a very big increase in the number of U.S. patents they have been granted in specific product areas. In office, computing, and accounting machines, the Japanese share of all patents granted in the U.S. increased from 15.4 percent in 1978 to almost 40 per cent in 1988. A little over 30 percent of U.S. patents granted in communications went to the Japanese in 1988, more than double since 1978. The growing penchant among American consumers to purchase Japanese-built cars is likely to worsen since U.S. patents granted to the Japanese for motor vehicles and other transportation equipments increased sharply between 1978 and 1988.[12] (See Figure 2).

Most threatening is that if the trend established by the Japanese in 1970 continues, within a decade or so they will be granted as many U.S. patents as will the Americans (Figure 1). It is fair enough to say now that the U.S. economy is increasingly being propelled by Japanese innovations. Should this established course prevail in the future, U.S. Japanese relations are sure to worsen even more than they are today.

Currently accompanying this pattern of increasing Japanese patents in the U.S. is the growing capital investments in the manufacturing sector. Japanese producers are taking advantage of the continued movement away from manufacturing in the U.S. Direct foreign investment by the Japanese in the U.S. manufacturing sector more than tripled between 1987 and 1990, exceeding $ 15 billion in the later year.[13]

Along with the big increase of direct investments by the Japanese in the U.S. manufacturing sector has been the growth of manufacturing assets. Between 1986 and 1989 the value of Japanese manufacturing assets in the U.S. increased by nearly 300 percent.[14] Although in 1989 Japanese manufacturing assets ranked third behind Great Britain and Canada, the average annual increase during the period 1986-1989 was much greater for Japan (59.7 percent) than it was for these other nations (34.5 percent and 11.5 percent respectively).

Japan's very recent penetration of the U.S. industrial economy is nothing more than an extension of what its manufacturers have been doing at home for years. The Japanese have long maintained the lead among industrialised nations in capital expenditures relative to gross national product. That they have increased this spending in the U.S. should be expected, given the size of the American market, and importantly, the continued movement toward a service economy.[15] While not conclusive because of some non-manufacturing investments, the magnitude of Japanese spending on new plant and equipment in the U.S. is a foreboding indicator for a nation now being unrealistically attracted to a "buy American" position. By 1987 the Japanese were spending more on plant and equipment in the U.S. than investors from any other foreign nation devoting about 6 percent more than second-place Great Britain. By 1989 theJapanese was investing 38 percent more than the British.[16]

Capital investments by the Japanese have permitted them to establish a manufacturing presence in America, thereby providing them with the opportunity to evade the increasing protectionism of the U.S. Indeed, their concern with U.S. protectionism is likely to be one of the contributing factors for the growing Japanese presence in the American economy.[17]

Supported by what is easily the most comprehensive and coordinated civilian S&T policy in the world, the Japanese are quick to patent new innovations at home and show a definite commitment to strong capital investment, all of which contributes favorably to rapid commercialization. Significantly, in very recent years this industrial activity is being transported to the U.S. coinciding more or less with their increasing investments in the U.S. manufacturing sector, the Japanese, as we have seen, have also been increasing their patent activity in America. Because of the perceived gain, what the Japanese patent at home they tend to patent in the US.[18]. Indeed, Japanese patent activity in the U.S. indicates their determination to produce commercial outcomes.

A direct consequence of this emergent industrial activity by Japanese producers in America has been more manufacturing jobs - something good for workers but not necessarily good for U.S. -- Japan relations. Because of the Japanese presence in U.S., more than 48 thousand manufacturing jobs became available between 1988 and 1989. Although the number of jobs stemming from Japanese manufacturing activity in the U.S. ranked third behind Great Britain and Canada in 1989, the rate of increase from the previous year for Japan (25 percent) was markedly higher than those for the other twocountries (17 percent and 4 percent respectively).[19]

Finding it more difficult than in the past to compete in foreign markets and specifically having experienced big trade imbalances with Japan in manufacturing, including high-technology areas, the U.S. must now contend with a major movement of capital resources by the Japanese to America. This penetration of the American market by Japan is threatening the relations between these two nations, especially since they have already been damaged by major trade problems.[20] Even though capital investments by the Japanese in the U.S. at this time are generally not as great as those of Great Britain and Canada, their more rapid growth rate will only compound the devastation already inflicted on American producers in trade. Making this matter even worse is the strong and growing technological presence of the Japanese in the U.S., as indicated by the patent data analyzed above.

**Industrial and Political Grounds for Converting S&T**

For years discussions centering on the declining U.S. competitiveness, job loss, relative technological decline, and the need to revitalize industry have attracted the attention of many analysts. These problems, currently exacerbated by the evolving patterns discussed above in regard to Japan, indicate a serious structural inadequacy of the U.S. industry in manufacturing and in civilian technological competence. The conversion of America's S&T resources to civilian use is now especially practical since it would help redress this structural inadequacy during a time when political conditions justify implementation.

**Industrial Needs and Conversion**

Converting the military S&T resources of the U.S., even before the conversion of the entire military network, should be a policy priority for several important economic reasons. The expeditious conversion of S&T resources to civilian use would be a major impetus to the revitalization of the U.S. industry, helping to restore and strengthen a manufacturing infrastructure that has been generally neglected by business behaviour, more oriented toward quick profits. Increased investments in the U.S. manufacturing (process) technologies are urgently needed to improve competitive chances in international trade, to maintain control over domestic markets and, significantly, to better employment prospects. Yielding similar outcomes, converting America's S&T complex to civilian use would also allow for the direct concentration of national resources on product improvement. New products that demonstrate "quality" to domestic and foreign consumers require a dependable and sufficient flow of R&D capital that supplies a steady stream of scientific and technological freshness to the productive process. This would give the corporate and government sectors more incentive to remove political and business obstacles standing in the way of commercialization.

Because of Japan's commitment to spend more on basic research in recent years, the scientific and technological challenges for U.S. industry are accelerating and are very unlikely to be met by the fruits of military supported work and by the efforts of the private sector. Major Japanese ministries and agencies have markedly sharpened their focus on important basic research programs. Only one of many undertakings, Japan's Ministry of International Trade and Industry has increased basic research support for its Next-Generation Fundamental Industrial Technology Researchand Development Program.[21] Begun in 1981 and relying on the familiar cooperative triad of government, business, and universities, this ongoing basic research program, which has an annual budget of about $ 50 million, concentrates on the innovative development of four areas of industrial technology-superconductivity, special materials, semiconductors, and biotechnology-that could easily provide Japanese manufacturers with enough of an edge to devastate U.S. manufacturers.

As a result of the deteriorating economic relations between Japan and the U.S., which are in the processes of degenerating further because of the emergent industrial trends analyzed above, these nations are moving closer to a trade and economics-based war that America is simply not prepared to wage.

Both the U.S. and Japan officially acknowledge the big contribution that S&T has made to each nation's real economic growth -- accounting between 40 to 50 percent of the increase.[22] The combined facts that Japanese industry-supported R&D is growing faster than that of the U.S. and that military R&D spending in America is increasing at about the same pace as industry-supported R&D (see Figure 3), suggest that defense-financed outcomes *must* produce positive commercial effects to offset the difference between these trends.

Is this likely to occur? Despite publicized efforts now under way in the U.S. that claim to expedite the transfer of technology from government to commercial use, a high degree of success is not imminent. Because of the frequent specialization requirements associated with defense technologies and in view of bureaucratic constraints such as the military's need for concealment, the prospect for a sufficient amount of technology transference to the civilian sector that is necessary to acquire a competitive advantage for U.S. industry is not good. Recent U.S.efforts to promote a dual-use policy for technology is not especially promising either. Not only is the military to commercial part of this policy not without problems, but also the civilian to military component is faced with similar problems. Civilian to military transference could work effectively if the U.S. industry produced the required amount of leading edge technology; if it did so, however, manufacturing and technology-based trade problems would not exist, at least not to the degree that they have been over the last decade. Moreover, to the extent that U.S. industry does not produce the required industrial technologies, the persistent efforts of American policy makers to try and access Japan's technologies in order to compensate for this deficiency will only worsen relations between these nations.[23] The relatively faster growth of industry-financed R&D in Japan and government's supporting S&T complex that has been concentrating more directly on basic research make it increasingly likely that it will be a more important source of many advanced technologies in the future than it is today. Already we see the attempts by the U.S. to gain access to Japan's industrial technologies as a major striking point in the relations between these nations.

Thus our argument is that the conversion of the U.S. military S&T complex to civilian use is an important first step in meeting the Japanese industrial challenge. To be effective, the conversion of these resources must be structured by a well coordinated and comprehensive civilian-oriented S&T policy. If properly organized and guided by appropriate legislation, there is absolutely no reason why a civilian S&T complex in the U.S. would not significantly improve its technology base and enhance manufacturing competence.

The nearly $ 50 billion that are annually consumed by America's military and related S&T programs and the many scientists,engineers and other technical workers supporting it represent a pool of productive resources that can be used to strengthen U.S. civilian industry. Considering a lag of three decades, this statement is not inconsistent with a key remark made in a U.N.-authorized report that appeared in 1962. This report pointed out that one of the most important social and economic consequences associated with the absence of military spending would be "a sizable release of resources for civilian research and development."[24]

Supporting the strengthened industrial base that would be revived by the conversion of America's military S&T complex to civilian use would be productivity gains in manufacturing, not by just the introduction of new process technologies, but also by improving the nation's badly damaged infrastructures. The U.S. productivity problems have been shown to be related to the condition of the nation's deteriorating infrastructure,[25] and converting America's military S&T complex would make an important contribution to the preparation of the technical blueprints necessary to begin this leviathan job. Because scientists, engineers, and other technical personnel now need extensive retraining to perform efficiently in the civilian sector, the conversion of the military S&T complex prior to the total conversion of the defense budget would provide an interim period for the retraining of these workers, along with the funds to do it. Thus although adequate funding for infrastructure repair would have to wait until capital is available from general conversion, the retraining of these workers, especially of engineers, needed to reconstruct and modernize America's highways, railroads, bridges, ports, and utilities systems, could begin with the conversion of the military S&T complex.

**Political Circumstances Supporting Conversion**

There are also several political reasons for converting America's military S&T complex. Transferring U.S. military S&T resources to the civilian sector would be a definite and perhaps the only genuine sign that the Cold War is really over. As long as the U.S. and other nations continue each year to devote billions to defence R&D, an arms race, though now at a standstill, could be provoked and easily rationalized from a military perspective. Moreover, by converting its defense S&T complex, the U.S. would be putting a great amount of pressure on other militarized and militarizing nations to do the same. Since the U.S. possesses the world's most advanced military S&T apparatus, taking the initial step would truly be symbolic of a "new world order".

Certainly the military S&T complex the U.S. now possesses much exaggerates the extent of international security problems. The currently much-improved political relations between the U.S. and the previous Soviet republics make it an opportune time to begin shifting global policing responsibilities to the United Nations and other international organizations. While the threat of military problems erupting in smaller nations may still exist, international organizations could become suitably prepared to contend with them, and with less military sophistication than America has today. Indeed, much of America's advanced weapons and equipments that were designed and built for dealing with tensions between the superpowers are now obsolete -- not, as was the case in the past, because of the counter technological measures taken by an opponent, but because the suspension of the competitive arms race has created the situation where these sophisticated systems no longer have a reasonable political explanation for existing.

At the same time that the conversion of America's S&T complex would help establish an international environment built on noncoercive diplomacy, it would also improve its political relations with Japan which have been weakened considerably by economic developments. This transition would prod U.S. policy makers away from their growing attraction to protectionism and Japan-bashing. Since a converted military S&T complex would unleash a large amount of capital and personnel resources to the civilian sector, this would strengthen the nation's weakened industrial base, thereby improving its competitive capabilities and eventually lead to a healthy employment growth.

S&T conversion could have another positive effect on U.S.--Japan relations. U.S. pressure on Japan to contribute to its own security persists in this decade.[26] Conventional wisdom in the U.S. is that, despite the diminished threat from the former Soviet republics, lingering problems in Asia from other countries mean that "Japan still needs protection."[27] Much of the Japanese public, however, is unlikely to share this conviction, since they never really held it in the past. In fact, even before the reduction of the claimed Soviet threat, the Japanese public saw the Reagan administration's military buildup as equally perilous as the growth of the Soviet's Pacific armada in the second half of the 1970s.[28] By converting its military S&T complex and envisioning complete conversion within a reasonable period, U.S. policy makers would have no justification for pressurizing Japan to contribute to its military security. This, in turn, would help smooth U.S.--Japanese relations.

**Implementing Conversion**

America's S&T and general conversion processes could be made politically less complex, as well as economically less painful for those adversely affected, by attending totwo matters in advance, both of which are now more possible because of the winding down of the Cold War. The first is the passage of detailed conversion legislation that would provide the planning necessary for the processes to be effective. Existing conversion legislation in the U.S. could serve as basis for this matter.[29] The second is the initiation of a general and complete disarmament process by international organizations that assemble the U.S., the major former Soviet republics, and other interested nations. If this were to occur, nations not part of the process would begin to feel pressures from the international community to comply. Nations deemed a military threat to world peace but refusing to comply with the disarmament process, could be subjected to strong economic sanctions imposed by an international organization such as the United Nations.[30]

Thus movement on conversion would benefit from renewed and heightened pressure on national governments by international organizations. In the past conversion was incorporated into the United Nations' dialogue on disarmament. Such a dialogue needs to begin in earnest today. Because of the size of its military S&T complex, most pressure should be directed at the U.S. to help structure a strategy that results in meaningful steps being taken towards general and complete disarmament, which could subsequently initiate the rapid conversion of its military S&T resources.

Given the continuing diminution of significant military conflicts, academicians from all scientific disciplines and analysts from policy-oriented organizations have the responsibility to examine fully and to publicize the merits of converting U.S. military S&T resources to immediate civilian use. This activity will lead to greater receptivity of conversion among the public and much improve the possibility of passing conversionlegislation. The media too has the responsibility to explore and publicize the positive outcomes of converting U.S. military S&T resources to civilian use.

In the final analysis the decision of whether the convert S&T in the U.S. rests in the hands of policy makers. They must choose either to continue hedging on conversion or to execute it with a conviction based on a reasonable assessment of worsening U.S. industrial conditions in a changed, and definitely less threatening, political environment. It is because of these very diverse political possibilities that pressure for S&T conversion in the U.S. must come from concerned academicians and analysts as well as from different institutional sources.

It is now important that U.S. policy makers "seize the moment." America's threat assessment of potential security issues should begin being shifted from defense agencies that create too many unlikely military scenarios to international organizations that consistently monitor public impression and perform independent inspections. With improved relations between the U.S. and the erstwhile Soviet republics, there is currently a visibly legitimate reason to challenge politically the recently developed position that America's military and economic security in matters involving advanced technology are intricately related.[31] Not only could the U.S. industry benefit from the technological resources set free by converting America's military S&T complex, but there no longer is a pressing political imperative to resist.

**Endnotes**

1. See, for example, Office of Technology Assessment, *Competing Economies: America, Europe and the Pacific Rim* (Washington, D.C.: U.S. GovernmentPrinting Office, October 1991); and Michael Dertouzos, Richard Lester, and Robert Solow, *Made In America: Regaining the Productive Edge* (New York: HarperPerennial, 1990).

2. Council on Competitiveness, *Competitiveness Index 1991* (Washington, D.C.: Council on Competitiveness, 1991).

3. Nathan Rosenberg and L.E. Birdzell, "Science, Technology and the Western Miracle", *Scientific American*, November 1990, p.43.

4. Murray Weidenbaum, "Filling in the Hollowed-Out Corporation: The Competitive Status of U.S. Manufacturing", *Business Economics*, January 1990, pp.18-22.

5. While the rate of increase for the U.S. during this period was 5.7 percent, for Japan its was 9.3 percent. Calculated from Organization for Economic Co-operation and Development, *Basic Science and Technology Statistics* (Paris: OECD, 1991), pp.197, 337.

6. National Science Foundation (NSF), unpublished data; and NSF,*Highlights*, NSF 90-307.

7. Murray Weidenbaum, *Small Wars, Big Defense: Paying for the Military After the Cold War* (New York: Oxford University Press, 1992), pp.89-106; and Department of Defense, *Critical Technologies Plan* (Washington, D.C.: Department of Defense, May 1, 1991).

8. *Economic Report of the President* (Washington, D.C.:U.S. Government Printing Office, 1991), pp.149-154.

9. see, for example, ibid., p.151.
10. Anthony DiFilippo, "Technology and Trade Problems in Post-Industrial Economy: Japan Versus the U.S." in *Perspectives and Issues in International Political Economy*, ed. Chronis Polychroniou (New York: Praeger, forthcoming).

11. Anthony DiFilippo, *From Industry to Arms: The Political Economy of High Technology* (Westport, CT: Greenwood Press, 1990), pp.37-43.

12. Already we see that as American automobile manufacturers experience depressed economic conditions and increased layoffs, Japanese auto production in the U.S. continues to grow. Both Toyota and Nissan are substantially increasing their annual production of cars manufactured in the U.S. (the former by 1,80,000 and the latter by 200,000). See "Japan's U.S. Plants Pose a Dilemma", *The New York Times*, February 6, 1992, p.D2.

13. See United States Department of Commerce, *Survey of Current Business*, August 1991, pp.51-54.

14. See United States Department of Commerce, *Survey of Current Business*, July 1989, July 1990, and July 1991 issues.

15. The Japanese have invested very heavily in the U.S. service sector. For example, Japanese banking assets in the U.S. are the highest among foreign nations. Two main functions of

Japanese banking activities in the U.S. are to support imports from Japan, and, currently of growing importance, exports to Japan. At the end of 1989 Japanese banking assets in the U.S. totaled $ 420.7 billion. See Japan Economic Institute, *Japanese Banks in the UnitedStates*, Report no. 22A (Washington, D.C.: Japan Economic Institute, June 8, 1990). Japanese non-bank investments in the U.S. service sector have grown very quickly in recent years. Japanese non-bank assets in the U.S. service sector (wholesale and retail trade, finance, insurance, real estate and services) increased from $ 84.2 billion in 1986 to $ 278.9 billion in 1989. U.S. Department of Commerce, *Survey of Business*, July 1989 and July 1991 issues.

16. See U.S. Department of Commerce, *Survey of Current Business*, July 1990, and July 1991 issues.

17. James Jackson, "Japan's Financial State in the United States: How Stable Is It?" in *Japan's Economic Challenge* (Washington, D.C.: U.S. Government Printing Office, 1990), p. 192. (Study Papers Submitted to the Joint Economic Committee, Congress of the United States).

18. Telephone conversation with an analyst from Computer Horizons Inc.

19. U.S. Department of Commerce, *Survey of Current Business*, July 1991, pp.86-87.

20. By 1990 the U.S. manufacturing trade deficit with Japan exceeded $ 55 billion. See U.S. Department of Commerce, International Trade Administration, *U.S. Trade on the Rebound*, June 1991, Appendix, Table 3.

21. See Science and Technology Agency, Japanese Government, *White Paper on Science and Technology, 1990*, pp.10, 283-284.

22. Ibid. p.23; and *Economic Report of the President* (1989), p.223.

23. See Anthony DiFilippo, "Political and Economic Contradictions of U.S. Science and Technology Policy" (paper presented at the meetings of the Allied Social Sciences Association, New Orleans, January 3-5, 1992).

24. United Nations, *Economic and Social Consequences of Disarmament*, United Nations publication, submitted February 28, 1962. Reprinted in Seymour Melman, ed. *Disarmament: Its Politics and Economics* (Boston: The American Academy of Arts and Sciences, 1962), pp.332-396.

25. See David Aschauer, "Is the Public Capital Stock too Low?" *Chicago Fed Letter*, October 1987, no.2.; Alicia Munnell, "Why Has Productivity Growth Declined? Productivity and Public Investment", *New England Economic Review*, January/February 1990, pp.3-22; and John Ullmann, "Building a Peacetime Economy", *Technology Review*, August/September 1991, pp.56-63.

26. Susumu Awanohara, "Security Burden Sharing: A Japanese View", in *The United States and Japan*, The Atlantic Council of the United States (Lanham, MD: University Press of America, 1990), p.85.

27. James Schlesinger, "New Instabilities, New Priorities", *Foreign Policy*, no.85, Winter 1991-92, p.17. See also Fred Ikle and Terumasa Nakanishi, "Japan's Grand Strategy," *Foreign Affairs*, Summer 1990, pp.83-86; and Roy Werner, "America and Japan in a World of Multipolar Tensions," in *The United States and Japan*, p.64.

28. Igarashi Takeshi, "Circumventing Japan U.S. Conflict", *Japan Quarterly*, January-March 1991, pp.16-17.

29. The most promising of which is that proposed by Congressman Ted Weiss (D.N.Y), The Defense Economic Adjustment Act.

30. On the importance of economic sanctions see David Alexander and Anthony DiFilippo, "Beyond INF: Toward Progressive Arms Reduction", *Defense Analysis*, vol.4, no.4 (December 1988), pp.398-400.

31. See, for example, Allan Bromley, *U.S. Technology Policy* (Washington, D.C.: Executive Office of the President, Office of Science and Technology Policy, September 26, 1990; and B.R. Inman and Daniel Burton, Jr., "Technology and Competitiveness: The New Policy Frontier", *Foreign Affairs*, Spring 1990.

FIGURE 1

## U.S. & Japanese Patent Shares
### 1970-2005

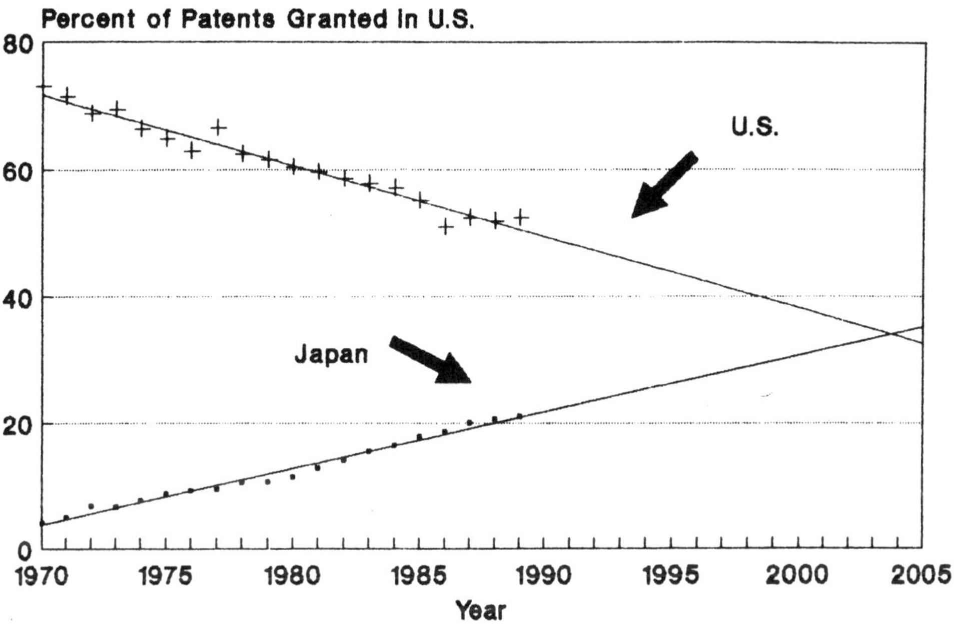


Source: Calculated from National Science Foundation, <u>International Science and Technology Data Update: 1991</u> (NSF 91-309), p. 101.

FIGURE 2

**Japanese Patents in U.S.**
**Selected Industry Areas, 1978 and 1988**

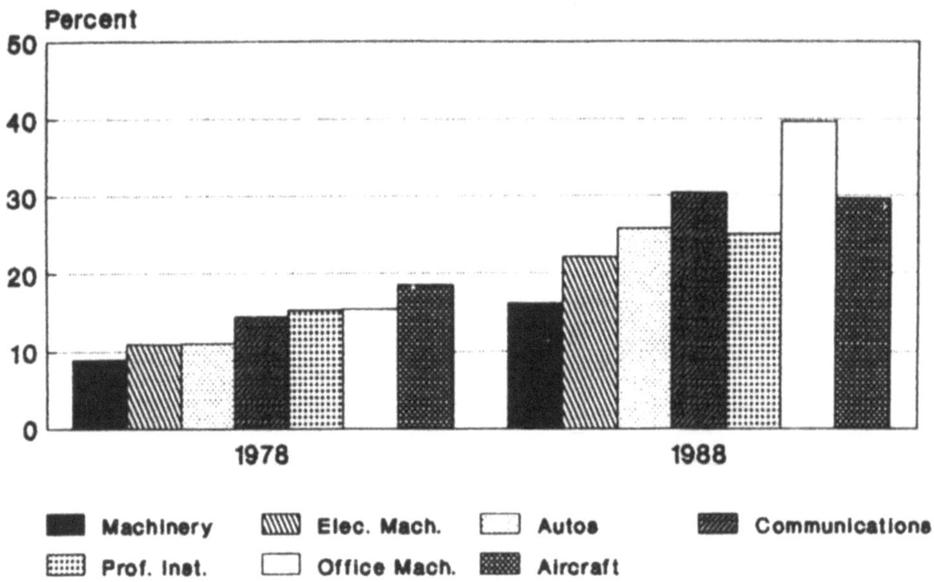


Source: Same as Figure 1, p. 103.

FIGURE 3

# Military and Industry R&D
## 1980-1989

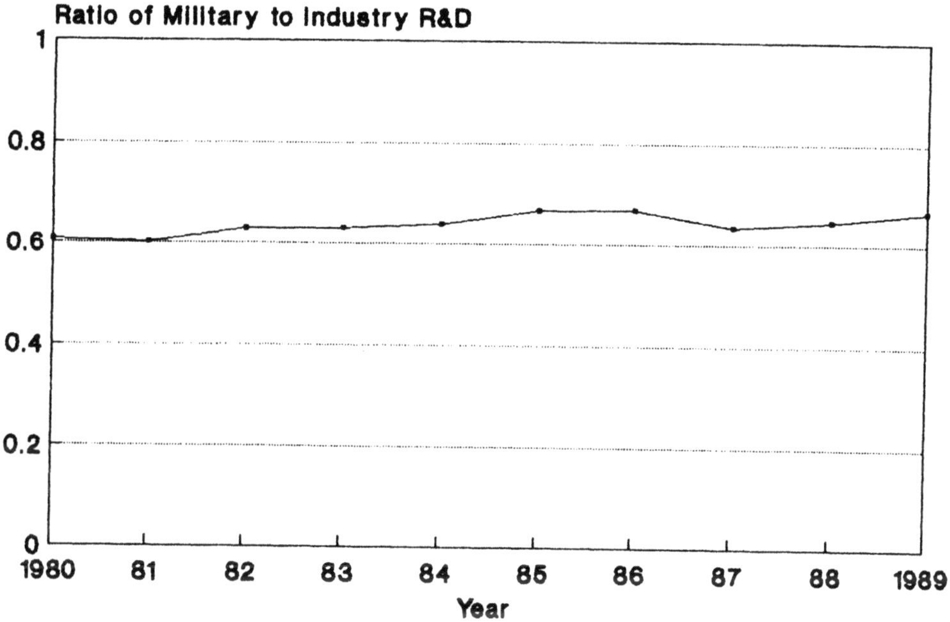


Source: Calculated from U.S. Bureau of the Census, *Statistical Abstract of the United States: 1991* (111th edition.) Washington, D.C., 1991, p. 588.

# MILITARY CONVERSION AND ECONOMIC DEVELOPMENT THE CHINA CASE

**Mel Gurtov**
*Director of International Studies, U.S.A.*

The cold war is over; but international violence, and preparations for it, are not. Despite major changes in global politics brought on by Soviet-American detente and the revitalization of the United Nations as a peacekeeper, the militarization of world politics continues. The Persian Gulf War, ethnic conflict in Yugoslavia, and fighting in Northern Ireland and Georgia attest to the importance of warfare in the post-cold war era. Global military spending has somewhat diminished; but at somewhere between $ 900 million and $1 trillion, it still is a major factor in the budgets and ambitions of most countries, large and small. As hopeful as the end of the cold war is, the best we can claim is that we are in an era of relative peace.

Nevertheless, this breathing space provides opportunities for states to re-examine their priorities and shift military resources to serve non-military purposes. China is one of the few states that has actually done so, seizing upon changes in international political economy to embark upon a major reorientation of its military industrial complex. In fact this "conversion" of military resources to civilian use began in 1979-80, well before the Gorbachev phenomenon.

The following analysis traces and evaluates China's conversion program, stressing its relationship to economic development planning. Unlike most western analyses, however, mine looks at China's program as *an economic strategy of military integration* rather than a program to convert "swords into plowshares". China, after all, continues to be very much a part of the prevailing system of arms competition and spending. But at the same time, it has nowhad over a decade of "conversion" experience, with impressive results that deserve study by rich and poor states alike.

## Origins

Two powerful forces combined at the end of the 1970s to prompt China's commitment to military conversion. One was the diminished sense of external threat to China; the other was the dramatic shift to a new model of economic development that emphasized "opening to the outside world".

The diminished sense of threat occurred in spite of China's involvement in Indo-China and the overly hostile relationship with the Soviet Union characterized by the "three major obstacles" to normalization of their relations.[1] Behind the scenes, the new PRC leadership was re-evaluating its onesidedly anti-Soviet position; by 1981 this took form in emphasis on an independent foreign policy and still later, in the "peace and development line", in world affairs.[2]

Among the specific elements of the new outlook in the 1980s were: steps to normalize relations with the Soviet Union, but without alienating the U.S.; a re-evaluation of the cold war and China's revolutionary line; and a positive appraisal of (and desire to emulate) the export-led economies of Asia. China gradually resumed high-level contacts with Soviet representatives in 1983-84. Sino-Soviet trade began to rise significantly in the first half of the 1980s. Gorbachev's Valdivostok speech in mid-1986 capitalized on this new openness on China's part to suggest further steps to normalize relations, leading eventually to his trip to Beijing in May 1989.[3] Meanwhile, relations between the PRCand the U.S., while occasionally disrupted by disputes over Taiwan and trade, continued to develop on the economic side.

The re-evaluation of China's foreign policy under Mao, and of the cold war in general, largely took place internally. It involved considerably more than the overarching concept of an "independent foreign policy" and equi-distance from the superpowers. Led, it would seem, by State Councillor Huan Xiang, but also including military leaders and defense intellectuals, the re-evaluation attacked a number of sacred cows in Mao's foreign policy without ever mentioning the Chairman directly. Not until the late 1980s did the key elements of the re-evaluation become public: criticism of various dogma on the collapse of capitalism and the vitality of socialist systems; questioning of the role of revolution in promoting global change; questioning also specific cold-war policies; and re-evaluation of the prospects of a world war.[4] The resulting redirection of Chinese foreign policy was toward openness to productive relations with all governments and social systems. Strategically, the two superpowers were said to be in a standoff; nuclear war had become unthinkable. Economically, they were in decline. The superpowers therefore needed to focus on domestic problems, helping create relative international tranquillity and opportunities for new interdependencies. China should capitalize on these changes and the emerging global multipolarity to move ahead in its economic development.[5]

The dynamism of Japan and the "four little dragons" of Asia also caught China's attention. One suspects this was not entirely due to their export successes, but was also related to their political systems.These states were able to take off economically while maintaining political stability. The attractiveness to China's leader lay in *how* that stability came about: the maintenance ofhighly centralized, one-party systems, with (Japan excepted) low levels of popular participation, severe restrictions on dissent, a strong military presence in and around government, repression of labor, and a controlled press. The specific time frame of those models may have been Korea under Pak Chung Hee in the early 1960s rather than Roh Dae Woo; Taiwan under Chiang Ching-kuo rather than under Li Teng-hui; and Japan as a one-party (LDP) system rather than as a parliamentary democracy. In any case, when Deng and other PRC authorities spoke of "catching up" with the rest of the world, they especially had in mind the Asian economies.[6]

The second major impetus to military conversion, and the one of primary interest in this paper, was economic development. China's "opening to the outside world" (*kaifang*) reflected not only Deng's ambitious goals for a more prosperous China[7], but also a conviction that China's economic expansion would have to fit within a single, capitalist global economy.[8] This would call for expanding exports to pay for technological and other necessary imports; sending large numbers of students and younger scholars abroad; inviting large numbers of foreign specialists to China to share their knowledge; gaining membership in multinational creditor organizations; attracting overseas Chinese and other funds for investment in coastal enterprises; and establishing laws and locales to induce foreign investment. Over arching all these revolutionary measures was the argument that a "peaceful and prosperous China" was in the West's self-interest to support.[9]

In this overall context of relaxation of international tensions and the objective of rapid economic development, China's leaders took a hard look at the country's vast network of military industries. They decided that these industries, especially the so-called "third line"industries that had been built in interior provinces for strategic reasons beginning in 1964[10] had become a serious drag on the state's resources. They represented an enormous outlay of money that may have significantly held back economic development in other regions.[11] Although third-line industries represented as much as one-third of state-run enterprises and workers,[12] they had become notorious for their inefficiency in production, management and labor, waste of funds, lax accounting, inappropriate sitting, and political interference.[13] Enterprise managers knew the state could always be relied on to bail out unprofitable industries ("eating from a big pot"); and workers knew they had permanent employment regardless of their efforts(the "iron rice bowl"). This set of circumstances was bound to intensify after 1979, when orders for military goods would be greatly reduced and many workers in military industries would be idle.

Yet military third-line plants were also recognized by central and local leaders as having uncommon assets: they are large in scale, form autonomous conglomerates that combine heavy industries with research institutes and transportation networks, and have more advanced technology and skilled personnel than most other industries in China.[14] Many of these resources could be shifted to produce for and, through technology transfer, directly support the civilian economy. Rather than dismantle military industries (although some were later dismantled), the various ministries responsible for them were told by Deng in 1978 to plan on employing one-half, and eventually two-thirds, of their workers in producing civilian goods.[15] This approach would not only satisfy pent-up consumer demand but also at a time of reduced military spending (see below) keep millions of workers productively employed, make use of idle plant capacity, and earn funds for the military.[16]

Still another motive of conversion that is linked to economic development is to achieve independence from foreign technology and capital as soon as possible. As one authority put it, China is "about 15 to 20 years" behind advanced world levels in the technical quality of high-tech products from defense industries. Until it can meet world standards, it will have to depend on technology imports from abroad that "always are subject to limitations by the developed countries..."[17] Conversion is seen as a method of promoting import substitution, in high-tech military items as well as civilian products.

## The Re-evaluation of China's Military Economy

By 1979 restructuring the military was not only desirable in its own right but it had also become an indispensable ingredient of economic modernization. Military modernization -- in the sense of bringing the military's technological level up to world standards -- ranked last among China's "four modernizations" as the PLA was basically told to wait its turn. This decision, favoured by Deng Xiaoping,[18] also helped resolve a high-level debate over the larger question "whether the defense industries should lead industrial development or vice versa." According to a western authority, that debate ended in 1979 with the decision to put military science and technology at the service of the other "modernizations", while also continuing to upgrade conventional and nuclear weapons".[19]

Once the key political decisions had been taken, the essential question was *how* most efficiently and effectively to organize and use military-industrial assets to serve the civilian sector and the national economy. The Chinese approach has had several components, each of which will now be briefly explored.

One element of conversion is the establishment of defense economics as a separate field of study. This important intellectual development led to the re-evaluation of military goods as commodities, although commodities "with special characteristics". This was a major shift in thinking that was a prerequisite to civil-military integration. Among the implications of this shift, which is spelled out in numerous

Chinese studies,[20] are a critique of the third-line strategy; recognition of the need for a more scientific, nonpolitical approach to military budgeting and planning;[21] cost-effective evaluation of a military enterprise's operations, in order to assess its relative potential contributions to the civilian and military economies; reconsideration of the role of value, pricing, and profit in military production; and reorganization of the military-industrial system so that its products, including weapons as well as civilian goods, would be made under contract and enterprises themselves would be subject to market forces.

One can see why, as two Chinese specialists have argued, this new view of defense economics "came as a shock" to traditionalists.[22] It meant that the usual treatment of the military-industrial system as a closed system, cut off from internal oversight and from the Chinese and world economies, would have to end. "For the first time," writes a U.S. analyst, "the [PRC] defense establishment feels the need to justify itself economically".[23] The new view meant subjecting military production to new standards that might compel greater efficiency and "economic linkages" (*jingji niudai*) with other parts of the economy and with the world.[24] Inevitably, a relaxation of internal censorship and secrecy would have to occur, since these are handicaps to exchanges of experience and technology, and to the financial success of enterprises.[25] Finally, defense economics requires studyingthe operations of defense departments and military-civilian industrial relations in other countries. This task has included examining others' conversion experiments and the role of economists in defense planning and budgeting.[26]

The "shock" extended to forcing PLA industrial managers to become entrepreneurs. They would have to find new ways to earn money in a period of austerity. Among other things, due recognition would have to be made of the true market value of their civilian products, and of the profit that could be earned from them. "Military enterprises always felt that the fixed profit of five percent was too small, that it was `eating from the same big pot' ..."[27] It was pointed out, for example, that five percent was far lower than profits in other industries.[28] Under the new military economics, the actual costs of production, as well as technological intensiveness and capital investment, would be taken into account in pricing.

The second crucial element in conversion was the reorganization of the bureaucracies concerned with military research and production. In the early 1960s China's defense industries were under eight ministries (numbered for secrecy) of machine building, six of which were charged with specific armaments production. Co-ordination of production was vested in the National Defense Industries Office (NDIO); but military R&D, including funding, training, and weapons testing, rested with the National Defense Science and Technology Commission. Both the NDIO and the NDSTC were responsible to the Central Military Commission (CMC, also known as the Military Affairs Commission).[29]

Although a re-organization occurred in May 1982 -- the ministries were now named, reduced to seven, and all placed under civilian leadership -- the military-industrialsystem still operated like "independent kingdoms". Each ministry's industrial network was "extremely self-contained, with their own equipment for processing raw materials and manufacturing components, their own schools, and their own research institutes."[30] Each ministry had its own set of plants, some exclusively making weapons, others only civilian goods, still others both. This vertically integrated system, isolated from the rest of the economy, worked well in the 1960s and 1970s in those areas -- nuclear weapons and aeronautics (missiles) -- that had high theoretical and political priority. But for the other sectors, it appears that isolation was detrimental to developing local sources of supplies and other inputs that can promote efficiency, innovation, and lower prices.[31]

The key organizational policy decision occurred later in August 1982 when NDSTC, NDIO, and one other unit were merged into the Commission on Science, Technology, and Industry for National Defense (COSTIND, abbreviated *Guofangkegongwei*).[32] COSTIND is the chief body responsible for co-ordinating military R & D, conversion, and new weapons and technologies. It is responsible to leading party and state bodies; the party central committee's CMC (whose membership has been identical to the state's CMC), and the State Science and Technology Commission and State Planning Commission under the State Council. From the outside, it appears that COSTIND is strategically situated to overcome "the longstanding conflict between the military research and development sector and production sector".[33] To the extent it can do this and maintain its independence-- i.e., avoid running afoul of party-state conflicts and pressures from the Ministry of National Defense and the PLA bureaucracy -- COSTIND may be able to overcome the perennial inefficiencies of military industries and promote military-to-civilian integration.

Various sources indicate that COSTIND's authority in fact goes beyond mere coordination. Ostrov maintains that COSTIND has "direct control over the ministries and factories relevant to its mission."[34] (In 1988 a further consolidation of military-industrial leadership took place that may reflect this centralisation of control. The seven ministries were reduced to what the Chinese call "two-and-a-half": the Ministry of Machine-Building and Electronics, the Ministry of Aeronautics and Aviation, and the China General Corporation of the Nuclear Industry). With specific reference to conversion, according to Beijing sources whom I interviewed, COSTIND is directly responsible to the premier of the State Council (LiPeng) and provides overall financial and administrative supervision of conversion R&D. COSTIND must accept or reject the conversion of individual enterprises to nonmilitary production. It also plays a leading role in directing technical research priorities.[35]

COSTIND's activities seem to extend to facilitating the proper functioning of military industry conversion itself. It has provided technical assistance to enterprises on planning, equipment, product development and worker training, drawing upon thousands of technical personnel from military plants and defense universities.[36] As one of the case studies below will note, COSTIND may also allocate funds to military industries as start-up grants or loans. Its province-level offices, as in the case of the Shaanxi COSTIND office, guide conversion efforts throughout the province, from production planning to marketing.[37] Assisting in technology transfers, such as by bringing together military factories, local governments, and industrial producers, very likely is one such liaison function.

The contract system that was introduced to military industries in 1986 is probably another example of COSTIND'S involvementin conversion. Its defense intellectuals had long urged adoption of the system as a crucial way to rationalize weapons production and conversion. According to a press report,[38] "nearly 1,000 models of weapons and equipment" were brought under the contract system., so that by 1988, 80 percent of this production was performed under contract, "Some" by competitive bidding.

Along with organizational change there has occurred a professionalization and "youthization" of the military-industrial leadership. Following instructions from the CMC,COSTIND in 1985 undertook the streamlining of its organization and the promotion of younger, more professionally competent leaders. The average age of the new leaders in the COSTIND is now 10.6 years younger than before. Also unprecedented in Chinese bureaucratic politics, all of the new leaders were graduates from colleges, including two who have finished graduate studies. Ding Henggao, COSTIND's director, graduated from Nanjing University in 1952 and studied abroad between 1957-61. He had been the Deputy Head of COSTIND's Science and Technology Section before being promoted to the Director's position.[38]

The fourth element of China's conversion was sharp cutbacks in military budget and manpower, and institution of a retraining program for demobilized soldiers and cadres. This step followed on a speech Deng Xiaoping is supposed to have delivered as early as 1975 about "restructuring" the Chinese military to make it less "bloated" than it traditionally was.[39] According to official PRC statistics, military spending before the reforms had risen from Y14.25 billion in 1975 to Y22.27 billion in 1979. By 1981, however, military spending had dropped to Y 16.8 billion, and by 1986 it was still only Y 20.08 billion.[40] Expressed as a percentage of the state budget, the dropoff of military spendingis perhaps more meaningful. Military spending had been 17.4 percent of the state budget in 1975 and 17.5 percent in 1979. But in 1981 it dropped down to 15.1 percent, and then to 8.6 percent in 1986.[41] As a percentage of GNP, China's military spending in 1986 was 6 percent, ranking it sixth in the world in military spending.[42] Yet the percentage of GNP also represents a considerable decline; it was over 10 percent before 1890, and 7.5 percent in 1985.[43] Although arguments are almost inevitable over official statistics on military spending in China (or anywhere else)- - the U.S. Central Intelligence Agency, for example, uses far higher figures-- it seems indisputable that during most of the 1980s the PLA experienced major cuts in its budget as the spending priorities of China's leadership changed.

One reason these budget cuts were possible is that the size of the armed forces was greatly reduced. Demobilization of army soldiers and cadres had been attempted, with unimpressive results, early in the 1980s.[44] A more determined effort was announced by Deng Xiaoping on June 4, ,1985: the PLA would be reduced by one-quarter, or about one million soldiers. This goal, said to have been achieved in mid-1987, was designed not only to save money but also to help regularize and professionalize the armed services.[45] It occurred simultaneously with other changes in the PRC military, including the retirement of many old cadres, reduction in the number of military regions, emphasis on higher education for army officers, and the reestablishment of ranks.[46]

The leadership's hope was that regularization and professionalization would make the PLA more attractive to educated younger people while offering demobilized soldiers and cadres opportunities for retraining. Over forty training centers were reported to have been established by 1987for cadre retraining.[47] For ordinary soldiers, PLA units and local authorities set up classes in civilian-use technology, with the aim of having each solider spend at least 100 days a year in them. The emphasis in training demobilized soldiers has been on developing" dual-use talent", that is, skills in both military and civilian life. Early in 1987 it was claimed that over 430,000 soldiers had undergone such training.[48] But problems have been reported. Press reports reveal that finding jobs, schooling, and housing for so many demobilized soldiers puts great pressure on the social system, and particularly on the retraining centres.[49] An internal government publication adds that unlike retired Western Officers, PLA officers usually leave the service without either retirement funds or an education. Many lack the motivation to work in other than urban bureaucracies that are already bloated with personnel.[50]

The fifth element of conversion has been the reorganization and in some cases relocation of third-line industries. Whether in Shaanxi, Sichuan, or any other third-line province, the problem of what to do with military-industrial assets that were in economic trouble or could not effectively convert to civilian production became acute in the post-Mao period. Between 1986 and 1990, 121 industries or "projects" (*xiangmu*) were "closed down, halted, merged, transformed, or moved" closer to cities.[51] As one province official said to me in explaining the restructuring or relocation of twenty-two military industries in Shaanxi, conversion poses challenges that some enterprises cannot handle. Foremost is that their leaders are reluctant to think beyond the military production of individual firms. Additionally, enterprises are isolated from energy sources and transportation, lack capital and advanced technology, and are in need of managers who are market-oriented and know what it takes to develop competitive products. Typically, however, these are problems associated withsmall-and medium-size firms, not large ones like Xian Aircraft Company (discussed below).[52]

Perhaps the key breakthrough was for military industries to establish "windows" -- outlet factories or partner enterprises -- in nearby or coastal cities beginning in the mid-1980s. For example, a leading third-line official has described events in Sichuan Province after party and state officials met in 1986 to ponder the fate of the province's military industries. Taking into account that 50 percent of the country's military - industrial productive capacity was located in Sichuan, the officials decided that the "patient" had to be saved. A policy of "leading the tiger down the mountain", or "putting wings on the tiger" was instituted. The task would be to determine which of the third-line enterprises could most effectively be linked with local or first-line (coastal region) plants, and to decide on closing or absorbing certain other plants.[53] Some specific experiences of enterprises in other provinces are recounted later in this paper.

Finally, conversion had meant the redirection of military resources to civilian needs and international markets. Here again, this element has been common to virtually all the writings of China's defense intellectuals. Their attention has been drawn to developments in high-technology such as micro electronics, nuclear energy, aerospace, lasers, and nautical science; to the shift from materials-based to information-based production; and to investment in advanced industrial countries in technology industry clusters (*Chun*), especially in electronics.[54] Investments in these fields promise high profits and hard-currency earnings, these analysts believe. They also expect that advances in technology will occur, with some spillover into new military applications.

Of course, only already technologically advanced enterprises are actually positioned to move in these directions. An example is China's nuclear industry, which employs about 300,00 people in over 200 enterprises. It has readily made the shift from overwhelmingly military to substantial civilian production. The industry's top official reports increasing profits and a nearly four-fold leap in the value of production between 1984 and 1986. He attributes these increases mainly to development of nuclear power at China's two plants (Qinshan in Zhejiang Province and Dayawan in Guangdong Province).[55] But the nuclear industry also produces seven other categories of products: nuclear technology and radioisotope products, fire alarms, rare-earth products, non-ferrous and precious metal products, metallurgical-chemical products, and machinery instruments. About 1,000 items from these groups have been put on the market.[56] Whereas 42 percent of the nuclear industry's output by value in 1989 was in civilian products, one industry official expects that figure to climb to 50 percent by 2000.[57]

For technologically limited military enterprises, manufacturing and marketing products require far more resourcefulness than in the nuclear, aerospace, and other large-scale industries. A tank factory, for example, can fairly easily adapt to produce tractors, and aircraft engine plants can switch to manufacture gas turbines. Moreover, enterprises with large assembly lines and a skilled workforce can respond fairly quickly to profitable opportunities. Small-and medium-size plants, however, not only lack the basic ingredients of large-scale plants (technology, capital, and skilled labor). They also do not have the network of contacts, research support, managerial knowhow, reputation, and international marketing representation that the large firms have. These comparative disadvantages make for difficulties in acquiring local or foreign partners, developingcompetitive products, and stimulating production. As a result, many smaller firms must be content to produce light industrial and household goods--televisions, cameras, bicycles, refrigerators -- that are already in abundance and are technologically inferior to what is available on the international (and in some cases also the local) market.

We can best see these changes at work by examining actual cases of conversion. Based on my three- year interview project in China, I have selected two types: a large aircraft company in the interior and electronics firms that have moved to the coast.

## Some Case Studies of Conversion

*Case 1 : Xian Aircraft Company.*[58]
Xian Aircraft (XAC), on the outskirts of Xian in Shaanxi Province, is a very large enterprise with about 17,000 workers. It is under the Ministry of Aeronautics and Aviation. In the post-World War II era the company produced only bombers; civilian aircraft had to be imported from the USSR and other countries. But in the 1970s XAC began developing the capability to replace the Boeing 707 with its own Y-7, and by 1984 it had produced the first of these. Now, besides mass-producing the Y-7 and other aircraft, XAC also produces radar, satellite and microwave antennae, diving boards, ferris wheels, hot food carts, and aluminium doors and windows.

The company, whose leadership looks to Boeing as something of a conversion model, is headed by engineering and management professionals, not by military people.[59] In fact, about one-fourth of its workers are categorized as "professional" a much higher figure than the usual 10 percent I encountered in other military industries. All the senior-level people I met in XAC had travelled abroad, which is also unusual.

XAC's leadership is pleased with the shift to civilian production. It meets national needs, and creates new challenges, they said. As an example they cited the hot food carts. Production was based on existing technology and human resources, requests from institutions needing the product, success in a competition run by the province, and market research. "All we need is a contract to go ahead", they said.

The contract, however, involves other organizations besides the producer and customer. XAC works closely with the Shaanxi Province Office for Defense Science, Technology, and Industry. This office, COSTIND's counterpart at province level, is the coordinator and overseer of all conversion activities. Its permission is apparently required before contracts and production can begin. In addition, other firms and research organizations may be involved in a particular task. The Xian Aircraft Design Research Center, next door to XAC, does all flight testing. the province office's coordinating function would comprehend the involvement of other units such as that one.

XAC's production is now 70 percent for the civilian market, which is roughly the industry average.[60] Twenty percent of production is still for the military (the Hong- 6 bomber and repairs), and 10 percent consists of components and assembly work for foreign firms (such as access doors and vertical fins for the Boeing 737, and airbus doors for the French and Italian A-626). The work remains highly labor intensive despite access to foreign designs, labor-saving technology and training (in the United States by Boeing), and some foreign capital, as well as a 16-20 percent profit margin on overall sales.

Unlike many third-line firms in Shaanxi, Xian Aircraft has been able to move out from reliance on significant government funding. XAC only calls upon government aid when itwants to launch a new project that requires more personnel or technology. About 90 percent of its financing comes from bank loans at interest. Profits are heavily taxed, moreover: real estate, income, and other taxes take about one-half of profits after expenses. Such funds are likely to be allocated by the Government to less successful firms.

Despite the strengths and sales successes of XAC , it acknowledges some problems. One is the weakness of overseas sales, which account for only about 3-4 percent of total sales. It has yet to conclude a joint venture with foreign firms, and its leaders express disappointment with the lack of technology turnover (a common complaint by Chinese state officials about doing business with other countries). Thus, revenue from conversion for export has so far been limited, although XAC looks to Third World markets (in particular in the Middle East) for purchases of its medium-size commercial aircraft.

XAC's capabilities are still not used to the fullest, its managers, said. A large domestic demand is not being satisfied. The firm is looking ahead to production of long-range aircraft for domestic flights. Walking through the assembly lines, it is apparent to the outside observer that XAC could indeed accommodate more aircraft projects. The aerospace industry was designated the "center" of national hi-technology development for the 1990s.[61] It has a healthy record of foreign sales to the Third World that are pushed by three export-promoting corporations under the Ministry of Aeronautics and Aviation.[62] Thus, although XAC leaders worry about the firmness of the aerospace market, the future seems reasonably bright.

*Case 2: Shenzhen's Electronics Firms*: The city of Shenzhen is the rapidly growingspecial economic zone (SEZ) next door to Hong Kong. It is one of the "windows" often chosen by military electronics industries that have shifted research and production to the coastal region. Since being declared an SEZ in 1980, its higher wages and better working conditions have attracted many of China's skilled (and degreed) younger people, senior engineers and administrators, and older professionals from the north and interior. Population, construction, communication, and transportation have all boomed in the last decade. Shenzhen's city government has offered reduced fees and paper work to military enterprises wanting to relocate. The proximity to Hong Kong is a major advantage for research and capital, especially as 1997 draws near. In all, Shenzehn has provided a near-perfect locale for export-minded military industries in the interior.

Electronics firms are under the Ministry of Machinery and Electronics, which reorganized in 1988 when five previously separate ministries were amalgamated. In view of their technical levels and export potential, it made sense to move electronics projects and many of their professional staff to Shenzhen and other coastal cities. There, they could become linked to the world economy as well as to each other in "groups" (*jituan*)-- the nonmilitary face of China's military-industrial complex. In Shenzhen, electronics firms have the incentive to create new consumer products in response to market conditions, and to develop or refine dual-use and strictly military products. They can also take advantage of modern communications and improving transportation, such as a new international airport and high-speed highway linking Shenzhen with Guanghou (Canton). By 1991 Shenzhen-based companies were leading the country in seven of ten electrical product categories. Electronics accounts for about 40 percent of the SEZ's industrial output by value.[63]

One example is the Huafa Electronics Corporation.[64] It was originally part of a third-line conglomerate founded in 1970 and located in the Southern junction of Guizhou, Yunnan, and Sichuan provinces. When the conversion decision was made in 1979, the company leadership (then composed of economists, managers, and a few retired PLA officers) decided to change its production line from electrical relays for the military to washing machines. This proved quite successful, and required no new technology. Subsequently, Huafa moved into color television and into Shenzhen. It set up two factories in Hong Kong, on which it relies for technical expertise, to join with the one in Shenzhen (which employs about 1,400 workers). These factories mainly produce for the export market (about 70 percent of sales), since the domestic market is considered too limited for high-quality TVs. PCBs and injected plastic parts are the main products for the Chinese market.

Huafa's leadership considers the company very competitive in a crowded electronics industry in Shenzhen. There are several hundred such firms there already, of which more than ten specialize in color television sets. But although lacking brand recognition, Huafa ("Happy" brand) has found an export niche in the third tier of television exporters, behind the United States and Japan (tier one) and Taiwan and South Korea (tier two).

Huafa is part of the Zhenhua group of electronics firms that includes Zhenhua Electronics Industry Corporation.[65] Zhenhua was established in 1966 in Guizhou Province to produce satellites, instruments for aircraft, and other hi-tech electronic products for the military. It was one of some thirty companies in the province under the Ministry of Electronics, as it was then known. These companies employed several thousand technical personnel from some of China'sbest universities. Long before Zhenhua moved to Shenzhen, its general manager told me, it had suggested and begun planning for conversion to take advantage of its superior technology and personnel. When the move finally was made in 1981, about 1,000 technicians went along, leaving the Guizhou plants to continue military production.

Now Zhenhua is established in over twenty coastal cities. It is an "export network", with branches in nearly all the specially designated economic and development zones. Total personnel in the enterprise group, including R & D, testing, and marketing, are over 30,000 of whom over 8,000 are considered technical. In 1987 the firm began developing advanced telecommunication products for the China Telephone and Telegraph Bureau. This set a pattern. Zhenhua now produces about US $ 1.5 million in nonmilitary products -- over 80 percent by value of total production -- and claims to be making a large profit that contrasts with many money-losing firms in Shenzhen. The city planning bureau has helped by cutting red tape and eliminating certain fees, while local banks have assisted Zhenhua with loans.

Zhenhua's production falls into three categories. The major product is ground satellite stations for outlying provinces to receive television via 52 relay stations. This line of work was implemented very rapidly and was widely reported in Chinese newspapers. Second is circuits and other telecommunication parts, to replace imports from Japan and elsewhere. (Circuits are the only product said to have received state support -- about US $1 million). Within a year of production, in 1990, Zhenhua circuits were being exported to Southeast Asia and other Third World regions. Some types of circuits are now used widely by state-run electrical machinery plants. Computer software and accessories fill out of the list of product lines.

A third Shenzhen electronics firm, Languang, further illustrates conversion by shifting production.[66] Languang was established in Gansu Province as Gansu Electrical Industry General Corporation. It then had over thirty plants, almost all military, manufacturing radar, telecommunications, missile technology, and military-use software. The decision to shift to civilian production was designed to maximize use of the technology and equipment. It also was spurred by the fact that "people had nothing to do", the general manager said. There were no incentives to produce new or higher quality products.

Beginning in 1982 twenty enterprises moved to the coastal region, leaving the rest behind to continue military R & D. A number of technical personnel who had initially trained in the United States or elsewhere abroad, and who had returned to do military work in Gansu only to find none, were tapped to go to Shenzhen. Once, again, the basic idea being the move was to make Languang and its sister coastal firms (such as Lantian, Lanhua, and Lanhai) "windows" for the interior parent company -- new settings for testing, training, concentrating technical capabilities, and attracting foreign technology to spur exports.

Languang Electronics is China's first company to produce stereo sets. Starting in 1983 with a total state investment of Y 1.1 million, the company is now worth Y 54 million (not including taxes on profits turned over to the state). Its products are sold about equally to domestic and international markets. Products are of five kinds: high-level electronics, such as stereos; electronic instruments; computers and office equipment; computer keyboards; and plastic covers for computers. Languang boasts high efficiency and sales (US$ 40 million in 1990), and acknowledges the role of state support in getting started by reducing taxes and fees.

Like Zhenhua, Languang reflects China's emerging strategy of rationalizing and concentrating technological assets in the coastal region. Its general manager is also chief executive officer of two other electronics firms (Guangda and Lanyun) that in turn are part of a nine -company group, each of which specializes in electronics products. One sees here a socialist version of the South Korean export conglomerates, such as Hyundai and Lucky-Gold Star. Evidently, PRC leaders are pleased with the results. In 1991 they decided that four cities -- Shenzhen; Baoji,, Guangxi Province; Xiangfan, Hubei Province; and Jianyang, Sichuan Province --would become test sites (*Shidian*) for the application of military technology to local economic development. Military resources, including operational resources and software, will be concentrated in these areas to push growth forward. Some 60 *xian* (counties) have also been designated for similar plans.[67]

**Assessing Conversion in China**

*The noteworthy aspect of conversion in China is that it has taken place in a strategic context*, that is, in a planned way, with specific economic objectives in mind, and on the basic of a fairly comprehensive and balanced critique of the military's role in the economy. Military resources have to a significant degree been integrated with civilian production simultaneously with a streamlining of the military establishment. Bureaucracy, budgets, manpower, and military-industrial firms have all undergone reorganization and reduction to facilitate integration.

What are the concrete results of these efforts thus far? Production of civilian products by military industry has significantly increased, helping satisfy consumer demand and increasing the amount and quality of exports. Overall, Chinese sources now claim that about 66 percent by value of military industrial products are for the civilianmarket.[68] This compares with a mere 6.9 percent by value in 1975[69] and 45 percent in 1981.[70] In all, according to a senior PLA official, between 1979 and 1988 "the output value of civilian products [by military industries} increased eight times. . ."[71]

The variety of new and spinoff products is very large, ranging, as we have noted, from high-tech to low-tech. The high-tech side is most interesting in as much as it reflects (in most cases) technology turnover from the military to the civilian sector. Chinese sources list and display hundreds of new products developed by the nuclear, chemical, and aeronautics industries (for example) that have application in medicine, metallurgy, communications, and transportation. These accomplishments do not demonstrate the argument in favor of military spinoffs as opposed to investments in non-state enterprises to produce the same products.[72] What they do demonstrate is that military industries have responded to the call to shift production to goods with greater social value than weapons.

Our study reveals several other ways in which China's conversion strategy has promoted economic development. It has contributed to the regional concentration of state and foreign investments in the coastal areas, as the section on Shenzhen indicates. The relocation of firms from third-line areas to the coastal region is an important reversal of a highly unbalanced and wasteful strategy in Mao's time. Conversion has also led to significant cost savings in the military budget that evidently have been transferred to development (or at least nonmilitary) programs. Use of idle military plant capacity has negated the need to build new enterprises.[73] Technical training for many demobilized soldiers, the professionalization of military-industrial management, and new standards for judging the performance ofindividual enterprises are other contributions to economic development.

The long-term nature of conversion planning indicates the important role of political and intellectual leadership. Breaking down the protective walls surrounding China's (or any country's) military-industrial complex takes a sustained political commitment. In China it has required discarding a number of taboos regarding state-run enterprises, going against the tide of traditional military interests, and creating a new set of economic rules (such as contracts) and incentives (such as more than five-percent profit).

Three sources of leadership have made possible these major changes in the way military industries do business. First is Deng Xiaoping himself: he has been leading the charge for a more streamlined and productive military establishment for over 15 years. Second is COSTIND, which has been directly involved in conversion efforts at national and local levels. Third is China's new breed of defense intellectuals. Although these people, mainly military and civilian analysts responsible to COSTIND, typically make their case in the context of Marxism and China's national interests, their analyses are fairly independent and sensitive to post-cold war trends. Their books and papers have proven instrumental in prompting "new thinking" (*xinsiwei*) about the economics of national defense, and have paved the way for many of the reforms discussed in this paper.

As impressive as China's strategy is, it is not "conversion" in the strict sense of the term. Two Chinese authorities call it a "transition" of military resources to use in both economic and military modernization.[74] "Transition" is a better term than "conversion". but it still does not fully describe what has actually been taking place. For one thing, civilian resources are used by military industries for weapons production, a kind of*reverse conversion*. This might occur, as many Chinese analysts state, in time of war or international tensions. It could also occur, and in fact seems to occur regularly, as the military's means of gaining access to foreign technologies such as electronics.[75]

Secondly integration of the military and civilian economies has put pressure on the PLA to make money in ways that have nothing to do with conversion--or economic development. Arms sales to the Third World are a prime example. Many of the same defense economists who have pushed for conversion have also argued that the commodity nature of military production means stepping up arms sales abroad.[76] They observe that the global arms competition has become more diverse, with a number of Third World States (for examples, Brazil) entering and profiting from it during the 1980s. China should join the competition, they say. Thanks to organizations like NORINCO (China North Industries Corporation), which is both an arms trading and a merchandise trade firm, Chinese arms sellers have found a niche in Third World markets, especially in the Middle East. They are not likely to yield to political pressure and abandon these markets--certainly not so long as the United States and Europe pursue arms deals of their own. Whether or not these arms sales actually buy political influences for China they are evidently quite profitable. One writer estimates that 25 percent of PLA profits from arms sales have been used to make up for budgets cuts.[77]

Another kind of business venture for the PLA unrelated to conversion is simply investments and ownership. PLA units have used capital to develop profitable business interests, such as owning a hotel and providing labor for a joint venture.[78] Some sources report that as much as ten percent of PLA soldiers are involved in commercial activities on the side, often illegally usingmilitary equipment and buildings for such purpose.[79]

Even while conversion is occurring, military resources, for strictly military purposes may increase. For example, allocations to the PLA have risen 15 percent and 12 percent in 1990 and 1991, pushing its share of the budget back up again after several years of decline or no change. Several reasons have been advanced by Chinese and Western sources to explain the increases.[80] The military's need of modern equipment has been long delayed, especially in the airforce and navy. Inflation has lowered the day-to-day living standards of ordinary soldiers and made a military career very unattractive. PLA recruitment is therefore well down. Soldiers living in the interior and the northeast military regions, which have not benefitted from the economic reforms the way the coastal region has, are especially anxious to find their fortunes in civilian life. Further more, as recent Chinese analyses of the global strategic picture have indicated, military modernization is still needed to take account of potential trouble sports. None is ever mentioned specifically; but the implication is that China's borders, although quiet of late, could again be tense. The Gulf War, moreover, has made an impact on Chinese thinking in this regard, by suggesting the decisive role of high-technology weapons and the need for technically skilled soldiers to handle them.

The unmentionable additional factor behind the PLA's increased budget is political: the need to reward the PLA for its loyalty during the student and worker demonstrations of May-June 1989. Since then-- and even more since Boris Yeltsin's ascendance after the August 1991 coup attempt in Moscow-- the CCP leadership has stressed ideological work in the armed forces in order to prevent China's "peaceful evolution" to a bourgeois state. Politicalloyalty has reemerged as the primary factor in assessing leadership.[81] The consequences for the conversion program of this emphasis on party discipline in the armed forces-- an old theme in Chinese politics-- remain to be seen. But given the large and growing importance of military-industrial production to the civilian economy, it is difficult to imagine that a strong dose of politics in the PLA will result in change of direction in defense economics.

Conversion in China is limited by several factors. Among the most important are the size and technical capabilities of the industries, the lack of adequate state support, the lack of managerial (including marketing) expertise, the low technical level and quality of many goods, unadaptable product structures, and the inadequacy of research staffs.[82] In comparison with nonmilitary plants, military industries on average have much lower labor productivity (Y 13,000, compared with Y17,000 in machinery enterprises nationwide) and lower average profits (about 50 percent lower in military and civilian enterprises of the same type)[83]. As a result, it seems fair to predict that the shakeout of noncompetitive, uneconomical medium-size and smaller firms will continue. Those that probably will fare best are in aerospace, electronics, shipbuilding, and the nuclear industry-- so long as they reinvest in higher-technology, higher-value products.[84]

Finally, our assessment should note that environmental factors have played no perceptible role in conversion planning. China (and, for that matter, most of the Third World) has had a longstanding resistance to balancing economic development against environmental protection. My research does not indicate any systematic effort to factor environmental protection into conversion planning, either with respect to the end-use of technologies or the cleanup and disposal of industrial wastes. On an ad hoc basis,military enterprises do produce environmental technologies.[85] But production for the market, rather than for sustainability, has left environmental considerations out in the cold. This is unfortunate in as much as China's environment has been heavily damaged and could surely benefit from the research and technology of military enterprises. Yet it is hard to imagine such enterprises, many of which contribute to the problem to move in the direction of environmental protection unless it is profitable to do so. In this sense, attitudes in China are similar to those in much of the rest of the world.

## Endnotes

1. `See Chi Su, "Sino-Soviet Relations of the 1980s: From Confrontation to Conciliation, " in Samuel S. Kim, Ed., *China and The World: New Directions in Chinese Foreign Relations,* Westview Boulder Co, 1989, 2ed., P.117.

2. Steven I. Levine, "Sino-American Relations: Renormalization and Beyond", ibid., pp.91-2; Deng Xiaoping *Fundamental Issue in present-day China*, Foreign Languages Press, Beijing, 1987, pp.4, 46-47; Deng Xiaoping, Speech of January 16, 1980, quoted in A. Doak Barnett, *China's Economy in Global Perspective*, Brookings Institution, Washington, D.C., 1981, p.72.

3. Chi Su in Kim, ed., pp. 112-18

4. Huan Xiang, "On Sino-U.S. Relations" *Foreign Affairs*, Vol. LX, No.1(Fall, 1981), P. 47; Yang Dezhi, "A strategic Decision on Strengthening the Building of Our Army in the New period", *Honqqi* (Red Flag), No.15 (August 1, 1985), trans. JPRS/CRF, No.85-020 (September 17, 1985), p. 4; Yang Dezhi, Huan Xiang *et al.*, *Guofangfazhan Zhanluesikao* (Thoughts on National DefenseDevelopment Strategy), PLA Press, Beijing, 1987.

5. See for example; Huan Siang's essay, "The Future International Environment and Our National Defense Construction", *Guofangfazhan Zhnluesikao* pp. 6-26, "Concentrate on the Four Modernization", *Honggi* No.3 (February 1, 1980), in JPRS, No. 75425 (April 2, 1980), pp.12-16; Xue Mouhong, "The New Situation in Our Country's Diplomacy", *Honggi* No.6, (March 16, 1986), pp. 19-24; interview of Huan Xiang, "A Tumultuous, Changing, Unquiet Year", *Renmin ribao* (People's Daily, Beijing, Overseas ed.), January 2, 1988, p. 6; Foreign Minister Qian Qichen, "A Year in which the International situation Seems to have improved" *Renmin ribao*, December 17, 1988, p.2.

6. Huan Xiang, "A Tumultuous, Changing, Unquie Year".

7. These goals have usually been put in terms of GNP ($1 trillion) and per capita income ($ 800) by the year 2000. For example, see Deng, p.47.

8. Bruce Cumings, "The Political Economy of China's Turn Outward", in Kim, Ed., pp. 204, 213.

9. Deng, pp. 70-71.

10. One source ("China Invents the Entrepreneurial Army" *The Economist,* May 14-20, 1988 p.67) gives a figure of 30,000 industries of all kinds that were eventually built in third-line areas. The interior development strategy and its economic costs are examined by Barry Naughton, "The Third Front : Defence Industrialization in the Chinese Interior" *The China Quarterly,* September 1988,351-86. "Third line" or "third front" (*Sanxian*), industries are located in all or parts of ten provinces in China's west and southwest, the main ones being Sichuan and Shaanxi.

11. Naughton, P. 379, calculates that "China's annual industrial output is currently 10-15 per cent below what it would have been if the Third Front had never been undertaken. . . "

12. Interview of Zhou Changqing, deputy director of the Third Front Office of the State Council, in *Renmin ribao*, June 6, 1987, p.3

13. Yu Dajiang, "On Third-line construction and Establishment of a Third-Line development Zone", *Junshijingji yanjiu*

(Military Economic Research), No.1 (1988) pp. 64-7; He Shanchang, ed., *Guofangjingjixue* (Defense Economics), Hubei People's Press, n.p., 1988 p. 162.

14. Jin Zhude and Guo Tiejun "An Exploration of the Multiplier Effect in National Defense", in Beijing Association of Defense Economic Research, ed., *Guofangjingfazhan Zhanlue lunwenji* (Essays on National Defense Economic Development Strategy), PLA Press, Beijing, 1987, p.46. Jin and Guo state that China's military-industrial complex includes nearly 1,000 enterprises that specialize in the production of weapons and military equipment, over 200 research institutes, over 300,000 engineers and technical personnel, and nearly 3 million workers.

15. He Shanchang, ed., p. 162

16. See *New York Times* August 27, 1981, I-B/p. 2

17. Wang Luyeh, "The Purpose of National Defense Technology Industries in the National Economy", *Renmin ribao*, December 29, 1989, p.2

18. See his speech to the Chinese Communist Party (CCP)'s Central Military Commission in December 1977, in Deng, p.94

19. David L. Shambaugh, "China's Defense Industries: Indigenous and Foreign Procurement", in Paul H.B. Godwin, ed., *The Chinese Defense Establishment Continuity and change in the 1980s* pp.50-51.

20. See, for example, Song Zhenduo and Ku Guisheng, *Guofanqjingjixue qailun* (outline of Defense Economics), Hunan People's Press, Changsha, 1986 pp. 141-42, 166-67; He Shanchang, ed., op.cit.; Jin Zhude and Chen Zaifang, "Several Issues in China's Defense Economic", in Preparatory Group of the Research committee on Defense Economics in China, ed., *Guofang jingjixue lunwenji* (Essays on Defense Economics), PLA Press, Beijing, 1986, pp.30.-39.

21. Note, for example, the following remark: "To put it a bit unkindly, so-called planning and programming in fact was simply a matter of increasing or reducing figures sent forward by the various departments". Jin and Chen, in Preparatory Group, ed., p. 31.

22. Song and Ku, p. 141.

23. Wendy Frieman, "China's Military R & D System: Reform and Reorientation", in Dennis Fred Simon and Merle Goldman, eds., *Science and Technology in Post-Mao China* Harvard University Press, Cambridge, Mass., 1989, p. 268.

24. Jin and Chen in Preparatory Group, ed., p.39.

25. Liu Yongen and Sang Zili, "Some of China's Military Industrial Technology Secrets are Turned over to Civilian Use", *Liaowanq* (Outlook Weekly), overseas ed., No.10, (March 7, 1988), pp. 14.15. According to this article, the basic decision on reducing excessive secrecy for conversion was made by the COSTIND in March 1987. Establishment of the newspaper *Jungongjishu-minyongbao* (Military to Civilian Weekly), and of the semi-official CAPUMIT (China Association to Promote the Peaceful Uses of Military-Industrial Technology), were among the vehicles chosen by COSTIND to speed up the application of military technology to civilian production

26. For example, see He Shanchang, ed., p.13

27. Song and Ku p.151

28. For petroleum, the profit margin is about 75 percent; for electricity, 72 percent; and for machinery, 21 percent. The year is not given. Yeh Zitong, "On the commodity character of Military Products and Reform of the Price System", in Preparatory Group, ed., pp. 243, 246.

29. Shambaugh, p. 46

30. Frieman, p.256.

31. Ibid., p. 266

32. The premier source on the background and activities of the COSTIND is Benjamin C. Ostrov, *Conquering Resources: The Growth and Decline of the PLA's Science and Technology Commission for National Defense,* M.E. Sharpe, Armonk, NY 1991.

33. Ibid., pp. 63-4

34. This includes a direct role, he writes, in budget formulation, allocations, and monitoring in areas such as weapons procurements and basic and applied research, Ibid., pp. 67-8.

35. On this last point, see *Renmin ribao* June 21, 1987, p. 4 The article reports that COSTIND for the first time issued a planning document, called the "Planning Directory" *(jihua zhinan)*,which specified the main items of technological research to be carried out by military industries. The areas covered were said to include lasers, computers, man and machine environment , the atmosphere and water power.

36. *Renmin ribao,* April 21, 1987, P.1

37. From interviews with the Shaanxi Province COSTIND office. See the case study below on Xian Aircraft company.

38. *Renmin Ribao*, December 12, 1988, P.4

38. Zhang Chunting, "The Youth Movement in the Top Leadership of the Chinese Military" *Liaowang*, No.43, (October 28, 1985) p.27.

39. Deng *Selected Works of Deng Xiaoping, 1975-1982* Foreign Languages Press, 1984 p. 154. He pinpointed five problems within the army -- "bloating, laxity, conceit, extravagance and inertia" -- and emphasized the need to "streamline the army establishment, and restructure it as a whole. . ."

40. Jiang Yuqi, ed., *Ahongguo guofang jingjifazhanchanlue yanjiu* (Studies in China's National Defense Economic Development Strategy), table 1 p.31.

41. Ibid., table 2, p.35

42. Ruth Leger Sivard ed., *World Military and Social Expenditures 1989* World priorities. Washington, D.C., 1990, p.52.

43. *Zhongguo tongji nianjian* (China Statistical Yearbook), quoted in *Zhengming* (Dissent, Hong Kong), No.95, (September 1, 1985) p.28 *Renmin ribao*, April 14, 1987, p.3 and Lester R. Brown *et al State of the World 1986*, Norton, Newyork, 1986, p. 207.

44. June Tuefel Dreyer., " The Demobilization of PLA Servicemen and Their Reintegration into Civilian Life", in Dreyer ed., *Chinese Defense and Foreign Policy*, Paragon House, New York, 1988 pp. 304-19. Dreyer found many instances of rebelliousness in the PLA that frustrated the political leadership's efforts to reduce its size by any substantial amount.

45. See Deng, *Selected Works* p. 373.

46. See for example, *Renmin ribao*, April 3, p.1 ; April 5, 1987, p. 1.

47. *Renmin ribao* July 24, 1987, p.1

48. *Renmin ribao,* February 10, 1987, p.4

49. *Renmin ribao,* August 5, 1987, p.2

50. State Council office, Small Group on Military Cadre Transfers and Relocations, ed., *Junduei Zhuanyeganbu anzhigongzuoyanjiu* (Research on Military Cadre Transfers and Relocations), Laodong renshi, Beijing, 1985, p.11 Still other problems are recounted by Dryer, pp. 321-23. Dreyer also raises questions about the claim that the 1-million target actually was reached.

51. See *China Daily*, August 19, 1989, P.2 (which refers to 121 military *industries*), and *Renmin ribao,* December 4, 1991, p. 1 (which refers to the same number of "projects").

52. For similar conclusions based on the Sichuan Province experience, see *Renmin ribao* July 9, 1987, p.3

53. *Renmin ribao*, June 8, 1987, p. 3, interview of Zhou Changquing, deputy director of the third-front office of the State Council.

54. See, for example, He Shangchang, ed., pp. 139-42.

55. *China Daily* July 31, 1988, p.1

56. Jiang Xinxiong, president of China National Nuclear Corporation, in State Planning Commission and COSTIND, eds., *Zhongguo guofanggongye junminjiehe shizhounian zhuanji* (Special Edition on the Tenth Anniversary of Military Conversion in China's Defense Industries,), Beijing, Ordnance Industry Publishing House, 1990, p.20

57. Li Dingfang, vice-chairman of the China National Nuclear Corporation in *China Daily*, August 8, 1990, P.1 Mr. Li's greatest concern is that civilian output can be developed fast enough to create jobs for 40,000 workers idled by reduced military production.

58. Based on a visit in August 1989 and interviews with the executive Vice-President and Chief Engineer.

59. Only a few older PLA cadres have been retrained, XAC's production manager said retraining is a difficult task in light of their age, prior training, and educational level.

60. According to Minister Lin Zongtaing, civilian products have increased from 16 percent in 1980 to 75 percent in 1988 as a proportion of output value. *Zhongguo guofang...zhuanji* p.13.

61. *Renmin ribao* December 9, 1988, p.5

62. Sales abroad by the industry were US $ 294 million in 1990. Best-selling in civilian items are the Y-12 (a 17 - seat Commercial plane), MD - 82 trunkline aircraft, helicopters, and Long March rockets to launch communication satellites. The export corporations are CATIC, Great Wall, and Precision. See *China Daily* August 11, p.1 and *Renmin ribao* October 16, 1991. p. 1.

63. Based on interviews in 1990 and 1991 with Shenzhen city officials and numerous enterprise heads.

64. Interview of the Vice-General Engineer, August 24, 1990.

65. Information from interviews in Shenzhen August 22, 1991.

66. Based on interviews in Shenzhen August 20, 1991.

67. Interviews in Beijing, August 1991, with COSTIND officials.

68. In the case of dual-use products (e.g., helicopters), whether they are counted as "civilian" or "military" depends, according to some Chinese managers I interviewed, on the end -use of the product. But other enterprise officials said it depended on the funding source.

69. Li Tao, "The Strategic Question of the Arms Industry's Developing the Production of Civilian Products", Foreign Broadcast Information Service-China, September 25, 1984, p. K12.

70. *New York Times* August 27, 1981, IB/p.2

71. Zhang Bin, Vice-head of the PLA General Logistics Department, in *Zhongguo guofang, ..zhuanji* p.22 Exactly how much money the state invested to produce these results is not clear, although Zhang does state that from 1985 to 1988 the total investment was over Y 500 million.

72. As noted earlier, China's conversion process has been consistently informed by the experiences of other countries, including the role of technology spinoffs from military research and production (e.g., in the Reagan- Bush Star Wars or SDI program).

    My view is that Chinese analyses exaggerate the importance of spinoffs. Perhaps this is deliberate, in order to lend further support to the value of military conversion. In any case, a closer analysis seems to be called for of the actual economic benefits from spinoffs when matched against the level of expenditures and alternative investments.

73. Jin Zhude and Chai Benliang. "China's Experience : A case Study ", paper for the UN Conference on Conversion, Moscow, August 13-17 , 1990 p. 8.

74. Ibid, p.2

75. This point is made by Wendy Frieman, pp. 272-73.

76. See, for example, Song and Ku, p. 174.
77. Paul H.B. Godwin, "Soldiers and Statesmen: Chinese Defense Policies in the 1990s,: in Kim,ed., p. 196.

78. "China Invents the Entrepreneurial Army" p.68.

79. *Kaifang yuekan* (opening Monthly, Hong Kong), September 1990: trans in *Inside China Monthly*, vol. XII No.11, (November 1990-) pp. 26-28.

80. Based on conversations with various authorities in Beijing over the past two years. See also Harlan W. Jencks, "The Military in China" *Current History* September 1989, pp. 265-68, 291.

81. Taj Ming Cheung "Basic Marxit Training" *Far Eastern Economic Review,* August 23, 1990, pp. 46-7.

82. These observations are from my own experience; but most of them are also recited by Chinese enterprise managers in conversations. In addition, see similar remarks by the general director of NORINCO, Jin Lailie, interviewed in *Jungongjishu minyongbao* (Military to Civilian Weekly), May 16, 1989, p. 3.

83. Speech by Huai Guomo, Deputy Director of COSTIND, in *Jungongjishu-minyongbao* (Military to Civilian Weekly), August 22, 1989, p.1

84. Mr. Huan Xiang, the late State Councillor, was among the internal critics of conversion efforts that simply produced money-making household goods rather than aimed at developing

higher technologies and a new generation of workers who could use them. See his "The Future International Environment and Our National Defense Construction", p.25.

85. For example, the Wuhan (Hubei Province) Instrument Factory under the Ministry of Aeronautics and Aviation produces waste water treatment equipment for export. The Beijing Electron Tube Factory, under what is now the Ministry of Machinery and Electronics, manufactures a variety of energy-saving lamps and bulbs for both domestic and international markets. The Nanfeng Machinery Factory, also under the Ministry of Aeronautics and Aviation in Shaanxi Province, advertises environmental testing equipment for export.

FIGURE 1. THE CHINESE MILITARY-INDUSTRIAL NETWORK

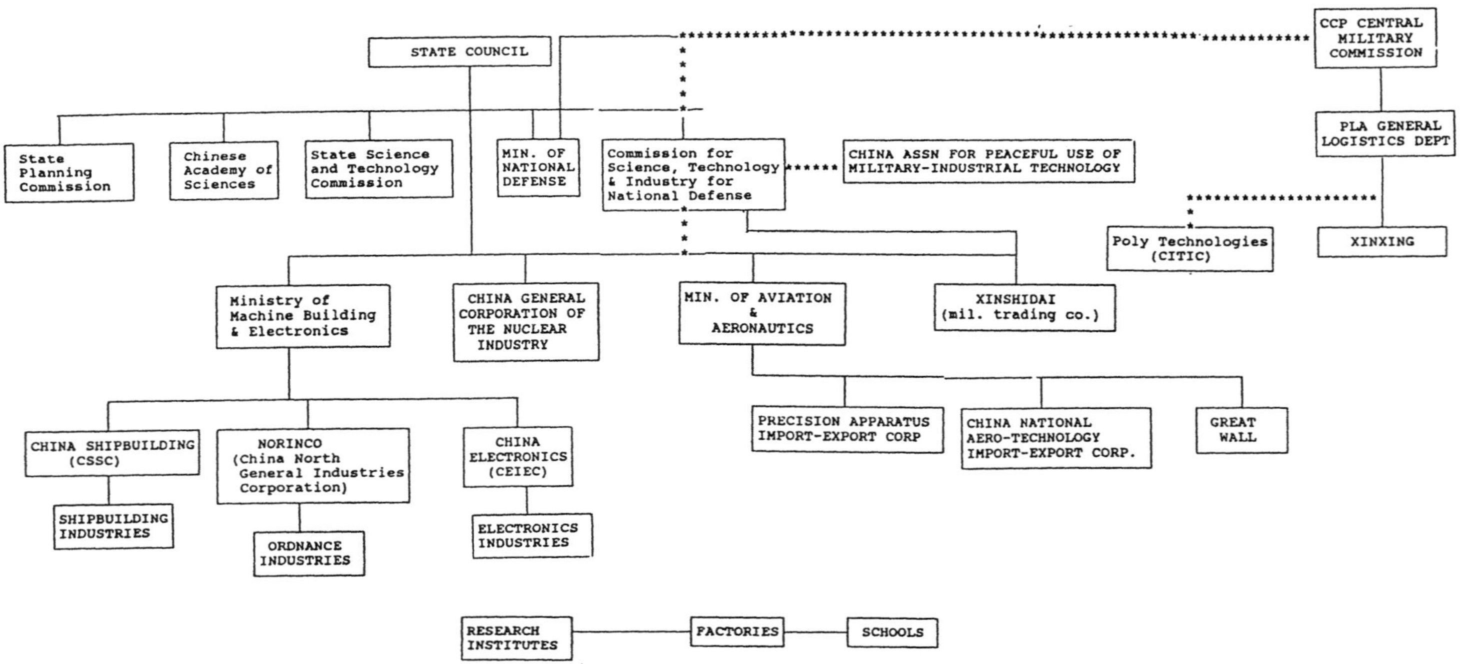


Sources: Frieman, p. 281; "Inside Mainland's War Machine," The Open Monthly, Hong Kong, Oct. 1990, translated in Inside China's Mainland, XII, No. 12 (Dec. 1990), pp. 21-22; interviews in China.

# PROBLEMS AND PROSPECTS OF CONVERSION

**Dr. Betty G. Lall**
*Council on Economic Priorities, U.S.A.*

## Introduction

For the first time in over 45 years, since World War II ended, the United States (Government and Citizens) can contemplate a substantial reduction in its military budget and military hardware, troops, and bases. The world has not become completely peaceful--many armed conflicts continue to rage, e.g., in Yugoslavia, Africa, South and Southeast Asia.

But the Cold War which consumed so much of the resources and manpower of the United States and its allies has ended. The collapse of the Soviet Union and the movement toward democratic governments in Europe means that the United States and other democratic states can devote attention to alleviating economic and social problems that have long been neglected.

The United States, throughout the 1980s, expended enormous sums to build a military establishment to prevent war. Its defense budget exceeded $305 billion by 1985 and although it has begun to decrease, the pace of reduction appears slow, given the lack of threat to the country and its immense back-log of needs to be met.

The U.S. defense establishment penetrated all parts of the globe. Over half of the national budget was expended overseas, primarily in Western Europe, but the U.S. has military installations in the Middle East, South Asia, SE Asia, South Korea and Japan. The Defense Department also located military facilities in the Western Hemisphere.

The DoD budget consists of five major functions: military personnel, operations and maintenance, research and development,procurement of weapons and military construction. A separate budget, lodged outside of the Defense Department, covers nuclear weapons programs. This is the responsibility in large part of the Department of Energy.

## Proposals to Cut the Military Budget

With the Cold War essentially over many members of the U.S. Congress and former officials of the Department of Defense have proposed that the U.S. military budget be scaled back. Such proposals began circulating in the U.S. Congress, the Pentagon, and in private research and action groups, particularly in Washington, D.C.

Two of the most detailed and prominent proposals were authoured by former U.S. Secretary of Defense Robert McNamara in the Kennedy Administration, and former Assistant Defense Secretary Lawrence Korb who currently is a visiting scholar at the Brookings Institution in Washington, D.C. For Fiscal Year 1991 Secretary of Defense Richard Cheney (a former U.S. Congressman from Wyoming) proposed a U.S. defense budget of $ 306 billion. This was a substantial increase over the 1990 budget of $ 296.3 billion.

McNamara proposed that the 1992-94 defense budget be reduced by $ 138 billion by the Fiscal year 1994. He also recommended that additional cuts be made and to do this the United States should conduct a complete reappraisal of its foreign policy, security commitments, threat assessment and military strategy.

### Views of Lawrence Korb

Korb's proposal specified that "the goal of a 50 percent after-inflation cut in the defense budget by the year 2,000 is a reasonable one". He spoke for a study group of the Committee for National Security (CNS) and the Defense Budget Project. The group would cancel several major weapons systems including the C-17 cargo aircraft, the advanced medium range air-to-air missile, SSN/21 attack submarine, rail garrison MX missile, and the B-2 bomber. Total savings from the cancellations of these five major weapons programs would total $ 23 billion according to the CNS study.

### Cuts Proposed by William Kaufmann

William Kaufmann of the Brookings Institution and M.I.T., proposed a plan to cut defense spending in three phases by $ 105 billion over a ten year period. He would reduce U.S. forces by at least 62,000 and strategic nuclear warheads to 4,100 for the U.S. and U.S.S.R.

### The U.S. Joint Chiefs of Staff

Former members of the U.S. Joint Chiefs of Staff (General David Jones, Admiral William Crowe, and General John Vessy) testified before the U.S. Senate Armed Services Committee in February 1990 that the U.S. and U.S.S.R. should reduce their strategic nuclear weapons to a very limited number on both sides. General Andrew Goodpaster, former head of NATO strategic forces, would cut NATO's current strength by 50 percent; reduce naval forces by substantial amounts, and cut strategic nuclear warheads to 500 to 1,000 for the U.S. and the Soviet Union.

The views expressed by former military officials about decreasing the defense budget suggest that these former military officialswould move much faster to cut weapons programs and expenditures than the President and many members of the U.S. Congress.

The American public has not yet focussed in any major way on the defense budget issue. But this lack of interest is changing. An example of increased interest concerning conversion of military expenditures and goods to peaceful and productive activities is the bill passed in the U.S. Congress in 1990 to provide $ 200 million for conversion efforts. Congresswoman Mary Rose Oakar of the House Manpower Stabilization Subcommittee successfully steered the legislation through the Congress with the special help of the House Majority Leader Richard Gephardt, a Democratic Congressman from the state of Missouri. He is well aware that his state receives the second largest amount of money from the Department of Defense prime contracts, a total of $ 4.548 billion. Gephardt used his influence to have the legislation passed.

### Eliminating and Converting Unnecessary Military Facilities

During the long years of the "Cold War" the U.S. Defense Department constructed a large number of military bases throughout the United States and in many countries. Such bases and related facilities helped local economies by purchases of goods from local merchants and employing local citizens to fulfill many military needs at these bases.

Servicemen and women enrolled their children in local schools causing the Pentagon to pay the school system for the extra burden placed on it as a result of the influx of children at the elementary and high school levels. Many communities benefitted from the influx of base personnel and thus when the Pentagon suggested a base might be closed local residents, civic leaders, andelected officials often lobbied their Congressmen and Senators to keep the base open.

During the long period of the Cold War the number of military bases and properties in the United States grew to 871. Another 375 such facilities were maintained overseas. Within local communities in the United States these military facilities became a way of life.

Local pressure to keep bases open became so great that the Pentagon was forced to yield to the community. In 1988 the Defense Department decided it needed to close some bases, but DoD officials knew of the pressures from members of Congress not to let a base close with the consequent losses of revenue for the tax base, merchants, and other affected groups.

A political solution was found to solve this touchy problem. Congress passed a base closing bill which established a high level commission to study and recommend which bases should be closed or realigned. The commission brought forth its recommendations to the Congress. The law establishing it specified that the Commission had only one vote to make in deciding whether all of the bases on the list should be closed or none of them. This plan worked with no political difficulties. Bases were closed or realigned and the community accepted the decision. Moreover, the land on which the bases were located was turned over to the local community to be used as the people desired.

The Pentagon's Office of Economic Adjustment (OEA) has operated for some thirty years to assist communities to adjust to the closure of military facilities and bases throughout the United States and abroad. Currently the OEA committee consists of twenty-two U.S. Government departments and agencies.

**The Nuclear Weapons Production Complex**

Among the many remnants of the Cold War perhaps the most dangerous is the enormous stockpile of nuclear weapons held by several states. In the United States these weapons are not only dangerous because of the enormous damage they can do if used in war, but they are also a threat to the public because their storage facilities are so inadequate that the lethal material stored throughout the United States is leaking into the soil and contaminating the area surrounding the facility.

In 1988 the Department of Energy announced that 16 sites were contaminated with radioactive and chemical pollution. One site alone, the Hanford reactor complex in the state of Washington, might require $ 39 billion for cleaning up. The U.S. General Accounting Office reported that 3,500 inactive waste sites needed to be cleaned up with the cost estimated at $ 25 to $ 65 billion.

# DEFENSE INDUSTRY CONVERSION: THE CASE OF CZECHOSLOVAKIA

**Milton Leitenberg**
*Center for International Security Studies at Maryland, U.S.A.*

In 1989, following Soviet President Gorbachev's address to the United Nations in December 1988, Czechoslovakia*, Poland and Hungary cut their military expenditures and armed forces sharply. They continued the cuts in 1990 and 1991, and with the dissolution of the Warsaw Treaty Organization, (WTO), these states were strongly motivated to reallocate resources away from their defense sectors. In addition to drastic cuts in their standing armed forces this also meant a program of defense industry conversion. They will attempt to convert a portion of their industrial infrastructure, which had been devoted to weapons production for the previous 40 years, to the production of civilian products. Most precisely defined, conversion is the utilization of existing defense industrial plant and personnel at the same factory site to produce non-military products. The conversion process is of greatest importance to the U.S.S.R. Its defense industrial sector was many times larger than that of any other state and its economy is in the direst condition and in greatest need of resources for its civilian sector. Czechoslovakia was the second largest weapons producer in the WTO and even before the loss of power by the communists the Czech government had announced that weapons production by its defense industrial sector would be drastically cut. In some of the earliest statements made by President Vaclav Havel and Foreign Minister Jiri Dienstbier after they assumed power in January 1990, they declared that Czechoslovakia would end its export of weapons. However by 1992 it was apparent that both the plans for extensive defense industry conversion and a cessation of weapons exports had been severely compromised by the domestic politics ofpotential Slovakian secession. It is therefore of some interest to see how this process played itself out in an economy in which the size of the defense industry was not altogether out of proportion, which is not in a state of total crisis and imminent collapse, and in which the government was accepted as being seriously interested in seeing the process succeed.

In the two years in question there was virtually no impact of the shift from a command to a market economy in Czechoslovakia in the process of defense industry conversion, since all of the industrial facilities concerned remained under state control. Some larger companies reportedly may be divided into sections in order to allow partial privatization, but the defense industry is to be privatized only in the last resort, and most of it is likely to remain under state control. The question of the marketization of the economy is therefore not otherwise addressed in the study.

The material which follows is organized in five sections:

1. The desirability of defense industry conversion

2. The similarities or dissimilarities between converting defense production facilities and converting or discontinuing capacity in any other industrial sector

3. Basic data on the Czechoslovak defense industry, and a description of the events in 1991 and 1992, among them the unsuccessful campaign to terminate arms sales

4. The possibilities for alternative production in Czechoslovak defense industries.

5. What Western governments or enterprises can do to facilitate such alternative production.

**The Desirability of Defense Industry Conversion**

The dismantling or conversion of defense industrial capacity is desirable when that capacity is greater than is necessary, and to the degree that it causes problems for the domestic economy and in international relations. The domestic economy can be weakened by the share of national expenditure allocated to the defense industry and by the amount of other resources required to support it. In international relations, threats arise both from the products of that industry -- for example, the massive tank production in the U.S.S.R. for some three decades -- and from the export of weapons to regions of the developing world involved in chronic or acute conflict. If defense-industrial capacity is reduced in the developed industrial states, fewer weapons will be produced and the pressures to sell weapons to developing nations will decline. As a corollary, the resources released can be used by societies for other needs which have been deferred.

The only real way to obtain the permanent reduction of excess defense capacity is through dismantling facilities or true conversion. Due to the historical accident of mimicking the Soviet model, many centrally planned economies developed a system whereby weapons production facilities also produced relatively large amounts of civilian goods. The purpose of this system was primarily to permit rapid expansion of defense production in case of war. For example, the Soviet and Chinese defense industries havelong manufactured very large quantities of civilian consumer goods in the same facilities that produce weapons. In the U.S.S.R., the ratio for the defense sector as a whole was 45 percent civilian production and 55 percent defense production. The same pattern of dual production of military and civilian goods occurred in all Czech weapon production facilities.[1]

In terms of arms control, conversion is particularly important when the goal is the true reduction in a nation's defense production capacity. That outcome is much more certain if defense industry resources are dismantled or truly converted than if plant and machinery are simply mothballed and the workforce used in alternative production lines at the same facility.

Beginning in the early 1980's China began reporting a substantial effort in what it described as defense industry conversion. However the actual nature of the process is ambiguous, and in some doubt. Although China apparently has been producing a large variety of civilian goods at military plants since the early 1980's, Chinese arms exports peaked in the early 1980's during the Iran-Iraq War because of sales to both combatant nations.[2] This combination of a sharp increase in arms exports and increased civilian production at weapons factories posed an obvious puzzle. The apparent answer is that the Chinese government maintained the old defense industrial capacity and built new production lines within the defense sector to enable it to expand civilian output. According to Chinese government officials some defense industrial plants have been permanently closed or converted, but there is no available evidence to assess how much of the total that is.[3]

It is equally impossible to provide an overall assessment of the degree of defense industry conversion that took place in theU.S.S.R in the three years between 1989 and 1992. In President Gorbachev's speech to the United Nations in December 1988, in addition to announcing cuts in military expenditure and force structure that the

U.S.S.R. would carry out in the following years, he also stated that a defense industry conversion plan would be elaborated, and in particular that "two or perhaps three" defense plants would be converted as an experiment.[4] For the next two years official Soviet agencies struggled to formulate and initiate, essentially unsuccessfully, such a "plan". What took place in the meantime was largely an ad-hoc process under the force of circumstances, and in only a small fraction of approximately five thousand Soviet defense plants.[5]

The assessment of a U.S. intelligence agency in mid-1991 read as follows:

Despite much fanfare, defense industry conversion has made scant progress, as the Soviets candidly admit. Of the several thousand plants producing for the military, only about 500 will undergo some degree of conversion. About 40 plants are to cease military production altogether. However, most of these belong to the civilian sector and devote only a small share of their output to defense.

While reductions in weapon production and defense spending are occurring, the defense industry, for the most part, is retaining a mobilization base. Military production lines are operating at reduced output levels or, in some cases, lines are being shut down but maintained in operating condition. Soviet economists and industrialists indicate that conversion, under current plans, will be achieved by expanding existing civilian production and adding new capacity at military plants rather than by switching weapon lines to civilian output. Although Soviet statistics show some increases inconsumer production by the defense sector, growth has been far short of plans.

Slow progress is not surprising. Conversion remains in the hands of government officials strongly tied to the defense industrial complex and interested in retaining its established priorities and skilled manpower.[6]

This was clearly not a serious program of defense-industry conversion. President Gorbachev's advisor, Alexander Yakovlev, appears to have been of a very similar opinion. He pointed out in September 1990 that the U.S.S.R.'s military industries had remained "almost untouched", and he ridiculed the claims of successes in conversion.[7]

**Structural Changes in Modern Economies and Defense-Industry Conversion**

Since the end of World War II, every industrialized state in the West has witnessed massive structural shifts in industrial employment. Employment in textiles, mining, forestry, steel, shipbuilding, automobile production, and electronics has fallen in turn by dramatic amounts in the United States, Great Britain, West Germany, and Sweden, as well as in other countries. The figures for structural shifts in employment in these countries over a period of one or two decades, say 1960 to 1970, or 1960 to 1980, are extremely high. Tens of millions of individuals have been involved, an order of magnitude greater than would be affected by any degree of defense-industry conversion that could reasonably be expected, either for the above-mentioned countries individually or as a group.

The problems of restructuring are essentially the same for defense industries as for any other industrial sector[8]

1. Can plant and machinery be applied directly or with relatively minor modifications to produce another marketable product?

2. Can the workforce be easily transferred to alternative production, or does it need extensive retraining?

3. Or must the plant be closed and the workforce released into the labor market?

In converting defence industries, in the most optimum circumstances, one would like to utilize the same workforce at its existing location, with existing skills or after retraining and employing the maximum amount of existing plant and machinery, on the assumption that the economic costs incurred for the transition are the lowest when this is possible.

Are there any characteristics of defense industries that make them more difficult to convert than other industries? There are two factors to be noted, but they are both political, and they bear no relation to the technological characteristics of defense industries and to arguments as to whether these characteristics make them more or less convertible than other industries. The first of these is that management or labour groups interested in continuing military production at current levels--and legislators representing the regions in which the plants are located --put forward the claim that continued production is necessary for reasons of "national security". The second is that the government may want to preserve the capacity as part of the defense-industrial base. In recent years, with the extensive rationalization that has already occurred in the different sectors of the defense industry in virtually all Western countries, defense departments are frequently concerned to retain portions of the defense-industrial base, should there be afuture need to reconstitute some capacity or an entire sector that appears to be facing extinction.

Finally, there is the question of whether the production technology and the work skills of the employees in defense industry are so highly specialized as to make their conversion to civilian production particularly difficult, as is frequently suggested. The skills of industrial workers on the factory floor producing tanks or artillery are not essentially different from mechanics or foundry workers producing civilian products. Military ship construction, excluding submarines, does not differ in any significant way from civilian ship construction. The same is true for military and civilian aircraft, military and civilian electronics, military and civilian communication satellites or military and civilian weather satellites. It is clear that there is a substantial amount of exaggeration in the view that the skills of defense industrial employees differ greatly from those producing civilian products.

Defense industry employs a higher percentage of R & D personnel than does civilian industry, but past experience with closures, at Boeing aircraft in the United States for example, in the early 1970s, indicates that some ninety percent of research personnel find other employment within a year. Scientific researchers should not face any overwhelming problem "converting". Civilian academic research scientists frequently face analogous problems in the course of a career, "retooling" as a matter of course, as a new approach to a problem is attempted, or as new methodologies or technologies are applied, or as new problems are tackled. A research scientist may prefer to continue in military R & D for reasons of prestige, relative salary, or other personal benefits, but that is another matter.

There is one characteristic of any structural industrial conversion effort, defense or any other, that always presents special problems when it exists and that is regional concentration. Regional concentration of the defense industry is not uncommon, and it nearly always makes the political pressure to maintain employment in the existing industry that much stronger. In the Czech case, the concentration in Slovakia as a whole as against Bohemia was not overwhelming, although the very largest plants were in a relatively restricted area within Slovakia, and other major industries were missing from that area. However the problem became greatly aggravated as soon as the central government began to face separatist and independence movements within Slovakia. The combination of the two factors was to have a profound and determining effect on the outcome of the defense industry conversion effort. The post- World War II political configuration of the Czech population had produced a highly organized and politicized labor force in the Slovakian weapon production factories, who were now still led by former communist functionaries. As separatist pressure in Slovakia grew during 1990, the central government in Prague became reluctant to press industrial issues, and to risk the dislocations that conversion might bring with it, even those that were planned by the previous communist government. The Czech government's decision on January, 1, 1991 to decentralize government operations gave policymakers in Slovakia authority to make industrial and economic policy decisions, and in particular, placed control of the heavy weapons factories in their hands. The result was that, at least for the time being, plans to convert the defense industry were greatly impeded.

## The Defence-Industrial Sector in Czechoslovakia and the Experience of 1989 to 1992

Czechoslovakia has been a major weapons producer for much of the twentieth century. The Czech firm of Skoda, along with Creusot-Loire of France and Krupp of Germany, were the three major European arms producers during World War I and in the inter-war period. During the Austro-Hungarian empire, however, arms production was located in Bohemia, in the Czech portion of the country. The Skoda works were in the city of Pilsen. It was only after the communists came to power that the primary location of the Czech defense industry shifted to Slovakia, closer to the Soviet border. The large tank works in Martin were built up between 1950 and 1955, and the large weapons plants in Dubnica and Adamov were constructed at the same time.

During the communist period, weapons production was called "special production" or "special technology", and was directed by a special section of the Federal Ministry of Metallurgy, Engineering and Electronics.[9] Between 1950 and 1968, the Czech government allocated between *35 and 91 percent* - depending on the year-of its total annual capital investment to the defense industrial sector.[10] One hundred and ten (or 111) enterprises made up the defense-production sector, and Czech defense industry workers earned twice the salary of the average Czech industrial worker. Czech weapons production peaked in 1988 at 8.3 percent of the Ministry's total production, equivalent to over 4 percent of the country's net industrial output,[11] and 2 percent of national product.

Data on employment levels in the Czech defense industry are somewhat contradictory. The first official figures released by Czech authorities in 1989 and 1990 appear to havebeen substantially inflated, perhaps with the purpose of exaggerating the employment problems that would attend conversion efforts. These stated that some 120,000 to 150,000 workers were directly employed in weapons production.(12) The expected changes in the early 1990s were also described as likely "to affect" between 200,000 and 250,000 workers, presumably including impacts on local shopkeepers, etc. Popper referred to "nearly 3 percent of the work force" as being employed in the defense sector, without giving the absolute figure that this would be equivalent to, while stating that the conversion program would cause the loss of 70,000 industrial jobs in Slovakia alone.[13] However in May 1991 Czech Economics Minister Vladimir Dlouhy reported that the number of people employed in arms production had peaked at 73,000 in 1987 and would be reduced to about 15,000 in 1992.[14] These figures were corroborated by Oldrich Cerny, a personal special representative of President Havel in his June 1991 testimony to a U.S. Senate committee. He reported that at its peak,". . . in 1987 the Czech defense industry employed 75,000 people...."[15] This would mean that the maximum number employed in the defense industrial sector had been only a bit over one percent of the entire labor force, and that the losses expected were about a half percent. Mr.Cerny added that "at the beginning of this year (1991) only 45,000 workers" were still employed in the defense sector. When President Havel visited Slovakia in the spring of 1991 he found that reports of job losses in Slovakia were "overstated" but there is no available information as to whether or how the 30,000 "reduced" employees have been reutilized, if they are unemployed, etc. As of early 1991, unemployment levels in the Czech region of Czechoslovakia were reported to be only in the range of 3-4 percent, and in the Slovakian region, according to different reports, between 4 and 7 percent. In February 1992 these figures were reported as reaching 15 percentin Slovakia, double the rate for the Czech region. Nevertheless even the higher rate was not much out of line with many parts of Western Europe at the same time. The 1989-90 Czech government figures also stated that 60 percent of defense industrial employment was in Slovakia and 40 percent in the Czech portion of the country; in 1991 Mr. Cerny gave this split as being 65 percent and 35 percent. Popper presents a figure of 16 percent of Slovakia's industrial work force as employed in the arms industry, with a very high concentration of these in the subregion of the Vah valley where some of the largest plants are situated.

Of all the major weapon producing states, Czechoslovakia may have had the highest weapons production to export ratio. *Seventy-five percent* of the weapons that Czechoslovakia produced - 80% according to Mr.Cerny - were exported to other Warsaw Treaty Organization states and to developing countries. The arms were made under Soviet license. All major sales to third world countries required the approval of the U.S.S.R., and the sales were managed by the Czech government owned Omnipol Company, which had a monopoly on all Czech arms sales. The co-ordinated U.S.S.R./Czech defense industry production plans - either annual or five year plans - were apparently independent of specific sales. It was assumed that the great majority of the annual production would be taken by Warsaw Pact states and the remainder, as directed by the U.S.S.R., would go to developing nations.

According to data published by the U.S. Arms Control and Disarmament Agency (ACDA), Czechoslovakia exported $ 18.5 billion worth of weapons between 1971 and 1988, sharing with China the ranking as the world's sixth largest arms exporter. (See table). (This is equivalent, however, to only 2.4 percent of the world market). Production consisted mainly of tanks, armored personnelcarriers, self-propelled artillery, small arms and ammunition, and a jet aircraft trainer. Czechoslovakia's major Third World arms clients included Iraq, Iran, Libya, Syria, Ethiopia, Afghanistan, India, Algeria,

Burma, Egypt, Nigeria and Pakistan. Another report claimed that arms sales provided an annual earning of $ 850 million per year, and that this was "half of Czechoslovakia's hard currency earnings".[17] This estimate appears to be somewhat high; arms sales apparently brought Czechoslovakia around $ 500 million annually until 1985, and represented one-third to one-half of export revenue. Particularly in the earlier years, weapons were sold on long term credits.

In 1989 Czechoslovakia reduced its military expenditure by 15 percent, and in 1990 by 13 percent. The first official government reports on weapon production cuts planned by the communist government, stated that weapon production was to have dropped by 40 percent in 1989 -- under the communist government -- and that by 1993 weapons production would be cut by 70 to 75 percent from the 1988 production levels. (The earliest report had predicted only a 50 percent overall cut). See Table 2.

In June 1991, Mr. Cerny stated that the Czech government plan was now to cut weapons production by *89 percent* of its peak year level, leaving only 11 percent of production.[18] In 1987 the value of Czech arms production was given as 29 billion Czech crowns -- around 1 billion $ U.S. -- whereas, defense industry contracts for 1992 were planned to be only 4 billion crowns. The original plan provided that some 30,000 workers would be released by the weapons-production facilities -- matching exactly the number that Mr. Cerny used -- and Czech government planners calculated that military equipment would then represent only 0.8-1 percent of national production. Given the reduction in the size of the Czech armedforces, and the large amount of equipment that it was retiring under the Treaty on Conventional Armed Forces in Europe (CFE), senior Czech military officials have stated that there would be no further modernization or purchases of Czech army equipment until 1994.[19]

When the new government came into power in January 1990,Foreign Minister Jiri Dienstbier stated that Czechoslovakia would end its weapon exports entirely, and that tank production would cease by the end of 1990.[20] Both of these decisions went through a tortuous evolution, and were heavily modified, if not reversed. The first qualification was made by a ministry spokesman within two days. In March the Foreign Ministry spokesman said that the country would "try to orient exports away from arms" but that "existing obligations" would have to be honored.[21] In the same week President Havel stated that Czechoslovakia was currently halting deliveries to "all totalitarian regimes and dictatorships", although it would continue to sell arms to democratic states "for a little while". In January 1991 Havel said that Czechoslovakia would halt arms exports "to areas of tension",[22], and in April 1991 that tanks could be exported if they already had been manufactured or are nearing completion, and if a country is found to whom they can be exported without any political risk.[23]. Arms sales continued in both 1990 and 1991. In February 1992, on a visit to Slovak weapons factories, "Economics Minister Dlouhy pledged a pragmatic new arms sales policy,....the government now condones the limited sale of arms, if such sales conform to Prague's foreign policy".[24]

Exactly a year after the new Czech government took office and Foreign Minister Dienstbier's original pronouncement, the power to make economic and industrial decisions for Slovakia was passed to the Slovak government. It immediately -- onJanuary 8,1991 -- decided to halt or to slow down the conversion program and to continue to export arms.[25] The Slovakian government claimed that the earnings from weapons exports would be used for the conversion program, a claim repeated by the Federal Premier, Marion Calfa (a Slovak by regional derivation), and that the budget did not provide sufficient funds for conversion, an argument somewhat belied by the fact that the continued arms production had to be heavily subsidized. Slovak Premier Meciar traveled to the USSR to see if continued arms deliveries could be arranged.[26]

In the spring of 1991 Czechoslovakia tabled data to the Treaty on Conventional Armed Forces in Europe (CFE) which indicated that it wanted to export more than 5,500 individual weapons under the export exemption of the Treaty.[27] These systems amounted to 43 percent of Czechoslovakia's inventory of major weapons, and over three-quarters of the equipment that Czechoslovakia would otherwise have had to destroy under the CFE Treaty. Article 3 of the Treaty allows an exemption from treaty limits for equipment "awaiting/or ...being refurbished for export or re-export...." Czechoslovakia had not indicated that it planned these exports when the CFE Treaty was signed in November 1990, but requested the exemptions during a 90-day period that was permitted for making amendments. Bulgaria also hoped to export a sizable amount of the equipment they would otherwise have had to destroy. Due to the number of weapons involved, the Czech request for export exemption went beyond the spirit if not the letter of the CFE Treaty. Since the Czech request was formally legal, and was not withdrawn, it was not challenged by the other CFE signatories.[28]

In March 1991, Mr.Dienstbier himself negotiated an oil purchase agreement with Iran for which Iran requested heavy arms andheavy machinery in payment. It subsequently transpired that a large shipment of tanks- variously reported as 300 or 1,500 older T-54 and T-55 tanks - was involved.[29] Iran had asked to buy T-72 tanks. Czechoslovakia also reportedly offered to produce ground support aircraft, under license, in Iran. In May 1991, a report followed that Czechoslovakia hoped to sell 300 T- 72 tanks to Syria.[30] The U.S. government pressured Czechoslovakia not to sell heavy weapons to either Iran or Syria. In mid-1991 Mr. Cerny explained the status of the two situations as follows; regarding Syria.

> "Virtually every country in the Third World and in the Middle East that the previous government supplied with arms owe Czechoslovakia money. The sum altogether exceeds 1 billion. . . Syria is our biggest debtor, and Czechoslovakia was promised by the Syrian government that it would try to do something about that debt, which exceeds $ 600 million"[31]

if Czechoslovakia sold it the additional 300 tanks. Czech authorities had stated that they would honor contracts for weapons purchases made by the previous communist government before the Havel government took office "-- unless they threaten to destabilize Czechoslovakia's good standing with the western countries."[32] This statement also seemed to indicate what the ultimate outcome of the sale to Syria would be. As for the requested sale to Iran, Mr. Cerny stated that "Czechoslovakia will do nothing to affect its integration with the west: such a sale will not be realized". To date, no contract has apparently been signed with Iran. The sale of tanks to Syria was made nevertheless, the Czech government arguing that the sale involved tanks in storage, and that no further sales to Syria would follow. Czech officials claimed that T-72 tanks intercepted by German officials being shipped to Syria in February 1991 were still part of the deliveryapproved in mid-1991.[33] Czechoslovakia is for the moment permitting the export of tanks already produced. The effort to establish a new arms sales policy "to countries that were not troublemaking" would not seem to have been a formulation that could easily have been applied to either Iran or Syria, or to any other state in the Middle East. Czech authorities subsequently expressed irritation that the U.S. had allegedly also prevented a proposed sale of 500 Czech tanks to Egypt and two months later sold 500 American made tanks to Egypt.[34]

In the fall of 1991 Czech Prime Minister Marion Calfa stated that "we have no other way to get the means for conversion to civilian production than to export the very weapons themselves once again".[35] Czech arms sales had reportedly supplied 3-4 billion crowns annually to the country's is treasury for over 20 years. 4 billion crowns were at first requested to aid conversion in Slovakia, and this was subsequently reduced to 1 billion crowns for the two years 1991-1992.[36] The claim that there is "no other way" is always dubious regarding both the political and economic decisions of governments, and is likely to be equally dubious in this instance. In Oldrich Cerny's words,

> "I think this is something that is closely connected with the overall political and economic situation in Czechoslovakia. As I pointed out in my statement, most of these large factories that need to be converted are situated in Slovakia. Now, Czech and Slovak relations at this time are a very, very sensitive issue. I mean the whole fate of our federation is at stake here, and if you close down the factories in Slovakia right now, there are no jobs, there are no places where these people can move to find new jobs.
>
> Therefore, it would be much easier just to close down the factory producingtanks and building something that would produce something else, but right now it is not politically feasible. I am not saying economically, but politically it is not feasible. As I say, this is a very sensitive issue".[37]

The Czech arms sales situation seems to have reverted to a somewhat thinned down version of its pre-1990 status. Czech Trade Minister Baksay reported arms sales totalling $ 326 million in 1990 and $ 186 million in 1991.[38] Between January and July 1990 deliveries went to Afghanistan, Algeria, Egypt, Ethiopia, India, Iran, Iraq, Libya, Nigeria, Pakistan and Syria.[39] 1,500 Czech technicians were reportedly employed at an ammunition factory construction site in Libya, and in 1992 it was reported that Omnipol was still operating a tank manufacturing plant in Libya, and had signed a contract in April 1991 for 600 Czech technicians to staff the factory.[40] Czech government officials themselves apparently suggested that $ 1.5 billion in weapons sales allegedly made to Nigeria utilized false end user certificates, with the weapons going to Croatia. In December 1991 the Czech Army Chief was in India offering to sell jet trainers, air defense radar, armored vehicles and spare parts.[41]

**Alternate Production for Czechoslovakia**

The major weapons produced by Czech industry are tanks, armored personnel carriers and artillery. Tank production involves essentially heavy metal foundry work, assembly and heavy welding. Artillery and gun barrels require machine tool work, such as boring, milling, drilling and other processes. The customary alternative production for tank plants are tractors, construction and road-building machinery (such as bulldozers and excavators), railroad flatcars, river barges, and, at times, locomotives. Many of these products have been, and are at present, civilian productsalready produced by Czech (and by Russian) tank factories. Tank engines can be redesigned as compressors or used to power construction machinery, and tank-transport vehicles could easily be redesigned for other heavy truck bodies and directly used for hauling heavy earth-moving equipment.

In mid-1990 Czech officials described the conversion prospects for Czech industry very optimistically. Lt. Gen. Jaroslav Kovacik, then responsible for the management of the Czech defense industrial sector, pointed out that *all* present Czech weapon procedures also produce civilian products. His prescription therefore was to increase the proportion of existing civilian goods already in production, and he listed an extremely wide assortment of these:

> They include consumer engineering products, consumer electronics, hydraulic elements and aggregates, trucks, robots and manipulators, combustion engines, machine-tools and single-purpose machines, tractors, building machines of all types, excavators, agricultural and textile machinery, printing and copying machines, airconditioning elements, laster technology, vacuum products, semi-conductor oscillators, sources for ion vacuum pumps, stroboscopes, communication exchanges and other technology, control systems for machine-tools, technology for electronics, labor safety means, special optical elements, as well as artificial kidneys, all sorts of equipment for ecology, shielding material for nuclear power stations, etc. The list could go on, for the range of products is wide.[42]

Kovacik further stated - in 1989 - that "It has so far proved that we are capable to master the technological, commercial and technical problems of the conversion", and that, of 110 defense-industrial productionenterprises, the government expected that *100*--over 90 percent - are capable to resolve their problems connected with the conversion within their own potentials and resources without the intervention of the Government." Thus, it was expected that only 10 percent of the plants, presumably the ten largest would need government assistance. By April 1990, several joint manufacturers production arrangements had already been set up with Western European manufacturers. Production was started by the Heavy Engineering Works in Martin on a new bulldozer in co-operation with Hanomag of Hanover, Germany, in September 1989, and Martin was also developing collaborative agreements with the French firm Poclain, and the Italian firm, Lombardini. By mid-1991, excavator components were being made for Caterpillar Corporation of the U.S.

As for new production alternatives, several factors clearly converge to recommend two related choices. Both are *greatly* needed not only in Czechoslovakia but all of the other Eastern European countries as well, and should be eminently marketable:

1. Czechoslovakia badly needs *energy* sources. Its oil imports must now be paid for in hard currency, and the quantities it previously obtained from the U.S.S.R. have been sharply cut.

2. Everyone, inside and outside Eastern Europe, agrees that the region is an ecological disaster area, and that *industrial pollution control equipment* is virtually nonexistent. This means a requirement for (a) such equipment to be retrofitted on industrial production facilities and (b) the clean-up of toxic waste sites.

3. Cities everywhere, including those in Eastern Europe, continue to producesolid waste, and landfill is already impossible or counterproductive.

4. One would like, where feasible and as much as possible, to make use of existing industrial plant and capabilities.

*It is possible to put these components together* : Materials-recycling plants, municipal waste-to-energy production plants, even compactors and dump trucks are all essentially heavy metal constructions. Any domestic production of energy from municipal waste would simultaneously reduce the Czech oil import requirements and save Czechoslovakia valuable foreign exchange. It would also decrease pressures for increased nuclear power production. These industrial products could be constructed in varying sizes to fit different community requirements. Every town and city in the country needs heat and electric power at the same time as it is producing solid waste. These products could also be exported to other East European countries, as well as countries in the Mediterranean area. That is one alternative. The second would be the production of industrial pollution control equipment, which also involves essentially large fabricated metal products. These too could certainly be export products.

Given the existing skilled labor in the Czech industrial workforce, the nature of the non-defense products already being produced at these sites and the prior assessments of the parent Ministry, there is every reason to expect that present Czech weapons producers should have no particular problems switching to these two lines of production. Existing machinery might in large part also be convertible to the new production requirements. That is the best one can do; beyond that additional investment would be necessary only for equipment needed to perform production tasks that existingequipment can not do. Absolutely the wrong course would be to seek to maintain Czech heavy weapon production facilities exactly as they are at present and to seek joint ventures with Western companies to do that -- as the U.S.S.R. has been attempting to do with portions of its aerospace industry - rather than to look for new industrial product lines to meet critical, unmet societal needs and new economic opportunities.

**What the West Should Do**

Early in May 1991, the Czech government asked the United States to assist it in converting its weapons plants. The request was made by Foreign Minister Dienstbier to the U.S. Department of State while on a visit to Washington. It was not clear if the request was primarily for technical assistance or for investment capital. It may have been for both. The request resulted several months later in the visit of a team led by U.S. Deputy Secretary of Defense Donald Atwood to Czechoslovakia (as well as to Poland and Hungary). The team included representatives from several U.S. weapons and automotive producers. It visited defense plants and the primary interest of its Czech hosts was to interest U.S. companies in investment and joint ventures. As best as can be ascertained not a single one of the weapons factories visited described any organized planning or research into alternative production to the visiting U.S. team. On the contrary, several months later information appeared claiming that the Slovak factories might take on subcontracting work for U.S. weapon producers, or that the plants might even be interested in going into the production of newer, more advanced weapon systems.(43) If this proved to be the nature of the joint ventures eventually formed "conversion", would be essentially disregarded, except perhaps to cover marginally released employees.

In line with the proposals made in the previous section of this paper it would appear far more desirable that the West should aid Czechoslovakia's heavy weapon industry in making the transition to the two new product lines recommended. There are two forms of technical assistance that could be supplied by the U.S.government. In the first, several U.S. cities have extensive experience with municipal waste-to-energy plants, St.Louis for example. U.S. government agencies could fund the exchange of Czech and U.S. industrial engineering experts to discuss the design of such facilities, as well as the production technology needed to build such plants. Other industrial specialists could assess the equipment in the Czech weapon plants to see how much of it could be directly converted for this purpose. The U.S. government could supply credits earmarked for the patents for relevant equipment and processes as a form of grant aid.

In the second, U.S. Environmental Protection Agency specialists with expertise concerning the pollution control technologies required by different industrial and chemical plants could also be sent to Czechoslovakia to inform Czech industrial managers as to what kinds of pollution control equipment is needed for various processes and installations that presently make up the country's industrial infrastructure. Western industry, for example United States and German corporations that produce and license such technologies, should be interested in joint ventures with and capital investment in Czech heavy industry for this purpose. Capital investment in Eastern Europe, just as in developing countries, previously omitted pollution control technology when plants were constructed, although the technology was always available by import if not produced domestically. Virtually *every* industrial site in Eastern Europe needs such technology. There is no question that products of this nature are marketable. Proposals for indirect U.S. support for industrial conversion in the U.S.S.R. could also be designed to involve Eastern European producers in a triangular relationship: U.S. credits could be given to the U.S.S.R. for purchases of industrial equipment from Czech or other Eastern European producers. In January 1992 Foreign Minister Dienstbier even suggested that "the west should buy and apply Central Europe's know-how in Russia in building grain silos."[44]

After two or three years there is no systematic data available to determine exactly what progress has been made - if any - in defense industry conversion in either the Czech or Slovak portions of the Czech and Slovak Federal Republic. No survey is available of the 110 defense plants and what has happened in them, or even in any representative or selective sample of them. The changes required were orders of magnitude less than those necessary in the U.S.S.R., and appeared to be within the capabilities of its industrial managers and strongly supported by the new non-communist government. The incidental reports available describe primarily inertia and resistance, although the government still announces the 1992 targets for weapon production cuts and defense industry employment levels. It is not possible to determine exactly what proportion of the defense industry work force may actually have been released since 1987. More recently bureaucratic resistance by local weapon factory managers has asserted itself, compounded by continuing strong communist affiliations of the Slovak management and work force. In Slovakia, which is the location of the heavy weapon and tank plants, developments became overwhelmingly affected by Slovakian nationalism, and prospects for conversion may depend highly on whether the Slovakian nationalist opposition wins the upcoming June 1992elections. It is not at all clear however, if any greater progress has been made in the Czech portion of the country.

## References and Notes


* The official name of the country is now the Czech and Slovak Federal Republic. For the sake of convenience, this paper continues to use "Czechoslovakia".

1. "Conversion Starting Off", Orbis Press Agency, 1539/8911, 10 pages, mimeographed (Interview with Lt. General Jaroslav Kovacik; see ref. #9 below)

2. - "Fact Sheet on Military Expenditure and Iraqi Arms Imports", Ian Anthony, et.al., SIPRI, Stockholm International Peace Research Institute, August 8, 1990, 10 pages, mimeographed.

   - Richard F Grimmett, *Trends in Conventional Arms Transfers to the Third World By Major Supplier, 1980-87,*

Congressional Research Service, Library of Congress. 88- 352F, May 9, 1988.

- Richard F. Grimmett, *Trends in Conventional Arms Transfers to the Third World By Major Supplier*, 1982-1989, Congressional Research Service, Library of Congress, 90-298F, June 19, 1990.

- Richard F, Grimmett, *Conventional Arms Transfers to the Third World, 1983-1990,* Congressional Research Service, Library of Congress, 91-578 F, August 2, 1991.

- Shirley Kan *China's Military: Roles And Implications For US Policy Toward China*, Congressional Research Service, Library of Congress, 91,731F, October 3, 1991.

3. Discussion with Huai Guomo, Deputy Minister in charge of Conversion, Peoples Republic of China, August 15, 1990.

4. "In Gorbachev's Words: 'To Preserve The Vitality of Civilization'....",*New York Times*, December 8, 1988.

5. Paradoxically, the first thing to occur under the auspices of the conversion program was a substantial increase in the size of the Soviet defense industrial complex in 1989- 1990, when several hundred factories belonging to two civilian industrial ministries were placed under its jurisdiction. Substantial new capital investment was made for new civilian production capacity at defense industrial sites, but the old capacity and machinery apparently has not been scrapped or converted. Only *six* plants allegedly were totally "converted" in a period *of two years.* What is more, these particular six plants apparently had relatively little defense industrial output to start with among the products that they produced, and they may even have been under the jurisdiction of civilian production ministries.

6. "Military-economic developments in the Soviet Union", submission to the Joint Economic committee, U.S. Congress, May 16, 1991, by C.P. Duecy, Defense Intelligence Agency.

7. Joseph Fitchett, "Perestroika Architect Bemoans Defense Industry", *International Herald Tribune*, September 27, 1990.

8. David Gardner, "Euro MPS Seek Help For Loss of Arms Jobs", *Financial Times*, September 12, 1991. For sources to the basic literature on the problems of defense industry conversion, see:
-*Convertibility Of space and Defense Resources to Civilian Needs: A Search For New Employment Potentials,* United States Senate, Committee on Labor and Public Welfare, 1964.

-*Economic Policies and Practices; Paper No.8, Programs For Relocating Workers Used by Governments of Selected Countries,* United States Congress, Joint Economic Committee, 1966.

- Seymour Melman (Ed.) *The Defense Economy ; Conversion of Industries and Occupations to Civilian Need* (New York, Praeger Publishers)1970.

- Bernard Dis, *Adjustment of High Technology Organizations to Reduced Military Spending: The Western European Experience,* National Science Foundation, October 1974.

- John Lynch (Ed.) *Economic Adjustment and Conversion of Defense Industries* (Boulder, Colorado, Westview Press), 1987.

- Marcus Breitschwerdt (Ed.) *Rustungskonversion, Facetten Einer Strukturfrage,* Wissenschaftsedition, SPD Baden- Wurttemberg, 1988/89.

- Klaus Schomacker et.al.,*Alternative Production Staff Rustung,* Bund Verlag, 1987.

9. Except where otherwise identified the statistical data on the Czech defense industrial sector quoted in this section is taken from three items either authoured by or reproduced in interviews with Lt. General Jaroslav Kovacik, "representative of the Federal Ministry of Metallurgy, Engineering, and Electrical Engineering for questions of special technology".
   a. "General Details Arms Production Phase Out", FBIS-EEU- 90-O56, March 22, 1990, Page 22.

   b. "Conversion Starting Off" Orbis Press Agency, 1539/89/1, 10 pages, mimeographed.

   c. Jaroslav Kovacik, "Problems of Conversion", Address to the UN conference on Conversion, Economic Adjustments in an Era of Arms Reduction, Moscow, August 13-17, 1990, 7 pages, mimeographed.

10. Stephen Tiedtke,"Czechoslovakia", in *The Structure of the Defense Industry - An International Survey,* ed. Nicole Ball and Milton Leitenberg, (London: Croom Helm), 1983, p 188.

11. Steven W.Popper, "Conversion of Arms Industries in Central Europe", *Trade in Conventional Weapons: The International Arms Bazaar,*Hearing , Committee on Governmental Affairs, U.S. Senate, 102nd Congress, First Session, June 12,1991, pgs.21 to 29.

12. See ref.9(a), above, repeated on Radio Prague, October 9, 1990, using the figure 150,000 to 200,000 which then found its way into *Report on Eastern Europe*, (RFE/RL Research Institute) 1:43 (October 26,1990) pg.15

13. Steven Popper, op.cit.

14. Vladimir Dlouhy, Prague CTK in English, May 30, 1991; FBIS - EEU #107, June 4, 1991, pp.12-13.

15. Oldrich Cerny (Presidential Advisor, Czech and Slovak Federal Republic), "Czechoslovakia; Problems withConversion,"in *Trade in Conventional Weapons....'* op.cit.

16. Mary Battiata, "Tank Seizure by Germany Spotlights Arms Trade by Czechoslovakia", *Washington Post,* February 2, 1992.

    In October 1991 a TV news report claimed that "Unemployment .. in the region around ZTS Martin (the largest tank plant)... is up to 30 percent". This seems dubious and would be difficult to correlate with any of the other available figures, unless it perhaps refers to the single plant and the single town. Report by Daniel Elias, McNeil-Lehrer News Hour, National Public Broadcasting Television Network, October 29, 1991, Manuscript text.

17. Daniel Elias, (McNeil-Lehrer), ibid.

18. Oldrich Cerny, op.cit. These figures were again first released by Economics Minister Dlouhy a week earlier on May 30, 1991;see ref. 14 above.

19. -Lubos Dohrovsky, Prague CTK in English, January 9, 1991; FBIS-EEU,#7, January 10,1991, pp.22-23.

    -Jan Obrman, "Interview with first Deputy Chief of the General Staff"

(Lt.Gen Josef Vincenc), *Report on Eastern Europe,* (RFE/RL Research Institute)1: (July 20, 1990) p.16.

20. Craig R. Whitney, "Prague Aide Says Country Will End Weapons Exports", *New York Times (and International Herald Tribune)*, January 25, 1990. Dienstbier stated that "... Czechoslovakia will simply end its trade in arms without taking into account what the pragmatists will say, that it will be a blow to the state coffers, that those people will get armsfrom somebody else anyway if we don't supply them. (The measures have) not been announced yet, but it simply will be done, and weapons just won't be sold anymore. We can stop that actually even without knowing how big the trade was".

21. Jan Obrman, "Czechoslovakia and the Arms Trade", *Report on Eastern Europe* (RFE/RL Research Institute), 1: (April 20, 1990) pp. 21-24 (quoting *Time,*March 26, 1990; AP, March 20,1990)

22. Ibid.

23. Prague CTK, in English, April 29, 1991, in FBIS -EEU; # 83, April 30, 1991, p. 19.

24. John Tagliabue, "Czechoslovaks Find Profit and Pain in Arms Sales", *New York Times* February 19, 1992.

25. "Arms and the Man in Slovakia", *Financial Times,* January 22, 1991; See also FBIS -EEU #12, January 17, pp. 19-20, FBIS -EEU, January 30,1991 p.19 and Stephen Greenhouse, `Slovaks Are Hurt By Arms-Sales Ban,' *New York Times,*August 29,1990.

26. Ana Goshko, "The Czech and Slovak Federal Republic in Transition", Congressional Research Service, 92-52F, December 27, 1991, p.6

27. *Vienna Fax*; News and Analysis From The Vienna Talks on Conventional Forces in Europe And Confidence - and Security- Building Measures", Institute for Defense and Disarmament Studies, Cambridge, Massachusetts : Issues of May 8, 1991, and May 28, 1991, (Vol. 2, #5).

28. *Vienna Fax,* Institute for Defense and Disarmament Studies, 2:10 & 11 (December 4, 1991) p. 10.

29. The figure of 300 appears in Obrman, op.cit., and the 1,500 in Goshko, op.cit. The 1,500 appears to be the correct number.

30. Leslie Colitt, "Prague and Iran Plan Oil-for-Arms Deal", *Financial Times*, March 12, 1991; "Hard Pressed Czechs Retain Arms Trade", *New York times*, May 3, 1991; Hugh Carnegy, "Stop Czech Arms Sales to Syria--Shamir", *Financial Times*, May 7, 1991. See also James L. Graff, "Czechoslovakia; A Tankless Task" *Time*, June 24, 1991, pp : 40-41.

31. Oldrich Cerny op.cit. Obrman (ref.21, above) writes that in 1989 Czech Prime Minister Adamec journeyed to Egypt and to Iraq in an effort to recover monies owed for past arms shipments, and that at the time Czechoslovakia was owed over 7 billion dollars, including payments from Warsaw Pact allies. If Cerny's figure of 1 billion dollars owed to Czechoslovakia by developing nations in mid-1991 is correct one must presume that most of the larger sum was paid to Czechoslovakia in the intervening two years.

32. Oldrich Cerny, op.cit.

33. Mary Battiata, op.cit. John Tagliabue, op.cit.

34. Daniel Elias, (McNeil-Lehrer) op.cit. It is possible that the U.S. *delivered* the tanks to Egypt at that time, but had made the sale earlier; this remains unclarified.

35. Daniel Elias, op.cit.

36. Steven Popper , op.cit.

37. Oldrich Cerny, op.cit. See also Henry Kamm "As Slovak Separatism Gains, Havel Faces the Unthinkable", *NewYork Times,* November 20, 1991; "Havel Advances Slovak Autonomy Claims", *New York Times,* November 11, 1991.

38. John Tagliabue , op.cit.

39. Jan Obrman, "Arms Trading Continues", *Report on Eastern Europe* (RFE/RL Research Institute), 1:43 (October 26, 1990) pp. 15-17. See also Jan Obrman, "From Idealism to Realism", *Report on Eastern Europe* (RFE/RL Research Institute), 2:51-52 (December 20,1991), pp: 9-13.

40. Mary Battiata, op.cit.

41. Steve Coll, "South Asian Nations Scramble For Arms After Soviet Crackup; End of Cold War Could Affect Regional Military Balance", *Washington Post*, January 5, 1992. It is impossible to tell from the press report whether this might be military surplus or newly produced equipment

42. Kovacik, 1989, op.cit.

43. Daniel Elias, op.cit.

44. Stephen S. Rosenfeld, "Central Europe's Escape", *Washington Post,* January 24, 1992.

Table 1

SELECTED OFFICIAL FIGURES ON CZECHOSLOVAK ARMS PRODUCTION,

1987 - 1990*

| | 1987 | 1988 | 1989 | 1990 |
|---|---|---|---|---|
| PRODUCTION | | | | |
| Total Arms Output, mn KCs | 29298 | 26737 | 18996 | 15107 |
| Index: 1987-100 | 100,0 | 91,3 | 64,8 | 51,6 |
| Share of Heavy Ind. Mfg. | 11,3 % | 10,0 % | 6,9 % | 5,7 % |
| REGIONAL BREAKDOWN | | | | |
| Czech Arms Output, mn KCs | 11557 | 12331 | 10587 | 7515 |
| Percentage of total | 39,4 % | 46,1 % | 55,7 % | 49,7 % |
| Slovak Arms Output, mn KCs | 17741 | 14406 | 8409 | 7592 |
| Percentage of total | 60,6 % | 53,9 % | 44,3 % | 50,3 % |
| SALES | | | | |
| Domestic | 22,4 % | 28,8 % | 35,8 % | 47,7 % |
| Export | 77,6 % | 71,2 % | 64,2 % | 52,3 % |
| of which: | | | | |
| Socialist | 75,0 % | 79,5 % | 91,6 % | 79,7 % |
| Other | 25,0 % | 20,5 % | 8,4 % | 20,3 % |

Source: Tydenik Hospodarskych Novin, 16 May 1991
(as reported by CTK-Ecoservice, 17 May 1991)

*These figures should be taken as representative rather than definitive, especially when comparing foreign and domestic "sales", because of the chronic valuation problems that were a sympton of the former Soviet-type economic system.

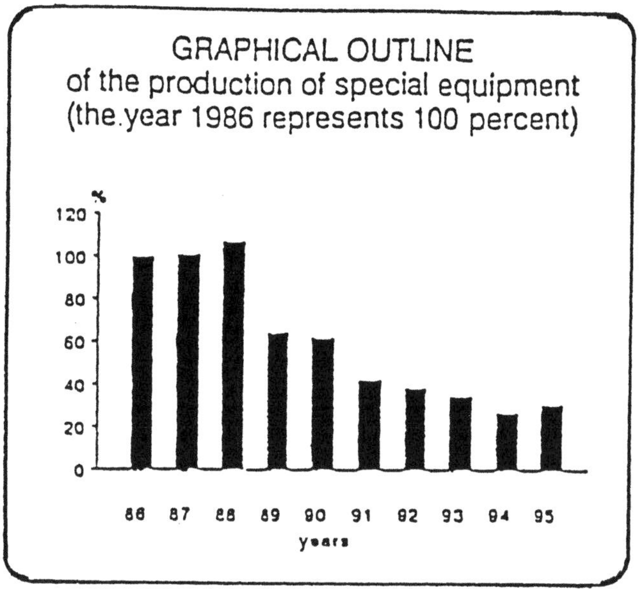


Source: Jaroslav Kovacik, August 1990.

Major Arms Exporters, 1971 - 1988

(1988 US $m.)

| Country/Region | Total | Percent World |
|---|---|---|
| World | 780,908 | 100.0 |
| Soviet Union | 310,711 | 39.8 |
| United States | 201,997 | 25.9 |
| France | 56,646 | 7.3 |
| Great Britain | 31,263 | 4.0 |
| West Germany | 20,490 | 2.6 |
| China | 19,116 | 2.4 |
| Czechoslovakia | 18,527 | 2.4 |
| Poland | 16.821 | 2.2 |
| Italy | 14,628 | 1.9 |
| Switzerland | 7,456 | 1.0 |
| Yugoslavia | 5,951 | 0.8 |
| Israel | 4,892 | 0.6 |
| Bulgaria | 4,663 | 0.6 |
| North Korea | 4,639 | 0.6 |
| Brazil | 4,539 | 0.6 |
| Spain | 4,464 | 0.6 |
| Canada | 4,343 | 0.6 |
| Romania | 4,273 | 0.5 |
| East Germany | 3,954 | 0.5 |
| Netherlands | 3,902 | 0.5 |
| 5 Major Exporters | 621,107 | 79.5a |
| 20 Major Exporters | 743,275 | 95,2a |

a = Figures may not add up to totals due to rounding.

Source: United States Arms Control and Disarmament Agency, World Military Expenditures and Arms Transfers, Washington, D.C.: U.S. Govt. Printing Office, 1983, 1987, 1989 and 1990. Table prepared by Nicole Ball.

# SOVIET RESOURCES IN THE DEFENSE SECTOR AND THEIR AVAILABILITY FOR ECONOMIC RECOVERY

**Milton Leitenberg**
*Center for International Security Studies at Maryland, U.S.A.*

The economy of the former USSR is in acute crisis. It was stagnant or in actual decline since the late 1970s[1]. It is now recognized that the sums expended by the USSR in military expenditure for four and half decades since the end of World War II represented approximately 25 percent - or more - of its national product. The reallocation of resources from the military to the civilian sector was always necessary. But there was no reallocation. There was none before March 1985 and Secretary Gorbachev's accession to power. There was none between 1985 and the first announcement of projected reductions in military expenditure in 1989, and there was hardly any between 1989 and 1991. The issues were the same before the August 1991 attempted coup, before the dissolution of the USSR at the turn of the year 1991-92, and they are the same now. Even in the period since 1989 the responses of senior soviet policymakers in the defense sector, as well as individual defense industrial plant managers, have been haphazard, ineffectual, and frequently overtly resistant to any change in the status quo of the defense industrial base.

In the Spring and Summer of 1991 there was an extensive public debate on the pros and cons regarding the "Grand Bargain" -- the proposal that Western nations supply $ 25-30 billion per year in economic assistance to the USSR forfive to six years in exchange for a process of change in the Soviet economic and political system. During this debate scarcely a word was mentioned regarding two of the most significant resources that the USSR has at hand and which it could use to rebuild portions of its domestic economy. These are its military expenditure and its unconverted defense industry.

Several of the most senior figures responsible for directing the Soviet defense-industrial complex participated in the August 1991 coup attempt in Moscow. Following the failure of the coup and the removal of these individuals from their positions, and with the accelerated devolution of economic decision making to the Republics, which also followed the coup, it could only be anticipated that weapons procurement in the USSR would drop more precipitously than it would have before August.

In late October 1991, Andrei Kokoshin, Deputy Director of the Academy of Sciences USA-Canada Institute in Moscow, stated that "the Soviet military industrial complex....will receive virtually no orders from the military next year"[2], and in January, the Russian parliament approved a budget that cut weapons procurement to one-seventh of the level of the previous year[3]. But for the bureaucratic and industrial interests that worked against such an event, any outsideobserver should have expected this change long ago. Given the reductions that they will be making under the START and CFE agreements, Russian military services will scarcely be in need of any new major equipment for some years to come, and a total procurement stop would not handicap them in any way. In addition, the Soviet Navy has been undergoing cuts in capital ships, independent of any treaty mandated reductions, and following the failed coup, the Soviet Army has also announced major additional reductions in force structure. Although there has been the same expectation before, it certainly seemed likely that the Soviet defense industry would now face conversion at a forced-march pace, in a way that it has not in the past three years. It would be useful to briefly look at the magnitude of these resources and the question of their rapid availability for other economic uses in the USSR.

On April 27, 1990, at a speech at the Soviet tank plant, in Nizhny Tagil, President Gorbachev revealed that Soviet military expenditure had grown at 9 percent per year during the 11th and 12th Five Year Plans, a solid decade during which overall Soviet economic growth had averaged around one percent per year. In the same speech he also stated that Soviet military expenditure was consuming 18 percent of the USSR's GNP - the same figure used by the U.S. Central Intelligence Agency. (Only a few weeks earlier the first Soviet submissions to the United Nations of its military expenditure reckoned these as equivalent to 9 percent of USSR GNP). At the 28th Party Conference in early July 1990, Gorbachev referred to the "inadmissible amounts being spent" for the Soviet defense program, fully half of which is devoted to weapons procurement. A day or two later, one of Gorbachev's closestassistants, Gyorgi Shaknazarov, in an interview on American television, stated that Soviet defense expenditure amounted to"...up to twenty percent of national income."[4] The very next day, Foreign Minister Shevardnadze said that "official estimates" of "past" Soviet defense expenditure indicated that these were twenty-five percent of GNP[5] Oleg Bogomolov, (Director of the Institute of Economics of the World Socialist System in Moscow), had already alluded to 27 percent cut in April 1990.[6] Figures of between 25 and 30 percent are now widely accepted,[7] virtually double the U.S. CIA estimates, and on a substantially *smaller* overall economic base than the CIA had estimated.

According to U.S. government estimates, Soviet defense expenditure fell by roughly six percent in 1989, and by the same amount in 1990. Procurement expenditure dropped by about ten percent in each of the two years and probably in 1991 as well. Nevertheless, for 1991, the Soviet defense budget appeared to remain at almost exactly the 1990 level in real terms. From its nominal 78 billion rouble level in 1990, it was raised to 98 billion roubles in 1991. The 25 percent increase was explained as compensation for inflation. However no other category of the 1990 U.S.S.R. state budget was compensated for inflation in establishing the 1991 expenditure disbursements.

Of greater significance was the statement by the chief scientific associate of the Scientific Research Economics Institute of GOSPLAN, that "the real Soviet defense budget" equivalent to the 1990 78 billion rouble figure was actually 140 to 150 billion roubles:

> "Rayzberg said that basic and applied research for the military by academicand scientific institutes, training of military industrial specialists, military preinduction training, and capital investment in the defense industries all come from the state budget, not the

military budget, where they belong. He also confirmed that prices paid by the military for equipment are artificially low due to hidden budget subsidies.."

Assuming a similar proportionality, the 1991 - 98 billion rouble sum would imply a "true" Soviet defense expenditure of 180 billion roubles. In other words, the U.S.S.R's announced defense expenditure was still missing various categories, and was still not a complete figure. A series of other statements in 1991 by members of the Supreme Soviet corroborated the above exclusions from the "defense" budget figures, and added several others as well.[9] The USSR's 1992 defense budget was announced in December 1991 as 529.4 billion roubles, with that of the Russian Republic as 323.5 billion roubles, the extraordinary four-fold increase presumably explained by its being calculated in 1992 prices[10].

If one steps back for a moment to look at the years before 1988 to 1991, it is clear that what must now take place has never before occurred in the U.S.S.R--at least not during the last forty years--and as indicated, it didn't occur during President Gorbachev's tenure either. All through the SALT and START negotiations, that is since 1968, and particularly between 1985 and 1988, the most senior Soviet spokesmen repeatedly claimed their desire to reallocate savings that could be obtained from strategic arms control agreements to the domestic needs of the U.S.S.R. The economic rationale for Soviet interest in these negotiations was uniformly repeated by western arms control specialists. However it is absolutely clear that between 1968 and 1985 *no* such savings or reallocations were made in the U.S.S.R.[11] Official U.S. estimates indicated that Soviet military expenditure also continued to increase in the 1985-1988 period of Secretary Gorbachev's tenure, which he corroborated in his April 1990 speech referred to above.

Two other pieces of evidence in the months preceding President Gorbachev's December address to the United Nations in 1988, are significant indicators of the approach of Soviet policymakers at the time. Between 1985 and 1988--during the period in which Soviet military expenditure continued to increase--the U.S.S.R. borrowed $ 9 billion from western banks. In September - October 1988 it borrowed an additional $ 8 billion, or $ 17 billion in all.[12] On the other hand, the highest estimates of savings to the U.S.S.R. from the INF Treaty signed in 1989 were approximately $ 10 billion, and that over a period of ten years. This strongly suggested that funds which the U.S.S.R. wanted for the purchase of consumer durable and consumer capital production goods were being obtained by borrowing from Western states, and not by the reallocation of its own military expenditure. More recently, it was learned that the U.S.S.R. sold off large portions of its gold reserves, in addition to incurring additional borrowing after 1988, for the same purpose.

The second rather surprising piece of evidence was the degree of interest that Soviet space authorities expressed in an extensive Mars space program during the first half of 1988. The cost of such a program would be enormous, and Soviet authorities pressed the U.S. to engage in a joint program. The estimated cost to the Unites States for the initial exploratory stages of aMars program is $ 2.5 billion over six years, and one can assume that the initial cost would be roughly the same for the U.S.S.R. Even if the incremental costs for early portions of a Mars mission are considered relatively small on top of the rest of the space program, they would undoubtedly serve as a justification for continued pressure for manned space missions. Estimates for the cost of a fully manned Mars program in 1988 were $ 200 billion (in current dollars), and in 1991 they had grown to $ 400 - 500 billion. The fact that this proposal was pressed by the U.S.S.R. in 1987 and during much of 1988 would strongly indicate that its leadership was not thinking first and foremost of applying savings gained from arms control to domestic civilian uses.[13]

In President Gorbachev's address to the United Nations in December 1988 he promised that the U.S.S.R. would develop "a plan" for defense industry conversion, to be completed within a year.[14] Two years of fitful struggle took place before the plan could be formulated, and like all previous Soviet economic plans, it was essentially a series of exhortations, guidelines, targets and goals. The plan was never publicly released, but the central government was to maintain control over the availability of all resources and inputs, as well as the process of placing state orders with individual plants. In the post-World War II years, the Soviet Union developed a system whereby defense-industrial facilities also produced relatively large amounts of civilian goods. The purpose of this system was primarily to permit rapid expansion of defense production in case of war. The 1988-1990 ratio of defense and civilian goods production within the defense industrial sector was 55 percent civilian products.[15] Under the "conversion" plan itwas to be inverted to 65 percent civilian and 35 percent defense in five years.[16]

Paradoxically, the first thing to occur under the auspices of the conversion program was a substantial nominal increase in the size of the Soviet defense-industrial complex in 1989-1990 when several hundred factories belonging to two civilian industrial ministries were placed under its jurisdiction. Only *six* plants were allegedly totally "converted" *in a period of two years.* What is more, these few plants apparently had relatively little defense-industrial output to start with among the products that they produced, and they may have been under the jurisdiction of civilian production ministries. According to testimony presented by US government agencies to Congress in May 1991:

> Despite much fanfare, defense industry conversion has made scant progress, as the Soviets candidly admit. Of the several thousand plants producing for the military, only about 500 will undergo some degree of conversion. About 40 plants are to cease military production altogether. However, most of these belong to the civilian sector and devote only a small share of their output to defense.
>
> While reductions in weapon production and defense spending are occurring, the defense industry, for the most part, is retaining a mobilization base. Military production lines are operating at reduced output levels or, in some cases, lines are being shut down but maintained in operating condition. Soviet economists and industrialists indicate that conversion, under current plans, will be achieved by expanding existing civilian production and addingnew capacity at military plants rather than by switching weapon lines to civilian output. Although Soviet statistics show some increases in consumer production by the defense sector, growth has been far short of plans. Slow progress is not surprising. Conversion remains in the hands of government officials strongly tied to the defense industrial complex and interested in retaining its established priorities and skilled manpower.[17]

This cannot be called defense-industry conversion. President Gorbachev's advisor, Alexander Yakovlev, pointed out in September 1990 that the U.S.S.R.'s military industries had remained "almost untouched", and he ridiculed the claims of successes in conversion.[18] Some "conversion"--very loosely defined--did taken place between 1989 to 1991 essentially on an *ad hoc* basis. What occurred for the most part, however, was not defense industry conversion. Rather *additional* production capacity for increased production of civilian goods was created--often involving *new* capital investment--at facilities that remain under the jurisdiction of the defense industrial ministries. The defense production lines remained intact and no conversion took place.

The original conversion plan prepared by the government proposed to *partially* convert some 422 defense ministry plants:

- 124 enterprises would reduce defense production by "up to 10 percent"

- 118 enterprises would reduce defense production by "up to 20 percent"

- 56 enterprises would reduce defense production by "up to 30 percent"

- 124 enterprises would reduce defense production by "over 30 percent."[19]

"Over 100" non-defense ministry plants were also to be converted, 34 of them totally. In September 1990 President Gorbachev used the figure of 550 as the number of plants that would be involved altogether.[20] The total number of Soviet defense plants is not known, but it is usually assumed to be around 5,000, including major plants as well as component producers, and the 550 would therefore only be around 10 percent of the total.[21] These figures are widely misunderstood and misreported in important western sources.[22]

The most significant aspect of the plan was that *after five years*, it was "... expected to make a relatively modest contribution to the growth of the defense sector civilian output". *By 1995, "conversion" was to have accounted for only ten percent of the total production of civilian goods by the defense sector, and a great majority of that due to newly installed production capacity rather than to true conversion.*[23]

Against this virtually trivial increment one should set three other figures concerning the Soviet economy. Speaking to the Supreme Soviet in the spring of 1990, Major General German Kirilenko quoted official Soviet government estimates of waste in three separate sectors as follows:

> According to information presented in the Supreme Soviet on 3 June 1989 by P. Mostov, chairman of the State Committee for Material and Technical Supply, half of the industrial product is lost in the country annually - this loss amounts to R 450 billion.
>
> According to figures adduced in their official speeches by M.S. Gorbachev and N.I. Rzyhkov, there is an annual loss of agricultural products to the extent of 20-40 percent, which constitutes a loss within the R 47-95 billion range.
>
> Losses of work time (U.S.S.R. Academy of Sciences figures) are put at 30 percent or R 83 billion.
>
> A summary estimate, however, made on the basis of the testimony of F.Fedorov, leader of a working group of the U.S.S.R. Supreme Soviet studying the problem of unproductive losses,

shows from far-from-complete data that total unproductive expenditure and losses in the national economy are put within the R 580-629 billion range.[24]

The estimate of 600 billion roubles *per year* lost due to waste is of course incredibly high. Nevertheless an economist on the staff of the Central Committee told this author in August 1990 that although he considered the above figures "too high", he thought "there might be 25 percent loss in the civilian industrial sector and 25 percent loss in the defense sector".[25] Since this reduced only the estimate for the first of the three factors and left intact the estimates for the loss of agricultural production and for loss of work time, the *annual* loss due to waste was reduced by somewhat less than half, to around 325 billion roubles. Even this lower estimate of waste in the Soviet economic system *for one year* would be about ten times the amount that the pre-coup Soviet defense industry conversion plan hoped to generate over five years.

There appear to be three major conclusions that result from the considerationof the above information. *First*, major reductions in the U.S.S.R's former defense production capacity are intrinsically desirable from the standpoint of arms control. The U.S.S.R. formerly produced weapons on a scale equal to or greater than all the NATO nations combined (including the United States), plus Japan and Australia.[26] Clearly it is in everyone's interest to see that production capacity *permanently* removed, not just set aside and mothballed. *Secondly*, short-term economic benefit is available at much higher levels and much sooner through the reduction of wasted resources in the U.S.S.R. A September 1991 press report noted that

"In most industrialized nations, harvest implies worthy silos and roads, but not here where a third or more of the crop rots or is lost before it even goes bumpily along to market...The idea of silos, of vast special rooms dedicated to storing food, seems a hallucinatory luxury..."[27]

The U.S.S.R. is after all the world's largest producer of both cement and steel, and silos and other storage facilities for agricultural products can be constructed of cement and steel. The caption for a photograph accompanying the article reads "Soviet farmers sitting idle last week at a co-operative in Leninship Lach while their crops rotted in the fields". Two months later Soviet officials reported a "shortfall" of between 20 and 30 percent in the 1991 grain harvest, although there is no information available to apportion the losses due to environment impacts in the growth cycle from those due to losses during harvest.[28]

The single specific investment recommendation made by the authors of "the Grand Bargain" was for "infrastructure". "Infrastructure" means roads, railroads, bridges, water systems, agricultural storage facilities and silos, etc. They are all made using cement, steel, crushed rock, wood, and local labor, all in abundant supply in the U.S.S.R. Neither foreign investment nor foreign technology are necessary. Local currency would be used for salaries for any national public works program. This is not a defense industry conversion issue; it is a national expenditure reallocation issue. The only thing necessary is the reallocation of government expenditure from the defense sector to these priorities.

Finally, in order to obtain major economic benefit from defense industry conversion, the U.S.S.R. would have to go far beyond the conversion plan that was drawn up by the government prior to the coup attempt in August 1991.

> The Soviet Union,...with something like 20 to 30 percent of their GNP in defense, if they are to get it down to the levels of western countries, they have to make cuts of about 5 percent or more per year in GNP; not 5 percent of the defense budget but 5 percent of the GNP has to be shifted in a given year out of defense into the civilian activities.[29]

What has the response of Russian policymakers been during 1991, and more particularly since August, following the coup attempt, and since January 1992 and the dissolution of the USSR?

As regards individual defense production facilities, orders have unquestionably gonedown. Leningrad shipyards will reportedly no longer produce military vessels. This should pose no problem, since the USSR has contracted out of the country for virtually all its civilian shipbuilding for decades:

- Poland and the DDR had built fishing fleets, factory vessels and freighters, as well as military cargo ships.

- Finland built heavy icebreakers, including those for Soviet military fleets.

- Sweden built one of the Soviet military fleet's largest capacity drydocks.

- Czechoslovakia built the Moscow River tour ships.

- Austria built refrigerated ships (on the Danube) for the USSR's Mediterranean fruit trade.

With one of the world's largest merchant fleets, former military shipyards in the USSR should obviously not lack for things to turn to. Saratov Aviation, whose production formerly consisted of 50 percent military aircraft and 50 percent civilian aircraft, has had its military production drop from 50 percent to 5 percent.[30] For the Red October Machine Buildup Corporation producing jet engines, whose defense output was 80 percent in 1989, it is expected to be 40 percent in 1992.[31]

Many other examples can be cited regarding individual defense industry firms and sharp drops in other military orders. But what of conversion and civilian production? In September 1991 there were very highly informed observers whose assessment was not very different from Alexander Yakovlev's in September 1990: little had yet takenplace.[32] A.I. Diduikov, Chief of the Department of Conversion of the State Committee of the USSR for Science and Technology, called for the selective production of the highest technology export products, (implicitly weapons and space systems).[33] President Garbochev, after having just previously referred in an interview in *Moscow News* to the Soviet defense industrial complex by saying "if this is not half of society, then it's at least a third", reportedly suggested to US Deputy Secretary of Defense, Atwood, on November 5, 1991 that the U.S. collaborate with the U.S.S.R. "on the development and production of advanced weapons..."[34] Not to be outdone, President Yeltsin informed a gathering of the heads of Russian state run industries that Russia could still make profits by selling arms overseas, and that he had just recently taken up with U.S. President Bush "the question of whether the arms market should be divided between the two countries".[35] On January 22, 1992, Gennady Burbulis, Mr.Yeltsin's first Deputy Prime Minister, presented an interpretation of "the transformation to civil manufacture" of "the enterprises of the Russian military-industrial complex" that can only be described as Neanderthal in its sentiments. The country, its political leadership and its factory managers could do *nothing* for themselves; *only* foreign investment and financial aid could alter an otherwise certain descent, and if the west did not finance R & D in the former Soviet defense research establishments,"...our highly trained nuclear specialists will be forced to look for a way to apply their abilities in various regions of the world".(36) The *old Soviet* habit of expression, in which all things were and are "forced" on a reluctant nation, with no alternative choices, was striking. Mr.Burbulis had in fact been preempted by several weeks by the Urkranian Minister of Defense VictorAntonov: the Ukraine would sell arms unless they were given Western economic assistance. They already had customers...[37] Israel, it turns out, had expressed interest, and Mr.Antonov told journalists some weeks later that the Ukraine also hoped to sell weapons to India and to France.[38] On February 4, 1992, Michael Malei, an adviser to President Yeltsin on military conversion (Russion State Councilor on Conversion) announced that "the small Russian autonomous republic of Udmurtia, where defense enterprises make up 82 percent of all production, would be a test bed for a new 'concept' which would downgrade conversion in favor of finding new markets for arms...Mr. Yeltsin had approved a scheme under which the republic would retain 70 percent of all hard currency receipts for weapons"[39] Yegar Gaydar, also Deputy Prime Minister and Yeltsin's chief economic policy maker ("deputy prime minister for economic reform") was made head of the government's new arms sales organization. Yeltsin announced that arms would be sold "...on a temporary basis while there are difficulties with the defense budget', and that a military housing program would be financed by several billion dollars from arms sales."[40] Presumably, given the sharp cuts in weapons procurement, virtually the entire Soviet defense budget in 1992 was supposed to have been available for civilian purposes and there should be no need for arms sales to fund it.

In September 1991, Vitalii Shlykov, Vice Chairman of the RFSR State Committee on defense, has written "...up to now the captains of our defense industry were convinced that by putting pressure on the government they can evade the inevitable".[42] The above rendition is more than a harbinger of the continuation of a long series of actsbetween December 1988 and August 1991 by Soviet authorities to avoid any serious defense industry conversion. It is an atmosphere in which one again hears calls for joint US Russian Mars missions,[43] and in which the Deputy Director of the Yuzhny Machine Building Factory (Yuzhmash) in Dnepropetrovsk in the Ukraine, which built the USSR's SS-18 ICBM and Zenith satellite launchers, dreams in an apotheosis of unrealism of ICBM missiles that could deliver "a peaceful payload of food or medical supplies and fired off to disaster zones20 on missions of mercy".[44] This ICBM plant appears to be one to which Israel has addressed queries concerning the purchase of technology, (which presumably would violate the Missile Technology Control Regime, but at the same time the plant can't seem to produce sausage-making machines successfully).

In striking contrast to all of the above, at least one detailed proposal has been prepared for the Russian Republic ("On Avenues for Converting Enterprises and Organizations in the Defense Complex") by the Engineering Academy of the RSFSR, which itemizes production equipment in the categories of (1) Agricultural Resources, (2) Energy Conservation, (3) Transport, (4)

Resource Conservation, (5) Ecology, (6) Medicine, (7) Information Processing, and (8) Construction, that could be produced in converted defense manufacturing facilities.[45] Detailed product sets are itemized under each of the eight categories. *All* would be products of the machine building sector. Many are high technology products. Many would introduce generous energy savings into manufacturing processes.

Soviet defense industry will only be converted to a significant degree if a policy ofproduction for domestic needs takes shape, as exemplified by the above proposal, as opposed to efforts to find high technology products for international export, space launch services, and so on. Virtually the entire industrial infrastructure (of the Commonwealth of Independent States) needs renewal and there are no lack of choices. Instead, the priorities and energies of Russian and Ukrainian policymakers have gone into fighting over the Black Sea Fleet and the allegiance of the military officers corp and not into implementing a defense industry conversion program. It has been simpler not to think about the hows and wherewithal of change, but to fall back once again on the continued production and export of weapons, or as most innocently expressed, in calls for foreign investment, joint ventures, high technology and export-oriented production. *These paths would neither convert much defense capacity nor produce the industrial products that the successor states of the USSR need.*

On April 25, 1991, Robert McNamara, former head of the World Bank, suggested that developing countries that spend less than a certain amount of their national resources in the military sector be given preference in aid allocations.[46] If that is reasonable for developing countries, it would certainly seem to be reasonable for the successor states of the Soviet Union, which, using the criteria of need, would not seem to be more worthy recipients of aid than the developing nations in general. Resources devoted to the military sector - annual defense expenditure and the enormous defense-industrial complex - should be reduced at least by half from their current levels before it should be considered reasonable to transfer billions of dollars in Western aid to Russia or to the Ukraine. Although Russian financial resources in itsown currency cannot now be used to purchase commodities on the international market, they can be used within the country to fund any chosen reordering of expenditure and production priorities: to harvest crops, build and rebuild transportation networks, build needed manufacturing capability, construct health facilities, and so on. Only with a massive shift of resources in this way will defense industry conversion be of significant economic benefit (as well), in addition to the benefits it achieves in conventional and strategic arms control.

## References and Notes

1. Henry S. Rowen and Charles Wolf, Jr. (Editors) *The Impoverished Superpower; Perestroika and the Soviet Military Burden*, (San Francisco, Institute For Contemporary Studies Press) 1990.

2. Joh Lloyd, "Sharp Cut in Soviet Defense Orders", *Financial times* , October 28, 1991.

3. Fredd Hiatt, "Russia Slashes Military Spending", *Washington post*, January 25, 1992; George Lardner Jr., "Republic's Procurement of Arms Said to Plunge", *Washington Post*, January 23, 1992; Elaine Sciolino, "CIA Chief Says Threat By Ex-Soviets is Small", *New York times*, January 23, 1992.

4. McNeil Lehrer News Hour, PBS-TV.

5. Scheverdnadze also claimed that some 700 billion roubles "had been wasted: in the USSR's military buildup "in the 1970's".

6. About 300 billion roubles per year, or three times the official figure, was alsoquoted by Pyotr Korokevich, a defense research official, in *Literaturnaya Gazeta*, quoted in Fred Hiatt, "Kremlin Official...Tactical A-Arms Cuts", October 3, 1991.

7. They are now uniformly used by Academy institute figures, such as Dr.Kokoshin.

8. The full text of Rayzberg's article appeared, in Russian, in *Argumenty i Fakty*, #13, March 1991, pg. 6. The quoted paragraph is taken from a short summary in *Armed Forces Journal*, (June 1991) page 78.

9. - Capt. V. Urban, "How Much Will Defense Cost?", *Krasnaya Zvezda*, January 12, 1991.

   - V. Romanyuk," In USSR Supreme Soviet Committees and Commissions; The Plan and Budget, Are the Predictions Realistic", zvesta, December 1, 1990.

   - Interview with Vladimir Lopatin, Moscow Radio, January 10, 1991.

   - Sergey Rogov, "Soviet Defense Budget Trends", *Kommunist* These disclosures supplying various additional portions of the USSR's military expenditure build on the official Soviet estimate of around 100 billion roubles, and notably are still only *half* as large as the Bogomolov and Korokevich estimates quoted above of 300 billion roubles, which *triple* the officially disclosed sum.

10. "RSFER Defense Budget", in *Report on the USSR* (RFE/RL Research Institute) 3:51-52 (December 20 1991).
11. Milton Leitenberg, "United States-Soviet Strategic Arms Control: The Decade of Detente 1970-1980, And a Look Ahead", *Arms Control*, 8:3 (December 1987) pp. 213-264.

12. "US Split on Allies Lending to Soviet", *New York Times*, October 21, 1988.

    In the same period, *net* borrowing by the USSR had only been $ 800 million; C.R. Neu, *Soviet International Finance in the Gorbachev Era*, Rand, R-4116-USDP, 1991.

13. Estimated reductions in Soviet expenditures for strategic missiles between 1975 and 1986 were more than compensated for by expenditures for space hardware. Adding the two sums together displayed a constant increase between 1975 and 1986. A Mars mission would have continued that trend. See "Soviet Budget Trends. Space and Strategic Missile Procurement". John Pike. *Public Interest Report*, (Federation of American Scientists), 40:9 (November 1987),7

14. "In Gorbachev's Words: `To Preserve The Vitality of Civilization' ...", *New York Times*, December 8, 1988.

15. Alexei Izyumov. "The National Experience of the USSR", United Nations Conference on Defense Industry

Conversion, Moscow, August 13-17, 1990.

16. U.I. Smyslov, "On the 03 Basic Directions of the Draft State Programme For the Conversion of Defense Industry in the USSR", United Nations Conference on Defense Industry Conversion, Moscow, August 13-17, 1990.

17. C. Patrick Duecy, US Defense Intelligence Agency, to the Joint Economic Committee, US Congress, May 16, 1991.

18. Alexander Yakovlev, quoted in Joseph Fitchett, "Perestroika Architect Bemoans Defense Industry", *International Herald Tribune*, September 27, 1990.

19. Yu. F. Glybin, "On Reduction of Military Expenditures and Some Approaches to Conversion in the U.S.S.R.", and V. Smyslov, op.cit., both to the United Nations Conference on Conversion, Moscow, August 13-17, 1990.

20. "Gorbachev to Transform Arms Plants", *Washington Post,* September 29, 1990.

21. Estimates of these numbers vary enormously: Russian-Vice President Alexander Rutskoit stated in September 1991 that "70 percent of Soviet Industry work for the Soviet Army", and the total number of Soviet Industrial plants is reported to be around 26,000. Different civic and industrial authorities for the city of Leningrad (St. Petersburg) give widely different estimates for the amount of defense industry in the city (going as high as 50 or 75%), For the Ukraine there are figures of "700-odd munitions plants" (January 25, 1992), 1,300 defense plants (Goshko, November 1991), while Ukrainian sources reported in October 1991 that there were around 2,000 defense industrial producers in the Ukraine, the figure also used by the *Washington Post* (September 10, 1991). Figures for the number of employees in the Soviet defense industry vary by equally large margins: Julian Cooper uses 6 million (ref. 23 below); Vitalii Shlykov, 9 million (September 1991) and Vladimir Kontorovich, 11 million (September 1991).

22. In a paper published in *Survival* by Daniel Nelson, "The Costs of Demilitarization in the U.S.S.R. and Europe", *Survival,* 33:4 (July-August 1991) 312-326 the author wrote: "The first steps of a five year conversion plan in the USSR were announced in June 1989. A year later, a more detailed programme was outlined. With economic conditions in the USSR worsening, however, the transformation of an additional 550 armament factories to civilian production was announced by President Mikhail Gorbachev in September 1990." They were not "an additional" - Gorbachev had just combined the "422" and the "over 100" - and, more importantly, the great majority of the total were to be only *partially* converted, as the percentages given in the next indicate.

23. Julian Cooper, "Military Cuts and Conversion in the Defense Sector, *"Soviet Economy"*, 7:2 (1991) 121-142.

24. Major General German Kirilenko; FBIS-SOV-90-071, April 12, 1990, pg. 69. Exactly what the industrial "loss" consists of is unexplained, but there is a rather large body of anecdotal evidence that has appeared in recent years suggesting what it might be; see, for

example, Anders Aslund, *Gorbachev's Struggle for Economic Reform*, Ithaca, Cornell University Press, 1989.

25. Alexei Kireyev, personal interview, Moscow, August 16, 1990.

26. Production Ratios of Selected Weapons, USSR to US and Warsaw Pact to NATO, 1974-1983, Richard DeLauer, FY 1985 Department of Defense Program for Research, Development, and Acquisition, to the 98th Congress, February 27, 1984, pp. 1-15.

27. Francis X. Clines, "Grim As Ever, The Thought of Winter Stalks Russian Spirit", *New York Times,* Sept. 22, 1991; Margaret Shapiro, "Soviet Grain Harvest Drops 30%, *Washington Post*, October 31, 1991.

28. John Yang and Guy Gugliota, "Bush Considering More Food Aid For Soviets", *Washington Post*, October 24, 1991.

29. Dr. William Perry, to the US Senate Armed Services Committee, September 19, 1991.

30. McNeil-Lehrer Newshour, PBS-TV, January 14, 1992.

31. Testimony on Military Conversion in the Russian Federation, by a visiting team of Soviet defense industrial managers. Committee on Armed Services, U.S. Senate, February 6, 1992.

32. Vitali Shlykov (Vice Chairman, REFSR State Committee on Defense), "The Point of View of the Russian Federation on Conversion", (undated, but apparently early September 1991).

- Vladimir Kontorovich, "Conversion of Military Industry in the USSR, A NewStage", Command Economies Research Inc., September 9, 1991.

33. A. I. Didiukov, "Conversion and Scientific and Technological Progress", International Conference on International Cooperation in Peaceful Use of Military Industrial Technology", Beijing, October 22, 1992.

34. Fred Hiatt, "Senior Pentagon Official Cautions on Aid for Soviet Military Conversion" *Washington Post,* November 6, 1991. The full text of the quotation reads as follows:

> "In a recent interview with the newspaper Moscow News, Gorbachev said: We cannot live a normal life in our supermilitarized economy..Why should the (the military-industrial complex) be given vast funds, the best resources, the most skilled personnel and the finest scientists? If this is not half of society, then its at least a third of it.'"

35. Daniel Sneider. "Soviet Swords Into Plowshares? Not Quite Yet", *Christian Science Monitor,* November 7, 1991.

36. Gennady Burbulis, "Come, Make Goods and Sell Them", *Washington Post,* January 22, 1992.

In a companion guest editorial, Alexei Izyumov, previously the Mayakovsky of young Russian Academy Institute analysts of the Soviet defense industry conversion question offered somewhat Milder but essentially similar sentiments: the Russians could solve none of the problems themselves, only outside investments would affect events: Alexeilzyumov", 16th Republic" Armed and Dangerous", -*Washington Post,* January 22, 1992.

37. Victor Antonov, interview, McNeil-Lehrer NewsHour, PBS-TV, January 27, 1992, p.58.

39. John Lloyd, "Under Financial Pressure, Some Republics Plan To Push Arms Exports; Military Chief says CIS has no Enemies", *Financial Times*, February 5, 1992. The article added that "Udmurtia production included chemical weapons---"although the USSR had declared in 1987 that it had ceased all chemical weapons production.

40. Peter Pringle and Ravel Felgenhauer, "Moscow Goes Down to the Arms Bazaar", *The Independent*, February 4, 1992; in FBIS-SOV-92-028-A, February 11, 192, pp. 3-4 as "Gaydar to Head Arms Sales Organization". See also "Udmurta may Produce Arms for World Market", FBIS-SOV-92-021, January 31, 1992. pg. 56.

42. Vitalii Shlykov, op.cit.

This presumably refers to Lev Zaikov, Oleg Baklavmov, Igor Belousov, the Central Committee secretaries responsible for the Military-Industrial Commission (VPK), the Aviation Ministry and managers of military aircraft and jet engine plants in particular, as well as many other defense plant managers.

43. William J. Broad, "Russian Scientists to Detail Plans for a Fast Nuclear Space Rocket," *New York Times*, January 13, 1992.

- William J. Broad, "In Space, and in East-West Politics, Its a Very Long Voyage to Mars", *New York Times*, February 5, 1992.

- Warren E. Leary, "US-Soviet Mars Flight is Proposed to Cut Costs", *New York Times*, June 27, 1991.

44. - Chrystia Freeland, "From Missile Maker to Sausage Stuffer; Ukraine's Colossus Still Builds SS-18's for Russia but Adds Umbrellas and Trolleys", *Washington Post*, January 30, 1992.

- Peter Maass, "Something New, From The Mind of Yuzhny; Huge Soviet Nuclear Missile Factory Now Hopes to Make Nice Things For Western Consumers", *Washington Post*, October 2, 1991.

45. "On Avenues for Converting Enterprises and Organizations in the Defense Complex", undated, Secretary of the Section of Engineering Ecology and Resource Conservation, Engineering Academy of the RSFSR.

46. Robert S. McNamara, "The Post-Cold War World and Its Implications For Military Expenditures in Developing Countries", World Bank Annual Conference on Developing Economics, 1991", April 25-26, 1991.

- Robert S. McNamara, "Reducing Military Expenditures in the Third World", *Finance and Development*, 28:3 (September 1991) 26-28.

- Nicole Ball and R.S. McNamara, "Link Economic Aid to Military Limits", *New York Times*, January 11, 1992.

# SUCCESSFUL CONVERSION EXPERIENCES

**Seymour Melman**
*Columbia University, U.S.A.*

## A Brief History of U.S. Economic Conversion Efforts

Since the Second World war, the United States has experienced several major wind-downs in defense spending, all of which have led to various degrees of conversion from military production to appropriate civilian lines of business.

### Post-World War II Experience

The most notable effort was following the second World War when the defense economy was reduced from about 40 percent of Gross National Product (GNP) to about 6 percent of GNP two years after the war. While national output fell somewhat due to the rapid decline in military spending, the level of employment and consumption was maintained through public policies instituted to counter the onset of recessions. Indeed, the Congress and Executive committed the nation to a full employment policy and passed the GI Bill which helped maintain consumer spending by providing returning GIs with money to go to college.

Contrary to popular belief, the *successful rapid transition to a civilian economy occurred because there was extensive planning* and policies were in place to cushion the economy from the shocks of demobilizing from the war[1]. Indeed, several planning boards such as the National Resources Planning Board and the Office of War Mobilization and Reconversion, among others, began to plan for reconversion before the war was over, with detailed plans developed as early as 1943. At the state and local levels, the Committee for Economic Development, a business group, had representatives working on local reconversion projects in over 1,800 communities by 1944. Furthermore, several major legislative initiatives were put in place to permit defense firms to begin civilian product development for the post war period and to set up procedures for the sale of government-owned plant and equipment to private firms.

As is well known, these sales of government-owned plant and equipment led to the creation of whole new businesses in such new sectors as the aluminum industry, and fostered new competition in transportation equipment and electronics[2] In short, the co-operative efforts of government and business to undertake advance planning and legislative initiative led to a successful conversion process.

### Post-Korean War Experience

Following the Korean War, several major aerospace companies undertook to develop commercial products which drew on their technological expertise in defense work. **Boeing Aircraft**, the most notable example, launched its conversion effort with the objective of developing a commercial passenger and cargo jet transport capability[3]. Boeing devoted a substantial percentage of its management, engineering and production talent to developing this commercial venture, which was viewed as a long term investment in an emerging market. As is well known, Boeing eventually became the worldwide leader in manufacturing jet passenger aircraft, and today the commercial division generates the lion's share of its revenues.

## Post-Vietnam Examples

During the wind-down in military spending after the Vietnam War, many defense contractors began to explore civilian market opportunities in the masstransportation market and in alternative energy research and development.

**Boeing-Vertol, Rohr Industries, Ingalls and Allied-Signal** corporation undertook to produce light rail cars and electronic control switches for the growing urban mass transit market in the 1970s.[4] These cases were only partial successes. They all produced new products and won contracts. But each contractor was plagued by technological and cost control problems which arose because they adopted the dubious military production and engineering practice of concurrency in their commercial work.

Concurrency is the simultaneous prototyping and full scale production of a product which has not yet been fully tested. This method has often been used by the military to accelerate the development and deployment of a new technology in order to gain a strategic advantage.[5] However, it is fraught with problems, since undetected errors in product design and production engineering usually require expensive redesign and retrofitting efforts to correct the defects in the product and the production process.

Boeing Vertol's concurrent testing and design practices resulted in major problems with its orders for the Massachusetts Bay Transit Authority. Eventually they worked out many of the problems, although they incurred rather large financial losses. Later, they learned from this experience, drawing on it to more efficiently produce and deliver orders for trolleys for the City of San Francisco and light rail cars for Chicago. While Boeing Vertol used mostly workers from their military production lines, it did not manage to re-employ a majority of its workforce in the trolley and rail car production.

Sudden changes in federal R&D and procurement policies during the 1970s and early '80s were principal factors confounding the conversion efforts of Boeing Vertol andother defense firms. Boeing-Vertol's failure to stay in the mass transportation market was to a significant degree a result of large reductions in federal urban mass transit discretionary spending during the Carter and Reagan Administrations. In addition, the failure of the Urban Mass Transit Administration to establish uniform standards for such vehicles contributed to the instability of the public market. Thus, public procurement and regulatory standards played a key role in creating and then destroying a conversion opportunity for aerospace firms.

Today, there is similar potential in the mass transportation market for aerospace firms, especially in Southern California where such a change is under active consideration. Yet because there is little commitment by the federal government to stimulate such a business development strategy, the market opportunity remains unexploited. It is noteworthy that today, the United States imports most of its mass transit rail car equipment from Japan, Italy and Canada.

A parallel conversion situation prevailed in alternative energy R&D. The oil price shocks of the 1970s stimulated large-scale increases in federal spending on alternatives. Several defense-related firms were attracted by this spending into alternative energy research, but they quickly exited when the money dried up in the early 1980s. Significantly, while the OEA's own study[6] covers many of these very examples, the OEA has failed to draw the obvious conclusion that linking and targeting federal R&D and procurement policies can create conversion opportunities for defense industries.

In this context, one can see the outlines of a sensible, R&D led conversion policy: it would permit a modest compensated shift of military R&D savings into critical civilian technology areas, creating important and lucrative conversion opportunities.

## Contemporary Examples

### The Curtis Universal Joint Company

Until recently, the Curtis Universal Joint Company of Springfield, Massachusetts sold most of its universal joints--an energy transmission device--to the military.[7] After the firm modified its production process and modernized its facility during 1989-90, however, costs decreased and quality increased. During the restructuring, it was discovered that much of the firm's machinery--including its machine tools--could be adapted to produce a commercially viable product.

### Kavlico Corp

Kavlico Corporation of Moorpark CA, an aerospace contractor, relied on military orders for almost 85 percent of its sales in 1985. Four years later, after it developed a sales organization and modified its manufacturing, half of the firm's sales are in the commercial market. The firm's overall sales and employment have doubled as well.

### Kaman Aircraft

Kaman Aircraft of Bloomfield, CT made helicopters exclusively for the Pentagon until the late 1960s, when it decided that the market was too undependable. By using the skills of its vibrational engineers, the company successfully developed and produced guitars, employing 100 people, including 30 people transferred from defense production.[8]

### York Naval Ordnance Plant

The York Naval Ordnance Plant in York, PA lost 1,092 civilian jobs following its closure in 1964. It was principally dedicated to the production of ammunition and ordnance for large bore naval guns. After its closure the site was redeveloped into an integrated production facility to produce precision equipment for Harley Davidson motorcycles. This case demonstrates that existingcapabilities within the military economy may be reused and redeveloped for appropriate civilian manufacturing. Today, 1,600 people are employed by Harley Davidson at the site.

### Frisby Airborne Hydraulics

In 1985, Frisby Airborne Hydraulics did 90 percent of its business with the Pentagon and military-firms. Five years later, this figure was reduced to 20 percent.[9] The firm's restructuring, which included establishing complete participative management in production and eliminating one layer of management, was accomplished without laying off any workers.

### Textron Aerostructures

Between 1988 and 1991, Textron Aerostructures of Nashville, Tennessee, a maker of airplane wings, turned an 80-20 percent dependence on military contracts into a 20-80 percent modest dependence. Management assigns a portion of the credit for the success of the transition to employee involvement in the necessary changes in accounting procedures and design. The company's loss of defense contracts led to significant lay-offs: the union work force fell from a peak of 5,400 to its current level of 1,200 employees. In addition to supporting the case for conversion, therefore, this example is also a cautionary tale, demonstrating the importance of advanced planning to avert layoffs.

**The Fail-Safe Technology Corporation**

James K. Lee, President and CEO of the Fail-Safe Technology Corporation, described his company's transition from the military to the civilian economy at a regional hearing sponsored by ECD in June 1991.[10] His firm, which makes self-repairing microcomputers, had been a subcontractor to the SDI program before launching a conversion process in 1987.

According to Lee, the transition to the civilian market was not easy. The company had considerable difficulty getting the capital necessary to reorient the firm. Nonetheless, Fail-Safe successfully developed a new product, the FS-99, which is a board level computer that is placed in larger systems. The FS-99 was licensed to Baker Hughes Oil Production and is used as an emergency shut down system in the oil and natural gas industry. It is now being used in Kuwait to repair the well-heads that the Iraqi army blew up during the recent war.

Mr. Lee noted that a Small Business Innovation Research (SBIR) Contract from the Federal government helped Fail-Safe make its transition. This contract enabled Fail-Safe to develop a computer that is plugged into other computers enabling them to repair themselves for 20 to 30 years. Lee emphasized, however, that the SBIR process consists of several phases, and in-between these phases, when the project is not going forward, "there's nothing to cover you". As a result, firms participating in the SBIR program have to continue paying the engineers working on the project during the program's lulls. Lee argued that a bill sponsored by California state Assemblyman Richard Polanco, which would create a state program to support the SBIR program, did not represent a special favor to business, because the entire economy would benefit. Highly paid and skilled employees pay taxes, he said, while "every one of them that's laid-off goes on the dole".

Mr. Lee believes that this nation needs "a national technology policy and industrial priorities". He said that labor, business and government "need to get more into co-operation and indeed into collusion, because if we don't, we're not going to be able to compete in the global economy".

**Air Industries**

Several years ago, Air Industries of Bay Shore New York sold 80 percent of itsproducts to several military contractors, including Grumman, Fairchild Republic and Sikorsky. By 1991, 80 percent of the company's business had become commercial, "much of it for Boeing Co.'s fleet of passenger planes," according to *New York Newsday*. Although profit margins for commercial contracts are lower than for military contracts, the company remains prosperous. Recently the state's Job Development Agency loaned the company $ 720,000 to expand its production facilities. They expect to hire an additional 35 to 40 workers as a result.[11]

**AM General**

The AM General facility in Mishawauka, Indiana is the manufacturer of the "Hummer" off-road vehicle made famous by the Gulf War. As part of its strategy for breaking out of its dependence on defense contracts, the company has begun adapting the Hummer for use as fire trucks and ambulances in remote areas, where the vehicles' air-and water-tight bodies will be useful in negotiating bad roads. Worker participation in both the production and testing processes are credited with saving hundreds of hours and substantially cutting costs. In addition it has prevented further layoffs and may create new jobs. Now the company has announced plans to produce an off-road recreational vehicle for the domestic civilian market.

**Job Retraining for Scientists and Engineers**

In 1971 the Technology Mobilization and Reemployment Program (TMRP) was designed in response to the contraction in the defense industry which occurred as the Vietnam War was ending.[12] The program was targeted to scientific and engineering workers who were laid off. It added staffing to address the needs of these workers through the "State Employment Service offices, job search grants, on-the-job training and short skill development courses," according to the Pentagon's Office of Economic Adjustment.Of the 532,000 people who participated in the program, 34,000 found new jobs through TMRP. The program cost $ 28 million and ran for two years.

**Base Closures and Redevelopment Experiences**

In a survey assessing the impacts of 100 major military base closures, the Pentagon's Office of Economic Adjustment claims that from 1961 to 1990 more new jobs were created (158,104) than were lost (93,424).[13] The "new" job figure, however, does not factor out pre-existing businesses and jobs which were relocated onto the base site. *Therefore, the "new job" figure is not an evaluation of "net" job creation for the community.* One should factor out jobs relocated onto the site from elsewhere in the community in order to determine whether there was net job creation. Despite this limitation there are several noteworthy cases that may provide models for base redevelopment and net job creation.

• The **Benicia Arsenal** in Benicia, California lost 2,321 civilian jobs when the facility was slated for closure in 1964. Ultimately, 5,700 new civilian jobs were created at the site. Exxon, Corey Construction Company, Huntway Refinery and Ace Hardware are among the firms active on the site. This case should be examined to determine if there were any regional economic advantages from redeveloping this site for these specific industries.

• **McCoy Air Force Base** in Orlando, Florida lost 395 civilian jobs after the Pentagon decided in 1974 to close the base. Today, 6,000 people are employed there by several concerns, including the Post Office, Federal Express, UPS and D.H.L. Airborne Express. This seems to be a clearer case of reusing an existing military facility for regional and industrial developments that complement the regional economy.

• **The New Iberia Naval Air Station** in New Iberia, Louisiana lost 85 civilian jobs following a 1965 Pentagon decision to close the facility. Today, 1,220 people are employed there by several firms including ERA Helicopters, the Otis Engineering Corporation and the Pelican Aviation Corporation. Again this is a more direct redevelopment example.

• **The Grenier Air Force Base** in Manchester, New Hampshire lost 138 jobs following a 1966 Pentagon decision to shut the facility. Today, 3,200 people are employed on the site by several firms, including Sanders Associates, Disogrin Industries and Summit Packaging.

• The closing of the **Charlestown Shipyard** in 1974 resulted in 3,015 lost jobs. The mayor established a representative economic planning council with authority to lease property and execute contracts. One of the first proposals accepted by the council established, at company expense, a vocational counselling and retraining project, in which the laid-off workers were given first priority. Within one year of the closing, all the jobs had been restored in six new, solid firms, and 500 students were enrolled in the vocational center. Ten years later the businesses and industries in the redeveloped area were employing 4,815 workers. This is a mixed use redevelopment example based on unique characteristics.

• **The Donaldson Air Force Base** in Greenville, South Carolina lost 672 civilian jobs following a 1963 Pentagon decision to close the facility. Today, several firms including the Woolworth Distribution Center, the 3M Company, General Electric and Proctor & Gamble Inc., employ 5,253 people at the site. This is an industrial park-type of redevelopment.

**The Kincheloe Air Force Base** in the upper peninsula of Michigan closed in 1978, eliminating 450 civilian jobs and a $ 40 million payroll. Today the site contains a 1850 -acrepark, a railroad, a power plant, a shopping mall, a waste treatment facility, a golf course and an airport. Now over 1000 local people work for 49 new firms, representing an increase of 10 percent per year in the county employment base. Yet this mixed use approach is not always feasible given the political and financial constraints of some militarily-dependent communities.

***Europe***

**Sweden**

The **Landskrona Shipbuilding Yard**. Facing the prospect of a shutdown of a large shipbuilding yard and large-scale job losses, labor and community groups in Landskrona Sweden successfully converted the government-owned Oresundsvarvet shipyard. Using the buildings already on site and existing workers' skills, supplemented by retooling and retraining, the production company created more than 40 new companies on the site. Among them are an oil spill cleanup company, a floating installation for raising fish on an industrial scale, and manufacturers of a wide variety of new products including hearing aids, aluminum windows and meatcutting tools. Of the 2,322 Landskrona workers laid off, 1,650 have found work in the area,fully half of them with these new businesses.

The potential in Sweden for conversion is enhanced by several government programs. One is the Swedish Industrial Development Fund, which was established by Parliament in 1979 with an initial capital of SKr 300 million. Its purpose is to support new product development by major Swedish firms. This support may take the form of loans or other forms of participation, such as cash grants in return for a royalty in future revenues.

A program sponsored jointly by the National Industrial Board and the Regional Development Funds supports new productdevelopment by small and medium-sized companies. In addition to technical advice, this support comes in the form of conditional loans and grants, repayable if the project succeeds.

**Germany**

Efforts on the part of three aerospace firms to increase their civilian production are still in the early stages.

MBB has created two new corporate sectors--Industrial Products and Industrial Energy and Process Technology -- to develop product lines outside of the company's traditional focus on aerospace. The first of these sectors has responsibility for explorations into the fields of data systems technology, industrial electronics, automation technology, transportation technology, medical technology, and high performance composite materials. The second will drive MBB work in bioengineering, thermal energy systems, wind power, vacuum superinsulation, photovoltaics and environmental protection. The emphasis in both sectors is on linking these diversification activities closely to the requirements of the market.

**Endnotes**

1 See Jack Stokes Ballard, *The Shock of Peace : Military and Economic Demobilization after WW II*, Washington, D.C., University Press of America, 1983.

2 Ibid, Also see "Conversion Planning: 1945 and 1990", by J. Davidson Alexander, December , 29, 1989, Allied Social Science Association meeting Atlanta, Georgia.

3 See Murray Weidenbaum, "Obstacles to Conversion," *Bulletin of the Atomic Scientists*, April 1964, for his discussion of the post-Korean War conversion efforts in Boeing and the aerospace industry. Curisously, Weidenbaum, who was a Boeing economist, plays down the success of Boeing's efforts because they did not immediately lead to a rapid growth in the commercial division's sales revenues. However, the key point is that the commercial division developed out of the transfer of resources from the defense work force and production capacity, and eventually led to the commercial division generating more sales revenues by the end of the 1960s.

4 A forthcoming study by the Office of Technology Assessment entitled *Technology Opportunities in Economic Conversion* treats these cases in Chapter 7. In addition, there is an examination of these cases in the OEA's study *Economic Conversion/Adjustment*, July 1985, Department of Defense, Chapter 7.

5 See *How the Military-Serving Firm Differs from the Rest* by Anthony Difilippo, ECD Briefing Paper #10, Washington DC, NCECD, March 1991, pp.4-5.

6 See *Economic Adjustment/Conversion*, op.cit. p.40

7 Dr. Frank Emspak, "Clearing Obstacles to Conversion: A Process Driven Model for Economic Conversion", paper presented at the American Academy of Arts and Sciences, September 10-11, 1990.

8 U.S. Department of Defense Office of Economic Adjustment and the President's Economic Adjustment committee, *Economic Adjustment/Conversion*, July 1985.

9 "CEO Reports Conversion Success Story", *The New Economy*, October 1990, p.3.

10 See James Raffel "Los Angeles Confronts Realities of Post Cold War Economy", *The New Economy*, Summer 1991, p.1.

11 James Bernstein, "A Switch in Gears", *New York Newsday*, November 1, 1991, p.53.

12 See U.S. Department of Defense Office of Economic Adjustment and the President's Economic Adjustment Committee, *Economic Adjustment/ Conversion*, July 1985, p.75.

13 See *Civilian Reuse of Former Military Bases, 1961-1990*, by the Office of Economic Adjustment, Department of Defense, Washington, D.C., June 1990.

# ENVIRONMENTAL DIMENSIONS OF DISARMAMENT AND CONVERSION

**Michael Renner**
*World Watch Institute, U.S.A.*

## Introduction : Toward a Peaceful and Sustainable Economy

The 1990s mark the beginnings of an era in which fundamental change occurs in two domains critical to the future of human existence. The first is the realm of national and international security. National security has been and still is invoked to justify the relentless pursuit of military prowess -- the maintenance of large armed forces, the deployment of ever deadlier weapons systems, and the frequent intervention in the affairs of weaker nations. But the end of the Cold War provides a unique opportunity to dismantle and convert to peaceful uses a large portion of national war-making capacities and to redirect society's priorities. The second area is the fate of the biosphere. The assault on the environment has reached crisis proportions, threatening to exhaust the earth's regenerative capacity. Humanity now stands at the threshold to irreversible degradation of many of the natural systems that underpin all of human life. At the same time, though, environmental awareness is soaring in many countries and the concept, of "ecological sustainability" is gaining ground as an alternative approach to economic development, suggesting that the current destructive path may yet be abandoned.[1]

The themes of disarmament/conversion and environment/sustainable development may seem separate and unrelated. Yet they interact in many ways, and solving one set of problems while disregarding the other will prove difficult. Reversing ecological destruction cannot succeed without halting the military's onslaught against theenvironment. Conversely, in this age of profound environmental crisis, military conversion needs to aim beyond a search for civilian labels and incorporate ecological guidelines into alternative use proposals. Put simply, producing more cars instead of tanks, for example, will not be part of the solution, but part of the problem.

The costs of warfare and permanent war preparation--human, economic, and ecological--are often justified by vague references to national security requirements. Yet, in the emerging post-Cold War era, the very definition of what constitutes true security needs reassessing. Arguably, social, economic, and environmental issues are at least as important to the security--that is, the well-being -- of nations and their citizens as safety from foreign attack on which traditional national security concerns are focussed.

This chapter begins by illustrating the ways in which environmental perils threaten national and global security. It then argues that military power is not only inadequate in this broader security perspective, but in addition the military economy pre-empts monetary and scientific resources needed to deal with the environmental crisis. Moreover, as the third and fourth sections document, the military sector--that is, the armed forces and the military industry - contributes significantly to environmental problems both during war and in peacetime. The environmental legacy of the war system joins a long list of solid reasons for moving towards serious disarmament. Unless humanity can find less violent ways of settling disputes, the fundamental incompatibility between the military and the environment will continue toconfront us. But environmental considerations also play a role once the disarmament process gets under way. The final section discusses three major dimensions: first, the disposal of weapons and military equipment; second, the clean-up of military bases and rehabilitation of land holdings before they can be turned over to civilian use; and third, the conversion of arms production facilities.

**Redefining National Security**

Over the course of history, and particularly during the half century of the Cold War, national security has come to be defined almost exclusively in military terms--the ability to deter or repel outside aggression. This growing reliance on force is counterproductive in at least three ways, however. First, with the spreading availability of weapons of mass destruction, the capability to truly defend a country's territory and population has been lost--and instead replaced by the capacity to destroy massively. Competitive national security policies have yielded international insecurity. Second, the unbridled pursuit of military power has meant a resource drain undermining economic vitality by retarding civilian R&D, slowing productivity growth, and contributing to indebtedness. And third, of central importance here, governments preoccupied with security threats of military origin have ignored the perils of environmental degradation.[2]

Environmental factors are often dismissed as irrelevant or unrelated to security issues. Yet military and environmental security share a fundamental objective: to permit people to go about their lives without undue harm. Although governments have appropriated the term, the dictionary definition of security contains not a hint of an exclusive military connotation: Webster's Collegiate Dictionary defines security as freedom from danger, from fear or anxiety, from want or deprivation. Thepotential for economic upheaval and dislocation inherent in the environmental crisis poses threats to nations' and individuals' well being and safety on a par with many wars.

National security is a meaningless concept if it does not encompass the preservation of livable conditions on earth. A reasonable definition of security needs to encompass breathable air and potable water, safety from toxic and radioactive hazards, an intact atmospheric zone layer, a stable climatic system, and protection against the loss of the topsoil that assures us of our daily bread. The well-being of nations and their individual citizens depends as well on economic vitality, social justice, and ecological stability as it does on safety from foreign attack. Pursuing military security at the cost of these other factors is akin to dismantling a house to salvage materials to erect a fence around it.

Because environmental pollution and degradation respect no human-drawn borders, they jeopardize not only the security of the country in which they originate, but also that of others, near and far. Border-transcending environmental degradation most immediately affects neighboring countries. Disputes over water use and quality, for example, simmer in virtually all parts of the world. These involve water diversion, industrial and agrochemical pollution, salinization of streams through heavy irrigation, siltation of rivers, and floods aggravated by deforestation and soil erosion. An estimated 40 percent of the world's population depends for drinking water, irrigation, or hydropower on the 214 major river systems shared by two or more countries. In the Middle East, a water scarce region, water has been a factor in previous wars and could play a prominent role in future conflicts.[3]

Deforestation seems an unlikely source of tension between nations. Yet by aggravating soil erosion and silt accumulation in river beds, it can lead to devastation floods. Denuded watersheds in the Ethiopian highlands have exacerbated floods in neighboring Sudan, devastating its capital, Khartoum, in 1988. The frequency and magnitude of flooding in Bangladesh has markedly increased, partly due to deforestation in the Himalayas since mid-century.[4]

The impact of pollution and ecological degradation originating in one country is not only felt in adjacent countries, as transboundary air pollution strikingly illustrates. Insecticides from Asia and southern Europe, for instance, are found in Arctic and Antarctic waters. All European countries are involved in an intensive exchange of air pollutants such as sulfur dioxide, a major component of acid rain. Norway and Sweden are large net importers of sulfur dioxide, originating primarily in Britain. Norway's foreign minister has said he considers the British reluctance to address the issue, a problem as serious for mutual relations as trade and defense.[5]

Spurred by a continuous stream of new scientific evidence, attention is shifting to those aspects of environmental degradation that have an all-encompassing, global effect: the depletion of the ozone layer and global warming pose problems from which no nation can insulate itself. As monitoring and scientific understanding grows, it is becoming clear that the thinning of the stratospheric ozone layer, which protects life on earth from harmful ultraviolet radiation, is accelerating. Initially, it was believed that the depletion was taking place only over the poles, but it is now significant in both the northern and southern hemispheres (record levels of chlorine monoxide, an zone-depleting compound, have been measured above the northernhemisphere in January 1992,leading NASA scientists to speculate that during the winter months, ozone loss there could be 20 to 30 percent). Also, ozone depletion was first registered only in wintertime, but it has now also been detected in summer, when ultraviolet rays pose the most serious hazard. Failure to rapidly phase out CFCs and other ozone-devouring chemicals - at a faster pace than current international agreements stipulate -presents a grave danger to human health, to agricultural productivity, and to marine fisheries. Already, significant crop damage and large number of skin cancers are inevitable,given the amounts of CFCs previously released into the atmosphere.[6]

The security of nations is similarly compromised by the "greenhouse effect" - the specter of global climate change brought on by the accumulation of carbon dioxide, methane, nitrous oxide, and other heat-trapping gases in the atmosphere. The prospect of an accelerated warming of the earth in coming decades puts the conditions essential to the stability of the biosphere in jeopardy. Shifting precipitation patterns and vegetation zones, and rising sea levels caused by global warming threaten to disrupt crop harvests, inundate heavily-populated low-lying coastal areas like the Bengal and Nile deltas, upset human settlement patterns, and undermine biological diversity.[7]

If the above observations are accepted, it follows that the means of dealing with security threats and solving conflicts need to undergo security. The key traditional instrument for maintaining national security-military power - is wholly inadequate to provide security as understood in this broader perspective. (This does not mean that military security and armed conflict are no longer of importance; the recent Gulf War and the Civil war in Yugoslavia serve as unpleasant reminders that the world community urgentlyneeds to develop effective methods of peaceful conflict resolution and prevention).

Obviously, military means cannot reverse resource depletion or restore lost ecological balance. Tanks, planes, and warships *might* be able to fend off a military attack, but no remedy exists to repel air or water-borne pollutants. Former Soviet Foreign Minister Eduard Shevardnaze recognized this when he said that "the biosphere recognizes no division into blocs, alliances, or systems. All share the same climatic system and no one is in a position to build his own isolated and independent line of environmental defense".

Environmental degradation is a problem very distinct from traditional conflicts. In the case of disputes over access, control and allocation of natural resources among competing countries, it is possible to have a "zero-sum game" ie.e, a winner and a loser. Environmental quality, however, is an indivisible goal. A zero-sum game is transformed into a no-win situation. Environmental security offers a more fruitful basis for cooperation and security among nations than military security because it is both a positive and inclusive concept. Whereas military security rests on the threat of annihilation, environmental security seeks to protect or to restore. While military security rests firmly on the competitive strength of individual countries (or coalitions of countries) at the direct expense of others, environmental security cannot be achieved unilaterally, but requires cooperative relationships among nations.

Security can no longer be achieved by exclusively national means; national security needs to be transformed into global security. Environmental reasons underscore the fact that nations are no longer the sole masters of their destinies. We have become accustomed to interdependence in economic affairs such as trade and investment, and we reluctantlyacknowledge an uneasy mutual dependence for survival in the nuclear age. Interdependence in economic, military, and environmental affairs has already begun to erode traditional notions of security and even national sovereignity, but the structures of the nation-state system--especially the war-making capacities--remain firmly in place. Because interdependence has come about by default rather than by design, it does not automatically generate the political will needed to adapt human institutions and thinking to new circumstances and requirements.

Military means not only are impotent in the face of environmental devastation, they themselves detract from environmental security in a variety of ways. The military sector (including the armed forces and the arms industry) consumes large resources and restricts society's ability to address environmental (as well as social and economic) problems and, as will be illustrated in subsequent sections of this paper, it directly contributes to environmental problems both in times of peace and war.

The military sector preempts a sizable portion of the financial resources and scientific and technological capacities needed to stem environmental degradation and to assist in creating a sustainable economy. According to an estimate by Lester Brown and Edward Wolf of the Worldwatch Institute, the global community would need to expend a cumulative sum of about $ 774 billion during the final decade of this century to reverse adverse trends in just four priority areas: protecting topsoil on croplands from further erosion, reforesting the earth, raising energy efficiency, and developing renewable sources of energy. While likely to be a conservative estimate and excluding other important areas, this sum is less than what the world currently spends for military purposes in one year.[8]

The resources poured into the military economy could go a long way toward meeting the environmental challenge (see tradeoff examples listed in Table 1). For example, the United States spends roughly $ 300 billion for military purposes, compared with about $ 115 billion (by government and industry) on pollution control. Similarly, in Western Europe military spending stands at roughly $ 150 billion, while pollution control outlays are estimated at anywhere from $ 50 billion to $ 100 billion.[9]

Perhaps an even more telling indicator of current priorities is the money devoted to research and development. At an estimated $ 100 billion annually, world military R&D exceeds the combined governmental research outlays on developing new energy technologies, improving human health, raising agricultural productivity, and controlling pollution. Worldwide, some 25 percent of all R&D funds, and of all scientists and engineers, are devoted to military ends and are thus unavailable to explore and develop ways to advance environmentally more benign production technologies. The share of governmental R&D outlays that goes to the military was as high as 70 to 75 percent in the United States and the Soviet Union during the 1980s, but lower in other countries (50 percent in Britain and 30 percent, in France, 12 percent in the former West Germany, and 3 percent in Japan). The end of the Cold War has begun to bring some change, but the Bush administration's budget proposals for 1993 still allocate 59 percent of all federal R&D spending to military projects. The United States alone, which accounts for at least 40 percent of the global total, has spent more than $1 trillion on military R&D since 1955. Civilian spinoffs have been limited, the opportunity costs tremendous.

**Warfare and the Environment : The Gulf War[11] :**

From the Punic Wars in the third century B.C. on, armies have poisoned wells, salted soils, and destroyed crops to foil the enemy. However, over time, the environmental impact of warfare has grown as sophisticated technology has boosted the firepower, range, and speed of weapons. In addition, modern industries present many high-profile targets whose destruction can wreak environmental devastation, as conflicts in Vietnam, Afghanistan, Central America, and the Persian Gulf amply demonstrate. Often, this is simply an inevitable consequence of the fighting, but at least in some cases environmental modification techniques have consciously been employed as a weapon.[12]

The recent Gulf War provides an unfortunate illustration that warfare and environmental integrity are incompatible. It ranks among the most ecologically destructive conflicts ever. While the fighting was brought to a swift conclusion, the ferocious onslaught against the environment continues. Before retreating from Kuwait, Iraqi troops set fire to hundreds of oil wells that severely polluted the atmosphere and imperiled human health and harvests in the region; wellheads that failed to ignite spilled huge amounts of oil into the ground that threatens to taint underground aquifers; oil deliberately released into the Gulf waters is poisoning marine life; attacks on refineries, petrochemical plants, and chemical and nuclear facilities have likely released substantial quantities of toxic materials; and the massive movement of troops and their heavy equipment has imperiled an already fragile desert ecology.

**Oil in the Atmosphere and the Desert**

It appears that 650 oil wells were set on fire (with an additional 100 damaged butunignited), and that as many as 6 million barrels of oil may have initially gone up in flames every day, almost four times Kuwait's production prior to the Iraqi invasion, or 9 percent of the world's petroleum consumption.After an agonizingly slow start, the fire fighting effort gathered momentum, as the number of team battling the inferno rose from 5 to 27 and the ranks of those involved in the effort swelled to almost 10,000 workers. The last fire was extinguished on November 6, 1991.[13]

Richard Golob, editor of *Oil Pollution Bulletin* in Cambridge, Massachusetts, estimates that at least one billion barrels of oil have been released. Kuwait's Oil Minister, Hamoud al-Ruqba, mentioned a figure of

600 million barrels in November 1991, yet also estimated that Huwait had lost 3 percent of its total oil reserves from fires and gushing oil wells which would put the total loss at roughly 2.7 billion barrels. Large amounts of oil gushed from damaged but unignited well heads - at least 30 million and perhaps as much as 70 million barrels of oil are still covering the surface of the Kuwaiti desert. Forming an estimated 200 lakes (some of them more than a mile wide and several feet deep), the oil is gradually seeping into the ground, threatening to poison underground water supplies.[14]

The clouds of oil smoke lingering over the Gulf region for eight months are likely to have contained large amounts of soot, sulfur dioxide and nitrogen oxides (the major components of acid rain), and a large variety of toxic and potentially carcinogenic chemicals and heavy metals. We may never know exactly what substances were present because governmental assessment efforts were woefully inadequate. The National Toxics Campagin, a U.S. grassroots group, found a number of toxic hydrocarbon compounds in the smoke in concentrations that exceeded U.S. health standards. Forexample, the NTC found high levels of 1,4 -dichlorobenzene, which can harm the liver, kidneys and respiratory system; and diethyl-phthalate and dimethyl phthalate, both of which attack the respiratory system. The U.S. Environmental Protection Agency estimated in March 1991, roughly 10 times as much air pollution was being emitted in Kuwait as by all U.S. industrial and power generating plants combined. Flying through the smoke clouds over Kuwait, scientists with the British Meteorological office found that the concentration of soot, sulfur dioxide, and nitrogen oxides at an altitude of 6,000 feet was about 30,20 and 10 times, respectively, the levels in a typical city plagued by air pollution.[15]

While the fires were burning the number of Kuwaities suffering from respiratory problems was reported to be up sharply. Although considerable uncertainty persists concerning the long-term toll on human health, many air pollutants are thought to cause or aggravate a wide range of conditions, including blood disorders, respiratory problems such as asthma and bronchitis, coronary ailments, cancer and possibly genetic damage. In independent assessments, George Thurston of New York University and Haluk Ozkaynak of Harvard University's Risk Assessment Group both predicted as many as 1,000 additional deaths in Kuwait - a 10 percent increase - over the course of one year.[16]

In parts of the Gulf region, the smoke plume significantly diminished solar radiation, thereby lowering day time temperatures and reducing the amount of rainfall. In April, daytime temperatures in parts of Kuwait were as much as 27 degrees Fahrenheit below normal. Bahrain experienced its coolest May temperatures in 35 years. Reduced photosynthesis, combined with the deposition of soot and other toxic materials, imperils crops. In fact, reports from parts of Iran andPakistan indicate that considerable crop damage has occurred. Saudi crop production has fallen by an average of 40 percent, according to the *Washington Post*.[17]

A huge area -- from the waters of the Nile to the snows of the Himalayas, from the Black Sea to the Indian Ocean -- was susceptible to acid rain and soot fallout ("black rain") from the Kuwaiti oil fires, according to the Max Planck Institute for Meteorology in Hamburg, Germany. Acid deposition is known to destroy forests and reduce crop yields. It can also activate several dangerous metals normally found in soil -- including aluminum, cadmium, and mercury -- making them more soluble and therefore more of a threat to water supplies and edible fish. The World Meteorological Organization reported in July 1991 that the environmental effects of the Kuwait oil fires have been felt as far away as Chad and the South Pacific. Kuwaiti soot has also been detected in Japan, Hawaii, and parts of the continental United states.[18]

Whether the atmospheric efforts could be of a global nature was a hotly contested issue among scientists even before the war started. While soot from Kuwait has evidently travelled around much of the globe, it appears that the quantity that was lifted into the upper atmosphere (at 35,000 feet or more, where it can remain for months or even years) is too small to set off any significant changes in global climate patterns. It turned out that, contrary to expectation, the soot particles easily bonded with water and most of them were therefore washed out of the sky. Still, this unhappy episode had all the trappings of a huge, uncontrollable game of roulette.[19]

**Oil in the Water**

The oil spilled into the Gulf waters is posing a severe test for marine ecosystems. Now estimated at some 6-8 million barrels bythe Saudi government, the spill is by far the largest in history -- 25 to 30 times the size of the 1989 Exxon Valdez accident in Alaska.[20]

Following "routine" spills (an average of 2 million barrels is lost every year during peacetime) and additional ones during the eight-year Iran-Iraq war, the Gulf was already a highly stressed environment, in poor condition to withstand additional ecological assaults. A relatively shallow sea, it is essentially a closed ecosystem with only a narrow outlet to the Arabian Sea through the Strait of Hormuz. Because Kuwaiti oil is of a "light" variety, up to 40 percent of it may have evaporated. The warm waters of the Gulf allow the remaining oil to decompose fairly rapidly, but significant amounts foul shorelines or poison the sea bottom. The Gulf waters are calm, with little wave action to break down the oil slick.[21]

From the beginning, the focus of the Saudi government's containment and cleanup effort was on protecting the country's desalination plants. Natural habitats were left to fend for themselves, and roughly 350 miles of Saudi coastline, from the Kuwaiti border to Abu Ali island, are now blanketed with oil. By September 1991, about 1.5 million barrels of oil had been recovered from the sea. Yet the oil has simply been dumped in pits dug in the desert; if it is left there - one option discussed - it will gradually seep into the soil and may contaminate the groundwater.[22]

Considerable harm to Gulf fish and other wildlife -- including porpoises, turtles, and seabirds -- is inevitable since many nesting and spawning grounds -- mud flats, seagrasses, salt marshes, and coral reefs -- have been soaked in oil. Up to 80 percent of Saudi Arabia's coastal mangrove swamps are dead or dying, according to a recent *Wall Street Journal* report. An estimated 20,000 to 30,000 birds were killed. Some areas are so contaminated that they had to be declaredoff-limits to fishing, threatening the livelihoods of commercial and subsistence fishes. The Saudi shrimp industry, for example, has been severely affected and is considered unlikely to recover before the end of the decade. If large quantities of plankton have been killed or poisoned, the entire food chain in the Gulf may be threatened. "It will take at least two decades for the marine life in the Gulf to recover from this war", according to Khalil Thaqfi, a Saudi Health Official.[23]

**Desert Wasteland**

The presence, during the Kuwait-Iraq crisis, of more than 1 million soldiers with their immense arsenals, has placed severe strains on the already fragile desert ecology of Kuwait, Saudi Arabia, and Iraq. Normally inhabited only by Bedouins, the desert of the Arabian peninsula cannot bear such a massive burden. Dessert vegetation is sparse, but it helps to stabilize and protect the soil. Tanks and other vehicles disrupted and compacted the soil and destroyed plants whose root systems are often close to the surface. As a result, the ground in many areas has been rendered susceptible to accelerated erosion. Seeds that lie dormant for large parts of the year, but which spark to life during spring rains, were likely to be affected. The incidence of sandstorms may well be increased. If a significant portion of the desert vegetation is destroyed, dry spells might be lengthened and the region's ecological balance could be tipped into long-term decline. It may take decades or even centuries for the desert to recover from the massive pre-war maneuvers and the tank battles, according to John Cloudsley-Thompson of the University of London.[24]

Another immediate and long-term peril stems from unexploded bombs and mines littering large parts of Iraq and Kuwait. Gar Smith reports in *Earth Island Journal* that as much as 20 percent of the one million landmines laid by Iraq may remain undetected after cleanup efforts. Of roughly 140,000 tons of explosives dropped by the allies on Iraq and Kuwait, perhaps 10 percent failed to explode and must be cleared. (British demolition specialists estimate that at least one quarter of the unexploded ordnance they have been disarming came from U.S. warplanes). Again, many bombs will likely remain undetected, and disarming those found - particularly the "Smart bombs" is extremely difficult. Since the end of the war, 200 people in Kuwait alone have been killed by land mines and more than 1,000 wounded.[25]

Kuwait is no longer an occupied country, but it has been transformed into a disaster, zone. With such results, it is difficult to distinguish between victor and vanquished. Indeed, the war's ecological impact extends far beyond the battlefield, blurring the distinction between the combatants and non-combatants, and affecting countries that were not party to the conflict and had no say over its course.

**The Military's Peacetime Assault on the Environment**

Whether advertently or inadvertently, warfare entails environmental devastation. Yet the military's war against nature begins before the first shot is fired and continues after a cease-fire. Even in "peacetime" preparing for war-the military contributes to resource depletion and environmental degradation, in some instances heavily. The production, testing, and maintenance of conventional, chemical, biological, and nuclear arms consumes substantial amounts of energy, generates enormous quantities of toxic and radioactive substances, and contaminates the earth's soil, air, and water. keeping troops in a state of readiness imposes a heavy toll on large tracts of often-fragile lands.[26]

The best information on military land use exists for the United States. The Department of Defense directly holds 100,000 square kilometers, an area equivalent to the state of Virginia.In addition, the armed services lease a total of about 80,000 square kilometers from other federal agencies, and an unknown amount of land owned by the states is devoted to military purposes.

The U.S. Department of Energy's nuclear weapons complex, meanwhile, spreads over 10,000 square kilometers. Altogether, at least 200,000 square kilometers -2 percent of total U.S. territory is devoted to military purposes. In addition, the pentagon controls about 8,100 square kilometers of land outside the United States, including an expanse of 700 square kilometers in the Philippines, which makes the U.S. military the single largest holder of agricultural land in that country.[27]

In what used to be the Soviet Union, the Defense Ministry controlled about 420,000 square kilometers, roughly 2 percent of total Soviet territory. In the republic of Kazkhstan, more land is devoted to military purposes than is used to grow wheat.[28]

Direct military land use in Western Europe is estimated at 1-3 percent of the total land mass. yet indirect or nonexclusive use tends to be much higher. In the Netherlands, for instance, direct use stands at 1.15 percent of the country's territory, but indirect uses of both land and water surfaces areas (and the airspace above) are believe to add another 10.5 percent. At least one third of the ecologically unique but fragile Waddensea and half the Dutch sector of the North Sea are used militarily. In Britain, in total land area controlled, leased, or held on license by the armed forces came to about 4,500 square kilometers in 1989. In the former West Germany, the national armed forces had exclusive access to about 2 percent of the total territory, or almost three times thecombined area of all nature preserves. Including indirect land use, the military's control came to 5.6 percent (U.S.Army forces stationed in West Germany occupied 1,000 square kilometers, or 0.4 percent of the territory). The Soviet armed forces stationed in what used to be East Germany controlled an estimated 2.5 to 4 percent of the land, while installations of the East German army covered another 1.7 percent. Soviet troops in Poland occupied some 1,730 square kilometers, or 0.5 percent of all the land.[29]

Global estimates of direct military land use are sketchy at best. It is believed to be around 1 percent of the earth's total land area (approximately 1.5 million square kilometers). This may sound small, but it represents an area roughly the size of Indonesia. it is unclear whether the estimate includes the former Soviet Union (where an area the size of Sweden was under direct control of the Defense Ministry). To arrive at a truly comprehensive measure, however, the space occupied by arms-producing enterprises would need to be added to the land controlled by the armed forces themselves - information that is unavailable.[30]

In a world dramatically short of productive land, any unproductive and destructive use of territory seems a misplaced priority. The military appetite for land increasingly collides with other needs, such as agriculture, wilderness protection, recreation, and housing. Beyond the land formally under the military's domain lie far larger areas that are used to varying degrees by the armed forces. While these may not be off-limits to civilians, they are often damaged or rendered dangerous or unusable even by occasional military use. Land used for war games is prone to suffer severe degradation. Maneuvers demolish the natural vegetation, disturb wildlife habitat, erode and compact soil, silt up streams, and cause flooding. In the former West Germany, maneuvers wereheld on as much as a quarter of the country's territory. Bombing ranges transform the land into a moon-like wasteland, pockmarked with craters. Shooting ranges for tanks and artillery contaminate soil and ground water with lead and other toxic residues.[31]

**Energy Consumption, Global Warming, and Ozone Depletion**

Whether it is jets roaring through the skies, tanks rumbling across the land, or warships navigating the high seas, the armed forces use large amounts of energy, and thus contribute to air pollution and global warming. Unfortunately, however, few statistics are available. Petroleum products account for roughly three quarters of all energy used by the armed forces worldwide, but by far the most important is jet fuel. Worldwide, nearly one quarter of all jet fuel some 42 million tons per year - is used for military purposes. The share rises to 27 percent in the United States, to 34 percent in the Soviet Union, and to 50 percent in West Germany (See Table 2.)[32]

Again, data are most readily available for the United States. The U.S. Department of Defense consumes about 37 million tons of oil equivalent (mtoe) a year. This works out to 2-3 percent of total U.S. energy demand and 3-4 percent of oil demand, but would be somewhat higher if military-related activities of the Department of Energy and the National Aeronautics and Space Administration were included. The Pentagon, perhaps the single largest consumer worldwide, uses enough energy in 12 months to run the entire U.S. urban mass transit system for almost 14 years.[33]

These figures would be considerably higher if energy used in manufacturing weapons were included. Up-to-date data in this regard, however, are virtually unavailable.[34] In 1971, producing arms in theUnited states used about 47. mtoe. An extrapolation for a more recent year 1989 can be made on the basis of 1971 and 1989 (inflation-adjusted) expenditures for U.S. domestic arms procurement and revenues from arms exports. Assuming that energy intensities have remained unchanged, this computation yields a figure of about 68 mtoe for 1989, almost twice the armed forces' direct use of energy. Assuming that energy efficiency in the arms industry improved on a par with civilian U.S. industry, the figure would be about 50 mtoe. Total U.S. military-related energy consumption would thus be at least as large as Australia's total consumption.[35]

Worldwide, energy use for weapons production may be assumed to be relatively lower because most countries do not have a significant arms industry. Still, the military sector's share of total oil and energy use may well be double the armed forces' direct share of 3-4 percent. If these assumptions are correct, the world uses about as many petroleum products for military purposes as Japan, the world, second largest economy, does for everything.[36]

There are virtually no data that would allow a detailed assessment of the military sector's contribution to air and water pollution. According to a 1983 estimate, emissions from the operations of the armed forces account for at least 6-10 percent of global air pollutants generated by all air traffic over that country's territory.[37]

Little work has been done to date on the military's contribution to global warming. In 1988, the U.S. armed forces contributed some 46 million tons of carbon emissions, or roughly 3.5 percent of the U.S. total. If the U.S. share of global military spending - just under a third - is any indication, then the U.S. armed forces' carbon emissions would have to be tripled to arrive at a rough globalestimate. The result of such a back-of-the-envelope calculation is about 150 million tons of carbon - nearly equal to the annual emissions of Britain. Again, if arms - producing industries were included, these numbers could well double.[38]

A similar dearth of information pervades the military use of ozone-depleting substances. The U.S. Defense Department is a major user of Halon -1211 and CFC-113, accounting for 76 percent and just under 50 percent of total U.S. use of these two compounds, respectively - which together are responsible for 13 percent of overall ozone depletion. (The military's share of other substances implicated in this global problem, however, is apparently much smaller). In addition, the combustion of solid rocket fuel, used in strategic and tactical missiles and on the U.S. Space Shuttle (some of the missions of which are military), releases hydrogen chloride gas which, transformed into hydrochloric acid, contributes to the depletion of the ozone layer. Scientists believe that the Shuttle's effect is relatively limited globally, but more significantly locally, in areas adjacent to the lunch site. However, Aleksandr Dunayev, former head of Glavkosmos, the Soviet space agency, estimated in 1989 that the ozone layer would be completely destroyed at an annual rate of 300 shuttle launches. The contribution of missile launches to ozone depletion, meanwhile, is difficult to calculate because no one knows how many rockets are fired each year (the U.S. Space

Command estimates and average of 500 to 600 during the 1980s). There are also no data on the amount of ozone-depleting chemicals used in the arms industry. A considerable part of the electronics industry's output, for example, goes to the military (20 percent in the United States in 1987), and presumably a comparable share of the industry's CFC consumption (used to clean semiconductors) is thus military-related.[39]

**Military Toxics[40]**

In their incessant pursuit of prowess and preparedness, the armed forces are poisoning the land and people they are supposed to protect. The production maintenance, and storage of conventional, chemical, and nuclear weapons and other pieces of military equipment generates vast amounts of materials inimical to human health and environmental quality. These wastes include fuels, paints, solvents, heavy metals, pesticides, poly-chlorinated biphenyls (PCBs), cyanides, phenols, acids, alkalies, propellants, and explosives.[41]

Knowledge about the health effects of these substances remains limited, but it is suspected that human exposure through drinking, skin absorption, or inhalation may cause cancer,birth defects, and chromosome damage or may seriously impair the function of a person's liver, kidneys, blood, and central nervous system. Having been dumping grounds for a lethal soup of hazardous materials for decades, military bases have become health time-bombs detonating in slow motion. Over time, the toxics migrate underground from their dump sites, tainting the soil and water of adjacent communities.[42]

The military is the largest generator of hazardous wastes in the United States and quite likely the world. In recent years, the Pentagon generated between 400,000 and 5000,000 tons of toxics annually, more than the top five U.S. chemical companies combined. Military purchases directly account for 13 to 16 percent of the final demand for chemicals in the United States. And these figures do not even include the large amounts of toxics spewing from the Department of Energy's nuclear weapons complex.[43]

Assessing the extent of toxic contamination is akin to opening Pandora's box: the number of U.S. sites on which problems have been spotted mushroomed from 3,526 on 529 military bases in 1986 to 17,482 on 1,855 installations in 1990. (See Table 3 for selected cases). In addition, more than 7,000 former military properties are being investigated, some 97 bases are so badly polluted that they are already on the Superfund National Priorities List, a number that could well double in coming years.[44]

Many sites owned or run by the Pentagon's contractors are similarly contaminated. According to a report by the National Toxics Campaign, the top 10 U.S. weapons contractors are listed 133 times by the EPA as "potentially responsible parties" at the nation's 100 most serious Superfund sites. Boeing, maker of missiles and aircraft, has dumped some 24 million gallons of toxic waste in two landfills in the Seattle area over the past 30 years.[45]

The extent of pollution at the Department of Defense's 375 bases outside the United States remains shrouded in considerable secrecy, although some evidence has emerged that suggests there are substantial problems (See Table 4). Publication of a 1986 and a 1990 General Accounting Office (GAO) study that identified significant contamination at U.S. bases in Italy, the United Kingdom, and West Germany was suppressed by the Pentagon and the State Department. U.S. military installations abroad are exempt from the 1970 U.S. National Environmental Policy Act. Although U.S. forces on foreign soil should in theory comply with pertinent host-nation laws, in practice they have been exempt, because the host governmental and the local population have no effective means of enforcement. But host governments are beginning to demand compliance with their environmental rules. As the U.S. military prepares to withdraw from some of its European bases, the question of who will shoulder the cleanup costs looms large. The U.S. Army has spent some moneyon cleanup, but plans to turn over vacated facilities in western Germany without a major rehabilitation effort, even though there are at least 358 polluted sites.[46]

As Soviet forces withdraw from Eastern Europe, it is becoming clear that the bases they occupied are in desperate need of cleanup. In addition, fuel, wastes, and unexploded ammunition were dumped in many unmarked off-base locations. In Vysoke Myto in central Bohemia, for example, groundwater tests reveal toxics in concentrations 30-50 times allowable levels. In Czechoslovakia, up to 8,000 square kilometers - 6 percent of the total territory - have been polluted or despoiled. In the former East Germany, some 700 Soviet installations are suspected to be contaminated, with at least 90 of them severely polluted; up to 10 percent of the territory sustained some kind of damage due to military activities. At Larz Air Base in Mecklenburg, for example, more than 50,000 tons of fuel leaked into the soil; at the Werneuchten airfield northeast of Berlin, 14 grams of kerosene per kilogram of soil were found. In Poland, the aquifer under Swinoujscie naval base near Szczecin has been polluted by fuel dumping; sewage from Chojno airfield flowed untreated into the Rurzyca river. The most severe cases of soil contamination in Hungary have been identified at air bases in Sarmellek, Tokol, and Kunamadaras, at a fuel depot at Petfurdo, and at the Huszar army post.[47]

Little is known about the toxics used in the arms industry and about workers' exposure to them. Military secrecy has hampered efforts to monitor and enforce safety at aerospace and other factories. Yet the manufacturer of explosives, composite materials, and electronic components can endanger human health. The production of explosives and propulsion systems, for example, entails potential exposure to suchhazardous emissions as chlorine gas, dibenzodioxins, and dibenzofurans. Chemicals used to bond composite materials like carbon fiber and glass fiber that make military aircraft lighter and stronger (including phenol formaldehyde and methylene di-aniline), are thought to cause cancer when inhaled or absorbed through the skin, while fiber fragments can damage the lungs and also cause cancer. In the manufacture of semiconductors, workers may be exposed to a multitude of substances that are suspected carcinogens, teratogen, or mutagens.[48]

### Nuclear Wasteland[49]

During the Cold War, staying ahead in the feverish nuclear arms race between the United states and the Soviet Union took precedence over the health and safety of those involved in producing the weapons. for decades, officials in the two countries knowingly subjected their unsuspecting citizens to the dangers of radioactivity in the name of national security. While the worst excesses belong to the past, the problems of nuclear waste will haunt future generations long after the last warhead is dismantled : plutonium, for example, has a half-life of 24,000 years.[50] Since the forties, the United States alone produced some 60,000 atomic warheads. Either stored or assembled in warheads are 90-100 tons of weapon-grade plutonium and 500 tons of highly enriched uranium. Stockpiles in the former Soviet Union are believed to be roughly of the same magnitude. Every step in the bomb-making process involves severe environmental threats. At the now-closed Purex plant at the Hanford Reservation in Washington, the production of a single kilogram of plutonium generates about 1,300 liters of liquid high-level radioactive waste laced with hazardous chemicals, more than 200,00 kilograms of low - to intermediate-level waste, and almost

10 million liters of contaminated cooling water. Military nuclear reactors are responsible foran estimated 97 percent of all high-level nuclear waste and 78 percent of all low-level nuclear waste (both by volume) in the United States. Measured in curies, however, the military portion comes to 6 and 74 percent, respectively. The military-related high-level waste inventory in the United States has been estimated at about 1.4 billion curies. (A curie measures the intensity of radiation and is equal to 3.7 x $10^{10}$ disintegrations per second; by comparison, about 50 million curies were released during the accident at Chernobyl.)[51]

By 1989, more than 3,200 sites in about 100 locations owned by the U.S. Department of Energy (DoE) had been identified as having tainted soil, groundwater, or both. (See Table 5). More than 50 Nagasaki-size bombs could be manufactured from the waste that has leaked just from Hanford's underground tanks. At Rocky flats in Colorado, enough plutonium has accumulated in ventilation ducts to make seven nuclear bombs. After a large 1969 fire there, investigators found the highest concentrations of plutonium ever measured near an urban area, including around Nagasaki.[52] Additional danger looms from wastes that have not escaped into the environment -- at least not yet. Some storage tanks at Savannah River and Hanford, holding plutonium by-products such as cesium, strontium, and iodine are apparently in danger of exploding, according to a government advisory pane. Should such a detonation occur at Savannah River, residents nearby could contract up to 20,000 additional cancers.[53]

Radiation is known to cause cancer, leukaemia, multiple myeloma, brain tumors, thyroid disorders, sterility, miscarriages, and birth defects. Damage to the human body depends on the size and type of the radiation does and on how fast it is absorbed. it is difficult to establish a causal link between a specific radiation exposure and averse healtheffects, but there is growing belief among scientists that no level of exposure is "safe". Recent studies conclude that the risks of ionizing radiation are three times higher than previously thought, and many scientists believe there is no "safe" level of radiation exposure.[54]

Some 300,000 people, or half of those who ever worked in the U.S. nuclear weapons complex, are believed to have been affected by exposure to radiation. The question of health effects on workers remains contentious, however, in part due to DOE's refusal to release sufficient data. Similarly, a February 1991 report by the Office of Technology Assessment notes the lack of sufficient data about off-site human exposure and concludes that "published reports and available data can neither demonstrate nor rule out the possibility that adverse public health impacts have occurred or will occur as a result of weapons site pollution."[55]

Mikhail Gorbachev's *Glasnost* gave the world first insights into the consequences of the Soviet nuclear weapons program. At Kyshtym in the eastern Urals (the Soviet counterpart of Hanford), perhaps over 6,000 workers were exposed to radiation of more than 100 rem--a dose that can produce cancer in the long run. Cesium, strontium, and other liquid radioactive wastes were dumped into the Techa river from the late 1940s to 1952. The river became so polluted that traces of radioactivity showed up in the Arctic Ocean, nearly 1,000 miles away. Those living along the Techa had to be evacuated. From 1952 on, nuclear waste was dumped into nearby Lake Karachay; the heat of the radionuclides began to dry out the 10-square-kilometer body of water. By 1988, it contained 120 million curies of strontium-90, cesium-137, residual plutonium, and other long-lived isotopes, two-and-a-half times more than was released at Chernobyl. The lake is now being covered by a thick layer ofconcrete to contain the radiation. In September 1957, a chemical explosion occurred at a high-level nuclear waste storage site at Kyshtym. The accident severely contaminated 15,000 square kilometers of land that was home to more than a quarter-million people. The explosion released about a third as much overall radiation as at Chernobyl and forced the evacuation of 10,000.[56]

Warhead testing is the final phase in the development of nuclear arms, but it was the activity that elicited the earliest health and environmental concerns. From 1945 to 1990, 1,908 bombs were exploded at more than 35 sites around the world. Roughly a quarter of all tests, most of them before 1963, were conducted in the atmosphere, injecting far more radioactive debris into the atmosphere than the Chernobyl accident did. The fallout may have caused as many as 86,000 birth defects worldwide and many eventually result in some 150,000 premature deaths. Although underground testing has cut down or radiation, some still escapes into the atmosphere (known as "venting") and is also suspected of leaching into groundwater.[57]

In the United States, individuals exposed to nuclear fallout include 400,000 "atomic veterans" (soldiers ordered to observe atmospheric testing), 100,000 test site workers, and a similar number of "downwinders" living in parts of Nevada, Arizona, and Utah. Yet millions more Americans are thought to bear trace amounts of plutonium in their tissues and organs. Between 1946 and 1958, the U.S. government conducted 66 atmospheric tests in the Marshall Islands. High levels of radioactivity in the soil and food crops have rendered Bikini Atoll in the Pacific uninhabitable since 1954. Many inhabitants of Rongelap -- downwind from Bikini -- have developed thyroid tumors. From the early 1980s on, increases in the number ofleukemia cases, brain tumors, and thyroid cancers have been registered in French Polynesia, the site of more than 160 French tests since 1966. In Australia, where Britain conducted nuclear tests from 1953 to 1963, 400 square kilometers of Aboriginal land has been contaminated. The Soviet Academy of Medical Sciences determined in 1989 that residents of Semipalatinsk, near the main test site in Kazakhstan, had experienced excess cancers, genetic diseases, and child mortality because of radiation exposure from pre-1963 atmospheric tests. A total of nearly 500,000 people were exposed to radiation. In 1988, the incidence of cancer was 70 percent above the national average. Kazakh activists claim that life expectancy in the republic has declined by four year over the past two decades, and that the number of people suffering from blood diseases has doubled since 1970. The Radiology Research Institute in Semipalatinsk has found that the 500,000 people living near the test site show incidence of cancer 40 percent higher, and leukemia 50 percent higher than in other regions of the Soviet Union.[58]

## Ecological Conversion

As the East-West confrontation fades away, and progress is made toward disarmament, a number of environmental challenges and opportunities present themselves in three key areas of conversion> 1) the disposal (and in some cases possible civilian reuse) of military equipment 2) the cleanup of military bases and land before they can be turned over to civilian uses; and 3) the conversion of arms production facilities.

## Military Equipment

Since Mikhail Gorbachev came to power in Moscow in 1985, a number of arms control and disarmament treaties have been concluded; the Soviet Union (and recently the United States, too) has also undertaken anumber of unilateral disarmament initiatives. Together, these measures mean that huge stockpiles of weaponry need to be disposed of. This disposal could take one of two forms: either the dismantling and destruction of weapons or the conversion and civilian reuse of military equipment (or the salvaging of parts and components thereof).[59]

Considering the enormous resources expended in producing these arms in the first place, even an effort to use some equipment for civilian purposes cannot recoup these investments. In any event, civilian reuse is likely to be difficult technically and marginal economically: most of the military hardware has no intrinsic civilian value. Nevertheless, such a reuse is being attempted, to a degree, by the former Soviet Union which needs to decommission large numbers of tanks, personnel carriers, and other equipment under the 1990 CFE Treaty. Already, several hundred SS-20 missile launchers prescribed by the 1987 INF Treaty have been transformed into self-propelled hoisting cranes.[60]

From a confidence-building point of view, destroying weapons is a better option than allowing for a civilian purpose configuration because the latter option cannot exclude altogether the possibility that converted equipment might later be refashioned again for military purposes. Arms reduction accords are likely to mandate the verifiable destruction of military equipment. The CFE Treaty, for example, offers four options for dealing with arsenals in excess of agreed deployment levels: destruction, withdrawal from the geographical area covered by the treaty (an option available only to the Soviet Union, the United States and Canada), conversion to civilian use, or foreign sales. The treaty permits both NATO and the former Warsaw Pact to each convert a relatively limited number of pieces of equipment: 750 tanks, 3,000 armored vehicles, many combattrainers, and most helicopters. (The Soviet Union needs to cut its arsenal by about 12,000 tanks and a similar number of armored vehicles).[61]

Yet the destruction of military hardware needs to proceed in an environmentally acceptable manner--an undertaking that no country or organization has yet much experience with. Of concern are the following areas: 1) solid-fuel missiles, 2) nuclear weapons-grade materials, 3) chemical warheads, and 4) stockpiles of ammunition and explosives.

### 1. Solid - Fuel Missiles

Under the terms of the 1987 INF Treaty, the United States and the Soviet Union had to destroy close to 2,700 rockets; another 2,000 long-range missiles will have to be disposed of under the START agreement, and unilateral reductions announced or offered in September/October 1991 and January 1992 consign large numbers of additional launchers to the scrap-heap. Yet, a careless demolition entails considerable air pollution. Missiles are, as *New York Times* science reporter William J. Broad put it, in essence "big tubes filled with exotic alloys, toxic chemicals and explosive fuels." Solid rocket fuel, when burned, can create clouds of dangerous chemicals and noxious fumes. U.S. Pershing missiles along, prescribed by the INF Treaty, contained almost 2,000 tons of solid propellants. The Pershings were burned on the ground, releasing aluminium oxide, titanium oxide, hydrochloric acid, and asbestos, and unleashing public protests in Utah and Colorado. The U.S. Air Force is now experimenting with a variety of ostensibly less polluting ways of disposing of solid rocket fuel.[62]

## 2. Nuclear Weapons-Grade Materials

Under both the INF and START treaties, plutonium and highly-enriched uranium from dismantled warheads was allowed to be "recycled", that is, used to produce new warheads. The uncertainty concerning future developments in the Commonwealth of Independent States--four of its members are in possession of strategic nuclear weapons -- has made control over nuclear warheads and materials an urgent issue. According to a recent report in *Komsomolskaya Pravda*, many of the older nuclear weapons in the former Soviet Union are in an reliable state; one weapons scientist believes that could soon be "hundreds of Chernobyls". Dismantling these weapons will likely pose a tremendous challenge: many of the relatively small number of specialists with intimate knowledge of the weapons design, faced with unemployment, are leaving to find work elsewhere, raising the specter that much of the know-how needed for dismantling may be lost.[63]

Together, the United States and the former Soviet Union have accumulated about 200 tons of plutonium and 1,000 tons of weapons-grade uranium 235. A substantial share of these stockpiles will be affected by recent disarmament measures. Short of releasing these fissile materials into the environment, there are four options of handling them: a) storage in their current form, which would permit their renewed military use at a later time; b) storage in diluted form, which would make such reuse more difficult; c) usage in submarine or other military reactors, which would call into question the demilitarization of these fuels; and d) usage in civilian nuclear power plants, which would consume the fuels but be objectionable on environmental and proliferation grounds. It appears that option b) is the most desirable one, provided that an acceptable international solution can be found to safeguard these materials.[64]

Long-term, safe storage of nuclear materials and wastes, however, remains an unresolved task in both the civilian andmilitary realms. For example, the U.S. Department of Energy's plans to open a repository for plutonium bomb wastes near Carlsbad, New Mexico experienced yet another setback in February 1992 when a federal judge ruled that the agency needed prior approval from Congress and the state of New Mexico. Much of the waste is currently stored in ways that are unsafe in the long run or perhaps even the short term.[65]

## 3. Chemical Warheads

The U.S. Congress passed legislation requiring the destruction of the U.S. chemical stockpile by 1997. The U.S. Army is planning to dispose of its stockpile of nerve and mustard gas by incinerating them in plants on the uninhabited Johnston Atoll in the South Pacific and at Tooele (Utah), and at smaller facilities to be built in seven other U.S. states. A 1984 study by the National Research Council concluded that this was the safest method available. Even under the best of circumstances, however, the incinerators would emit some toxins, and according to Greenpeace, "in a `worst-case-scenario' prepared in 1987, the Army projects 8,896 to 62,773 deaths, depending on the site". A large proportion of toxic byproducts of incineration are captured by filters which, however, must themselves be safely disposed of. Originally, the U.S. Army had planned to dump them in the Pacific; current plans call for their shipment to U.S. landfills. Greenpeace has called for placing the weapons in carefully monitored storage until a safer disposal method is developed. Yet, many of these weapons are aging and deteriorating: the longer they are stored, the more likely it is that leaks will occur, with grave environmental consequences.[66]

The Soviet Union signed a bilateral chemical arms destruction agreement with the United States in 1990, but intense public concern has, for the time being, derailed plans to incinerate 40,000 tons of chemical weapons. In September 1989, the Soviet government stopped construction of its only facility for that purpose, located near Chapayevsk on the Volga River. The Russian Federation (the only one of the successor states of the Soviet Union that possesses chemical warheads) has acknowledged that it has not made any progress on that question. Sergei Batsanov, head of the Russian delegation to the Geneva Conference on Disarmament, said in early 1992 that "Russia as of now has no project ready for destruction". He acknowledged that foreign "help might be important for us".[67]

**4. Ammunition**

Both the production and the destruction of ammunition constitutes a hazardous undertaking. The production of explosives entails potential exposure to emissions of chlorine gas, dibenzofurans, and dibenzodioxins. Particularly the countries of the former Warsaw Pact are now faced with the need to dispose of vast stockpiles. For example, the now-dissolved East German army had stocks 290,000 tons of ammunition, and the Soviet troops now withdrawing from the former East German territory are believed to have stored at least another 1 million tons (though some estimates speak of 2.5 million tons). Soviet stocks are generally too old and fragile to be transported, and must be destroyed on site. Experience with destroying ammunition--there is a bewildering variety of types, each of which requires different procedures of destruction--is very limited: in West Germany, no more than 1,000 to 2,000 tons of obsolete ammunitions were destroyed annually. Incineration generates highly toxic combustion gases which must be contained.[68]

The costs arising out of the physical destruction of arms--including verification measures and environmental safeguards--are substantial. Destroying chemical weapons, for example, is reckoned to cost more than their production. Estimates for the U.S. stockpile range from $ 4-10 billion, while General Vladimir Pikalov, the former chief of the Soviet chemical forces, has put potential Soviet expenses at up to $ 20 billion. The cost to develop appropriate technology and to build plants to destroy stocks of ammunition in the former East Germany are estimated at anywhere from 500 million Deutsche Marks (DM) to 3 billion DM ($ 300-1,800 million at current exchange rates).[69]

**Base Cleanup**

As the previous section demonstrated, many military facilities and militarily-used tracts of land are severely contaminated; before they can be seriously considered for civilian re-use, they will need to undergo a meticulous cleanup process. The cost of repairing the damage wrought will be staggering. For the United States, nuclear decontamination costs may run to $ 200 billion, or $ 3.3 million for every nuclear warhead the nation has ever produced. Coping with toxic wastes at U.S. domestic military bases could reach between $ 100 and $ 200 billion, according to estimates from the Department of Defense Inspector General's Office. Environmental damage caused by Soviet troops in Hungary and Czechoslovakia has been estimated at $810 million and $ 120 million, respectively. In Poland, the damage, including losses from the inability of using sites for commercial purposes, has been put at $ 4.8 billion. In the former East Germany, the rehabilitation of fuel-contaminated soils is estimated to cost 300 DM (180) per cubic meter. Cleaning up the Maralinga nuclear test site in Australia may cost anywhere from $ 75 million to $ 450 million.[70]

But the financial cost is only part of the story. The most severely poisoned areas may prove impossible to "clean up" or otherwise restore. Fenced-off and unsuitable for any use, they may become "national sacrifice zones", ghastly monuments to the Cold war.

Another enduring and perilous legacy of war preparation is found in the large tracts of land strewn with unexploded bombs and ammunition. For instance, the Jefferson Proving Ground's 90 square miles in Indiana have been saturated with 23 million artillery, tank, and mortar shells since 1941, 1.4 million of which failed to explode. It might cost as much as $ 5 billion to restore the area. But to make the land safe for civilian use, one would probably have to remove a 10-meter layer of soil.[71]

Not only military land proper is affected, however. In November 1989, for example, the Pentagon had to close off some 275 square kilometers of public land in Nevada after discovering 1,389 live bombs, 123,375 pounds of shrapnel, and 28,136 rounds of ammunition accidentally dropped outside an Air Force bombing range. It costs an estimated $ 250 million to rehabilitate a single square kilometer of strafed and bombed land. Robert Stone, Deputy Assistant Defense Secretary and the Pentagon's top land manager, acknowledged that land used for bombing practice often has to be closed to human use permanently because even an intensive effort will fail to locate all unexploded bombs.[72]

## Plant Conversion and the Requirements of a Sustainable Economy

Arms production facilities, R&D labs, and design bureaus need to be switched to civilian use. This is the task that has received by far the most attention in the conversion literature. Plant-based conversion involves the following components: (a) making an inventory of existing skills and equipment; (b) identifying alternative civilian products; (c) conducting engineering studies to determine the feasibility of producing these goods with present capabilities; (d) retraining employees or upgrading their skills; (e) refashioning machinery and production layouts where needed; (f) preparing marketing studies to ascertain whether alternative products selected can generate sufficient demand; (g) making new arrangements for raw material supplies and other inputs; and (h) erecting civilian marketing networks.

Arguments for a conversion of the military economy to civilian use focus on the adverse impact of military-led economic development on productivity and on the fact that military industry corporations tend to be incapable of functioning effectively in the commercial market place. Yet, in an era of unparalleled environmental crisis, this emphasis alone is misplaced. If conversion is not to simply result in reinforcing the civilian protectionist system, such a switchover should involve not just a crossing of the military/civilian threshold, but put much greater emphasis on ecological sustainability, i.e.

* boosting conservation and efficiency
* relying more on renewable energy
* extracting nominally renewable resources only to the degree that they can regenerate themselves
* minimizing waste and maximizing reuse and recycling
* avoiding the use of hazardous materials

* developing production technologies that are more environmentally benign and less hazardous to the health and safety of workers and
* designing products to be more durable and repairable.[73]

Could military contractors lead the way in moving toward sustainability? They frequently pride themselves at being at the cutting edge of technological development, and in response to recent military budget cuts a number of them have come forward with claims of spectacular potential spinoffs from their military research, including lasers and sensors that could be used for environmental, medical, or other purposes. Yet problems pervasive in the military economic--cost-maximizing and the kind of overdesign that Mary Kaldor at the University Sussex has labeled "baroque"--should inject a measure of caution into overblown expectations. There is also growing evidence that they are no longer a technological advantage. A number of studies by the U.S. Defense Science Board, for example, concluded that the Pentagon purchases electronics products that are often 5 to 10 years behind the state of the area and up to 8 times more expensive than civilian models.[74] Finally, Michael Closson, executive director of the Center for Economic Conversion in California sharply questions whether the "corporate culture" present in the arms industry can ever be amenable to environmentally-sensitive approaches"

> "The General Dynamics and Lockheeds of the world are authoritarian, hierarchial, patriarchal, secretive and not particularly innovative. They are hardly the foundation upon which one could build a sustainable peace economy. [...] The sort of projects that they are best suited to are multi-billion dollar manned space stations and voyages to Mars, exactly the type of activity we should not bepursuing at this point in time when we have so many Earth-bound problems to address."[75]

Still, there is no single, typical "arms producer." The companies making up the military industry belong to a wide variety of industrial sectors (aerospace, electronics, motor vehicles, shipbuilding, chemicals, machine tools, and others); they include huge conglomerates and small, specialized firms and ventures; some of them are highly innovative, while others are structurally or even technologically conservative; for some, military contracting is the mainstay or even exclusive focus, but for others it is a business of limited significance; some have a long history of building weapons, while others have only a fleeting involvement; some produce equipment that has only military applications, while others manufacture civilian-like products that may be built to military specifications or is indistinguishable from civilian goods.

In short, the ability of military contractors to convert-and to do so in an environmentally responsible manner-varies enormously. Thus, some companies may be well placed to respond in imaginative ways to the environmental challenge. A number of them (such as Grumman and Rockwell in the United States and Eurometal in the Netherlands) have in the past taken an interest in solar technology, for instance, though with mixed success. Some (such as the tank producer Krupp Mak in Germany) have expertise in building locomotives and other mass transit vehicles. Others (such as Diehl in Germany) are trying to apply their know-how in sensor and process technology to the needs of pollution control equipment.

Among the conversion groups initiated by arms industry workers, a number have focused specifically on environmental criteria in identifying possible alternative uses of theircompany's equipment and know-how (see Table 6). Alternative use committees as envisioned in conversion legislation drafted by Representative Ted Weiss in the U.S. Congress (composed of representatives from management, labor, and the local community) might indeed be an ideal vehicle to identify civilian alternatives to military contracting that meet environmental needs. With the help of environmental experts, such committees could refine and strengthen the conversion concept by developing guidelines for an environmentally sound conversion process and perhaps preparing "eco-conversion audits" that assess a company's specific options.

An ecological conversion process could be either supply-driven or demand-oriented. In the first case, the process would be focused on plant conversion, and the companies concerned would most likely seek to stay as close as possible to the production technologies and products that they are familiar with. The success of an environmentally-inspired conversion process would, as discussed above, hinge primarily on the extent to which military contractors' know-how is useful is applicable to solving environmental problems. The process would thus be shaped by the capabilities of military contractors and, in a sense, be a captive of their interests. The--perhaps preferable--alternative is a demand-oriented approach. It would channel government spending from military budgets to R&D and procurement program that help develop, and generate demand for, environmentally sound technologies and products. While such environmental markets would certainly be open to former military contractors, the conditions and parameters of success would not be (exclusively) defined by them.

**Conclusions**

This paper has presented the argument that significant detrimental environmental impacts are a virtually inevitable consequence of both warfare and war preparation, and documented the environmental legacy of the war system. Environmental considerations present both challenges and opportunities in the disarmament and conversion process.

The environmental legacy of the war system is most discernable in the wake of armed conflicts. The recent Gulf war provides tragic evidence. Though it has been one of the most ecologically damaging conflicts ever, it is only one in a long series of highly destructive wars. With some prominent exceptions, the environmental effects of producing arms and keeping armies in a state of readiness--the permanent state of war preparation--are not as dramatic as the flaming inferno in Kuwait that could be seen on television screens around the world. But they are no less serious; cumulatively, they may be even more serious. Even less noticeable--to the general public, anyway--are the indirect ways in which military research and spending priorities detract from solving environmental (and manifold other) problems.

With the end of the Cold war, the environmental challenges of disarmament are becoming evident. Stocks of ammunition and weaponry need to be disarmed and disposed of in the least objectionable manner environmentally. Arms factories and military bases need to undergo a meticulous, lengthy, and costly cleanup. New technologies need to be developed and tried out, and large sums of money budgeted. The challenges are tremendous, but so are the opportunities.

Rehabilitation efforts and an ecological conversion process--oriented toward the development of environmentally-sound products and production technologies--mean that immense alternative markets are emerging.

Coping with the environmental legacy of the Cold War is an unavoidable necessity. But if ecological considerations are taken seriously, there are implications not just for the past, but the future as well. Not only has the radical political transformation in the former Warsaw Pact countries eclipsed military power as a prime determining factor in the East-West relationship, but there is a also a nascent discussion about the need to conceptualize "security" in much broader terms than traditional definitions suggest. Central to these new concepts is the recognition that security can increasingly be obtained only globally, not on an exclusively national basis. Such a redefinition of security implies that military power is less appropriate and acceptable as a means of settling conflicts.

**Table 1. Trade-offs Between Military and Environmental Priorities**

| Military Priority | Cost | Environmental Priority |
|---|---|---|
| Trident II submarine F-16 jet fighter programs | $ 100,000,000,000 | Estimated clean-up cost for the 3,000 worst hazardous waste dumps in the United States |
| Stealth bomber program | $ 65,000,000,000 | Two thirds of est. costs to meet U.S. clean water goals by 2000 |
| 4 days of global military spending | $ 8,000,000,000 | Action Plan over 5 years to save the world's tropical forests |
| Development cost for Midgetman ICBM | $6,000,000,000 | Annual cost to cut sulfur dioxide emissions by 8-12 million tons/year in the U.S. to combat acid rain |
| 2 days of global military spending | $ 4,800,000,000 | Annual cost of proposed U.N Action Plan to halt Third World desertification, over 20 years |
| 6 months of U.S. outlays for nuclear warheads, FY 1986 | $ 4,000,000,000 | U.S. government outlays for energy efficiency, FY 1980-87 |
| 10 days of European Community military spending | $2,000,000,000 | Annual cost of clean up hazardous waste sites in 10 European Community countries by the year 2000 |
| 3 B-1B bombers | $ 680,000,000 | U.S. government spending on renewable energy, Fy 1983-85 |
| 2 months of Ethiopian military spending | $ 50,000,000 | Annual cost of proposed U.N. Antidesertification Plan for Ethiopia |

*Source:* Michael Renner, *National Security: The Economic and Environmental Dimensions*, Worldwatch Paper 89 (Washington, D.C.: Worldwatch Institute, May (1989), Table 8.

**Table 2. Use of Petroleum Products by the Armed Forces, United States and the Soviet Union, 1987**

| Petroleum Product | United States | | Soviet Union | |
|---|---|---|---|---|
| | Military Use (Million Tons) | Share of Total Domestic Use (Percent) | Military Use (Million Tons) | Share of Total Domestic Use (Percent) |
| Jet Fuel | 18.6 | 26.9 | 11.8 | 33.9 |
| Diesel | 4.4 | 3.0 | 3.8 | 4.9 |
| Distillate | 0.5 | 0.9 | n.a. | n.a. |
| Residual Fuel | 0.3 | 0.1 | 1.4 | 1.7 |
| oil | 0.7 | 0.4 | 0.3 | 0.4 |
| Gasoline | | | | 3.9 |
| Other[1] | 28.1[2] | 3.4 | 17.4[3] | |
| Total, Military | | | | |

1 Includes lubricants, greases, refined products, and refinery fuel.
2 Includes 3.6 million tons purchased for stockpiling. At least one third of the total is used on U.S. bases abroad.
3 Total does not add up due to rounding.

*Source*: Author's calculations, based on Tom Cutler, "Myths of Military Oil Supply Vulnerability, *Armed Forces Journal International*, July 1989.

**Table 3. United States: Selected Military Hazardous Waste Sites**

| Location | Observation |
|---|---|
| Otis Air Force Base, Mass. | Ground Water contaminated with trichloroethylene (TCE), a known carcinogen, and other toxins. In adjacent towns, lung cancer and leukemia rates 80 percent above state average. |
| Picatinny Arsenal, N.J. | Ground water at the site shows TCE levels at 5,000 times Environmental Protection Agency (EPA) standards; polluted with lead, cadmium, polychlorinated biphenyls (used in radar installations and to insulate electrical equipment), phenols, furans, chromium, selenium, toluene, and cyanide. Region's major aquifer contaminated. |
| Aberdeen Proving Groujd, Md. | Water pollution could threaten a national wildlife refuge and habitats critical to endangered species. |
| Norfolk Naval Shipyard, Va. | High levels of copper, zinc and chromium discharges. Contamination of Elizabeth River, and of Willoughby and Chesapeake bays. |
| Tinker Air Force Base, Okla. | Concentrations of tetrachloroethylene and methylene chloride in drinking water at levels far exceeding EPA limits. TCE concentration highest ever recorded in U.S. surface waters. |
| Rocky Mountain Arsenal, Colo. | 125 chemicals dumped over 30 years of nerve gas and pesticide production. The largest of all seriously contaminated sites, called "the most contaminated square mile on earth" by the Army Corps of Engineers. |
| Hill Air Force Base, Utah | Heavy on-base groundwater contamination, including volatile organic compounds up to 27,000 parts per billion (ppb); TCE upto 1.7 million ppb; chromium up to 1,900 ppb; lead up to 3,000 ppb. |
| McClellan Air Force Base, Calif. | Unacceptable levels of TCE, arsenic, barium, cadmium, chromium, and lead found in municipal well system serving 23,000 people |
| McChord Air Force Base, Wash. | Benzene, a carcinogen, found on-base in concentrations as high as 503 ppb, nearly 1,000 times the state's limit of 0.6 ppb. |

*Source:* Michael Renner, "Assessing the Military's War on the Environment", in Lester R.Brown *et al.*, *State of the World 1991* (New York: W.W.Norton & Co., 1991), Table 8-6.

**Table 4. Toxic Waste Problems at Selected U.S. Military Bases in West Germany**

| Location (by State) | Observation |
|---|---|
| Hessen: Rhein-Main Air Base, Frankfurt | Massive and fast-spreading plume of jet fuel contaminating the subsurface water table. Decontamination is expected to take at least 5 years. |
| Babenhausen Barracks, Darmstadt | Chlorinated hydrocarbons from motor vehicle maintenance shops have polluted soil and ground water. Cleanup was to begin in 1991, at $ 1.2 million cost. |
| Rheinland-Pflaz: Bitburg Air Base | Kallenbach river and feeder streams contaminated by organic waste, solvents, etc., due to inadequate sewage treatment and leaking underground fuel system. New treatment plant completed, in 1989 after protests by West German authorities. urea[1] was found at a concentration of 163 parts per million (ppm). Concentrations of 1 ppm are toxic to fish, which have been killed of massively in the Kallenbach. |
| Gonzenheim Army Depot, Mainz | Considered one of the U.S. Army's most serious pollution problems, though the full extent of ground water contamination (solvents and other wastes) remains to be determined. Est. costs: $10 million. |
| Marshall and Rose Barracks, Bad Kreuznach | A dry-cleaning plant leaked carcinogenic chemicals into drinking water supplies of this spa town for 25 years. Cleanup began in 1987. Gasoline, oil, and lubricants from a tank farm and from motor vehicle maintenance shops have polluted soil and ground water. Total estimated clean up costs: $ 7.1 million. |
| Rhine Ordinance Barracks and Army Depot, Kaiserslautern | Soil and ground water have been contaminated by improper use and storage of gasoline, oil, lubricants and chlorinated hydrocarbons. Cleanup costs estimated at $ 2.5 million. |
| Baden-Württemberg: Taylor Barracks, Manheim | The most polluted U.S. Army site in West Germany. TCE and other chlorinated hydrocarbons have polluted a nearby protected water area. Cleanup under way since 1985, is estimated to cost $ 10 million. |
| Germersheim Army Depot | A leaking heating oil tank spilled 1 million liters of heavy oil, polluting an aquifer underneath the base and adjacent land. $ 9 million has been spent to remove contaminated soil and restore the aquifer (total costs are estimated at $ 16 million). |
| Cooke Barracks, Göppingen | Soil contaminated with gasoline, oil and lubricants. Estimated cleaup costs: $ 1.2 million. |

| | |
|---|---|
| Neurut Tank Farm, Karlsruhe | Soil contaminated with gasoline, oil, and lubricants. Estimated cleanup costs: $ 1.6 million. |
| Tompkins Barracks, Schwetzingen | Soil and groundwater were contaminated by a motor vehicle depot operated in the middle of a protected water area. Estimated cost: $ 2 million. |
| Bayern: Grafenwöhr | Discharge from a military waste dump has contaminated soil and ground water. The ground at a firing range is contaminated with lead from ammunition residue. Some cleanup work has begun. Total costs estimated at $ 17.4 million. |
| Hohenfels | Pollution from military waste dump. Estimated clean up costs: $ 8 million. |
| Johnson Barracks, Nürnberg | Chemical cleaning carried out until 1980 contaminated Fürth drinking water supplies with chlorinated hydrocarbons. Estimated costs: $2.1 million. |
| Faulenberg Barracks, Würzburg | Leaching from military dump has reached the ground water. Chlorinated hydrocarbons contaminated a Würzburg drinking water well. Total cleanup costs estimated at $ 2.5 million. |

1 Urea is used in de-icing compounds applied on military airfield runways. Urea laden runoff robs rivers and lakes of oxygen and can lead to severe nitrate pollution of ground water.

*Sources:* John M.Broder, "Pollution `Hot Spots' Taint water Sources", *Los Angeles Times*, June 18, 1990; "U.S. Army Faulted for Environmental Damage", *JPRS-TEN-91-002*, West Europe, January 29, 1991, pp.81-82; Petra Lösch, "When the Sun Shines, We Have War in the Skies", *Earth Island Journal*, Summer 1989.

**Table 5. United States: Radioactive and Toxic Contamination at Major Nuclear Weapons Production Facilities**

| Facility (Task) | Observation |
|---|---|
| Feed Materials Production Center, Fernald, Ohio (converts uranium into metal ingots) | Since plant's opening, at least 250 tons of uranium oxide (and perhaps six times as much) released into the air. Off-site surface and groundwater contaminated with uranium, cesium, thorium. High levels of radon gas emitted. |
| Hanford Reservation, Wash. (Recycles uranium and extracts plutonium) | Since 1944, 760 billion liters of contaminated water (enough to create a 12-meter-deep lake the size of Manhattan) have entered ground water and Columbia River; 4.5 million liters of high-level radioactive waste leaked from underground tanks. Officials knowingly and sometimes deliberately exposed the public to large amounts of airborne radiation from 1943 to 1956. |
| Savannah River, S.C. (Produces plutonium and tritium) | Radioactive substances and chemicals found in the Tuscaloosa aquifer at levels 400 times greater than government considers safe. Released millions of curies of tritium gas into atmosphere since 1954. |
| Rocky Flats, Colo. (assembles plutonium triggers) | Since 1952, 200 fires have contaminate the Denver region with unknown amount of plutonium. Strontium, cesium, and cancer-causing chemicals leaked into underground water. |
| Oak Ridge Reservation, Tenn.. (Produces lithium -deuteride and highly enriched uranium). | Since 1943, thousands of pounds of uranium emitted into atmosphere. Radioactive and hazardous wastes have severely polluted local streams flowing into the Clinch River. Watts Bar Reservoir, a recreational lake, is contaminated with at least 175, 000 tons of mercury and cesium. |

*Source*: "Status of Major Nuclear Weapons Production Facilities: 1990", *PSR Monitor*, September 1990; Robert Alvarez and Arjun Makhijani, "Hidden Legacy of the Arms Race: Radioactive Waste", *Technology Review*, August/September 1989.

**Table 6. Environmental - Related Conversion Proposals by Selected Workers' Alternative Product Working Groups, United States, Britain and West Germany**

| Location, Year started | Alternative Product Proposals/Management Response |
|---|---|
| United States: Unysis, St.Paul/Minneapolis, Minnesota, 1989 | IBEW local conducted skill audit and gathered over 40 alternative product ideas, focussing on pollution control equipment, energy-efficiency technologies, and water-conserving irrigation systems. Management has refused to cooperate. |
| Britain: Vickers Shipyard, Barrow Alternative Employment (BAEC), 1984 | 1987 report `Oceans of Work' assessed alternatives to Trident submarine production in marine technology and renewable energy. |
| West Germany : Blohm & Voss Shipyard, Hamburg, November 1980 | Water and waste treatment, desalinization, biogas, wind power for energy generation and shipping, and energy-saving technologies. The working group's activities put pressure on the company to engage in (relatively successful) diversification measures. |
| Krupp MaK (Machine Tools), Kiel, January 1981 | Innovative rail systems, co-generation, wind power. Management mostly uncooperative, but pursued some ideas originally proposed by the working group, including a windenergy plant on the North Sea island of Helgoland. |
| MBB (Aerospace) Bremen, March 1982 | Some 62 alternative product ideas, including solar collectors, biogas, wind energy, energy storage and transport systems, and high-tech zeppelins (for surveillance of North Sea Environment). Management showed interest in pursuing wind energy market. |
| MBB, Augsburg, 1988 | Worker's group was catalyst in bringing about a project in partnership between MBB and the city government to explore ways in which MBB's know-how can be applied to solving environmental problems. Initial focus on boosting energy efficiency and reducing water pollution in Augsburg's textile industry; later, technologies for an environmentally more benign urban transportation system will be explored. |

*Source:* Michael Renner *Swords Into Plowshares: Converting to a Peace Economy*, Worldwatch Paper 96 (Washington, D.C.: Worldwatch Institute, June 1990). Tables 4 and 5.

## Endnotes

1. A major milestone in the sustainable development discussion was reached with the publication of what is widely referred to as the Brundtland report: see The World Commission on Environment and Development (WCED), *Our Common Future* (New York and Oxford: Oxford University Press, 1987).

2. This section draws heavily on my earlier work. See Michael Renner, *National Security: The Economic and Environmental Dimensions*, Worldwatch Paper 89 (Washington, D.C.: Worldwatch Institute, May 1989).

3. "Environment and Conflict", Earthscan Briefing Document 40, International Institute for Environment and Development, London, November 1984.

4. Martin Wright, "Mixed Blessings of the Flooding in Sudan", *New Scientist*, September 22, 1988; "Bangladesh Blames Neighbors for its Floods", *Panoscope*, December 1987.

5. Pesticides from Michael Weisskopf, "Toxic Clouds' Can Carry Pollutants Far and Wide", *Washington Post*, March 16, 1988. Sulfur dioxide from Hilary F. French, *Clearing the Air" A Global Agenda*, Worldwatch paper 94 (Washington, D.C.: Worldwatch Institute, January 1990), and from Tomasz Zylica, "European Airborne Pollution: Economic, Social and Political Aspects of a Possible Reduction Program", Warsaw University, unpublished manuscript, 1986. Norwegian foreign minister from Norman Myers, *Not Far Afield: U.S. Interests and the Global Environment* (Washington, D.C.: World Resources Institute, June 1987).

6. For most recent findings, see William K. Stevens, "Summertime Harm to ozone Detected Over Broader Area", *New York Times*, October 23, 1991, and Warren E. Leary, "Ozone-Harming Agents Reach a record", *New York Times*, february 4, 1992.

7. For a review of the major issues with regard to climate change, see Christopher Flavin, *Slowing Global Warming: A Worldwide Strategy*, Worldwatch Paper 91 (Washington, D.C.: Worldwatch Institute, October 1989).

8. Lester R. Brown and Edward C.Wolf, "Reclaiming the Future", in Lester R. Brown et al., *State of the World 1988* (New York: W.W. Norton, 1988).

9. U.S. military spending from Executive Office of the President, Office of Management and Budget, *Historical Tables: Budget of the U.S. Government, Fiscal year 1992* (Washington, D.C.: U.S. Government Printing Office, 1991); U.S. pollution control spending from Management Information Services Inc., "PABCO Expenditure", Washington D.C., May 1990, and from Gary Rutledge and Nikolaos A. Stergioulas, "Plant and Equipment Expenditures by Business for Pollution Abatement, 1987 and Planned 1988," *Survey of Current Business,* November 1998. West European military spending from U.S. Arms Control and Disarmament Agency, *World Military Expenditures and Arms Transfers 1989* (Washington D.C.: U.S. Government Printing Office, October 1990). West European pollution control expenditures from Mark Magnier, "Market Abroad for Pollution Control Beckons", *Journal of Commerce*, December 8, 1989, and from Michael E.Porter, "America's Green Strategy", *Scientific American*, April 1991.

10. Comparison of military and selected civilian R&D from WCED, *Our Common Future*. An estimated 27 percent of all U.S. Scientists and more than 30 percent of British Scientists are involved in military projects. See Michael Dee Oden, *A Military Dollar Really is different: The Economic Impacts of Military Spending Reconsidered* (Lansing, Mic.: Employment Research Associates, 1988), and Mary Kaldor et al., "Industrial Competitiveness and Britain's Defense"*Lloyds Bank Review*, October 1986. Share of governmental R&D in the United States and the Soviet Union devoted to military purposes during 1980s from, respectively, Renner, *National Security: The Economic and Environmental Dimensions*, and from Alexei Izyumov, "The National Experience of the USSR",

paper presented at United Nations Conference on Conversion: Economic Adjustments in an Era of Arms Reduction", Moscow, August 13-17, 1990 Other countries from Organisation for Economic Co-operation and Development, *Main Science and Technology Indicators 1981-1987* (paris, 1988), Table 39. 1992/1993 U.S.spending from Andrew Polack, "Businesses Want Broader measures for Research", *New York Times*, January 31, 1992. U.S. spending since 1955 from William J. Broad, "Defining the New Plowshares Those Old Swords Will make", *New York Times*, February 5, 1992.

11. This section draws heavily on my earlier work: see Michael g. Renner, "Military Victory, Ecological Defeat", *World Watch*, July/August 1991, pp.27-3.

12. The subject of the environmental impact of warfare has been treated in the following publications, among others: Arthur Westing, *Ecological Consequences of the Second Indochina War* (Stockholm: Almqvist & Wiksell, 1976); Arthur Westing, *Weapons of Mass Destruction and the Environment* (London: Taylor and Francis, 1977); Arthur Westing, Warfar in a fragile World (London: Taylor and Francis, 1980); Ralph Ostermann, "Umwelt also Waffe-Das Zerstörungaspotential okologischer Kriegsfuhrung", in Arnim Bechmann (ed.), *Umwelt Braucht Frieden* (Fischer Verlag: Frankfurt, 1983); Arthur Westing (ed.), *Herbicides in War: The Long-Term Ecological and Human Consequences* (London : Taylor and Francis, 1984); Arthur Westing (ed.), *Environmental Warfare A Technical, Legal and Policy Appraisal* (London: Taylor and Francis, 1984); The Environmental Project on Central America, *Militarization: The Environmental Impact,* DPOCA/Green Paper No.3, San Francisco, 1990; Arthur Westing (ed.) *Environmental Hazards of War: Releasing Dangerous Forces in an Industrialized World* (London: Sage Publications, 1990).

13. "Desert Storm Still Rages", *Friends of the Earth*, September 1991, p.3; Youssef M. Ibrahim, "Most Oil Fires Are Out in Kuwait, But Its Environment Is Devastasted", *New York Times*, October 19, 1991; Matthew L. Wald, "Amid Ceremony and Ingenuity, Kuwait's Oil-Well Fires Are Declared Out", *New York Times*, November 7, 1991.

14. Andre Carothers, "Oceans of Oil", *Greenpeace Magazine*, October-December 1991, p.17; Ibrahim, "Most Oil Fires Are Out in Kuwait, Buts Its Environment Is Devastasted"; Wald, "Amid Ceremony and Ingenuity, Kuwait's Oil-Well Fires Are Declared Out; Jennifer Parmelee, "Environmentalists Survey the Blackened Wasteland that Was Kuwait", *Washington Post*, December 20, 1991.

15. Matthew L. Wald, "High Levels of Toxic Substances Found in Kuwait Oil Fire Smoke", *New York Times*, July 16, 1991; Michael Weisskopf, "Oil Fire Pollution Assessed", *Washington Post*, April 4, 1991; Marlise Simmons, "British Study Disputes Lengthy Climatic Role for Kuwait Oil Fires", *New York Times*, April 16, 1991.

16. Michelle Hoffman, "Taking Stock of Saddam's Fiery Legacy in Kuwait", *Research news*< August 30, 1991, p.971.

17. Peter V. Hobbs, "Summary of the University of Washington's Preliminary Findings on the Effects of the Smoke from the Kuwait Oil Fires on the Atmosphere", Atmospheric Sciences Department, University of Washington, Seattle, June 21, 1991, p.6; Tom Wicker, "Kuwait Still Burns", *New York Times*, July 28, 1991; Parmelee, "Environmentalists Survey the Blackened Wasteland that Was Kuwait".

18. "U.N.Virtually out of Money to Pay to Combat Pollution from Gulf Oil Fires", *International Environment Reporter*, July 17, 1991, p.401; Peter Hadfield, "Soot from Burning Wells `Will Not Upset Climate' ... but Falls on Japan", *New Scientist*, July 6, 1991, p.16; Marlise Simmons, "Beyond Kuwait, Black Rain and Acid-Filled Clouds", *New York Times*, March 16, 1991; John Horgan, "Burning Questions", *Scientific American*, July 1991, p.17.

19. John Travis, "As the Gulf Region Chokes, the Global Climate Breathes Easier", *Science News*, Vol.140 (1991), p.25; "Black Smoke from About 3 Million Barrels per Day of Burning Kuwait Oil Likely to Fall to Earth Quickly, Scientists Say", U.S. National Science Foundation, News Release, June 24, 1991; "NCAR Scientists Release preliminary Resultsfrom Kuwait Oil Fires Study", National Center for Atmospheric Research, Boulder, Colorado, News Release, June 24, 1991; Stephen H. Schneider, "Smoke Alarm", *World Monitor*, March 1991, pp.50-51.

20. U.S. Environmental Protection Agency, Gulf Task Force, "Summary of the Gulf Task Force/Non-Governmental Organization Meeting to Discuss U.S. Activities to Address Oil Well Fires in Kuwait", Washington, D.C., October 1991.

21. Louise Kehoe and David Thomas, "Difficulties with Oil in Narrow Waters", *Financial Times*, January 38, 1991; William Booth, "War's Oil Spill Still Sullies Gulf Shore", *Washington Post*, April 6, 1991.

22. Thomas Y. Canby, "After the Storm", *National Geographic*, August 1991, p.28; Ken Wells, "The Battles Are Over, But Gulf Environment Still Fights for Its Life", *Wall Street Journal*, October 15, 1991.

23. Charles Sheppard and Andrew Price, "Will Marine Life Survive the Gulf War?", *New Scientist*, March 9, 1991, pp.38-39; John Horgan, "The Muddled Cleanup in the Persian Gulf", *Scientific American*, October 1991, p.107; Wells, "The Battles Are Over, But Gulf Environment Still Fights for its Life"; Robert D. McFadden, "Oil Threatens Fishing and Water Supply", *New York Times*, January 26, 1991; Matthew L. Wald, "Gulf Oil Spill Vexing Cleanup Efforts", *New York Times*, April 7, 1991. Thaqfi quoted by Parmelee, "Environmentalists Survey the Blackened Wasteland that Was Kuwait".

24. J.L. Cloudsley-Thompson, "The Destructive Effects of Warfare on the Desert Environment", *Environmental Awareness*, Vol.13, No.2, (1990), pp.43-48.

25. Gar Smith, "Nuremberg Trials for Eco-Terrorism?", *Earth Island Journal*, Spring 1991, p.50; Parmelee, "Environmentalists Survey the Blackened Wasteland that Was Kuwait".

26. Center for Disarmament, *The Relationship Between Disarmament and Development*, Disarmament Study Series No.5 (New York: United Nations, 1982).

27. Pentagon-controlled and - leased land from U.S. Department of Defense (DOD), *Our Nation's Defense and the Environment. A Department of Defense Initiative* (Washington, D.C.: 1990), and from Edward McGlinn, "The Military Land Grab", *The Riverwatch* (Anglers of the Au Sable River, Grayling, Mich), Winter 1990. Energy Department land from Thomas B.Cochran et al., *Nuclear Weapons Databook, Vol.II: U.S. Nuclear Warshed Production* (Cambridge, Mass.: Ballinger, 1987). Philippines from Joseph Collins, "Cory's Broken Promise", *The Nation,* November 14, 1987.

28. Alexei Izyumov, "Conversion in the Soviet Union - and Possibilities for Cooperation in the Baltic Region", *Bulletin of Peace Proposals*, Vol.22, No.3 (September 1991), p.274. Kazakhstan from Paul Quinn-Judge, "Soviet Writers Blast Nuclear Testing", *Christian Science Monitor*, March 14, 1989.

29. Netherlands from Paul J.M.Vertegaal, "Environmental Impact of Dutch Military Activities", *Environmental Conservation*, Spring 1989. British figure includes about 1,000 square kilometers used for agricultural purposes. See Secretary of State for Defence, *Statement on the Defence Estimates 1990*, Volume 2 (London: HMSO, April 1990), Table 5.12. West Germany from Olaf Achilles, "Der preis der Freiheit", in Olaf Achilles, ed. *nature Ohne Frieden* (Munich: Kanur, 1988), and from "Environmental Damage from Military forces Detailed", *JPRS-TEN-91-001*, June 17, 1991, p.107. East Germany from Frank Marczinek, "Conversion of the Armed Forces in the GDR", paper presented at

United Nations Conference on Economic Adjustments in an Era of Arms Reduction, Moscow, August 13-17, 1990, from "Alles Zerwuhlt and Kaputt", *Der Spiegel*, October 1, 1990, and from "Die US-amerikanischen und sowjetischen Streitkrafte hinterlassen in Deutschland ein grobes Altastenproblem", *Okologische Briefe*, No.11, March 13, 1991, p.13. poland from Bogdan Turek, "Soviet Troops are Devastating Polish Environment", UPI wire story, February 27, 1991. A recent Polish newspaper report puts the land occupied by Soviet troops at 700 square kilometers, or 1 percent of Polish territory. See "Environmental Damage Caused bySoviet Forces Assessed", *JPRS-TEN-91-014*, July 90, 1991 p.45.

30. Global estimates are from: Wolfgang Schwegler-Rohmeis, "Rustungskonversionals Scherheitspolitik", in Marcus Breitschwartdt, ed. *Rustunqskonversion. Facetten einer Strukturfrage* (Stuttgart: Wissenschaftsedition SPD Baden Wurttemberg, 1988); Samuel S. Kim, *The Quest for a Just World Order* (Boulder, Colo.: Westview Press, 1984); Nicolai N. Smirnov, "The Impact of Conventional War on Natural Areas of the USSR", *Environmental Conservation*, Winter 1989.

31. Bruno Jerlitschka, "Umweltzerstorung durch Truppenstandorte und Manover", in Arnim Bechmann, ed., *Umwelt Braucht Frieden* (Frankfurt: Fischer Verlag, 1983); Olaf Achilles, "Bodenbelastung und Flachenverbrauch druch Militar", *Forum Wissenschaft*, No.1, 1989; Peter Grier, "Defense Contractors Go for `Green' Look", *Christian Science Monitor*, May 22, 1990. West German maneuver space from Schwegler-Rohmeis, "Rüstungskonversion als Sicherheitspolitik".

32. Share of petroleum products from Gunar Seitz, "Ressourcenvergeudung durch Rustung", in Bechmann, *Umwelt Braucht Frieden* global share of military jet fuel use from Olaf Achilles, "Militär, Rustung und Klima", MOP Studie VII, Arbeitsund Forschungsselle Militär, Okologie und Planung, Bonn, West Germany, June 1990; U.S. and Soviet shares from Tom Cutler, "Myths of Military Oil Supply Vulnerability", *Armed Forces Journal International*, July 1989; West German figure from Schwegler-Rohmeis, "Rüstungskonversionals Scherheit-spolitik".

33. DoD share of total U.S. Oil and energy consumption calculated from Energy Information Administration, Department of Energy (EIA/DOE), *Annual energy Review 1988* (Washington, D.C.: U.S. Government Printing office, 1989); comparison with urban mass transit it author's calculation based on American public Transit Association, *1989 Transit Fact Book* (Washington, D.C. 1989).

34. The tri-annual *Manufacturing Energy Consumption Survey* compiled by theU.S. Energy Information Administration (EIA/Department of Energy) provides data on the two-digit SIC code level, with more desegregated information (on the fourdigit level) only for four industry sub-groups that are not of prime importance for the military industry. The two-digit level is too aggregated for a meaningful comparison that would allow the preparation of an estimate of total energy consumption in arms production. The EIA is planning, however, to conduct a more detailed consumption survey for the next edition, to be published in 1994.

35. 1971 figure from Helge Hveem, "Militarization of Nature: Conflict and Control over Strategic Resources and Some Implications for Peace Policies", *Journal of peace Research*, Vol.16, No.1, 1979; 1989 figure is author's estimate, based on data in Stockholm International peace Research Institute (SIPRI), *SIPRI Yearbook 1990: World Armament and Disarmament* (Oxford: Oxford University Press, 1990), Tables 5.4 and 7.1., and in Paul Quigley, "Arms Exports: The Stop-Gap Alternative to Pentagon Contracts?", *Bulletin of Peace proposals*, Vol.19, No.1, 1988. Australian energy consumption 90.8 mtoe in 1988 - from British Petroleum Co.(BP), *BP Statistical Review of World Energy* (London, July 1990).

36. Estimates for armed forces' energy use from Centre for Disarmament, *The Relationship Between Disarmament and Development*; Achilles, "Militar, Rustung

und Klima", Comparison with Japanese oil consumption from *BP Statistical Review of World Energy.*

37. Global air pollution estimate from Seitz, "Ressourcenvergeudung durch Rüstung". West German jets' pollution is author's estimate based on date provided in Achilles, "Militar, Rustung und Klima".

38. U.S. military carbon emission is author's estimate, based on EIA/DOE, *Annual Energy Review 1988*, on Gregg marland et al., *Estimates of $Co_2$ Emissions from Fossil fuel Burning and Cement Manufacturing, Based on the United Nations Energy Statistics and the U.S. Bureau of Mines Cement Manufacturing Data* (Oak Ridge, Tenn.: Oak Ridge National Laboratory, 1989), and on BP, *BP Statistical Review.*

39. Casey Bukro, "Military Faces Difficult Task in Ending `War' on Environment", *Journal of Commerce*, March 22, 1989; depletion potential of CFC-113 and Halon 1211 from Cynthia Pollock Shea,*Protecting Life on Earth: Steps to Save the Ozone Layer*, Worldwatch paper 87 (Washington, D.C.: World-Watch Institute, December 1988). Rocket and shuttle contribution to ozone depletion from Lenny Siegel, "No Free Launch", *Mother Jones*, September/October 1990, from Steven Aftergood, "Environmental Impacts of Solid Rocket Propellants", *FAS Public Interest Report,* September/October 1991, p.6, from Steven Aftergood, "Poisoned Plumes", *New Scientist*, September 7, 1991, pp. 34-38, and from Warren E. Leary, "Study Calls for Effort to Limit Rockets' Pollution", *New York Times*, October 18, 1991; William J. Broad, "Some Say the Rockets' Red Glare is Eating Away at the ozone layer", *New York Time*, May 14, 1991.

40. This section draws on my earlier work: Michael Renner, "Assessing the Military's War on the Environment", in Lester R. Brown et al., *State of the World 1991* (New York: W.W. Norton and Co., 1991), pp. 141-145, and Michael G. Renner, "War on Nature", *World Watch*, May/June 1991, pp.18-21.

41. Heavy metals include beryllium, cadmium, chromium, mercury, lead, copper, zinc, iron, and others. They attack and weaken the cellular tissues of most living things. Solvents include tri- and di-chloro ethylenes, ethanes and ethenes, carbon tetrachloride, methyl chloride, and trichlorofluoromethane (Freon). *Uncle Sam's hidden Poisons*, reprint from a *Sacramento Bee* series of articles published September 30-October 5, 1984.

42. Ibid.

43. Waste generation data for recent years from Lenny Siegel, "The Growing Nightmare of Military Toxics", *Nuclear Times*, Spring 1990; "Turning the Wastes of War into a War on Waste", *Toxic Times*, Summer 1990; Will Collette, *Dealing with Military Toxics* (Arlington, Va.: Citizen's Clearinghouse for Hazardous Wastes, Inc., 1987); Emissions of top five chemical firms in 1988 from John Holusha, "Ed Woolard Walks Du Pont's Tightrope", *New York Times*, October 14, 1990. Military share of final demand for chemicals from Lenny Siegel et al., *The U.S. Military'sToxic Legacy: America's Worst Environmental Enemy* (Boston: The National Toxic Campaign Fund, January 1991), p.10.

44. Number of sites and bases and Superfund listing from DOD, *Defense Environmental Restoration Program*, and from Michael Weisskopf, "3,000 Military Base Sites Added to Toxic Cleanup List", *Washington Post,* March 29, 1991; former military sites and slow cleanup pace from Seth Shulman, "Toxic Travels: Inside the Military's Environmental Nightmare", *Nuclear Times*, Autumn 1990.

45. Siegel et al., *The U.S. Military's Toxic Legacy: America's Worst Environmental Enemy*, pp.8-9.

46. John M. Broder, "U.S. Military Leaves Toxic Trail Overseas", *Los Angeles Times*, June 18, 1990. For reports of contamination at U.S. Overseas bases, see "Japanese Group Plans to Sue U.S. Navy for Violating NEPA", *Multinational Environmental Outlook*, October 31, 1989; Jorge Emmanuel, "Environmental

Destruction Caused by U.S. Military Bases and the Serious Implications for the Philippines", Paper presented at "Crossroad 1991: Towards a Nuclear Free, Bases Free Philippines", Manila, May 14-16, 1990; and :U.S. Defense Department lax in Environmental protection at Overseas Bases, Official Charges", *World Environmental Report*, July 23, 1987. U.S. Army plan from John G. Roos, "US Army Plans to Turn Over Facilities in Germany in `As Is' Condition", *Armed Forces Journal International*, September 1990. "U.S. Army Faulted for Environmental Damage", *JPRS-TEN-91-002*, West Europe, January 29, 1991, p.79.

47. Henry Kamm. "Americans Help Czechs Clean Up After the Soviets", *New York Times*, July 24, 1990; Peter S. Green, "Cleaning up After the Soviet Army", *U.S. News and World Report*, May 28, 1990; Weva Rich, "Departing Red Army Leaves Its Rubbish Behind", *New Scientist*, June 2, 1990; "Alles Zerwuhlt und Kaputt", "Das wird eine Gratwanderung", and "Die Fliegenals Erste Raus", all in *Der Spieqel* July 16, 1990; "Environmental Inspectors to Visit Soviet Bases", Foreign BroadcastInformation Service (FBIS) Daily Report/Soviet Union, Rosslyn, Va., August 30, 1990; "Pollution Problems in Soviet Army Units Viewed", Foreign Broadcast Information Service (FBIS) Daily Report/Soviet Union, Rosslyn, Va., August 30, 1990; "Pollution Problems in Soviet Army Units Viewed", FBIS Daily Report/Soviet Union, Rosslyn, Va., September 5, 1990; "Survey on Environmental Damage Left by Soviet Army", *JPRS-TEN-91-017*, September 13, 1991, p.32; "Environmental Damage from Military Forces Detailed", *JPRS-TEN-91-001*, West Europe, June 17, 1991, p.109.

48. "Secrecy Hides the Hazards of Working in the Aerospace Industry", *New Scientist,* April 8, 1989; Kenneth B. Noble, "Health Troubles at Military Plant Add Mystery to Top-Secret Project", *New York Times*, September 18, 1988. K. Geiser, "health hazards in the Micro-Electronics Industry", *International Journal of Health Services*, Vol.16, No.1, 1986.

49. This section draws on Renner, "Assessing the Military's war on the Environment", pp.145-150, and Renner, "War on Nature", pp.21-24.

50. "Defending the Environment? The Record of the U.S.Military", *The defense Monitor*, Vol.18, No.6, 1989; Howard Ball, "Downwind from the Bomb", *New York Times Magazine*, February 9, 1986; Keith Schneider, "Atom Tests' Legacy of Grief: Schneider, "U.S. Admits Peril of 40's Emissions at A-Bomb Plant", *New York Times*, July 12, 1990; Francies X. Clines, "Soviets Now Admit '57 Nuclear Blast" *New York Times,* June 18, 1989.

51. U.S. Warhead production and plutonium and enriched uranium stockpiles from Cochran et al., *Nuclear Weapons Databook, Vol.II*; Cochran et al., *Nuclear Weapons Databook, Vol. IV: Soviet Nuclear Weapons* (New York: Ballinger, 1989), and from "Status of Major Nuclear Weapons production facilities: 1990", *PSR Monitor*, September 1990. Waste products per kilogram of plutonium from Karen Dorn Steele, "Hanford: America's Nuclear portion by volume from "Defending the Environment? The Record of the U.S. Military"; military portion by curies from Scott Salaska *etal., Nuclear Legacy: An Overview of the Places, Problems and Politics of Radioactive Waste in the United States* (Washington, D.C.: Public Citizen Critical Mass Energy Project, 1989). U.S.high-level inventory from Radioactive Waste Campaign, *Deadly Defense* (New York: 1988).

52. Number of contaminated sites from Committee to Private Interim Oversight of the DOE Nuclear Weapons Complex, Commission on Physical Sciences, Mathematics, and Resources, National Research Council, *The Nuclear Weapons Complex: Management for Health, Safety, and the Environment* (Washington, D.C.: National Academy Press, 1989); number of locations from Keith Schneider, "In the Trail of the Nuclear Arms Industry", *New York Times*, August 26, 1990; Nagasaki bom comparison from Michael Satchell, "Uncle Sam's Toxic Folly", *U.S.News and World Report*, March 27, 1989; plutonium in Rocky Flats Ventilation

ducts from Mathew L.Wald, "Doubt on Safety at Weapon Plant", *New York Times*, June 21, 1990; 1969 fire from Fox Butterfield, "Dispute on Wastes Poses Threat to Weapons Plant", *New YorK Times*, October 21, 1988.

53. Mathew, L.Wald, "Hanford's Atom Waste Tanks Could Explode, Panel Warns", *New York Times*, July 31, 1990; Alvarez and Makhijani, "Hidden Legacy of the Arms Race".

54. Larry Thompson, "Scientists Reassess the Long-term Impact of Radiation", *Washington Post*, August 15, 1990; "ICRP to Recommend More Stringent Human Radiation Exposure Limits", *Multinational Environmental Outlook*, October 16, 1990.

55. Total number of workers from Martin Tolchin, "U.S. to Release Health Data on Nuclear Plant Workers", *New York Times*, May 24, 1989. "New Agreement Could Continue Department of Energy's Monopoly Over Radaiation health Research", *News Alert*, Physicians for Social Responsibility (PSR), Washington, D.C., October 26, 1990; Daryl Kimball, Associate Director for Policy, PSR, Washington, D.C., Private Communi-cation, November 2, 1990; "Health and Energy Agree, PSR Disagrees", *Bulletinof the Atomic Scientists*, April 1991, p.6. Office of Technology Assessment, *Complex Cleanup: The Environmental Legacy of Nuclear Weapons Production* (Washington, D.C., Feburary 1991).

56. Mathew, L.Wald, "High Radiation Doses Seen for Soviety Arms Workers, *"New York Times*, August 16, 1990; Bill Keller, "Soviet City, Home of the A-Bomb, is Haunted by its Past and Future", *New York Times*, July 10, 1989; Zhores A. Medvedev, "The Environmental Destruction of the Soviet Union", *The Ecologist* January/February 1990; Clines, "Soviets Now Admit '57 Nuclear Blast"; Vera Rich, "Thirty-Year Secret Revealed", *Nature*, June 22, 1989; David Dickson, "Kyshtym Almost as Bad as Chernobyl", *New Scientist*, December 23/30, 1989; Arjun Makhijani, Institute for Energy and Environmental Research, Takoma Park Md., Private Communication, October 23, 1990.

57. The United States, the Soviet Union, and the United Kingdom stopped atmospheric tests in 1963 but France and China continued until 1974 and 1980, respectively. Number of tests from Stockholm International Peace Research Institute (SIPRI), *SIPRI Year Book 1991: World Armaments and Disarmament*, Tables 2A.2 and 2A.3. Birth defects from Barry Commoner, "Do Nuclear Plants Make Deadly Neighbors?" *Congress Record*, March 23, 1972. Estimate of premature deaths from Bernd W.Kubbig, "Atomtests: Gefährdung fur Mensch and Umwelt," in Bechmann, *Umwelt Braucht Frieden*. "Venting" from John Hanrahan, "Testing Ground, "*Common Cause Magazine*, January/February 1989, and from "A Nuclear Unthreat", *Economist*, March 29, 1986.

58. Solider from Kubbig, "Atomtests, Gefährdung fur Mensch und Umwelt", Kim, *The Quest for a Just World Order*, reports a number of 250,000 soldiers; workers from Schneider, "U.S. Fund set up to pay civilians Injured by Atomic Arms Program"; downwinders and their health effects from Ball, "Downwind from the Bomb"; Richard L.Miller, "Let's Not Forget Radiation in the U.S." (op ed), *New York Times*, June 27, 1986; Americans' exposure from Keith Schneider, "Senate Panel Describes Data on Nuclear Risks", *New YorkTimes*, August 3, 1989. Results of U.S. tests in Pacific from Ian Anderson, "Potassium Could Cover Up Bikini's Radioactivity, *"New Scientist*, December 10, 1988; Eliot Marshall, "Fallout from Pacific Tests Reaches Congress", *Science*, July 14, 1989. French tests from Bengt Danielsson, "Poisoned Pacific: The Legacy of French Nuclear Testing", *Bulletin of the Atomic Scientists*, March 1990. Australia from Ian Anderson, "Britain Semipalatinsk from Quinn-Judge, "Soviet Writers Blast Nuclear Testing", from Olzhas Suleymenov, "We Cannot Be Silent", *Earth Island Journal*, Summer 1989, from Judith Perera, "Soviet Environmentalists Cite Health Problems Around Nuclear Testing Ground", *Multinational Environmental Outlook*, September 5, 1989, from "Soviet Official Reveals Increased Radiation

Hazard Near Semipalatinsk", *JPRS - TEN-90-017*, Society Union, December 6, 1990, p.79, and from Susan E.Reed, "Atomic Lake", *The New Republic*, October 28, 1991, p.12.

59. Treaties include the 1987 Intermediate-Range Nuclear Forces (INF) Treaty between the United States and the Soviet Union, the multilateral Conventional Forces in Europe (CFE) Treaty signed in November 1990, and the 1991 Strategic Arms Reduction Talks (START) Treaty between the United States and the Soviet Union.

60. SS-20 missile launchers into cranes from Alexandri Remisov, "Disarmament Treaties and Conversion of Military Production in USSR", *IDDC Internazionale*, September/October 1988.

61. *The Arms Control Reporter 1990*, p.407. B.410.

62. William Broad, "New Methods Sought to Dispose of Rockets with No Harm to Earth", *New York Times*, September 17, 1991. Broad briefly discusses some of the alternatives examined. Amount of Pershing solid propellants from Stephen Iwan Griffiths, "The Implementation of the INF Treaty", SIPRI, *SIPRI Yearbook 1990: World Armaments and Disarmament* (Oxford: Oxford University Press, 1990) p.457.

63. John Rettie, "Soviet Nuclear Nightmare", *The Guardian* (Manchester), January 29, 1992.

64. Paul Rogers, "Ploughshare Option for Warheads", *The Guardian* (Manchester), February 1, 1992.

65. Mathew, L.Wald, "Judge Sets up New Hurdles for U.S.Plutonium Dump", *New York Times*, February 4, 1992. For a general discussion of problems relating to nuclear waste disposal, see Nicholas Lenssen, *Nuclear Waste: The Problem that won't go away*, Worldwatch paper 106 (Washington, D.C.: World watch Institute, December 1991).

66. David Clark Scott, "US Plan to Incinerate Chemical-Arms Stocks Rouses Opposition", *Christian Science Monitor*, September 26, 1989; Keitch Schneider, "U.S.Plan to Burn Chemical Weapons Stirs Public Fear", *New York Times*, April 29, 1991; "Chemical Weapons Burn", *Greenpeace Magazine*, January/February 1991, p.7; "Arms Controllers V.Greenpeace", *The Bulletin of the Atomic Scientists*, November 1990, pp.4-6.

67. Scott, "US Plan to Incinerate Chemical-Arms Stocks Rouses Opposition". Batsanov quoted in "Russia Not Able to Destroy Chemical Arms", *Chemical and Engineering News*, January 20, 1992, p.17.

68. Hazards of ammunition production from Vertegall, "Environmental Impact of Dutch Military Activities". Ammunition stockpiles in the former East Germany from Burckhardt J.Huck, "Abrustung und Konversion in Deutchland", and from Wener Hänsel and Heinz Michael, "Rustungskonversion in den Neuen Bundesländern", both in *Informations dienst Wissenschaft and Frieden*, December 1990, p.62, from "Ecological Damage Caused by Withdrawing Soviet Soldiers,", *JRPS-UPA-91-010*, West Europe, June 11, 19912, p.91, and from "Environmental Damage from Military Forces Detailed", *JPRS-TEN-91-011*, West Europe, June 17, 1991, p.109.

69. Chemical weapons destruction costs from "U.S.Chemical Weapons Production: Poisoning the Atmosphere", *The Defence Monitor*, Vol.18, No.3(1989), p.5, from Schneider, "U.S. Plan to Burn Chemical Weapons Stirs Public Fear", from Keith Schneider, "Military Has New Strategic Goal in Cleanup of Vast Toxic Waste", *New York Times*, August 5, 1991, and from Walter Goodman, "The Special Horror of Chemical War", *New York Times*, October 23, 1990. Ammunition destruction costs in the former East Germany from Huck, "Abrüstung and Konversion in Deutschland", and from "Die US-ameikanischen and sowjetischen Stritkräfte hinterlassen in Deutschland ein großes Altlastenproblem, "Ökologische Briefe, No.11, March 13, 1991, p.13.

70. Nuclear decontamination costs from "Problems Persist at Weapons Plants",

*PSR Monitor*, September 1990. Toxic cleanup estimate from Siegel, "The Growing Nightmare of Military Toxics". Hungarian damage estimate from "Hungary, Soviet Union in Negotiations on Environmental Damage Caused by Troops", *International Environmental Reporter*, August 14, 1991, p.437. Czechoslovak estimate from "Cost of Ecological Damage by Soviet Troops", *JPRS-TEN-91-015*, August 7, 1991, p.32. Poland from "Environmental Damage Caused by Soviet Forces Assessed", *JRPS-TEN-91-014*, July 9, 1991. East German estimate from "Alles Zerwühlt and Kaputt", p.121. Australia from Anderson, "Britain Asked to Foot its Nuclear Cleaning Bill".

71. Tom Turner, "Ground: Zero. The American Military vs. The American Land", *Wilderness*, Fall 1991, pp.12, 14: Keith Schneider, "Military Has New Strategic Goal in Cleanup of Vast Toxic Waste", *New York Times*, August 5, 1991.

72. Nevada example from Bert Lindler, "Foes Unite to Fight Military Proposal", *High Country News*, February 12, 1990. Land rehabilitation cost from Edwar McGlinn, Anglers of the Au Sable River, Grayling, Mich., Private Communication, February 18, 1990. Robert Stone from Andrew Melnykovich, "Torn Between Cows and Jets, *High Country News*, February 12, 1990.

73. For a more detailed discussion of the requirements of sustainability, see Michael Renner, *Jobs in a Sustainable Economy*, Worldwatch paper 104 (Washington, D.C.: Worldwatch Institute, September 1991). I have explored a number of related questions in an earlier essay: "Conversion to a Peaceful Economy: Criteria, Objectives, and Constituencies, in Lloyd J.Dumas and Marek Thee (eds.), *Making Peace Possible: The Promise of Economic Conversion* (Oxford and New York: Pergamon Press, 1989). See especially Table 1.

74. For spinoff claims, see for example William J.Broad, "Defence Industry Goes Hustling to make a Buck without the Bang", *New York Times*, April 8, 1990. Mary Kaldor, *The Baroque Arsenal* (New York: Hill and Wang, 1981). Defence Science Board from "A Healthy Military-Industrial Complex", *U.S. News and World Report*, February 12, 1990, p.43.

75. Michael Closson, "Expanded Horizons", *Plowshare Press*, Vol.15, No.2 (Spring 1990), p.4.

# CONVERSION AND THE DEATH OF THE SOVIET MILITARY - INDUSTRIAL COMPLEX

**Christopher Smart**
*Hudson Institute, U.S.A.*

As the most demanding sector in the Soviet economy, as one of the most privileged elements of Soviet Society, and as the very foundation of Soviet military power, the military-industrial complex shaped much of the Soviet System's evolution over the last seven decades. The effort to convert military enterprises to civilian production, therefore, has meant more than mere industrial reform. Because the military-industrial complex has played such an important role in the country's evolution, its demise offers a rare perspective on the economic, political and military decay of the Soviet Union. Meanwhile, the transformation of the country since August 1991 has shaken this crucial foundation of the Soviet Union's international stature beyond reasonable chance of repair.

Economically, the shape of Soviet manufacturing has changed. The tight central planning that reserved the very best industrial resources for military purposes has been undermined by the dispersed and uncoordinated decisions of enterprise managers themselves. The collapsing Soviet budget has forced defense industrialists into a kind of military "conversion by default" as they have scrambled to find new sources of income with civilian products. Defense industrialists may use their new economic position to press for further decentralization and a weapons industry based on free markets. Alternatively, the disintegration of their military-industrial base may frighten them into forming new cartels that shadow their former production ministries. This will depend partly on their own choices and partly on the relationships that develop between the successor regimes.

Politically, the struggle to guide the reform of the country's military-industrial resources has been a critical test of strength among the factions contending over the broad agenda of reform. At least until August, the Gorbachev government mounted a carefully managed effort to bring the magic of Soviet defense industry to the rescue of the civilian sector, while the republics and municipalities attempted to wrest control over these resources from the center. Since the coup, the political contest has evolved from one between center and republics to one among the republics themselves over security cooperation, industrial ownership and trade obligations. If a viable economic and political community emerges, profitable inter-republican trade and lucrative Western investment may sway the newly-empowered defense industrialists to back further free-market reforms. If, however, mistrust between Soviet nations shatters the all-Union military-industrial base into rival autarkies, security concerns are more likely to encourage command methods of military production.

Finally, whether the market or the plan eventually dominates, the current reconstruction of the military-industrial complex has seriously damaged the very strengths that made the Soviet Union a superpower. Moreover, defense industrialists now have had their attentions focused on finding Western investors and selling to the profitable international market rather than answering the calls of Soviet military strategists for advanced technology systems. It may be too soon to judge the strength of the future military-industrial complex that may back, say, Russia's military power. But certainly the mightyindustrial locomotive that deranged a backward and defeated Bolshevik regime into the company of Great Powers is now sputtering to a halt.

## A. THE ECONOMICS OF CONVERSION

The logic of military conversion is as tempting as it is false: an economy that has been drained of resources by extensive military production need only re-direct these resources to the civilian sphere. The problem is that many of these resources cannot be "re-directed" profitably, if at all. A central government can derive benefit rather fast in the savings that come from radically slashing military spending. Most of the original investment, however, is irretrievable. And additional benefits must await the work of a market that guides land, labor and capital to new purposes. A government can help alleviate this transition by facilitating contacts between producers or offering incentives to develop key regions and industries, but the retooling of existing factories is rarely cost-effective.[2]

While it is not clear that Soviet officials have understood that the promise of conversion is largely hollow, they have been forced into making substantial cuts in military expenditure as the all-Union budget deficit deepens and as inflation rages into three-digits. The result has been a kind of "conversion by default," as weapons lines simply grind to a halt, and defense industrialists scramble to come up with something -- anything -- to replace traditional lines of production. The result will not produce converted factories that are any more cost effective, but the shape of the military economy now depends much more on the preferences of these enterprise directors than on the orders of central planners.

### The Centrality of Military Spending

Understanding the importance of Soviet military conversion to *perestroika* as a whole requires above all an appreciation of the central role of these industries in the Soviet economy. To be sure, Russia's age-old sense of insecurity and the Western intervention following the Bolshevik Revolution, contributed substantially to the decisions to build strong defensive capabilities. The feverish pace of weapons manufacture under the young Communist regime, however, was fueled substantially by an ideology that saw enemies everywhere and sought to nudge along the inevitable course of history at the point of a gun. Initially, it was Lenin's interpretation of Marx that called for a one-million soldier army in 1918 (and a three-million soldier army soon thereafter) to defend the Bolshevik coup and spread the revolution into Poland. This was to be a temporary force that would evolve into a territorial militia. Later, however, it was Stalin's interpretation of Lenin that harnessed the best of the country's resources to the requirements of military strength permanently.

In the face of the classic planner's dilemma between guns and butter, therefore, the Soviet leadership first chose guns. When Stalin launched his campaign to rebuild Soviet industry with the First Five Year Plan in 1928, military production was a top priority. By 1933, the people's Commissioner for the Army and Navy, Kliment E. Voroshilov, could declare that one of the plan's main aims had been "the technical re-equipment of the Red Army on the basis of the successful growth of our industry."[3]

By the 1980s, there were nine ministries for defense production.[4] Coordinating this production and that of all other defense-related industries was the powerful State Commission on Military-Industries. Headed by a deputy prime minister, the commissionensured that the military production lines had priority access to supplies and skilled employees, and that they fulfilled their planned obligations on time.[5] The State Planning Commission (Gosplan) and the State Committee for Material-Technical Supply (Gossnab) were also called upon to ensure that the military needs would be met first. At the regional and local level, the Communist Party organizations were responsible for smoothing any production hitches on their own territory.

Determining the size of Soviet military industries has always been a highly contentious undertaking in the West. Through the 1970s and early 1980s, the Central Intelligence Agency placed Soviet military spending at about 12-14 percent of Gross National Product. Last year, Gorbachev himself admitted that military spending approached 18-20 percent of 'national revenue."[6] Using the Central Intelligence Agency's tally of Soviet military spending and the GNP figures of maverick Soviet economist Grigori Khanin, the real burden of defense industry approaches 25 or 30 percent.[7] A fuller understanding of the Soviet defense burden, however, comes from a variety of other measures. Defense industries have had a lock on one third of the country's industrial workforce,[8] some 60 percent of the machine-building industry,[9] 55 percent of Russia's industrial production,[10] 80 percent of research and development personnel,[11] and 20 percent of Soviet energy output.[12] Such is the crushing burden of defense spending that Soviet leaders are now attempting to unload.

**The Soviet Experience with Conversion**

Soviet officials like to speak about three previous attempts at conversion. The first was in the early 1920s, following the Civil war, when the Red Army was cut from 5 million to 562,000. This, however, did not involve a great deal of industrial retooling. The second was said to have come after World War II, although, strictly speaking, this was mostly reconversion of industries that had only been mobilized for wartime production.[13] The third effort at conversion was said to have come under Nikita S. Khrushchev, when more than 2 million members of the Armed Forces were demobilized and military hardware was scrapped from 1958-64. Many of these reforms, however, were derailed early on. From time to time, defense industries were charged with increasing their contributions to the civilian sector, but their own grip on resources was never seriously challenged.[14]

What makes the current effort at conversion different in an economic sense is that it has become subsumed in the far broader transformation of the economy towards what Soviet leaders have euphemistically called "market relations" and what the rest of the world knows as capitalism. It also comes at a time when budgetary stringency has forced spending cuts that have left defense enterprises as independent economic actors for the first time.

Initially, the central ministries relinquished some of their traditionally tight control deliberately. In January 1989, in an effort to encourage greater efficiency in production, they and their civilian counterparts were forced to adopt a new system of self-financing and cost accounting. Long protected from real costs by generous government subsidies and artificially low prices, military industries were charged with covering all their own costs with income from their sales. By this year, however, the center's loss of control quickened as its own sources of money started running dry. In January 1989, Gorbachev announced that the Soviet defense budget would be cut 14.2 percent by 1991, and perhaps by a much asone third to one half by 1995. Production of military hardware would fall by 19.5 percent.[15] The 1991 Defense Budget was announced early this year at 96.6 billion roubles, up from the 71 billion roubles for 1990, with procurement rising from 31 billion roubles to 39.7 billion. Given the spiralling inflation rate, however, U.S. intelligence agencies estimate that procurement outlays fell by about 10 percent in each of the last two years, and predict that the real decline in 1991 will reach some 20 percent.[16] Still more cuts are on the way.[17]

Anecdotal evidence suggests that the effect on defense managers has been dramatic. "Saturn," an electronics firm based in Kiev, depended on state orders to fill 98 percent of its capacity in 1985. By 1990, this had fallen to 82 percent, but it meant a real loss in receipts of some 45 percent because customary subsidies were cut with the orders.[18] Igor' D. Spasskii, of Leningrad's "Rubin" design bureau asserts that while the shipbuilding ministry controlled all of his design work as recently as three years ago, by this year it has had nothing to do with some 60 percent of his work.[19] Moreover, the orders have been unpredictable. "Signal," a defense electronics enterprise in Leningrad, has had its military orders cut 30 percent in one year, and only learned of 12 percent of the cuts in January of this year. Valentin. P. Zanin, the general director, says he expects the same routine this year. The initial order will go out, but suddenly "they will say, we're sorry we don't have the money."[20]

Defense industrialists have also faced increasingly chaotic production circumstances. The lack of clear legal frameworks left enterprises caught in the middle of disputes between the republics and the central government about laws and taxes. The weakening of administrative controls, meanwhile, has meant that the habitual hunt for supplies has turned into a darkening nightmare. Well before the coup, the central ministries had even stopped guaranteeing supplies for some state orders, and Party officials had stopped intervening to reroute supplies destined for civilian production. Enforcement mechanisms to punish suppliers who failed to meet their obligations had disappeared.

Beyond this, the lucrative cooperative sector began to draw away some for the defense industry's most talented designers and engineers. A Leningrad city official estimates that over the past year, some 30,000 specialists left the city's 150 defense plants. This, he said, was a serious drain on the city's grand total of 600,000 industrial workers.[21] Beyond those leaving voluntarily, however, the additional pressure of falling defense contracts has begun to force out many workers. This spring, Deputy Prime Minister Scherbakov revealed Goskomstat data that some 300,000 workers were relieved of work in defense plants in 1990. Seventy-six percent were reassigned to other production, while the rest joined the ranks of unemployed. Trends projected another 300,000 relieved in 1991 with some 70,000 becoming unemployed.[22] More worrying still, those employees who remain at the defense plants have begun to organize politically. Even before the August coup, workers collectives threw out their factory's traditional Party cells and began making ominous calls for strikes.

**Results to Date**

None of these difficulties stopped government officials from finding positive signs to advertise. A senior staff member of the Military-Industrial Commission bristles at the suggestion that conversion had yet to yield results. `People don't know what they are talking about," says Albert F. Trifonov. "It's all speculation."[23] Still, even the official results are not encouraging. By the end of1989, only 23 of the 120 new types of civilian goods due from the defense sector had been produced, according to scholar Aleksei Iziumov.[24] By 1991, it had become clear that only six of the 400 arms enterprises targeted for conversion would be fully converted, and only 34 of the 100 ancillary plants targeted.[25] In early August, Deputy Prime Minister Vladimir I. Scherbakov declared that some 600 industrial plants were involved in conversion, but it was still not clear just how far along any of them were or what proportion of military production they represented.[26]

And specific reports on conversion offer plenty of cause for skepticism. One military plant produced an automatic sausage machine that turned out green hot dogs. ("Well, not entirely green, of course, but kind of grayish-greenish," according to one journalist). The Ministry of Atomic Energy and Industry copied Western vacuum-evaporating devices, but failed to account for the higher acidity of Soviet milk which left it sticking to the machinery.[27] In a spectacular display of over engineering, one design bureau developed a new washing machine with 20 different cycles.[28]

Though not all attempts are so absurd, most show little planning or research and rely on considerable improvisation. The travails of Leningrad's "Rubin" Central Maritime Design Bureau, which specializes in submarine and other underwater technologies, has pursued two news lines near their traditional expertise: oil platforms that can be built under Arctic ice conditions and a tourist submarine for underwater excursions. The rest is a curious mix that was apparently developed at random: stationary power stations with natural gas turbines, restaurant equipment, medical and sporting equipment, and a new hotel under construction on Rubin's property in central Leningrad.[29]

### Conversion: the Good News and the Bad

In their search for alternative lines of production, the defense industrialists hold a few advantages, but face a far longer list of obstacles.[30]

First, there is as yet almost no industrial competition in the Soviet Union, which makes markets easy to enter.[31] Meanwhile, the consumer market is so desperate for goods of almost any kind, that the producer need not conduct very sophisticated market surveys to land on a saleable item. Second, Soviet workers are generally well-educated. This is an important advantage in the colossal task of reorganizing the country's workforce. Third, there are widespread signs that a commercial culture is beginning to take root in the Soviet Union in spite of Russia's egalitarian traditions. While many opinion polls suggest that the idea of a free market is gaining legitimacy, a great many islands of entrepreneurial activity have appeared ranging from small cooperative shops and restaurants to fledgling commodities exchanges and breakaway state enterprises.[32]

That is the goods news. The bad news it somewhat more daunting. Above all, there is a serious deficiency in basic market knowledge. This is a problem for all of Soviet industry, but it is particularly debilitating in the defense industries where many managers have only just begun to travel abroad recently, and few have ever had the chance to place their products on the international market.[33] Thus, one finds otherwise bright and ambitious directors of large defense enterprises insisting that the shift to a market economy can be accomplished without massive unemployment, that the defense industrial ministries function much the way large Western Corporations do, or that a freely convertible rouble would rapidly stabilize on apar with the dollar.[34] As they actually try moving into the market, their ignorance grows still more glaring. "Saturn," the Kiev telecommunications firm, for example, spent many months trying to secure a western partner to create a direct telephone satellite link between Ukraine and the United States. Its directors were naively unaware, however, of international conventions and Western agreements that restricted transmission over a Soviet satellite or prohibited Soviet leasing of earth stations in the U.S.[35]

Compounding this lack of education is a more specific information problem: the deeply-rooted penchant for secrecy. Lacking a healthy class of brokers and traders, the Soviet market as a whole suffers from a lack of freely available information about who can make what for how much. Aggravating this problem still further in the military sector, however, is the stifling secrecy that has surrounded nearly every form of weapons production.[36] This has not only blocked technological spin-offs from reaching the civilian market, but it has concealed what suppliers and customers may find in this prize military sector. Frank press accounts and interviews with Soviet officials suggest that this habit is fading; still it will take some time for such a transformation to bear fruit.[37]

Third, much like Western military contractors who do not face the pressures of competition to keep their prices low, Soviet defense enterprises make even less effort than Soviet civilian enterprises to keep their costs down. Moreover, their elaborate production facilities frequently produce goods priced far beyond the reasonable reach of civilian consumers.[38] And because the rouble is not convertible and because Soviet managers have no real appreciation for their real production costs, they cannot set rational prices. Asked how he prices his trawlers for export, the director of the "Leninskaia Kuznitsa" shipyard in Kiev answers simply: "Intuition."[39]

Fourth, there is the small matter of money. Any kind of conversion, whether the actual retooling of an existing production line or the construction of a new civilian facility, requires money. On any list of shortage items in the Soviet Union, such capital ranks high. Indeed, the Leninskaia Kuznaitsa shipyard, one of three plants selected to showcase conversion methods in the early stages of the government program, has yet to be allocated the money it needs to retool. Even for such a prominent project, which only needs to convert some 10 percent of its overall production from military to civilian, the plant managers have had to scramble for money.[40]

Finally, successful conversion involves, more than anything else, a vast reallocation of labor. In this matter, the Soviet Union faces a number of hurdles. The need for residential permits has long prevented workers from seeking suitable work in a different part of the country. Even without such official limitations, however, the severe lack of housing will make it difficult for workers to move to regions where the economy most needs them. This problem will be acutely felt in cities where residents work almost entirely in military industries. They will not be able to count on other parts of the country to draw off their unemployed workforce.[41]

**Entering the Political Fray**

As their traditional world collapses around them, the defense industrialists have sought new foundations of support. On the one hand, they have broken from the traditional bureaucratic structures, setting up independent "concerns" either along or in conjunction with other enterprises in their industry. In November, for example, theMinistry of Shipbuilding and several prominent shipyards and design bureaus announced the formation of "Sudo-promimpeks," an unwieldy corporate entity charged with launching them on the international market.[42] On the other hand, they have formed industrial associations with their counterparts in the civilian sector in order to defend their political and economic interests.[43]

Their influence stemmed initially from their former Party connections, but their continuing power now rests with their crucial role in keeping the Soviet economy working. "If directors wanted *perestroika* to end, we could do it in a week," says Georgi S. Khizha, who is the director of Leningrad's "Svetlana" plant and chairman of both the Leningrad Association of Industrial

Enterprises and the Union of Associations of Leningrad Enterprises. The current economic difficulties require enterprise managers to work harder than ever, just to keep the system functioning. Relaxing their effort only slightly, he argues, could bring it all to a grinding halt, and send workers to the streets to demand order. "If there is *perestroika* today," says Khizha, "it is only thanks to the efforts of the directors."[44] Few others have made the point in quite the same way; indeed, prominent reformers have long accused enterprise managers of blocking any real economic reform. The fact remains, however, that these managers represent a formidable force in guiding the shape of the economy.

**An Ambiguous Presence**

Just what kind of force this emerging class will represent remains ambiguous. Traditionally, military enterprises might resist the military's calls for technical innovation because they threatened to derail their annual plan targets.[45] Nevertheless, ministries and Party superiors held unquestioned financial and political authority and could enforce central decisions. Now, their primary concern is to keep their enterprises operating in conditions of economic disintegration. On the one hand, if the political collapse continues, these managers could emerge as quasi-owners of their enterprises. This might launch them into a nascent form of competition, but it will also lead to further degradation of the industrial base. On the other hand, the sheer chaos may lead to new forms of central control and planning. This may protect crucial industries but the prospects for economic reform in this or any other sector will be snuffed out.

Many defense industrialists appear loath to give up the freedom that they now enjoy. In many respects they have become the *de facto* owners of these factories and they seem to like it. By all accounts, industry chiefs split in this spring's Russian presidential election, and many of them backed Boris Yeltsin and his bolder call for market economics. Some also opposed the coup. The president of the Scientific-Industrial Association, Arkadii Vol'skii, appears to have opposed the coup, and has spoken at length about the need for a slow, stable evolution towards a market.[46]

Moreover, for a class of managers with little experience in the market and little market to experience, there are signs that they are seeking to make it work anyway. In interviews, most Leningrad directors back the notion of a "free-enterprise zone," and speak excitedly about the prospects for destatization of industrial properties and Western investment. In Ukraine, directors who had made their careers in the Party or ministerial apparat grow positively misty when they speak of Ukrainian sovereignty and independence. Finally, several self-help regional data-banks have been formed to help enterprise directors exchange information about potential customers and partners. A Military-Industrial Exchange haseven been formed at the Flight Control Center near Moscow, to help these very industrialists develop a kind of wholesale market. Included, is a secret department that will collect information on enterprises still veiled in secrecy that cannot take advantage of conventional exchanges.[47]

Many are also finding that in spite of all the difficulties of operating in the civilian market, it has proven more reliable and profitable than their traditional military buyers.[48] Some have even been reported to be turning down defense orders altogether.[49] Others have been looking to the international arms market for new sources of income.[50] The lure of foreign investment has also encouraged considerable independence on the part of these enterprise directors, creating the basis for a military-industrial complex based much more on market principles and independent decision making.[51] If the trend continues, a few lucky defense industrialists may happen on a formula that offers a profitable new use for their facilities. For the most part, however, they will care more about keeping their plants running and their workers paid, than about the most efficient use of their plants. The result will be deeper debts, falling investment and a steady degradation of their factories.[52]

The crumbling union and the deepening economic chaos, however, could encourage many defense industrialists to back a restoration of order that would choke off democratic and market reforms. Many have done it before.

In an open letter in September 1990, defense enterprise directors complained of a flurry of confusing laws and resolutions from various levels of government and the "massive" loss of skilled workers, called for the government to reclaim control over the economy.[53] In February 1991, defense industrialists warned that a lack of planningand funding for any production of new civilian equipment assigned to them might force defense institutes and design bureaus to disband.[54] The Union Treaty that Gorbachev was preparing to sign this August would have essentially left defense industries at the mercy of the republics. So there was little surprise that some part of the military-industrial complex led the coup.[55]

Moreover, if market forces are allowed to prevail, even those enterprises that have found some successful alternate line of production will be forced to close. This offers a powerful incentive to preserve some central control over production that can keep factories operating and workers working. Indeed, there have been a variety of reports of defense enterprises continuing to pay workers to show up for work and do nothing. If central ministries do not reemerge to take up the call, the enterprise managers themselves could aggregate into cartels or production associations that would coordinate the economy. Indeed, in September, several industrial associations announced plans for an Enterprise Council under the State Council of the U.S.S.R. that would promote market reforms within a unified economic space and protect the rights of businessmen.[56] Specific reforms for the management of military industries were said to be under consideration.[57] Such a reassertion of central control might help protect some strategically important military industries from disintegration, but it will effectively derail any chance for real economic growth or technological advance. It will also end any chance of real democratic reforms.

## B. THE POLITICS OF CONVERSION

Just what kind of military sector emerges in the new Soviet Union, of course, will depend heavily on the kind of political order that emerges with it. Until the August coup, this effort was shaped mainly by the powerstruggle between the republics and the center. Since the coup, however, the contest has evolved into a series of disputes among the republics themselves about common defense needs, property rights and trade arrangements. The republican governments that now drive the political agenda will doubtless pursue the largely futile attempt to squeeze some civilian benefit out of their substantial military-industrial base. The size and shape of these bases, however, will depend ultimately on the strength of the economic and political union that emerges.

### Basis for Conversion

Soviet military conversion in its current incarnation has advanced largely on the momentum of three separate political agendas: one technological, one diplomatic and one economic.

Well before the full scale of the Soviet economic crisis became clear, it was dawning on members of the senior leadership just how far their country lagged technologically. By the mid 1970s, the major challenge to Soviet military and industry appeared mainly in two families of technology: micro-circuitry and semiconductors, and lasers.[58] Marshal Nikolai Ogarkov, then chief of the general staff, was among the earliest and most prominent officials to recognize the value of such weapons and to understand the importance of a strong economy to support such technological advances.[59] From the earliest months of his tenure, Gorbachev took up the call for economic reform based on the technological advances that the military wanted most.[60] While it was hardly clear just what kind of economic reforms he intended --let alone how radically they might evolve --the military leadership seemed to back him.[61]

This initial technological motivation dovetailed nicely with a second agenda to break through mounting Western fear of and hostility toward the Soviet Union. Proclaiming an era of "New Thinking," Gorbachev simply insisted that the West's growing relative military strength was irrelevant, and that Moscow viewed peaceful political solutions as more important than any of its traditional goals of class struggle.[62] The culmination of this campaign came at the United Nations in December 1988, when Gorbachev reiterated his new international philosophy, and sealed it with a promise to cut the Soviet army by 500,000 soldiers, and convert military enterprises for civilian use. "A breach has appeared in the heretofore apparently unbreachable wall of suspicion and enmity," he declared.[63]

Meanwhile, as these lofty thoughts of microcircuits and world peace bubbled away in the official rhetoric, the Soviet consumer market went from bad to worse. According to official Soviet estimates, the general availability of staple foods (in volume as well as in kind) dropped from 90 percent of demand in 1983 to 22 percent in 1989 to 11 percent by the middle of last year. Some 98 percent of 1,200 basic consumer goods were essentially never available.[64] With such dwindling supplies, the government began to look to converted military industries for help. Officials promised that these prized resources would soon be able to crank out desperately needed consumer goods. "Who can provide for making television sets at the quality of Sony, stereo systems at the level of Sharp, cameras at the level of Nikon and motorcycles at the level of Honda?" boasted Igor S. Belousov, Chairman of the Military-Industrial Commission. "None other than the defense industry can really provide for this in the next few years!"[65]

**The Gorbachev Program**

Gorbachev's conversion strategy was as simple as it was traditional: he turned to the formidable resources of the defenseindustries for help. "Here we are not lagging behind in anything," Gorbachev said. "This means we know how to work."[66] In part, his efforts were incoherent because they faced fierce opposition from other interests that would have them go far faster as slower. In part, his plan was timid and contradictory both from a lack of political will and from the disintegration of traditional bureaucratic levers.

Essentially, he sought three kinds of help: people, management techniques and technology. For the first he conjured up a steady stream of senior military industrial managers and placed them in key economic posts. By the end of 1988, one third of the 58 newly-appointed ministers had backgrounds in the defense industries (including 6 of 13 first deputy and deputy chairmen of the council of ministers).[67] Second, Gorbachev patterned his agricultural reforms on the powerful Military-Industrial Commission. His establishment of Gosagroprom was intended to coordinate food cultivation, processing and distribution much as the Military-Industrial Commission did to ensure reliable military production. The defense industry's quality control program was also copied. Third, Gorbachev sought to transfer the technological advantages of the military branches to the civilian sector. "The experience of the defense branches must be used in full measure," Gorbachev declared in 1985. "We have begun this work. It has to be continued actively".[68]

In the spring of 1987, large chunks of ailing civilian industry were entrusted to the military-industrial complex. The Ministry of Machine Building for Light Food and Industry and Household Appliances was abolished and its industrial facilities were merged with the Ministry of Medium Machine Building Industry, which was in turn folded with the Ministry of Atomic Energy into a new Ministry of Atomic Energy Industry.[69] In addition to producing nuclear weapons components, this ministry was assigned several conversion tasks, including the modernization of the Soviet dairy industry.[70]

When the official policy of conversion was announced at the United Nations, therefore, many defense industries had already been summoned into action. By February 1989, Military-Industrial Commission Chairman Belousov was boasting that 345 defense industry enterprises were "involved' in the production of equipment for the light and food industries, accounting for one fifth of such output annually.[71] Overall defense industries would boost their production of civilian goods from 40 percent to 60 percent or more, concentrating on ten troubled sectors ranging from agricultural processing and medical equipment to computers and civil aviation.[72]

**"The Latest Whim"**

These promises notwithstanding, Gorbachev's U.N. announcement appears to have caught Soviet industrialists by surprise. Many simply interpreted it as "the latest in a long series of the government's whims," according to Oleg I. Gapanovich, Chairman of the Leningrad City Soviet's Commission on the Military Industrial Complex and Conversion. By and large, when enterprise directors were told to convert, they just decided to produce more of whatever civilian product they were already producing without touching their military lines. "This program was just a joke," he says.[73] The coordination of the conversion program grew even worse as the government began to feel increasing pressure to fill the consumer market. Military enterprises and design bureaus were assigned new production tasks that only coincidentally corresponded either to their specialty or to what was needed.[74]

Other critics attacked the program, meanwhile, for not being nearly bold enough. The government had announced plans to invest some 40 billion roubles in conversion from 1991-1995, but less than a quarter of this sum was to be spent on retooling military production lines. The rest was for entirely new production lines within the defense complex.[75] The government appeared to be using the banner of conversion to conceal its effort to amass more resources. The civilian production that was assigned to the military sector, mean while, suddenly fell under its traditional veil of secrecy.[76] Even the government's revised conversion plan, announced in September 1991, was met with derisive howls. The plan's details remained vague and its implementation was to be left with the same central managers who had been using conversion as an excuse to accumulate more resources. "It wasn't a plan for conversion, it was a typical conservative product," says military analyst Sergei E. Blagovolin. "In reality, it was a plan of anti-conversion."[77] While Gorbachev backed the official plan, the publicly railed against resistance to conversion too. "Many believe that further conversion will make the country weak and put in jeopardy the industry's defense sector," he declared. "This is all nonsense. I believe that somebody wants to play on the sacred feelings of the people."[78]

## Deadlock and Disintegration

By this time, however, the government plan was caught in the political deadlock that had blocked other efforts at reform, even as the center's power was crumbling. If the military-industrial complex had the strength to block widespread market reform in the fall of 1990, however, its leaders did not have the strength to push the government's conversion plan through the Supreme Soviet. The political deadlock was such that the government withdrew its draft from consideration and claimed to be revising it.

Meanwhile, the government was rapidly losing control over the military-industrial complex that it intended to reshape. Not only was its budgetary leverage disappearing, but the traditional ministerial and Communist Party structures that once watched over defense industries was dissolving. Well before the coup, the Party, which was once the central facilitator and enforcer of military production plans, was falling apart.[79] The defense industry ministries, meanwhile, faced dramatic reductions in their own staffs and budgets too.[80] Thus, when the republics had the job of managing military industries thrust upon them in the wake of the coup, the central government's conversion efforts had effectively stalled.

## The Military

Just ad the center was losing control over the process of military industrial management, the military appeared to be losing whatever influence it enjoyed as well. Before August, the senior leadership more or less agreed that economic revitalization would be crucial to any Soviet effort to keep up with the West. If the theoretical musings of Marshal Ogarkov served as a warning of the Soviet Union's slipping qualitative edge in military affairs, the daily experience of soldiers and sailors burdened by shoddy equipment bore eloquent witness to the slide in a much more concrete fashion. A series of aircraft and submarine accidents--most notably the loss of the Mike-class submarine *Komsomolets* in 1979--fanned the military's anger still further.[81] Former Chief of the General Staff Mikhail Moiseev bluntly accused defense manufacturers of a *"diktat* of the producer over the consumer and technocratic egotism."[82]

Yet the military's support for economic reform faltered as soon as the Soviet economy began to enter an accelerating slide during 1990. "It is not yet possible to shiftcompletely to direct ties with suppliers," said Lieutenant General A. Trepalin, deputy commander for rear services of the district forces. "There will be breakdowns in the still-weak market and this is impermissible for the vital activities of the troops."[83] Other officials stressed that conversion efforts were blunting whatever edges the country enjoyed in high-technology production. "[I]s it sensible when aviation works, in the Ministry of the Aviation Industry, manufacture saucepans?" asked Deputy Defense Minister Colonel General Iuri A. Iashin. "Is it sensible when individual factories produce mousetraps? Is it sensible when rocket works, that is space rocket works, manufacture some kind of bedsteads or something else?".[84]

Just what role the Ministry of Defense will play in continued conversion remains deeply unclear. Again, it remains largely upto the republics to determine the nature and structure of their new union, and to define the shape and responsibilities of any joint defense ministry. If the remnants of the armed forces carry any political weight at all, however, they are not likely to resist further conversion significantly. Even before the coup, Defense Minister Evgenii Shaposhnikov had been a strong proponent of market reforms in the military industry.[85]

## Republics and Municipalities

With the center and the military essentially stripped of their traditional power on matters of state defense, the republics

and the municipalities are now making policy. Even before the coup, both Russia and Ukraine, where the bulk of Soviet military industries are located, claimed varying degrees of responsibility for military industries during the past year.[86] While Ukrainian officials backed away from an announcement in May that they would assume control of their military-industrial complex, the coup forced the job onto them. Dwindling budgets
will likely force the republics to continue and perhaps even accelerate the central government's efforts.

Moreover, within each republic, the municipalities have set forth their own ideas about how to best draw benefit from these industries. In Moscow, for instance, the non-Communist city council has made strenuous efforts to put their military industries to some consumer use.[87]

Meanwhile, Leningrad has set ambitious goals of its own. "More than 75 percent of Leningrad's industry works for the military-industrial complex, and only 25 percent makes consumer goods," said Mayor Anatolii A. Sobchak this spring. "I would like these figures to trade places".[88] On top of all the other obstacles hindering conversion, city officials must find a way to replace the handsome tax base that their local military-industrial complex provides. "I am accused of defending the military complex," says Nikolai N. Konstantinov, deputy chairman of the planning commission of the Leningrad City Executive Committee, who has questioned some efforts to eliminate military production. "I want to defend the city budget and the city budget is suffering".[89]

While the economic changes in the Soviet economy now give defense enterprise directors broad new influence, the ultimate shape of the military-industrial complex will also depend greatly on the shape of the political and economic union that emerges among the republics. Especially important, of course, will be the relationship between Russia and Ukraine where the bulk of these enterprises lie. A healthy economic and political relationship among them could conceivably encourage the defense industrialists to increase trade, diversify their production in the civilian realm and back further market reforms. A less predictable or trusting relationship, however, could force theregimes of the newly-sovereign republics to reassert some central control over their defense industrial base so that they are assured the proper military supplies.

Many republican leaders have expressed the desire to preserve some joint responsibilities for defense. Indeed, even as the Soviet Union was coming apart, Russia and Ukraine signed an agreement by which each pledged not to take any unilateral measures on military strategic issues, but they have so far failed to enter any significant economic relationship. Defense cooperation will require political coherence and trust, neither of which is in abundant supply. Even with a modicum of trust, it is unlikely that Russia will continue to rely on Ukraine for half its tanks and missiles.[90] The recent moves by various republics to establish their own independent militia or defence forces do not bode well.[91]

## C. THE MILITARY DIMENSIONS OF CONVERSION

While Gorbachev's conversion program assumed an economic dimension to supply a starving consumer market, and a political dimension in the historic struggle over reform, it remained at heart an effort to close the Soviet Union's qualitative military lag behind the West. The evidence is mounting, however, that by their very efforts to tinker with the traditional industrial system, conversion has seriously damaged their current capabilities and all but undermined any chance of closing their military technological gap behind the West. The conversion program as it has developed in the broader context of economic and political reform has destroyed the unique strengths that enabled the Soviet Union to keep pace this far.

**Traditional Strengths**

The Soviet defense industrial base has enjoyed several unique strengths in its race to keep pace with Western military technology: political primacy, administrative control and Western borrowing.[92] Each of these is now under seige.

First, the military sector has enjoyed unquestioned political primacy in the allocation of resources, both in the amount of money spent and in the priority granted to military producers in their quest for resources. High salaries, generous benefits and a healthy dose of domestic prestige ensured that the best engineers and cadres all found their way to defense enterprises.[93] With the loss of its unquestioned priority within the Soviet system, the military industries of the country now find themselves desperately short of key resources: money, people and supplies. "Even before the supply sectors used to let us down, but now the difficulties are even more acute...," complained the first deputy minister of machine building, F. Kotov. "Out of an order of 5,000 items of equipment, the USSR State Planning Committee has allocated only 2,700 for 1990."[94] Unreliable supplies are undermining not only existing production lines, but also the introduction of new products.[95]

The Soviet Union has also lost the tight administrative management that permitted the defense complex to respond to special technological challenges by mustering the country's very best resources to address a problem.[96] Not only has the central ministerial system been sharply weakened, but the disciplinary function of the Communist Party has disappeared altogether. With enterprises losing their state orders, they are no longer entirely dependent on the center, and cannot be forced to respond as before. "If we had begun work on developing the atomic bomb today, with the present state ofour economy and society," a Soviet nuclear scientist has been quoted as saying, "more likely than not we should never have built it."[97] A specific manifestation of this systemic decay appears in the fraying network of military representatives (or *voenpredy*) in defense plants. Once accountable only to their military superiors, these quality inspectors have gradually come under the thumb of the factory bosses they are meant to supervise. Frequently, for example, they have depended on plant managers for their housing privileges.[98]

Finally, the Soviet military industrial base benefitted tremendously from technological know-how acquired from the West, either through open Western licensing of production and trade or through illicit acquisition of know-how and technology.[99] Paradoxically, just as the Soviets have opened up to Western trade, the technologies that it needs are becoming more and more difficult to incorporate.

Thane Gustafson outlines five differences that distinguish the technologies that the Soviet military would like to assimilate now from those it managed to absorb in the past. First, the current generation of technological challenges has a broader and closer dependence on new areas of science where the Soviet Union lags. Second, they are increasingly dependent on civilian innovation, which has been largely barren in the Soviet experience. Third, new weapons systems are really interdependent systems of technological applications, making any simple reproduction exceedingly difficult. Fourth, there is a growing importance of software, where the Soviets are weaker. Fifth, the Soviet system cannot take advantage of new opportunities for reliability, ruggedness, custom design, and flexibility that are now so important in advanced Western designs.[100]

## Military Consequences

The military consequences of the Soviet military conversion are difficult to assess. From the available evidence, however, the complaints from both the military industrialists and their uniformed customers leave little doubt that the economic and political reforms of recent years have disrupted the Soviet Union's defense supply base significantly. Moreover, if current capabilities are in question, any efforts to introduce new technologies to fight what strategists envision as future wars cannot be considered more than wishful thinking.

From defense industrialists and central planners, both the force and specific nature of their warnings suggest a mechanism that is badly fraying. "[A]s part and parcel of our country's economic and industrial complex," announced Defense Industry Minister Boris M. Belousov, "our sector is feeling the full impact of all the collisions of our times associated with the reorganization of the economic mechanism or the economy's management, the attempts to organize market relations and the democratization of production and social interrelations, which is aggravated by the negative process which is occurring in a number of the country's regions".[101] In a remarkably candid conversation, Communist Party Secretary Baklanov declared: "Unfortunately, we are now losing our scientific-industrial potential, let alone other things, and this is dangerous... the backbone of our military industry is being broken".[102] Others in and around the industry complain about how conversion has set some of the country's best resources on tasks that fall well beneath their traditional capability. "It would not occur to anyone to drive a nail with a Japanese tape recorder," complained Boris M.Bocharov, deputy director of the "Energiia" Scientific-Production Association, where the Soviet space shuttle is produced, "but we force highly skilled specialists to screw nutson bolts." Conversion plans reportedly have his plant making a compressor for blowing flour onto a conveyer, surgical instruments, prostheses, a candy-wrapping machine, a cabbage planting system and a dough kneader.[103]

## Technical Prospects

Soviet military officials continue to clamor for rapid technological innovation in their defense industries, but the task of modernizing Soviet production lines amid such political chaos and industrial decay has become more elusive than ever.[104] Each of the elements that enabled Soviet military industries to keep up with Western technology has disappeared or is seriously threatened, whether it is money and skilled labor or administrative priority and special access to supplies. Engineers and specialists have been leaving the plants altogether to find more lucrative work in cooperatives.

Military enterprise directors now say that they count on the introduction of a market economy to help the Soviet Union make up its technological lag behind the West. They also hope that Western investment in their plant will help them procure modern production technology, which is likely to eventually find its way back into military applications. While reasonably optimistic about the prospects for his own shop in the years ahead, Spasskii of Leningrad's "Rubin" Design Bureau is decidedly pessimistic about the prospects for any significant scientific-technical revolution in the Soviet Union. There just haven't been enough resources, especially money, to make it work. "We needed to have a stimulus for that revolution to proceed", Spasskii laments. "We hope that when the market starts working, the new technologies will come with it. "Even with good, intensive work and a thoughtful approach", however, he does not foresee any significant advance for another 15 or 20 years. Some openly admit they do not expect any real progress for a decade or two. While it is conceivable that an authoritarian regime might reimpose discipline on the system of state mandated supply, the damage already done to the infrastructure of military supply will not be easily restored.

## Conclusions

Wherever danger appeared for Moscow, whether in the vague form of "imperialist aggression" or the more specific form of the U.S. atomic bomb, this massive aggregation of scientific and industrial resources was ordered to answer the call. With the disintegration of the Soviet economy, planners struggling to supply a desperate consumer market rather than match the armed strength of the West, these same prize industrial assets have been summoned forth again. Attempts at conversion so far, however, have been more destructive than constructive. The dual strains of a collapsing political order and a disintegrating economic system are sending unprecedented tremors through this crucial national foundation.

Economic forces have been driving whatever military conversion there has been so far as weapons orders have fallen and enterprise directors have been forced to improvise with new lines of production to keep their workers occupied. This type of conversion can be expected to yield rational civilian production only randomly. Meanwhile, the breakup of central economic planning mechanisms leaves military enterprise directors with a crucial role in coordinating and protecting much of the country's industrial activity. They have incentives to back the transition to a market. This may lead randomly to some useful conversion, but ultimately it will mean the degradation of the country's military-industrial base. They also have incentives, therefore, to reimpose some central control. This may preserve somemilitary industries, but it will end any hope for economic or political reform.

Whether they continue to participate in the further introduction of market reforms will depend heavily on the extent and kind on political co-operation that arises among the newly sovereign republics, and especially between Russia and Ukraine. It is possible that a reasonably healthy economic and political atmosphere will offer enough incentives for these enterprise directors to back further market reforms. Far more likely, however, are a series of tenuous and mistrustful links that will leave each republic to rely essentially on its own industrial base for its security needs, and force the re-establishment of some centralized industrial management.

The combined effects of these new economic choices and these emerging political trends point unambiguously to the steady disintegration of the once mighty Soviet military-industrial complex. Even if some successor regime suddenly chose to crank up massive military production anew, the facilities appear incapable of responding for quite some time. Military industries face the same disintegrative pressures that threaten all Soviet institutions. Without Herculean efforts to restore traditional planning methods, this complex of industries will have great difficulty meeting the most basic needs of the Soviet Military. Still less will it be up to the task of introducing the advanced weaponry that Soviet strategists insist will be crucial in fighting future wars.

## Endnotes

1. The author is a Research Fellow at Hudson Institute, 1015 18th Street NW, Suite 200, Washington D.C.20008; Tel. 202-223, 7770.

2. In the United States, the city of St.Louis has managed admirably to cushion the blow of recent defense cuts on its economy. City officials have helped retrain workers and lured new enterprises to the area, but individual firms have found it all but impossible to make the transition. See Richard W.

Stephenson, "So Far, St.Luois Handles Arms Cuts," *the New York Times*, August 8, 1992.

3. K.E. Voroshilov, "Strengthening the Defence of the USSR," in *From the First to the Second Five Year Plan,* (Moscow and Leningrad: Cooperative Publishing Society of Foreign Workers in the USSR, 1933), p.350. Quoted in David Holloway, "The Soviet Union," in Nicole Ball and Milton Leitenberg, *The Structure of the Defense Industry: An International Survey* (New York: St.Martin's Press, 1983), p.51

4. Aviation (aircraft & helicopters), communications equipment (electronic warfare equipment), defense (conventional armaments), electronics (computers), general machine-building (rockets & space equipment), machine-building (munitions), medium machine-building (nuclear warheads), radio (radar), and shipbuilding (naval ships). Some industries of the electrical equipment industry, chemical industry, and petroleum-refining industry also belong to the defense industry and operate under the supervision of the Military-Industrial Commission. See Christopher Davis, "The High Priority Military Sector in a Shortage Economy," in *The Impoverished Superpower: Perestroika and the Soviet Military Burden*, edited by Henry S. Rowen and Charles Wolf. Jr. (San Francisco: Institute for Contemporary Studies, 1990), pp.162-163. For a more detailedaccount of their evolution, see Holloway, "The Soviet Union," pp.64-65.

5. For a full account of the procurement process, see Julian Cooper, "The Defense Industry and Civil Military Relations," in *Soldiers and the Soviet State: Civil Relations from Brezhnev to Gorbachev*, eds. Timothy Colton and Thane Gustafson, (Princeton: Princeton University Press, 1990), pp.165-66.

6. "Great Consultation with the Working Class," *Sovetskaia Rossiia*, April 28, 1990, pp.1,2 (reprinted in Foreign Broadcast Information Service *Daily Report: Soviet Union*, hereafter FBIS-SOV-90-084, pp.106-107).

7. Richard Ericson writes: "Thus, the military places a greater burden on the development of the economic system than is generally believed in the West, further aggravating the now widely acknowledged economic slowdown. Moreover the slowdown must have a greater impact on the military, as there are fewer resources to shift in order to maintain its rate of growth. The demands of the military may be adding urgency to the drive toward radical reconstruction as it strives to maintain parity with the United States on a smaller economic base." See "The Soviet Statistical Debate: Khanin versus TSSU, in *The Impoverished Superpower*, p.91.

8. S.E. Blagovolin, "Geopolitisheskie aspekty oboronitel'noi dostatochnosti," *Kommunist*, No.4 (1990), p. 115.

9. This estimate comes from the Institute of Economic Prognosis. See Andrei I. Kondakov, "Konversiia v SSSR. Chetvertaia popytka," *Mezhdunarodnaia Zhyzn* (September 1991): p.49.

10. This estimate came from Ivan Silaev, former premier of the Russian Federation. RFE/RL, *Daily Report*, November 12, 1990.

11. Russia's Vice President Aleksandr Rutskoi made this estimate. "poslednii Bal VPK," *Moskovskie Novosti* No.39, 29 September 1991, p.13.

12. D.Derk Swain, "The Soviet Military Sector," in *The Impoverished Superpower*, p.104.

13. Moreover, as soon as the war had ended, Stalin announced that three more Five Year Plans of investment in heavy industry would be required for the Soviet Union to be prepared to meet "all contingencies." What retooling did occur led to a 17 percent drop in industrial output as plants adjusted to much more complicated forms of production. Only in 1947 did it pick up again. Holloway, "The Soviet Union," p.53.

14. In 1971, Leonid I. Brezhnev told the XXIVth Party Congress that some 42 percent of the military economy was devoted to military production.

15. Colonel-General Nikolai Chervov of the General Staff elaborated in an interview in January 1990 that a 20 percent reduction in weapons procurement would include a reduction in the output of tanks of 50 percent, ammunition of 30 percent, combat helicopters by 25 percent and armored vehicles by 20 percent. Moscow World Service, January 11, 1990 (FBIS-SOV-90-010, p.1).

16. See "Beyond *Perestroyka*: The Soviet Economy in Crisis," A paper presented by the Central Intelligence Agency and the Defense Intelligence Agency to theTechnology and National Security Subcommittee of the Joint Economic Committee, Congress of the United States, (June 1991), p.6, 18,. But with an increase in Soviet industrial wholesale prices rose of some 67 percent on January 1, 1991, John Tedstrom estimates a procurement drop of 25-27 percent. "Glasnost" and the Soviet Defense Budget," *Report on the USSR*, July 19, 1991, pp.6-14. Research and development, meanwhile posted a real drop from 13.2 to 12.4 billion, which will likely have dramatic effects on any technological developments.

17. Arkadii Vol'skii, member of the Committee for the Operational Management of the National Economy, announced that expenditures on defense industries would be cut 5.6 billion roubles. Radio Rossii Network, September 28, 1991 (FBIS-SOV-91-189, p.43).

18. Christopher Kedzie and Regina Mayor, "Soviet Defense Facilities Conversion: Implications for Western Assistance" Unpublished Paper, Harvard University, April 19, 1991.

19. Interview in leningrad.

20. Interview in Leningrad.

21. Interview in Leningrad.

22. RFE/RL, *Daily Report*, August 8, 1991. In 1989 and 1990, some 300,0--00 specialists were reported to have left. "if this process is not stopped, the government-run industry will collapse," warned Party Secretary Oleg Baklanov. "Is it right that workers at a state-run plant receive only 10 to 15 kopeks per rouble's worth of output, while a workersof a cooperative receives upto 50 kopeks?" "Problems of Conversion," *Glasnost*, 4 October, 1990, (FBIS-SOV-90-212, pp.65-66).

23. Interview in Moscow.

24. And only 15 percent meet international standards. Aleksei Iziumov, "The National Experience of the USSR." Disarmament: A periodic review by the United Nations, vol.XIV, No.1 (1991): pp.63-64.

25. "Conversion Needs Reliable Guidelines, *"Narodnyi Deputat*, No.4, (FBIS-SOV 91-120, pp.38-40). Also "Beyond *Perestroyka,"* p.9.

26. TASS, August 6, 1991 (FBIS-SOV-91-152, p.33). In July 1991, Shcherbakov had announced that conversion could well expand to as much as 80 percent of the defense complex. Coming as it did on the eve of Gorbachev's pitch to the London summit of industrial nations, however, the promise was vulnerable to skepticism. RFE-RL, *Daily Report*, July 16, 1991, p.1991.

27. See "The Shoemaker Bakes Pies..." *Trud*, February 20, 1990, p.2(JPRS-UMA-90-007, p.103-104).

28. Peter Almquist and Kevin O'Prey, "Beating Swords into Agricultural Complexes: Soviet Economic Conversion," *Arms Control Today*, December 1990, p.19.

29. Conversion, economic uncertainty and the recent introduction of computers has forced "Rubin" to reduce its workforce of "several thousand' by some 20 percent. Many, says director Igor' D. Spasskii, are more than happy to be retrained for the hotel business. Interview in leningrad.

30. For an excellent case study, see William S. Ritter, Jr., "Soviet Defense Conversion: The Votkinsk Machine-Building Plant," *Problems of Communism* (September-October, 1991): 45-61.

31. Most plants are monopolies or new monopolies in their line of production. A recent survey in the Soviet press found that 77 percent of the production lines in the machine building, metallurgy, chemical, timber and construction industries were supplied by only one producer. The machine-building industry alone tallied 87 percent. See Phillip Hanson, "Dimensions of the Monopoly Problem, *Report on the USSR*, April 26, 1991, pp.5-6.

32. For a profile of a commodities exchange in Moscow, see "Birzha: Dom svidanii rynochnoi ekonomiki s komandnoi," Moskovskie Novosti, No.37, September 15, 1991, pp.8-9. An *Izvestiia* poll in June 1990 showed 66 percent in favour of a market economy with 19 percent against. A *Wall Street Journal* poll in February 1991 showed 85 percent favoured private ownership of land in RSFSR. Meanwhile, by the end of 1990, 6.2 million were employed in cooperatives, 3,000 joint ventures had been established, 1,400 commercial banks and 40,600 private farms. See Anders Aslund, "Prospects for Economic Reform in the U.S.S.R." Paper prepared for the World Bank's Annual Conference on Development Economics, Washington D.C., April 25-26, 1991.

33. Indeed, Mikhail Agursky has argued that this isolation has led to widespread mediocrity within the defense sector. See *The Soviet Military-Industrial Complex*, Jerusalem Papers on Peace Problems, No.31 (Jerusalem, The Magnes Press, 1980), p.14.

34. Such ideas were expressed a number of times in conversations with defense enterprise managers and planners in the Soviet Union this summer. Some realize that their market understanding is primitive and have signed up for such training as Black Sea cruises featuring lectures by Western business experts.

35. For a full case study of Saturn, See Kedzie and Mayor, "Soviet Defense Facilities Conversion."

36. For a fascinating look at how crippling secrecy can be to Soviet military industries, see Agursky, *The Soviet Military-Industrial Complex*, pp.12-16.

37. Describing a recent trade fair in Italy put on by Soviet military industrialists, one reporter noted a clear evolution in the

tradition of secrecy. "If you are buying a pig in a poke, then (the catalog) offers a preliminary description of both the pig and the poke." See "Poslednii Bal VPK," *Moskovskie Novosti*, No.39, September 29, 1991, p.13.

38. When Soviet military industrial managers took control of the Ministry of Machine Building for Light Food and Industry and Household Appliances in 1987, the costs of food processing equipment soared. Partly, this was habit. Military enterprises are monopoly producers and are not very attentive to controlling costs. Partly, they tend to use more sophisticated equipment in production. See Judyth. L. Twigg, "Balancing the Soviet Military and Civilian Economies," in *Science and International Security*, Eric H. Arnett, ed., (Washington, d.C.: American Association for the Advancement of Science, 1990), p.256.

39. Interview with Dmitri A. Chernenko, director of 'Leninskaia Kuznitsa" Shipyard in Kiev.

40. Interview in Kiev.

41. The Umdurt ASSR in Russia's Ural region, for example, has some 83.5 percent of its industry engaged in military production. "Roll Up Your Sleeves in Order Not to Have to Tighten Your Belt-This is How the Defense Workers of Umdurtia View Conversion," *Pravda*, March 29, 1991, p.3 (FBIS-USR-91-009, p.51).

42. Their Swiss-based partner is called Inter-Maritime Management.

43. Among other things, they have used their financial weight to take control of alternative television outlets. In May, a group of defense industrialists in Leningrad was reported to be exploring the possibility of buying the local television company, which had long been feuding with authorities in Moscow over their more progressive programming schedule. RFE/RL *Daily Report*, June 20, 1991. In June, a group of some 120 defense managers in Moscow called "Konversiia" helped pull the Russian Federation's television network out of debt with a $6 million loan. "Rossiiskoe TV zhivet na deng'i Borisa Iosifovicha" *Kommersant*, No.20, 1991, p.10.

44. Interview in Leningrad.

45. Traditionally, while senior military officials have written extensively on questions of security policy, the leaders of defense industries have not played a visible political role. When they ventured to go public, their concerns tended to vary little from those of their civilian industrial colleagues: sloppy planning, unrealistic production targets, inadequate supplies, etc... Special claims based on their special role in defending the Motherland were rare. See Peter Almquist, *Red. Forge: Soviet Military Industry since 1965* (New York: Columbia University Press, 1990), pp.46-65 and 126-129.

46. The Scientific-Industrial Union's members include directors of state enterprises and the fledgling private sector and claims to account for 60 percent of the country's gross industrial sales. See "Topical Interview," Moscow Central Television, July 3, 1991, (FBIS-SOV-91-129, pp.41-43. See also, A.I. Vol'skii, "Rynok i ekonomicheskaia stabilizatsiia," *Kommunist*, No.6 (April) 1991, pp.46-57.

47. See "`Oboronka' budet torgovat", *Argumenty i Fakty*, No.23, p.4, and "Cosmonautics' Second Profession" *Izvestiia*, July 5, 1991 p.2 (FBIS-SOV-91-134, p.77).

48. Interviews in Kiev, Moscow and Leningrad. Those factories that do not have far to search for civilian output (for example, aviation and shipbuilding) will enjoy the most success.

49. Viktor Protasov, Chief designer at a defense plant, complains that the Finance Ministry has "set certain norms of profitability for defense industries," while suppliers have been allowed to set their prices at will. "As a result about 500 defense plants have come out as losing, some of them up to 200-300 million roubles in debt. So they begin refusing to accept defense orders." Interfax, October 4, 1991 (FBIS-SOV-91-194, p.6).

50. Soviet officials were aggressively drumming up business at the Paris Air Show this year.

51. The Boston-based investment firm Batterymarch Financial Management has been accorded extraordinary access to traditionally secret Soviet military enterprises with its talk of an investment fund of some $ 1 billion.

52. See "Don't Give Up Now: A Survey of Business in Eastern Europe,"*The Economist*, September 21, 1991 pp.10-12.

53. They further complained of "destructive criticism and attacks on the defense sector as well as attempts by authorities at all levels to meddle in the management of defense enterprises." See "Status `oboronke'" *Pravda*, September 6, 1992, p.2.

54. "Open Letter to USSR President M.S. Gorbachev and USSR Prime Minister V.S.Pavlov," *Sovetskaia Rossiia*, February 26, 1991, p. 2 (FBIS-SOV-91-041, pp.43-44).

55. Indeed, two members of the State Committee on the State of Emergency were Oleg Baklanov, the Communist Party secretary for defense industries, and Aleksandr I. Tiziakov, the President of the Association of State Enterprises and Industrial, Construction, Transport and Communications Facilities. Pavlov also joined the coup attempt.

56. RFE/RL *Daily Report*, September 27, 1991. Also, Interfax, September 17, 1991 (FBIS-SOV-91-181, p.34).

57. Interfax, October 9, 1991 (FBIS-SOV-91-197, p.42).

58. William E. Odom, "Soviet Military Doctrine," *Foreign Affairs*, (Winter 1988/89), pp.118-131.

59. He quoted Engels that "Nothing depends on economic conditions as much as the armed forces." Rusell Bova, "The Soviet Military and Economic Reform." *Soviet Studies*, Vol.XL, No.3 (July 1988), p.393. See also Thane Gustafson, "The Response to Technological Change," in *Soldiers and the Soviet State: Civil Relations from Brezhnev to Gorbachev*, eds. Timothy Colton and Thane Gustafson, (Princeton: Princeton University Press, 1990, p.217.

60. Mary C. FitzGerald, "The Soviet Image of Future War: Through the Prism of the Persian Gulf," *Comparative Strategy*, Vol.10, No.4 (Fall 1991).

61. As Admiral A.I. Sorokin, first deputy chief of MPA of Army and Navy, declared at Central Committee Plenum in June 1987: "The combat power of the Army and Navy depends on the level of economic development of our country and the conditions of the economy." Admiral N. Smirnov addressed plenum in similar terms, noting ow its economic program was vital to the maintenance and renewal of the large, modern and

complex Soviet Navy. See Bova, "The Soviet Military and Economic Reform," pp.398-399.

62. This new philosophy was outlined most completely in Gorbachev's book *Perestroika i novoe myshlenie dlia nashei strany i dlia vsego mira* (Moscow: Izdatel'stvo politicheskoi literatury, 1988).

63. M.S. Gorbachev, *Izbrannye rechi i stat'i*, vol. 7, Moscow: Izdatel'stvo politicheskoi literatury, 1990, p.198.

64. These are the surveys of the All-Union Scientific Research Institute for the study of the Population's Demand for Consumer Goods and Trade Cycles, attached to the Ministry of (Domestic) Trade. Anders Aslund, "Gorbachev, Perestroyka, and Economic Crisis," *Problems of Communism* (January - April 1991), p.19.

65. "Conversion: The Moscow Radius of Action," *Moskovskaia Pravda*, October 24, 1989, pp.1,2. (FBIS-SOV-89-215, pp.93-96).

66. Cited Almquist and O'Prey, "Beating Swords Into Agricultural complexes," p.18.

67. Cooper, "The Defense Industry and Civil Military Relations," pp.177-79. Senior leaders such as Lev Zaikov, Egor Ligachev, Nikolai Ryzhkov and Iuri Masliukov, who became central figures in the early Gorbachev leadership, had all risen through the privileged ranks of the military industries.

68. Ibid., p.180.

69. William H. Kincaide and T. Keith Thomson, "Economic Conversion in the USSR: Its Role in Perestroika," *Problems of Communism*, (January-February, 1990), p.90.

70. In addition, the Ministry of Machine Building merged with the Ministry of Defense Industry, and the Ministry of Communications Equipment Industry was folded into the Ministry of Radio Industry. Finally, the Military-Industrial Commission was entrusted with the reform and management of the State Committee on Computer Technology and Information science and the Ministry of Civil Aviation.

71. Belousov also declared that the defense complex was contributing "2,000 categories of goods for cultural, consumer service and domestic purposes," and by 1995 would be offering some 3,000 "categories of new equipment" to agro-industrial production. "Our Dialogue, *"Sovetskaia Rosiia*, February 10, 1989 p.1 (FBIS-SOV-89-028, pp.82-84).

72. Moscow Domestic Service, June 28, 1989. (FBIS-SOV-89-124, pp.36-37). Some of the specific promises were tantalyzing indeed: stereo television sets, general purpose cookers, refrigerators, freezers, vacuum cleaners, electric meat grinders, clothes driers, irons, dishwashers, compact disc players, camcorders, pocketwatches, voice synthesizers, and more. "Goods for the People off the Military Production Line," *Sovetskaya Rossiya*, February, 10, 1989, p.1 (FBIS-SOV-89-028, pp.82-84).

73. Interview in Leningrad.

74. "There was tremendous desire to improve our country's situation... the chief factor that drove (conversion) was how to quickly feed our market with consumer goods," says Igor' D. Spasskii, director general of the "Rubin" Central Maritime Design Bureau in Leningrad. Interview in Leningrad.

75. "Beyond *Perestroyka*," p. 9. The Soviet government invested about 9 billion roubles in metallurgy during 1986-1990.

76. Cooper, "The Defense Industry and Civil Military Relations." p.182.

77. Interview in Moscow. For examples of other detailed critiques of the government's approach, see Aleksei Kireev, "Conversion in the SovietDimension," *International Affairs* (May 1990), pp.92-102 or Aleksei Iziumov, "Conversion? Conversion! Conversion..." *Literaturnaia Gazeta*, July 12, 1989, p.11 (FBIS-SOV-89-141, pp.99-103).

78. Moscow World Service, November 28, 1990 (FBIS-SOV-90-230, p.45).

79. During the past two years, one fifth of the Party's membership left. Of those who remained, large portions refused to pay the two percent of their salary in membership dues. By summer 1991, this left the vast apparatus with a budget deficit approaching 1.1 billion roubles this year. Francis X. Clines, "Bankruptcy Seen as Threatening the Soviet Party," *The New York Times*, August 9, 1991.

80. The Council of Ministers staff itself was to be cut by 32 percent, with some ministries facing reductions of as much as 50 percent. RFE/RL, *Daily Report*, June 20, 1991.

81. Shipbuilding ministry has incurred fines of more than 3 million roubles for delivering substandard parts for submarines alone in 1985-1989. See Almquist and O'Prey, "Beating Swords Into Agricultural Complexes," p.20.

82. Acknowledging that weapons would cost much more under new conditions of cost-accounting, he continued: ".... (U)ltimately it is not a question of money. It is very important that the new equipment should be of the highest quality, that ships should not sink, that aircraft should not fall apart, that people should not die. That is what causes the most anguish." "Arguments of General Staff," *Izvestiia*, February 23, 1990, p.3 (FBIS-SOV-90-038, pp.35-36).

83. "Entering the Market," *Krasnaia Zvezda*, Jan. 19, 1991, p.2 (JPRS-UMA-91-007, p.36-37).

84. Moscow Domestic Service, February 22, 1990, (FBIS-SOV-90-039, pp.76-78). General Moiseev himself grumbled that for conversion there was no "well considered concept of this most important state matter. It is probably no wonder that at first we almost fouled it up." See "Military Reform: Realities and Prospects," *Krasnaia Zvezda*, June 12, 1991, pp.1-2 (FBIS-SOV-91-115, pp.35-40).

85. "Under the conditions of a market we will arrive at more businesslike and democratic relations. Life will compel it... We hope that a similar approach will compel aviation firm to be more enterprising, that it will create competition between them, that it will stimulate their creative search, and that, in the final analysis, it will have an effect on the quality of the product." See "Money Likes to be Looked After," *Krasnaia Zvezda*, September 15, 1990, p.2 (FBIS-SOV-90-183, pp.54-56).

86. Some 80 percent of the U.S.S.R.'s defense factories lie within Russian Republic, and they are heavily concentrated in areas of outspoken political opposition to the center: 20 percent in Urals, 17 percent in Central Russia, 10 percent around Leningrad. One third of all industries in Moscow work are defense-related. Iziumov and Kortunov, "The Monster: A Profile of the Military-Industrial Complex." About 13

percent of defense industries are in the Ukraine. RFE/RL *Daily Report*, May 29, 1991, p.8.

87. "Konversiia: Stolichnaia versiia, "*Moskovskie Novosti* No.28, July 4, 1991.

88. "Krizis vlasti ili vlast' v usloviiakh krizisa?" *Argumenti i Fakti*, No.12, March 1991, p.2.

89. Interview in Leningrad.

90. John Tedstrom, "Industrial Conversion in Ukraine: Policies and Prospects," *Report on the USSR*, August 23, 1991, pp.12-16.

91. Azerbaijan, for example, recently claimed ownership of all Soviet military equipment on its territory. Ukraine, meanwhile, announced its intention to form its own 400,000-strong army.

92. For a fuller discussion of the particular advantages of Soviet military industries, See Bova, "The Soviet Military and Economic Reform," p.386 and Weickhardt, "The Soviet Military-Industrial Complex and Economic Reform," pp.195-196.

93. The importance of prestige is not to be underestimated in the success of military industries. As Ed A. Hewitt writes: "It conveys a sense of urgency and importance to the work of the entire sector, clothed in an aura of patriotism, which is absent from the civilian sector. With that comes high priority in procurement of materials, investment and access to talented labor.... As a result, the military industries, unlike their civilian counterparts have the wherewithal and the incentive to produce high-quality products and to constantly search for ways to improve them." See *Reforming the Soviet Economy: Equality vs. Efficiency*, (Washington, D.C.: The Brookings Institution, 1988), pp.219-220.

94. "Goods Instead of Shells," *Izvestiia*, May 1, 1989, p.2 (FBIS-SOV-89-086, p.83-84).

95. Said P. Romanov, director of a machine building plant: "We are trying to make maximum use of the existing range of equipment and raw materials. To be honest I would never try to assimilate an article for which I do not have everything necessary at my disposal." *Ibid.*

96. As Defense Minister Dmitri F. Ustinov said in 1977: "Those who count on attaining military superiority over the Soviet Union with the help of such weapons (new types of weapons of mass destruction) should remember that the economy, science and technology in our country are now at such a high level that we are in a position to create in the shortest time any type of weapons on which the enemies of peace might wish to count." Cited in Holloway, *The Soviet Union and the Arms Race*, p.146.

97. "The Keys from the Nuclear Arsenal," *Pravitelstvennyi Vestnik*, March 91, p.12. (JPRS-UMA-91-013), p. 39-41. Even officials of the once-powerful Military-Industrial Commissions acknowledge that their influence has been sharply reduced. Interviews in Moscow.

98. Wrote one inspector: "As a result, the chief of military representation at times was forced to resort to doubtful compromises of conscience and accept production, let us say with unclear results from so-called periodic testing." "The Military Representative Under Dual Control: What Compels Him to Close His Eyes to Waste," *Izvestiia*, Jan 24, 1990, p.2 (JPRS-UMA-90-008, p.45).

99. For history of Soviet technological development, and a status of how SU lags in key military technologies, see Holloway, *The Soviet Union and the Arms Race*, pp.134-135. See also, Julian Cooper, "Western Technology and the Soviet Defense Industry," in *Trade, Technology and Soviet-American Relations*, ed. Bruce Parott (Bloomington: Indiana University Press, 1985), pp.170-175.

100. Thane Gustafson, "The Response to Technological Change," in *Soldiers and the Soviet State*, pp.207-13. "Created to cope with industrial backwardness and constant external change, the existing system has succeeded by concentrating priority and limiting technological risk-precisely the behavior it is now supposed to abandon," Gustafson writes. "The military-industrial system of weapons acquisition was Stalin's proudest creation, and it expressed the very essence of the Stalinist strategy of development-but it is now being asked to lead the way in a dismantling of the Stalinist order." *Ibid.*, p.238.

101. "Debate speech by B.M. Belousov, "*Pravda* February 7, 1990, p.2. (FBIS-SOV-900-027, February 8, 1990, 69-71).

102. In the same conversion, General Igor Rodionov responded: "That means I was right. Conversion from political, demagogic considerations is a crime, the murder of the army." "V gostiakh u generala Rodionova," *Den'*, No.9, (May).

103. "Come Now! Who Is Directing the Parade?" *Nedelia*, No.15, April 9, 1990, p.9 (JPRS-UMA-90-017, pp.36-38).

104. Again, see FitzGerald, "The Soviet Image of Future War".

# TECHNOLOGY POLICY & ECONOMIC CONVERSION IN THE UNITED STATES: BEYOND "DUAL-USE - TARGETING NATIONAL NEEDS FOR ECONOMIC DEVELOPMENT

**Joel Yudken, Ph.D**
*Project on Regional and Industrial Economics, Rutgers University, U.S.A.*

## Introduction : Barriers to Conversion

In the early eighties, aerospace giant McDonnell Douglas Corp. (MDC) was in a serious quandary. Its Douglas Aircraft plant in Long Beach, California was operating at only 30 percent capacity and as many as 10,000 of its employees had been laid off. MDC is the product of a 1967 merger between the St.Louis-based McDonnell Aircraft Co., a highly successful builder of military fighter aircraft, and the California-based Douglas Aircraft Company, known mostly for its commercial jetliners (the DC-9 and DC-10). Cash-rich McDonnell had hoped to establish a counter-cyclical ballast for its boom-and-bust military business, while Douglas was looking to be bailed out of its financial woes. But the marriage failed to produce the desired results for either partner. Soft civilian aircraft markets, accidents involving its planes, mismanagement, and bungled sales all contributed to Douglas' predicament.

Bob Berghoff, president of United Auto Workers Local 148 was also worried. Most of the unemployed Douglas workers were members of his unit. With the aid of an economic adjustment specialist from the California Dept. of Economic and Business Development, and consultants from the Silicon Valley-based Midpeninsula Conversion Project (now the Centre for Economic Conversion), Berghoff took the initiative. He succeeded in setting up a series of meetings with plant managers to explore alternative product options -- mass transit vehicles, commuter aircraft, and cogeneration equipment -- for utilizing idle plant capacity and reemploying some of his workers.[1]

At the time, MDC had few other options. It was unsuccessful in finding a foreign or domestic partner to invest in a joint design and production agreement for making a new fuel-efficient commercial widebody jetliner. The R&D costs alone for such a plane were estimated to be as high as $2 billion. The alternative was to invest monies out of its own pocket for the new product. The company also embarked on aggressive efforts to sell stretch versions of its MD-80 (the old DC-9 Super 80). While MDC eventually was somewhat successful with the MD-80s, these sales and those of its KC-10 military tanker (a converted DC-10) and some navy trainers, were not enough to offset the decline in Douglas' fortunes. In the end good business sense dictated yet another option. Despite signs of serious interest in the union-initiated proposal, MDC's upper management in St.Louis instead chose to militarize the Douglas plant. Douglas won a multi-billion R&D contract from the air force for the C-17 supercargo plane--with follow-on production contracts likely--flipping around its traditional 30/70 percent mix of military to civilian business to 70/30.

The MDC experience exemplifies some of the principal obstacles standing in the way of a successful economic conversion program for military-dependent firms, workers, and communities in the United States. One barrier is the "wall of separation" between military and civilian businesses.[2] Because military-dependent and civilian commercial firms operate in very different market environments, their respective organizational structures and cultures, and engineering, design and marketing practices tend to diverge, as well.In the MDC case, for example, McDonnell's military-oriented top executives, mostly members of the founding McDonnell family, consistently clashed with the Douglas leadership, and made many mistakes in their commercial marketing efforts. Their solution was to turn Douglas into a predominantly military firm; this was a type of business they understood.

An equally important barrier--one, in fact, that defines the larger context that has fostered this `wall of separation'—is the lack of coherent national policies that provide incentives and markets for converting military facilities. On the contrary, since World War II, the U.S. government has conducted a concerted military-led industrial policy that created a highly concentrated complex of military-dependent industries with substantial exit barriers institutionalized by government policies: the lack of effective markets for new alternative technologies and products; the lack of technical, marketing, and organizational experience and competence to produce and market these products; and the lack of incentives and supports for military firms to invest in ways to overcome these limitations. As the MDC case illustrates, military industrial companies will not be motivated to invest in new civilian technologies as long as powerful incentives remain in the military market, where both R&D and production costs are underwritten by the Pentagon.

The winding down of the Cold War and the growing competitiveness in global markets, however, present the United States with a new and very complex challenge for the 1990s: how can it most efficiently redirect the enormous economic, technological and human resources tied to military production to needed civilian applications, simultaneously strengthening economic performance and improving the quality of life, while mitigating the economic hardship for military-dependentworkers, businesses, and communities caused by the military build down? In this paper, it will be argued that to meet this challenge the tasks of converting military resources to civilian use and instituting new industrial policies for promoting economic development and meeting basic human needs must be linked in a common economic agenda implemented through a new national policy strategy.

First, the paper examines the nature, roots and costs of the military's industrial policy in the U.S., which has favored the development of a select set of industries (aerospace, communications and electronics equipment), while it neglected, ignored, or short changed others. By contrast, other advanced industrial nations have targeted civilian economic and social concerns for investment and technical development. Hence, European and Japanese companies enjoying the backing of their respective governments, through R&D subsidies, tax breaks, trade protections and other institutional supports have rapidly increased their share of civilian product markets in industry after industry. While U.S. companies lead the world in producing advanced military aircraft, missiles, and satellites systems, Airbus Industries, subsidized by four European-governments, has been capturing wider and wider shares of the commercial aircraft market, even threatening the predominant position of Boeing.[3] While U.S. electronics firms excel in producing a wide-range of ("milspec") electronic devices and components for military command, control, and communications, guidance, and electronic warfare systems, Japanese and European firms, with help from their respective governments, have captured most of the consumer electronics and components markets, and are rapidly chipping away in the few remaining high-tech areas of U.S. commercial superiority, i.e., computers and semiconductors. Similar stories can be toldfor advanced materials, robotics and machine tools, supercomputers, and other "critical technologies".[4]

Second, the paper critically assesses an important new policy initiative, dubbed "dual-use technology policy", which has emerged in reaction to the failure of traditional laissez-faire, cold war era policies of the Reagan/Bush administrations to reverse the decline in U.S. economic performance. Championed by a loose coalition of high-tech industrial, military, academic, and government officials, dual-use policy would give the government, particularly the Defense Department, a more active role in supporting the development of specific areas of advanced technology considered useful for both military and civilian ends. Motivated largely by industrial concerns over the declining U.S. position in civilian manufacturing, and Pentagon worries about a comparable deterioration of the defense industrial base, it seeks to reconcile the increasingly divergent needs of military and civilian technology, to bolster simultaneously U.S. industrial competitiveness and military superiority. This proposal constitutes a new de facto industrial and economic conversion policy for the nation, which has garnered strong bipartisan support in the Congress and, despite general White House opposition, even the backing of important Bush administration officials.

Third, the paper evaluates the limits of both the traditional macroeconomic and dual-use policies, first, with respect to their stated objective of improving U.S. economic competitiveness, and then, more importantly, with respect to meeting broad national needs. It contends that laissez-faire policies, which rely solely on market forces, will be inadequate for stimulating new technical and economic progress in the coming decade, and that dual-use policy would perpetuate the influence of the military's industrial policy onU.S. technical development, creating too few demand inducements in the civilian sector to truly improve the nation's sagging economic performance.

Fourth, the paper examines the prospects of economic conversion of military industry in the light of this policy debate, arguing that neither the laissez-faire nor the dual-use policy fosters the necessary demand and supply supports for enabling a successful conversion of military technological and economic resources at the national, regional or local levels. Nor do they make adequate provisions for short-term economic adjustment assistance. By the same token, conventional approaches to conversion and economic adjustment are also shown to be insufficient for either easing the pain of transition for military-dependent firms, workers, and communities, or making the shift to new productive activities and markets.

Finally, the paper proposes an alternative policy strategy which links economic conversion and adjustment to new investments in science, technology, and economic development keyed to critical civilian economic, social, and environmental needs. This strategy would seek to establish new national priorities--economic stability, the environment, and health--to replace the current emphases on national security and `competitiveness'. The new national goal areas would serve as the primary filters or `focusing devices'--using Stanford economist Nathan Rosenberg's term[5]--for guiding new public investments, new initiatives in science and technology, and economic adjustment and technical assistance programs. The paper further claims that this policy strategy would be the most preferable and beneficial approach to economic development and industrial restructuring, while creating the optimal conditions for implementing effective economic conversion and adjustment policies at the national, regional and local levels.

**One, Two, Three Many Industrial Policies**

Industrial policy has an even more controversial, if shorter, history than economic conversion. It has been defined generally as an "attempt by the government to guide or shape industrial trends through the use of tax incentives, subsidies and loans".[6] Ann Markusen identifies several typical features of an industrial policy: the targeting of particular sectors as growth leaders; substantial support for R&D through incentives or direct subsidies; provision of capital for plant, equipment, and operating expenses; encouragement for various forms of industrial collaboration and planning; the monitoring and shaping of competition, e.g., by spreading around business or providing emergency assistance to ailing firms whose survival is deemed critical; government guaranteed markets for industry outputs, especially in the early stages of development, which helps to ensure volume for growth and cost recovery; trade policies which promote exports of the industries involved and protects them from the rigors of international competition; adjustment assistance for firms, workers, and communities affected by closing of facilities.[7]

The concept of industrial policy gained currency in the early 1980s. The Democrats made it a key platform in their 1984 campaign, pitching it as an antidote to the decimation of heartland basic industries--auto, steel, textiles--that had reached its apogee in the beginning of the decade. Through a number of ploys, President Reagan successfully aborted this effort. In line with the deregulation and privatization movements of that period, White House laissez-faire ideologues argues strenuously against anything that smacked of `excessive' government intervention in the private sector. Their rallying cry was that government had no business `picking industrial winners and losers' in the market place.

In reality, this neoclassical handwringing over industrial policy is something of a red herring, if not hypocritical, given the numerous ways in which the federal government has historically played an active role, often with the blessings of free marketeers and in partnership with the private sector and in guiding industrial development in the U.S. Indeed, it can be argued that not only have there been several examples of direct federal intervention in promoting special industrial interests, but various types of industrial policies have been employed for many years by the federal government in pursuit of national, political, economic, and social goals.

Over the years, many of the policy instruments mentioned above have been used to build up industries deemed important to the nation's well-being (which usually had strong lobbies to champion their cause). An important early example, was the creation in 1915 of the National Advisory Committee for Aeronautics (NACA), later absorbed into the National Aeronautics and Space Administration (NASA, created in 1958), to further the science and technology of aeronautics and advise the military and other government agencies on aeronautical research.[8] NACA was a response to the nation's perceived inadequacy in aviation research and production--by World War I, the United States had only 23 aircraft compared to 1,400 in France and 1,000 in Germany.[9] It conducted and funded research on airframe and propulsion technologies, pioneered in the construction and use of large wind tunnels, and served as an important source of performance and other test data in aeronautics. As David Mowery and Nathan Rosenberg point out, NACA (and its successor NASA) "did more than simply support research yield results that were diffused widely within the industry; they underwrote a portion of the costs of theresearch infrastructure associated with innovation in airframes and engines".[10]

Linda Cohen and Roger Noll observe that every administration since World War II has pursued policies that directly benefitted select areas of industrial and technological development: the Roosevelt and Truman administrations funded the beginnings of the computer industry; the Eisenhower administration spawned the semiconductor and nuclear power industries; the Kennedy-Johnson administration gave the nation the Apollo program, the supersonic transport (SST) and the war on cancer; Nixon, Ford and Carter spent billions of federal dollars on energy R&D in response to the 1970s energy crises. Even President Reagan, who came to power on the promise of reducing government's role in the economy, "supported programs to develop commercially attractive breeder reactors, nuclear fissions, orbital manufacturing facilities (Space Lab) and rocket planes (the Orient Express)".[11]

The federal government has also actively engaged industrial policy tools in the areas of health (medical R&D subsidies, market creation through Medicare), agriculture (land-grant colleges, R&D subsidies, experimental stations and technical extension services, crop subsidies,), energy (R&D subsidies and other measures to spin-off a civilian nuclear power industry from the military's nuclear program, price regulation, subsidies for oil, coal, and gas); and space (the concerted national effort in the 1960s to put a man on the Moon, building up U.S. rocket technology for both military and civilian purposes).

For example, the U.S. for many years has conducted a medical industrial policy, creating a medical-industrial-academic-government complex every bit as entrenched, and in some ways more massive--medical care accounts for 12 percent of the GNP, twice the share of the military economy--thanthe military-industrial complex. This has garnered a powerful set of vested interests--medical practitioners and their professional organizations (e.g., the American Medical Association), hospitals and clinics, suppliers of pharmaceutical and medical products, research centers at universities, hospitals, government, and private firms, and last but not least, the insurance industry--who have a stake in maintaining the current system.[12]

In all cases, government intervention helped to spawn and/or nurture a select set of industries and develop the S&T and industrial base needed to pursue national objectives. Correspondingly, these policies have helped to shape specialized market and industrial structures, specific lines of science and technology development, and patterns of investment in these areas by the private as well as public sector, with specific social, economic, political, and cultural consequences. But, while these `industrial policies' molded activity in their respective arenas, none has been as comprehensive and vigorous, with as great an impact on economic, industrial and S&T development in the U.S. as the military-led industrial policy pursued since WW II.

***The Military's `Closet' Industrial Policy***

Every element of Markusen's list above has been employed in the military's closet industrial policy. It is called a `closet' policy because it has rarely been acknowledged as an industrial policy, as it was pursued under the rubric of national security, and not as a means to stimulate economic activity. Since WW II, the Pentagon has actively nurtured an enormous industrial base, with a massive science and technology enterprise geared to the specialized needs of strategic and tactical warfare. This military-led industrial policy spawned a powerful aerospace, communications and electronics (ACE) industrial complex and spun-off severalcommercially-important technical innovations and industries, which for many years contributed to U.S. technological advantage.[13] Not incidentally, the ACE industries are among the few in the U.S. manufacturing sector with a positive trade balance. Aerospace, for example, registered nearly $37 billion in exports in 1990, contributing $26 billion to the U.S. trade balance.[14]

Munitions and shipbuilding have also benefitted from the military's largesse over the years, but ACE industries have consistently grabbed the lion's share. Aircraft, shipbuilding, missiles and spacecraft, ordnance, tanks, and communication equipment rank as the most defense-dependent industries; each obtained at least 40 percent of its sales from the military. In 1985, missiles ranked the highest in defense-dependency getting almost 100 percent of its business from the military. Shipbuilding and ordnance each got over 90 percent, while aircraft and communications received 79 percent and 60 percent, respectively. But, of the 40 percent of the military budget that goes to the manufacturing sector for production and R&D contracts, the aerospace, communications and electronics equipment industries got 79 percent in 1990, or $62 billion of a total $78 billion in Defense Department outlays; this is far greater than the $11 billion received by shipbuilders.[15]

The ACE industries also both receive the largest portion of the government/DOD R&D spending and are the largest performers of industrial R&D.[16] The military has long been a dominant actor in U.S. science and technology. Since World War II technological superiority has been considered by military policymakers to be the key for maintaining military superiority. Providing more than 80 percent of all federal funds for R&D up through the early 1960s, the military currently accounts for nearly two-thirds of all federal R&D and about one-third of the total, publicand private, (and fully a third of all industrial) R&D spending. By the same token, private industry performs approximately two-thirds--$24.7 billion in 1990--of the Pentagon's R&D. In 1989, the aerospace industry ranked number one in R&D spending, with $19.1 billion, accounting more than a fifth of all manufacturing R&D. About 82 percent of these funds came from the federal government, primarily from the DOD and NASA. Electronics and communications, together, followed closely behind, with $18.5 billion in overall R&D expenditures. Of this, 43 percent came from federal sources. Not surprisingly, the ACE industries together garnered over 80 percent of all federal R&D funding disbursed to all manufacturing firms in the U.S.[17]

Government policies have directly aided these industries in other ways as well. This includes providing much of the capital for facilities and equipments used by defense contractors. In the early 1980s the Pentagon invested about $18 billion in plant and equipment, with a replacement value of $100 billion. The government's industrial holdings include 146 plants and a half million items of equipment, 60% of which were in the hands of contractors. It still owns a major portion of the facilities used by contractors including most of the plant space and equipment in the munitions and strategic missiles industries and about one-third of that of the aircraft industry.

The government, furthermore, has not been adverse to giving special breaks to ACE firms when they got into trouble, ostensibly on the grounds of preserving the industrial base for national defense. Between 1958 and 1973 the government undertook some 3,652 rescue operations to help financially troubled firms. A major example in the early 1970s was the $350 million government loan guarantee given to Lockheed to help it stay in business, after getting into serious financialtrouble over its C-5A military transport and commercial L-1011 jetliner projects.[18] Underlying such bailouts is a form of economic `blackmail', aptly expressed in the quip by Representative William Moorhead (D-PA) in 1970, in which he compared Lockheed's threat of bankruptcy to that of "an 80-ton dinosaur who comes to your door and says, `If you don's feed me, I will die'. And what are you going to do with 80 tons of dead, stinking dinosaur in your yard?"[19]

The Pentagon also attempts to spread around its contracts to ensure the well-being of its principal contractors. James Kurth calls this practice the "follow-on imperative," in which the big companies get "turns" in being granted the latest contract in the sequence of new weapons systems.[20] Kurth recently predicted that Lockheed, which has been having financial difficulties, would be the next to receive a follow-on contract. In early 1990 the Lockheed-led team won the bid for the lucrative Advanced Tactical Fighter contract.[21]

In addition to these direct forms of aid, the Pentagon has provided trade support to its favored industries for decades and operates a large Foreign Military Sales program. The Export-Import bank has also favored the commercial side of the aircraft business, perhaps more than most industries, with preferential financial packages for the export of aircrafts comparable to those given by European governments. Up through 1971, for example, Boeing received $600 million in loan guarantees from the Export-Import bank, which helped induce foreign airline purchases of its products[22]. The winding down of the cold war and corresponding reductions in the defense budget have given foreign sales new importance for the military industrial sector. The Persian Gulf War has given further impetus to U.S. military exports, which, as one aerospace executive has remarked, is "the only game in town right now"[23] Shortly after hostilities were over, the Bushadministration asked Congress to appropriate $1 billion in loan guarantees for overseas customers of U.S. military contractors, to help the latter compete in an increasingly crowded market; the first such request for aid since the 1970s.[24]

In addition to building, subsidizing, and bailing out industries, the Pentagon has made modest efforts to alleviate the worst impacts of military cutbacks, particularly base closures, on defense-dependent workers and communities. The Office of Economic Adjustment (OEA), established by the DoD in 1961, has provided guidance and adjustment assistance to several communities in redeveloping closed bases, with mixed success. Overall, the extent and quality of the programs available to aid stricken communities and workers have not been, by and large, adequate to the task. Nevertheless, there is at least tacit recognition by the Pentagon of a responsibility to provide such aid.

**Roots and Costs**

Some of the roots of the military-led industrial policy can be traced to the first decades of the century, the role of NACA in nurturing this nascent aircraft industry was noted earlier. But, a full-blown commitment by the Pentagon to build and foster a supporting industrial base was an outgrowth of the World War II. The U.S. emerged out of the war as the premier military and economic power on the planet, with a new sense of its `manifest destiny' as the guardian (and promoter) of the capitalist "free-world". The federal government embarked on an aggressive program of building up its military arsenal to counter the rise of Soviet power and back its growing foreign commitments and interests in the post-war era. The war convinced policymakers of the primacy of air power for carrying out the new U.S. military doctrine; later military doctrines would be builtaround missile capabilities, and even later, around achieving technological superiority.

To implement these doctrines the Pentagon recognized that it needed a compliant and robust industrial and technological base which it could mold to meet its specialized requirements. At the same time, an ailing aircraft industry lobbied hard in the late 1940s for direct forms of government support. Not coincidentally many of its top executives were in the forefront calling for an air power-centered military force. Building on the successful collaborative experiences of the war, the convergent interests of military, industrial and other government policymakers (not to mention many academic and labor leaders) subsequently cemented a new post-war partnership between government and the private sector, which spawned the emergent aerospace, electronics, communications and computer industries.

Over the first post-war decades, this policy seemed to serve U.S. foreign and economic needs well. It bolstered U.S. corporate interests internationally, while providing direct stimulation for economic development domestically on the regional level, forging a prosperous "gunbelt" -- described by Ann Markusen and Joel Yudken as "a patchwork of cities and towns strung out along an arc from Alaska to Boston, sweeping down through Seattle, Silicon Valley, and Los Angeles in the West, across the more southerly mountain and plains states, through Texas and Florida, and up to the eastern seaboard through Newport News and Long Island to Massachusetts and Connecticut."[25] Finally, direct-buys and spin-offs from the military-led industrial policy fueled the development of commercially significant high-technology products such as computers, jet planes, lasers, semiconductors, and robotics, and the industries that produced them.

But this early prosperity was also a product of certain advantages that the U.S. had over other industrial nations right after the war. While the U.S. industrial base was left largely intact- in fact, it was greatly stimulated and built up during the war- and benefitted from the massive pent-up capacity and demand for consumer goods in the post-war era, the economies of most of the other combatant nations were badly damaged, if not shattered, and needed time to be rebuilt. In short,the U.S. was able to maintain its economic and military preeminence in the world during the first post-war decades mainly because, with the exception of the Soviet Union in the military sphere, there simply were no viable competitors.

The military's industrial policy has not been without its costs,however. One group of Pentagon critics, the "depletionists" - its best known proponents include Seymour Melman, John Ullmann, and Llyod J. Dumas - has long contended that military spending is a major drain on the civilian economy, diverting vital capital, technical, and human resources away from productive uses. Critics of the depletionist view, such as Gordon Adams, director of the Washington - based Defense Budget Project and former White House economic adviser Murray Weidenbaum, on the other hand, downplay these impacts primarily on aggregate economic grounds. They like to point out that the defense budget accounts for only 5-6 percent of the GNP, for example.

There clearly is not enough space for a full discussion of this debate here. Markusen and Yudken present elsewhere in some depth a different set of arguments which shows how, as the U.S. military industrial policy directly and indirectly built up certain sets of industries, regions, and occupations, other were shortchanged over the past forty years[26] In short, while certainly not all the nation's economic ills can be blamed on the militaryeconomy, the actual pattern of U.S. economic development manifested locally, regionally and nationally in the postwar era owes a great deal to the Pentagon policies.

To a degree, there does indeed appear to be a diversion of resources, especially in the realm of science and technology, from the civilian to the military sector, as claimed by depletionist scholars. The greatest costs to the nation's economic and social well-being derive from the tradeoffs among competing national priorities forced upon the nation by the military industrial policy. The decision to build up the massive military-industrial system in the U.S. entailed significant "opportunity costs" in terms of various possible public and private investments in economic and social development that had to be foregone in the name of national security. Consider the illustrative example of steel. To make their industries competitive,European and Japanese firms and governments poured money into steel R & D, especially in the 1970s. In contrast, U.S. steel firms were cutting back, and the federal government preferred to concentrate its R & D resources on military industrial objectives. Hence, while the White House pushed the Strategic Defense Initiative program which has been bankrolled at the rate of $ 3-6 billion per year since the early 1980s, the steel industry was not able to get $ 15 million for its Leapfrog Technology initiative fully funded, despite support from President Reagan's Science Advisor George Keyworth.[27]

Unfortunately, measuring opportunity costs is very difficult, as it entails evaluating alternative futures that we cannot really know. On the other hand, the rise of European and Asian Economic powers in the last couple of decades, gives us another basis for comparison. The United States is no longer without major economic competitors. Its war-torn allies and former enemies-ironically, the latter have fared the best- are now fullyrecovered and modernized, and pursuing aggressive industrial policies of their own, built, however, around civilian goals. It is increasingly apparent, even to many government, business and academic leaders, that the U.S.'s military-led industrial policy -not to mention its lack of policies for addressing its environmental, health care, infrastructure, and urban crises - is much less successful than the policies of its trading partners in generating technological innovation, productivity growth and enhanced industrial performance.

**The Rise and Limits of "Dual-Use" Policy**

**Out of the Closet**. The collapse of the Cold War has forced a reconsideration of U.S. military doctrine, which is bringing the DOD's industrial policy out of the closet, with greater cries from both within and outside the Pentagon for a more explicit military role in strengthening domestic industries. Military planners are especially worried about the inability of U.S. firms to supply many vital components and systems, which has resulted in greater reliance on foreign sources. Once on the leading-edge,military technology has fallen behind commercial innovation in many areas. Moreover, the needs and requirements of military and civilian technology development have diverged significantly over the years.[28]

Meanwhile, a growing chorus of high-tech managers, academics and political officials has called for greater government attention to the nation's faltering civilian manufacturing sector.[29] This group wants greater government investment in commercially-relevant "critical technologies", along with more trade protections, R & D tax credits, relaxation of anti-trust laws, reduced regulation, strengthened intellectual property rights and new patent laws.

Opposition to these proposals has come from powerful White House officials who are against any form of "industrial policy" which targets industries and technologies, as they fear it would interfere with the workings of the market.[30] Their position, one that Reagan/Bush laissez-faire ideologues have argued ever since industrial policy became a burning issue during the 1984 presidential campaign, is that government should not be in the 'business of picking winners and losers' in the markets place. To circumvent this resistance, high-tech advocates turned to the one governmental entity seemingly immune from this objection : the military.

Military R & D planners and high-tech industrialists are now promoting the doctrine of "dual-use" which gives priority to technologies which serve both military and commercial needs, such as semiconductor manufacturing, high-performance computing, and advanced materials[31] all of which receive significant military subsides. The Pentagon's Defense Advanced Research Projects Agency (DARPA) has been elected as the principal federal agency to manage this effort.[32] DARPA is the main sponsor of "dual-use" projects, including $100 million given yearly to Sematech, a consortium of semiconductor firms, and $50 million for the recently initiated pre-competitive technologies program.[33]

**Through the Back Door** The "dual - use" cause has not yet prevailed, [34] but the need to do something about the nation's flagging industrial performance will inevitably require some sort of accommodation between the "dual-use" and laissez-faire forces. High-tech advocates have embarked on a piece-meal, `back-door' strategy, to promote the notions of a "technology policy", which has met with some success in Congress, and even grudging acceptance by the White House. Backed by the Office of Science and Technology Policy (OSTP) director D. Allan Bromley, the Bush administration hasendorsed so called "generic" pre-competitive technology R & D programs favored by the high-tech supporters, such as the multi-agency high-performance computing program and advanced manufacturing technology programs of the commerce Department's National Institute for Standards and Technology (NIST)[35] "Dual-use" policy proponents also take heart from a recent OSTP report to the Congress entitled "The U.S. Technology Policy"[36] Leading "dual-use" advocate Lewis Branscomb, from Harvard's Kennedy School of Government, quips that `Building a consensus in the White House for any document with the words"technology policy" in the title was no small achievement." Nevertheless, the Bush administration's 1992 budget request asked for only miniscule levels of funding for commercial technology programs. National defense R & D programs, especially for large weapons systems development will continue to receive most of the federal R & D largesse.

The pressure for a "dual-use" policy continues to intensify, however, with new proposals from influential private sector groups. A report by the Carnegie Commission on Science, Technology and Government, for example, urges the transformation of DARPA into a National Advanced Research Projects Agency (NARPA) to oversee the 'national technology base', under Pentagon's auspices.[37] Meanwhile, in a political end-run, bipartisan Congressional backers of "dual-use" policy tagged a major technology policy amendment onto the 1992-1993 Senate National Defense Authorization bill, which substantially institutionalizes the DARPA's lead role in managing the nation's "dual-use" program.[38]

**Limits of "Dual- Use",** A virtue of the "dual-use" proposal is that it demands a stronger government role in the nation's technical and industrial development. Its proponents correctly reject the laissez-fairepremise that market forces alone are sufficient for stimulating and directing technical innovation; a position, infact, supported by a growing body of empirical evidence. Much of this evidence is presented in the writings of economists and economic historians critical of traditional equilibrium models of technical innovation and economic change, such as Richard Nelson, Sidney Winter, Christopher Freeman, Giovanni Dosi, and Nathan Roseberg, among others. According to this school of thought--which has roots in the earlier works of Joseph Schumpeter-technological development is seen as both a product and cause of dynamic economic change. In their view, a complex set of economic, institutional and social forces, which operate as 'selection' and 'focusing' mechanisms, are what actually shape the patterns of technical and industrial development.[39]

Some analysts argue that there are serious deficiencies in the "dual-use" strategy, as well.[40] First by maintaining the Pentagon's persistent role as the main federal sponsor of high - tech development, technology transfer in significant areas of commercial and national need will inevitably be hindered. Not all critical areas of civilian science and technology meet the test of military relevancy. Moreover, the difficulties of bridging the widening gap between military and civilian paradigms in technological development may be an underlying reason why, despite substantial, long -term DOD funding for "dual-use" technologies - such as robotics, semiconductors, supercomputers, superconductors, airframes, advanced ceramic propulsion, and advanced U.S. commercial firms in these areas continue to lose ground to Japanese competitors. In the final analysis, DARPA is a military agency, and its programs of necessity favor the needs of military over civilian users. In any event "dual-use" initiatives make up only 3 percent of the DOD's total R & D budget; the massivemilitary R & D commitment will continue to crowd out investments in civilian research.

Second "dual-use" policy unduly, limits, -and it may even narrow further- participation in science and technology policymaking to an elite group of industrial, governmental, and professional leaders and experts. There are few provisions to involve other stakeholders in the national policy process, despite recognition, for example, that worker participation in technology decision making improves productivity. Hence, R & D managers have little incentive to incorporate broader social or environmental concerns into their agendas. The lack of such inputs into the processes that guide technical innovation will subsequently reinforce patterns of technological development in the economy that do not contribute, and may even be antithetical to social and environmental goas.

Finally, "dual-use" policy does not provide adequate demand -side inducements for promoting new directions for technical change. "Dual-use" policy assumes that global market forces will provide the necessary incentives for commercial innovation, if only the federal government will help out on the supply-side. Given the intensifying world competition in advanced technology at a time of global recession and international instability, this seems a dubious premise. Demand-side inducements are also needed for promoting new directions for technical change. By signalling new market opportunities, government-created demands help to stimulate private sector investments in new areas of technical development. This was particularly the case for the military-led industrial policy in the first postwar decades, which not only spawned new generations of sophisticated weapons technologies, but generated a number of commercially significant innovations. As economists David Mowery and Nathan Rosenberg observe, "the benefits that are sometimes perceived to flowfrom military R&D are in fact the product of military R&D plus frequently massive military procurement... Without the pull of defense procurement in such sectors as jet aircraft, integrated circuitry, and computers--especially in the critical early years of development of these technologies- the impact of military R & D spending alone would have been far smaller."[41] Similarly, without comparable public investments in civilian needs, such as alternative energy generation and mass transportation, the military will remain the demand creator or last resort for advanced technologies.

In short, a "dual-use" agenda which targets only a handful of so-called "critical" technologies for investment, based on narrow economic and military criteria, will limit opportunities for developing other economically and socially vital technologies. Although, public investments in rebuilding the public infrastructure, renewable energy sources, and cleaning up the environment, would enhance the nation's economic performance and stimulate broad new species of scientific and technical innovation, such programs would not be supported by"dual-use" policy. In fact given tightening federal budget constrains, "dual-use" programs may very well preclude public in vestments in these other critical areas of national need.

## Economic Conversion Redefined

Another shortcoming of both the dual-use technology and laissez-faire policies is the little attention paid to problems of economic adjustment, diversification and conversion for workers and communities suffering from dislocations due to military cutbacks. The conversion problem looms large at this time, not just because of the question of whether the U.S. adjustment capacity is sufficient to handle the mounting military-induced dislocation, but because of the *strategic*importance of the military sector, despite its small share of the total GNP, to the nation's industrial performance.

First, military industry is a much larger part of the nation's durable goods manufacturing sector than of the economy as a whole. The defense industry accounts for about 15-17 percent of the nation's manufacturing employment and an even greater share of these jobs in key industrial regions. In 1980, for example, *before* the 1980s build-up, defense-related jobs accounted for at least one-fifth of the manufacturing employment in California's Santa Clara County, according to an Office of Economic Adjustment-sponsored report prepared by SRI International.[42] Changes in military demand also have significant impacts on overall national and regional manufacturing outputs and exports in the U.S.

Second, military spending dominates U.S. public investment priorities. Hence, it has precluded investments -- in public infrastructure rebuilding, environmental controls, alternative energy sources, and commercial R&D -- which have significant economic pay-offs for U.S. industries.

Third, the most direct way the military impinges on U.S. civilian economic performance is through control over the major portion of the nation's vital science and technology resources. The military and commercial high-tech industrial sectors are increasingly in competition for many of the same scarce technical resources; a fact which gave rise to the dual-use policy discussed earlier.

***Limited Successes***. Economic conversion can therefore be defined as the *optimum reallocation of public and private economic assets, from the military to civilian industrial sectors, that enhances national economic performance, environmentalsustainability and social well-being.* A crucial criterion is that this be accomplished with minimal dislocation of workers and communities. Conversion of military facilities ideally would preserve the bulk of the jobs, reemploy much of the plant and equipment, and maintain as large a part of the local economic base as possible. Corporate diversification, in which a firm shuts down existing facilities while it buys new product divisions, leaving workers and communities high-and-dry, clearly does not meet this criterion. Nor do community-initiated diversification efforts -- including many high-tech development strategies -- which do not aid local workers and businesses most directly affected by cutbacks and closures.

Thee have been several corporate and employee-community initiated conversion efforts in the United States since 1970. Among the most notable efforts by firms-mainly for their lack of success-are attempts by Boeing-Vertol, Grumman and Rohr to shift military divisions to produce mass transit vehicles (rail cars, buses) in the 1970s. On a smaller scale, Kaman Corp. and Frisby Airborne Hydraulics have achieved limited success in converting military-based technologies to commercializable products.[43]

The best-known employee-community-led conversion projects in the 1980s involved Douglas Aircraft (see earlier), Blaw-Knox, Unisys, and General Dynamics' Quincy Shipyards.[44] Learning from their lack of success at the plant level -- noting the lack of incentives and resources to assist conversion, especially by small and medium-sized firms --community-worker groups have since then turned much of their attention to promoting municipal and state conversion policies. Minnesota, for example, with one of the best state-initiated programs, was first advanced by and includes extensive participation of labor and community conversion activists.

Critics point to the limited success of conversion as evidence to show that it lacks viability. In actuality, this is a classic "catch-22" situation. Inadequate government commitment and institutional support for conversion has been a major reason as to why it is both an unattractive option of corporate managements and very difficult to implement under current economic conditions. At the same time, the lack of conversion successes has been used to undermine political efforts to establish such mechanisms. Federal and state support might include assistance for the following : new product R&D; planning and feasibility studies; employee adjustment; retooling; finding new buyers; establishing ESOPs, worker buy-outs and the like; and arranging financing. If such institutional aid had been available on a larger scale and in a more coordinated fashion -- say, if the Defense Economic Adjustment bill sponsored year after year by Congressman Ted Weiss (D-NY) was enacted -- the conversion success rate might have been greater.

***"Wall of Separation."*** On the other hand, as the earlier McDonnell Douglas example illustrates, the problem of making conversion viable goes beyond providing economic adjustment and technical assistance. In general, many conversion advocates have tended to underestimate the significant institutional and deep structural economic factors which underlay the conversion failures, to-date. In practice, economic conversion is a very complex and difficult process. Conversion proponents and critics alike have noted the many barriers that military-industrial firms confront when trying to shift to civilian pursuits, whether through diversification or conversion.

In terms of the `neo-schumpeterian' writers above, the conversion problem should be understood as a switch from the technological `paradigm' (military) to another(civilian). Military-related industry has evolved its own unique paradigms of technical and organizational development, which have been fostered and institutionalized by the Pentagon's industrial policy for over forty years. A substantial divergence -- a "wall of separation" -- between military and civilian firms has emerged over time in their modes of operating, engineering, and marketing. These are the technical and organizational `exit barriers' that make it difficult for military firms to convert to civilian production. The `reconversion' from wartime to civilian production after WW II, which entailed primarily civilian firms switching back to their original modes of activity -- in reality, somewhat reconfigured by their years of war work -- was therefore a far easier task than the current problem of converting industries that have become increasingly entrenched in military-driven `paradigms' over the cold war period.

While, in theory, it is possible to identify many civilian products and technologies to which military firms may be capable of switching their production activities -- in fact, lists of such products have been developed in various studies and actual conversion efforts -- the problems of technology transfer involved are formidable. As one corporate executive sated in response to a U.S.I Senate questionnaire:

> Personally, I happen to question the ability of a company devoted solely to defense-space to succeed in the commercial field. They have a tendency to underrate the complexity of the problems and the capabilities of their future competitors.[45]

The Boeing-Vertol conversion experience supports this perception. Boeing officials admitted that they had failed to anticipate the engineering challenges presented in theproject.[46] The requirements of civilian technical development generally are substantially different, even in similar production areas, such as aircraft, than those in the military industry. Over the Cold War period, the skills of military-dependent engineers and scientists have correspondingly diverged from those required in civilian industries. It is hard -- though, not impossible -- for those trained in the design of military hardware and software, who are used to specialized military performance specifications, to transfer their experience to the civilian arena where costs or very different performance criteria are emphasized.

***Demand and Supply-Side Inducements***. Successful conversion therefore requires long-lead times and substantial planning, retooling, and retraining of many sectors of the work force, particularly managers and engineers, to enable firms to adjust to different cycles of technical innovation and product development.[47] Conversion legislation and policies would provide various forms of government assistance for these tasks. Support for R&D would also be needed, not only for short-term product development, but to help firms evolve new civilian-oriented paradigms of technology development over a long-term.

Supply-side inducements such as these, which help companies cross over the hurdles of technology transfer, are necessary, but not sufficient conditions, for successful conversions. Success is also contingent on there being effective demand for products that plants can convert their production to. For example, it can be argued that the Boeing-Vertol experience was in reality not a total failure. The company had moved quite far along the learning curve in the design and production of transit cars. The primary problem was a lack of stable markets for their product to allow them to follow that curve further, and actually begin to make a profit.Boeing officials themselves noted that a major obstacle was a lack of coherent national mass transportation policy. More generally, the comment of Boeing Chairman William M. Allen, in response to the Senate query, illustrates the corporate sensitivity to the market signals provided by government:

> The company has examined many of the civil systems markets during the past five years. These have included in-depth studies of surface transportation, water management, waste disposal and security systems. We find that each of these have elements consistent with our technical and systems management capabilities. However, we do not see either established national goals in these areas, consistent commitment to adequate funding, reasonable-size contracts of adequate duration, or contracting modes consistent with these (civilian markets).[48]

The general decline in U.S. civilian manufacturing performance, therefore, presents a serious obstacle to defense companies shifting their production to new areas, even if they do all the right things internally. As the U.S. loses ground in one civilian industrial arena after another, because of both a lack of technical competence and losses of market share, military firms seeking to diversify into civilian markets confront a dilemma when contemplating the difficulties of converting their facilities. At the heart of this dilemma, note Markusen and Yudken, is the fact that "no other product area" offers "the guarantee of federal commitment, public resources, and luxuriously long lead times that producing defense hardware did."[49]

But given the massive cuts in military hardware systems expected over the next few years, the problem devolves to this :

what kinds of new national `focusing devices' will be needed -- or possible -- to enable militaryfirms to overcome exit barriers and smoothly move into new areas of civilian technology, with minimum costs to the individuals, communities, and firms involved? It is obvious that neither the laissez-faire economic policies of the Bush administration nor the proposed dual-use technology policies will be able to do the job.

The former will do little to provide the kinds of targeted supply and demand inducements that would be needed. Moreover, an administration reluctant to extend unemployment benefits at the heart of a recession, is unlikely to do much to bolster an already overburdened system for aiding displaced workers, much less target defense workers for special consideration. The latter is too narrowly focused on helping a limited number of technologies and industries. Only a few types of military-dependent firms, mainly high-tech component suppliers, would directly benefit from dual-use policy. In addition, dual-use policy will not create new civilian markets for military firms to convert into. As noted earlier, it may not be even successful in helping domestic commercial high-tech firms compete in existing markets.

### Targeting Science & Technology for National Needs

What is required is a different kind of strategy, one that links conversion of defense resources to rebuilding the nation's economic capacity and addresses pressing national needs. The nation will have to redirect vital technical and economic assets now locked into the military economy -- the so-called `peace dividend' -- towards these ends. The new strategy would include mechanisms which minimize the costs of adjustment for workers and communities. What would such a strategy look like?

***New National Goals***. First, we need new national goals which would guide public andprivate sector commitments in the coming decades, just as national security served as the primary focus for the U.S. industrial policy during the Cold War. Ann Markusen and Joel Yudken maintain that three priorities should top the list: economic stability, the environment, and health.[50] Indeed, these goals imply a shift in national policies from the narrow objectives of national security and economic competitiveness to a broader concept of *global security* - "global" in both a generic and geographical sense.

Policies for economic stability would build and maintain the *economic security* and well-being of workers, businesses and their communities. This implies investments that promote vigorous and diversified economic growth, create new and decent jobs, enhance the productivity of domestic industries and the quality of their goods and services, increase the standard of living, and revitalize moribund industries and local economies (e.g., inner-city areas). At the same time, there would be an emphasis on regulating this growth to preserve the economic and social integrity of communities and maximize job security for the local work force. All too often, Markusen and Yudken observe, "champions of the market and capital mobility ignore the destructive effects on workers, households, local governments, and regional businesses when a local economic environment changes."[51] Hence, provisions for assisting those affected by such impacts would be mandatory in any new economic development program.

Environmental sustainability is a critical condition for achieving *environmental security* for all living things. Environmental problems cut across all geographical boundaries: pollution of the air, water, and soil; toxics in our homes and work places; depletion of scarce natural resources; global warming and a depleting ozone layer; threats to the biosphere and species diversity on the planet. Governments are being pushed to invest more in solutions, and strengthen controls and regulations. However, as most environmental problems derive from uncontrolled production practices, we need policies that emphasize sustainability, in which environmental objectives and criteria are built into economic development.

Providing for the health and safety of all citizens is a condition for achieving *physical and mental security*, and also a part of global security. As medical costs escalate, more and more U.S. citizens lack access to adequate health-care services and a means to pay for them. We need a new health policy that would tie public investments, health care delivery programs, regulation, and research and development, to new goals of preventive care, equal access to inexpensive, high quality medical care, and reduced threats to health and safety in our jobs and communities.

Finally, we must address these problems, not from a narrow nationalistic vantage point, but from a place of concern for guaranteeing the security and well-being of all peoples in the world. The United States and the other industrial powers have a responsibility and the resources to helpless advantaged societies in the world achieve global security in all its meanings. The rhetoric of global co-operation must replace the rhetoric of national competitiveness. MIT economist Paul Krugman notes that "there is surprisingly little coherent discussion of what `competitiveness' means," and when applied to nations it is not a relevant notion. Competitiveness raises again the spectra of nationalistic competition, that, Krugman observes, "like national defense.... can be easily used as a patriotic cloak for special interest politics."[52]

***Public Investment and Adjustment Assistance***. Achieving these goals willrequire the employment of three major policy instruments. First, a series of public investments are needed for rebuilding the public infrastructure, environmental protection, renewable energy, education, health care, mass transportation and affordable housing. Many economists agree that such public investments would stimulate the productive base of our declining economy -- increasing private investment, productivity and profits -- and put it back on track. Jeff Faux and Todd Schafer of the Economic Policy Institute report that "the United States could not have successfully developed a powerful private economy without large and sustained government outlays on transportation, education, and the generation of new technologies." However, over past years the U.S. invested less, while other industrial nations invested more in these areas, "setting the stage for further declines in living standards and competitiveness."[53]

Second, economic adjustment and technical assistance programs are needed to aid communities, workers, and businesses hurt by military cutbacks, technological change, and international competition.[54] A new economic development strategy would redirect vital technical and economic assets from the military sector to rebuilding the nation's economic capacity and addressing pressing social and environmental problems. Economic conversion and adjustment assistance would therefore need to be targeted to help military-dependent communities and workers make this transition. This includes providing adequate capital and technical resources for helping workers, communities and managements convert facilities and diversify local and regional economies. Measures that institutionalize early warning, rapid response, labor-management cooperation, and worker participation, as key features of all economic assistance programs, are also needed. Finally, there need to be stronger incentivesfor companies planning to close a plant or make major lay-offs to cooperate with worker/community initiated conversion, diversification and assistance effort. Such programs represent an investment in the efficient reuse of economic resources, as well as a means to reduce the costs to society of human hardship.

***Science and Technology Initiatives.*** Finally, a series of science and technology initiatives are needed to help achieve the new goals. Civilian R&D investments coupled to demand-creating social programs would shift the patterns of U.S. technical innovation in directions compatible with producing socially needed goods and services.[55] "Dual-use" advocates want to target "critical technologies" for R&D investment that are considered essential by government and private sector groups for both national security and commercial competitiveness; these include advanced materials, microchip manufacturing, computer-aided manufacturing, robotics, high performance computing, advanced propulsion, and biotechnology.[56] Many of these same areas probably would also be considered important in the new strategy.

However, the criteria for prioritizing technologies would be quite different in a national needs-driven S&T agenda. Concerns about the environment, energy efficiency, quality of work, health and safety, autonomy and participation, and public service would be considered as important as efficiency (or cost), product performance, or marketability in selecting R&D projects. Such criteria would tend to focus even so-called `generic' technologies along different paths of development, leading to different kinds of innovation. A machine tool technology or agricultural process developed to optimize these criteria, for example, might differ substantially in their specifications from thosedesigned according to military or narrow commercial objectives.

In addition, a national needs agenda would select areas of science and technology that have not been singled out in the government and private sector "critical technologies" lists. Only the OSTP-appointed National Critical Technologies Panel, unlike the DoD and Commerce Department, applied national needs criteria -- energy security and quality of life -- beyond industrial competitiveness and defense in its selection of 22 "critical technologies," though the latter were given higher priority.[57] Energy technologies (renewable and environmentally improved energy technologies, energy conservation, energy storage), pollution minimization, remediation and waste management, and surface transportation technologies were among those that appeared to address broader national goals.

In crafting a new agenda, it will therefore be necessary, first, to evaluate current R&D programs (e.g., environment, transportation, energy, health) sponsored by the federal agencies, to determine which kinds of science and technology programs not currently funded, or supported at too low levels, should be added to or supplemented. This would entail rethinking and redefining the missions of existing federal R&D agencies, such as the DOE, NASA, NSF and NIH, and even those of the DOD. This includes changing the mandates and priorities governing their in-house and extramural R&D programs. For example, if the DOE changes from its current primary emphasis on nuclear energy and weapons to producing alternative, renewable energy sources, the national laboratories under its direction -- especially, the weapons labs at Livermore, Sandia, and Los Alamos --would more easily be able to convert their research program to the new agenda.

Meanwhile, other agencies now at relatively low levels of R&D funding, such as Transportation, Commerce, Labor and the Environmental Protection Agency might find their R&D budgets beefed up. Proposals for creating new agencies, such as the National Institutes for the Environment promoted by ecologists and environments, or a civilian advanced technology agency, also may have a better chance of success. Undoubtedly, the new agenda would result in a downplaying of DARPA's role in civilian technology. Although the agency would continue to emphasize "dual-use" technologies, it may no longer be the primary sponsor of research in these areas.

Many of our most important problems, however, are cross-disciplinary in nature and too complex to fall under the missions of a single agency; they would require coordination and co-operation across many agencies and institutions. Each would target for public investment a unique set of social, industrial and R&D programs which reacquire the participation of federal, state and local governments, business, labor, academia, and the public interest community in their solution. Such cross-cutting S&T initiatives can be thought of as new "grand challenges" in science and technology, similar in scope to our efforts to reach the moon, cure cancer, or produce a strategic defense system. Candidates for such initiatives include the following:

***Sustainable development in industry and agriculture***. Can we produce energy efficient, environmentally clean industrial and agricultural technologies? Ken Geiser, professor of environmental policy at the University of Massachusetts-Lowell, points out that the notion of sustainability first emerged with regard to agriculture and is now being applied to manufacturing. "Sustainable farming," Geiser says "aims to lessen the need for chemical pesticides and fertilizers infavor of practices that work with natural ecological cycles to improve the soil and increase the pest resistance of plants." Sustainable manufacturing would include the following features: appropriate technologies for the desired ends; safe and environmentally compatible materials; production processes that minimize waste products; energy efficiency; safe and skill-enhancing work environments; and resource conservation.[58] The emphasis would be on designing pollution control and waste reduction technologies into the front-end of production. This would reduce, if not eliminate, the back-end costs of pollution clean-up and waste management for firms, benefitting both industry and society.

***Quality of worklife and productiveness***. Can we design technologies that enhance job security and the quality of worklife -- health and safety, skill, participation, and autonomy in the workplace -- which at the same time improve productivity and product quality? New information and materials technologies are causing a fundamental restructuring of the workplace. The goals of this initiative would be to develop new technologies used in manufacturing and services which enhance rather than diminish quality of worklife. It would emphasize research on concepts such as "participatory design" and "skill-based automation," and on reducing the hazards and stress-related work injuries associated with high-tech workplaces.

***Transportation infrastructure***. Can we build safe, efficient, environmentally benign large-scale transportation systems both between and within major metropolitan areas? We need to invest much more in building new surface mass-transportation systems, and improve our air and maritime transport capabilities that not only handle larger quantities of people and cargo, and go faster, but which emphasize such values asfuel-efficiency, pollution and noise reduction, and safety.

***Urban infrastructure development***. What technologies do we need to rebuild the transportation, communication, sewage, housing, and other public service systems of our metropolitan areas? We need to reevaluate how our metropolitan areas are constructed and operate from a wholistic planning and environmental perspective, examining the interaction of its infrastructure systems with its economic and social functions.

***Revitalizing the manufacturing base***. Can we introduce new `flexible' technologies, as well as incremental innovations and technical assistance which, while satisfying the criteria of sustainability and human productiveness, strengthen domestic high-tech manufacturing and revitalizes traditional industries such as steel and textiles?

***Benefits***. A strong argument can be made that this strategy would yield greater economic and social benefits than the current laissez-faire or "dual-use" policies; it would satisfy the needs and concerns of a broad cross-section of constituencies and interests in the nation. An underlying premise is that new public investments, coupled with targeted S&T initiatives and economic assistance programs, would guide the private sector in shifting its resources into activities that result in significant economic and social payoffs.

Not only would it address pressing problems in the society and the environment -- reducing so-called externality costs of environmental degradation, occupational and environmental illness and injuries, and unemployment -- the proposed public investment strategy would directly stimulate both economic and technological progress and improve the economic performance of domestic industries. It would create andpreserve jobs, strengthen the economies of communities and regions, and raise the standard of living, while meeting environmental, health and quality of life needs. By linking economic development to national needs, powerful, new inducements would be established in the national economy, in the form of new effective markets, that would stimulate technical innovation in the production of goods and services that directly serve human needs. New `virtuous circles' would be created: as the private sector is induced to reorient activities in response to the new government-provided market signals, it would increase its own investments in these same areas.

These initiatives and investments would subsequently generate new classes of manufacturing technologies, capital goods and consumer products, create new industries and types of high-skilled jobs, and stimulate new directions for scientific discoveries and new opportunities for technical innovation. Scientists, who are feeling the pinch of a zero-sum budget game and pork-barrel politics, would also gain, despite the goal-oriented nature of these programs. A national needs-driven strategy will actually diversify and expand their portfolios of funding sources as new problems in basic science are defined within the S&T initiatives. In addition, many of the so-called `critical technologies' for high-tech competitiveness will also be well-supported. For example, new computing, advanced materials, and microelectronic technologies would find numerous applications in the S&T initiatives listed above.[59]

Finally, the new strategy creates the optimum conditions for an efficient conversion of military economic and S&T assets to civilian use, and reemployment of displaced workers. National policies with new goals for guiding public investments and S&T resources, would create new markets forproducts and business opportunities, and provide new sources of capital, making both plant-site conversions and community diversifications efforts more viable, and adjustment for workers and communities easier.

***Getting there from here***. Democratizing the U.S. S&T system and implementing a new economic development strategy mentioned above are closely linked problems. A crucial condition for promoting the new strategy would be to expand the democratic participation of broad sectors of society in economic and S&T policies and decisions in both the public and private sectors. The participation of broader range of interests in the society is vital for crafting industrial policies which are both economically effective and socially equitable. Representation of labor unions, environmental organizations or community groups in this process would lead to the different criteria in the technology design and resource allocation processes than currently exists. Moreover, active promotion by national, regional, and local grass-roots, labor and public interest constituencies would be essential for making the new economic strategy a serious political issue at the national level.

Only recently have there been signs of concern among public interest, labor and other grass-roots constituencies, particularly among those working on conversion, environmental and workplace technology issues. This includes the emergence of community-labor-public interest-led initiatives to influence industrial and S&T policies, such as the Campaign for Responsible Technology, the 21st century Project, the National Toxics Coalition, the Federation for Industrial Renewal, the National Commission on Economic Conversion and Disarmament, and several state and local conversion projects in Minnesota, Maine, Washington, and California, among others. Labor unionsand technical professional societies have also begun to articulate their own concerns and agendas regarding these issues.[60]

Industrial renewal and economic conversion groups around the U.S. have for years fought to involve community-labor participation in economic and technological planning in responses to plant closures and military cutbacks. There are many examples of such efforts since the 1980s. One of the most prominent is the Unisys project in Minnesota, which has involved labor and grass-roots participation in designing an alternative use plan for that company's threatened military division.[61] Similarly, the Cleveland Coalition Against Plant Closings spawned the Regional Industry Center which seeks to integrate several different, primarily state-funded initiatives and programs in the Cleveland area into a cohesive industrial planning strategy, guided by the grass-roots. This includes early warning signals training and projects which help local businesses adopt new technologies. A common denominator in these efforts is that they seek to expand the notion of citizen participation to be more proactive in setting the S&T policy agenda, and not just react to one set by the powers that be.

Probably the leading representative of this genre, however, is the Campaign for Responsible Technology (CRT). Comprising of community, environmental and labor groups from around the nation, concerned with environmental and occupational health issues in high-tech manufacturing, CRT is trying to get the joint government-private sector funded Sematech consortium to broaden its mandate to include environmental, labor and community concerns.[62] In early February 1992, CRT met with public interest activists, labor leaders, and Congressional aides to draft a revision of Sematech's budget authorization legislation, which is coming up for renewal later in theyear. Their proposal would expand Sematech's advisory council to include experts on environmental and occupational issues, and local community representatives. It would also earmark government funds for R&D on pollution and toxic reduction in semiconductor manufacturing, and expand public access to information.

Working closely with CRT, is the 21st Century Project, which is an explicit program of democratizing technological investment decisions so that solutions to global and national problems become paramount in United States government policy. The Project is a program of the Computer Professionals for Social Responsibility (CPSR), the nation's only public interest organization of information technology professionals. It will attempt to integrate a national perspective with the work of citizens at the local level. Its national presence will be developed through an ongoing research process and a sustained effort of public outreach.[63]

The Project's research program will provide the intellectual content of a new agenda for science and technology policy. It is explicitly designed to include citizen and public interest, as well as "expert", involvement in the conceptualization and research process for constructing an alternative agenda. This will take the form of four working groups convened to address specific issues in science and technology policy: Democratic Management of Science and Technology; Sustainable Development; Quality of Work in the Information Age; Computers and Communication Infrastructure. Participants will include leading public interest activists, labor leaders, academics, policy experts, scientists, and engineers, among others. Each working group will examine the problem of developing an R&D agenda with regard to a particular theme, with a special focus on designing an alternative program in the area of information technology. A productof this study will be a model R&D program built around a set of key societal problems and potential relevant applications of information technology, taken as an example.

The growth of grass-roots activism around issues of science and technology, economic conversion and industrial development are encouraging indications that such concerns are finally beginning to be perceived as serious `bread-and-butter' problems that should not be left in the hands of managerial elites and scientific-technical experts. Only time will tell if this new momentum will be able to grow into a force that can create the kinds of shifts needed at the level of national policy.

* This paper is an adaptation of two earlier papers by the author: "Technology, Defense Conversion and Development: A National Policy Perspective," chapter in Kevin Cassidy (ed.), *Real Security-Converting the Military Economy and Building Peace*, (SUNY Press, in press); "Beyond `Dual-Use' Policy: Crafting a National Needs Investment Strategy for Science and Technology," paper prepared for the Conference on The Triple Revolution, One Generation Later: Harnessing Science and Technology for the Common Good, Institute for Policy Studies, Washington, D.C., October 4-5, 1991.

## Endnotes

1. For a discussion and evaluation of this effort see: Joel S. Yudken, "Conversion in the Aerospace Industry: The McDonnell Douglas-Project," in Suzanne Gordon and Dave McFadden (eds), *Economic Conversion, Revitalizing America's Economy* (Cambridge, M.A: Balligner Publishing, 1984), pp.130-143; and Catherine Hill, Sabina Deitrick, and Ann Markusen, "Converting the MilitaryIndustrial Economy, *Journal of Planning Education and Research* (forthcoming in 1992).

2. For discussion of the "wall of separation," see Ann Markusen and Joel Yudken, *Dismantling the Cold War Economy* (New York: Basic Books, 1992), especially chapter 4.

3. Steven Greenhouse, "There's No Stopping Europe's Airbus Now," *New York Times*, 23 June, 1991, section 3, pp.1,6.

4. For discussion of "critical technologies," see *Report of the National Critical Technologies Panel*, (Washington, D.C.: U.S. Government Printing Office, March 1991); Council on Competiveness, *Picking up the Pace: The Commercial Challenge of American Innovation* (Washington, D.C., 1988), and *Gaining New Ground, Technology Priorities for America's Future,* (Washington, D.C., March 1991). In its reports the Council

on Competitiveness, a private sector industry-labor coalition (not to be confused with the White House's Council on Competitiveness headed by Vice-President Dan Quayle) sounds the alarm over the decline in U.S. high-tech industrial competitiveness and argue for governmental policies to help bail out economy.

5. Nathan Rosenberg, *Perspectives on Technology*, (New York: Cambridge University Press, 1976), p.123.

6. Kim Moody, "Industrial Policy," *Labor Notes*, July 27, 1983.

7. See Ann Markusen, "Defense Spending: A Successful Industrial Policy?" *International Journal of Urban and Regional Research* Vol.10, No.1 (1986):105-22; and, Markusen and Yudken, *Dismantling*, especially chapter 3.

8. Markusen and Yudken, *Dismantling*, p.14.

9. Daniel S. Greenberg, The *Politics of Pure Science* (New York: New American Library, 1967), p.8.

10. David C. Mowery and Nathan Rosenberg, *Technology and the Pursuit of Economic Growth* (New York: Cambridge University Press, 1991), p.184.

11. Linda R. Cohen and Roger G. Noll, *The Technology Pork Barrel* (Washington, D.C.: The Brookings Institution, 1991), pp.1-2.

12. In what he aptly terms the "medical technology `arms race'," New York Times reporter Andrew Pollack reports that there is "a growing recognition that the uncontrolled use of high-technology medical equipment and procedures... helps drive the relentless increase in medical cost." Pollack observes that medical practitioners are rewarded more for using sophisticated treatments than for routine health care. Hence, there is a tendency to overuse advanced, expensive medical procedures. Bolstering this up is a "huge medical technology industry," which is "spewing out streams of innovations and marketing them heavily, with profit rather than social utility often its prime motivation," Andrew Pollack, "Medical Technology `Arms Race' Adds Billion to the Nation's Bills," *New York Times*, April 29, 1991.

13. This point and the discussion below is elaborated in Markusen and Yudken, *Dismantling*, chapter 3.

14. Aerospace Industries Association of America, "1990 Year-End Review and Forecast -- An Analysis," mimeo (Washington D.C., December 12, 1990), table VI.

15. Markusen and Yudken, *Dismantling*, p.35 and table 3.1

16. Military industry R&D funding is highly concentrated among certain industries and firms, which in turn, are much more R&D intensive than in the manufacturing sector as a whole. During the first cold war buildup, by 1956, the federal government financed 87 percent of the aircraft and parts and 56 percent of the electrical equipment industries' R&D expenditures, while shares for scientific instruments and chemicals were 25% and 3% respectively. In the same year R&D spending accounted for almost 20% of the sales of the aircraft and parts industry, compared to only 3% for all U.S. industry; by 1960 this share rose to 30%. See Markusen and Yudken, *Dismantling*, especially chapter 3 and chapter 5.

17. Ibid.

18. When Douglas was in serious trouble with its DC-8 and DC-9 in 1967 it got $75 million from the federal government. More recently, General Dynamics and others have benefitted from similar government supported recoveries. Markusen and Yudken, *Dismantling*, ch.3.

19. Quoted in *The Washington Post*, 6 March, 1970. Cited in Berkeley Rice, *The C-5A Scandal* (Boston: Houghton Mifflin Co., 1971), p.183.

20. James R. Kurth, "Aerospace Production Lines and American Defense Spending,"in Stephen Rosen (ed.), *Testing Theory of the Military-Industrial Complex* (Lexington, M.A: D.C. Heath, 1973), pp.135-156.

21. James R. Kurth, `The Follow-On Imperative in American Weapons Procurement, 1960-1990", Paper presented at the Conference on the Economics of Disarmament, Economists Against the Arms Race, University of Notre Dame (December 1, 1990).

22. Markusen and Yudken, *Dismantling*, p.54. See also Daniel Todd and Jamie Simpson, *The World Aircarft Industry* (London: Croom Helm, 1986), pp.59, 187; Robert Reich, *Minding America's Business* (New York: Vintage, 1983), p.236.

23. Quoted in Charles Lane, et. al., "Arms for Sale," *Newsweek*, 8 April, 1991, pp.22-27:23.

24. Clyde Farnsworth, "White House Seeks to Revive Credits for Arms Exports," *New York Times*, 18 March, 1991, p.D6.

25. See Markusen and Yudken, *Dismantling*, chapter 6. The concept of a `gunbelt' was first developed and elaborated in Ann Markusen, Peter Hall, Sabina Deitrick and Scott Campbell, *The Rise of the Gunbelt* (New York: Oxford University Press, 1991).

26. See Markusen and Yudken, *Dismantling*, especially chapters 3-6.

27. For discussion on the steel industry see Ann Markusen, *"Steel and Southeast Chicago: Reasons and Remedies for Industrial Renewal"*., Report to the Mayor's Task Force on Steel and Southeast Chicago, (Evanston: Center for Urban Affairs and Policy Research,Northwestern University, 1985) and "Planning for Communities in Decline: Lessons from Steel Communities," *Journal of Planning Education and Research*., Vol.7, No.3 (1988): 173-184; Seymour Melman, *Profits Without Production* (New York: Alfred Knopf, 1983), pp.188-189; Robert Reich, *Minding America's Business* (New York: Vintage, 1983), pp.155-168.

28. Military planners are especially worried about the inability of U.S. firms to supply many vital components and systems, which has resulted in greater reliance on foreign sources. Once on the leading-edge, military technology has resulted in grater reliance on foreign sources. Once on the leading-edge, military technology has fallen behind commercial innovation in many areas. Moreover, the needs and requirements of military and civilian technology development have diverged significantly over the years. See Jay Stowsky, "Beating Plowshares into Double-Edged Swords: the Impact of Pentagon Policies on the Commercialization of Advanced Technologies," (Berkeley Roundtable on the International Economy (BRIE), April 1986). Leslie Bruechner and Michael Borrus, "Assessing the Commercial Impact of the VHSIC (Very High Speed Integrated Circuit) Program," BRIE Working Paper (Berkeley, CA: BRIE,

December 1984); Jacques S. Gansler, *Affording Defense* (Cambridge, MA: The MIT Press, 1989), pp.215-238, and endnotes. For discussions about problems of the defense technology base, see U.S. Congress, office of Technology Assessment *The Defense Technology Base: Introduction and Overview - A Special Report* , OTA-ISC-374 (Washington, D.C.: U.S. Government Printing Office, March 1988), and, - *Holding the Edge:Maintaining the Defense Technology Base, OTA-ISC-420* (Washington, D.C.:U.S. Government Printing Office, April, 1989).

29. Since the creation of the President's Commission on Industrial Competitiveness, by President Reagan in 1984, headed by Hewlett-Packard CEO John Young, a number of government organizations and private sector commissions, such as the Council of Competitiveness (1988, 1991), a business-trade organization started by Young, Cuomo Commission on Trade and Competitiveness (1988) M.I.T. Commission on Industrial competitiveness (Dertouzos et al 1988), Carnegie Commission on Science, Technology and Government (1991), National Academy of Engineering, National Advisory Committee on Semiconductors, Department of Commerce (1990) and the Office of Technology Assessment (1990b, 1988a), among others, have issued a series of findings and recommendations that are highly consistent. Robert M. White, formerly president of the National Academy of Engineering, and currently Undersecretary of Commerce for Technology, summarized the commonality into four points: (1) the principal responsibility for harnessing technology for economic growth and remainology competitive in the global economy lies with industry; (2) there is a key role for government in providing the proper economic environment; (3)government at all levels must take a more prominent role in advancing education;and (4) there needs to be a stronger role for government participating with industry in the development of generic technologies when forces make it difficult for industry to capture the benefits of research and development.Robert M. White, "Technology Policy in an Interdependent World," in Sauer, *Science and Technology and the Changing World Order*, pp.273-278: p.274.

30. The most important of these officials has included former White House Chief of Staff John Sununu, Office of Management and Budget Richard Darman, and Council of Economic Advisers Chief Michael Boskin. It remains to be seen whether replacement of Sununu by the more moderate Samuel K. Skinner will lead to a toning-down of the hard-core laissez-faire position in this arena, especially in the light of the current economic problems which have put a damper on President Bush's re-election hopes.

31. Other so-called dual use technologies include high-definition display, robotics, optoelectronics, biotechnology and advanced machine tools. For discussion of dual-use technology see Lewis M. Branscomb, "The Case for a Dual-Use National Technology Policy", *The Aspen Quarterly* Vol.2, No.3 (Summer 1990) : 33-52 --, "Toward a U.S. Technology Policy", *Issues in Science and Technology* Vol.VII, No.4 (Summer 1991) : 50-55.

32. DARPA Officials in the late 1980s have even proudly equated the role of their agency to that of Japan's Ministry of International Trade and Industry (MITI). See Yudken and Black "Targeting National Needs", which outlines and *critiques* DARPA's commercial high-tech

role. See also Joel Yudken, "Technology, Defense Conversion and Economic Development : A National Policy Perspective", in Kevin Cassidy (ed.), *Real Security - Converting the MilitaryEconomy and Building Peace* (SUNY Press, in progress).

33. The DoD also supports a large manufacturing technology program and even funds the Office of Science and Technology Policy's (OSTP) newly - formed Critical Technologies Institute. David P. Hamilton, "Technology Policy : Congress Takes the Reins", *Science* Vol.250 (9 November 1990) : 747.

34. See Eliot Marshall, "Beating Swords into ... Chips?" (box in "U.S. Technology Strategy Emerges"), *Science* Vol.252 (5 April 1991) : 22; and, William J. Broad, "Pentagon Wizards of the Technology Eye Wider Civilian Role", *New York Times,* 22 October, 1991, pp.C1,C11.

35. In 1990, the Commerce Department also issued a report on "emerging technologies", with potential to advance "productivity and quality", and create a "multitude of new products and services". See U.S. Department of Commerce, Technology Administration, *Emerging Technologies, A Survey of Technical and Economic Opportunities* (Washington, D.C., Spring 1990).

36. Lewis M. Branscomb, "Toward a U.S. Technology Policy", p.51.

37. The report also proposes changes in the roles of the National Security Council and other Executive offices that would make technology policy a central concern of both national security and economic policy at the highest levels of the executive branch. See Carnegie Commission on Science, Technology and Government, *Technology and Economic Performance : Organizing the Executive Branch for a Stronger NationalTechnology Base* (New York : CCST&G, September, 1991).

38. The recently passed bill, the FY 1992 Defense Authorization Act, includes appropriations of $ 281 million for Manufacturing Technology Development and $60 million for Critical Technologies Partnership under DARPA. Senators Hollings and Bingaman had earlier tried to get bills passed to set up a civilian technology program, and failed. They then pushed several aspects of their program through the military route because of the better chances of getting through Congress. Their judgement was rewarded.

39. See, for example, Richard Nelson and Sidney Winter *An Evolutionary Theory of Economic Change*, (Cambridge, M.A : Belnap Press of Harvard University Press, 1982); Giovanni Dosi et. al. (eds.), *Technical Change and Economic Theory*, (London: Pinter Publishers, 1990); David C. Mowery and Nathan Rosenberg, *Technology and the Pursuit of Economic Growth* (New York, Cambridge University Press, 1991).

40. For a critical elaboration of the weaknesses of the high-tech "dual use" industrial policy, see Yudken and Black, "Tergeting National Needs".

41. Mowery and Rosenberg, p.144.

42. President's Economic Adjustment Committee, Office of Economic Adjustment (OEA), Office of the Assistant Secretary of Defense (Manpower, Reserve Affairs and Logistics), *The Role of Defense in Santa Clara County's Economy* (Washington D.C. : The Pentagon, August 1980), p.66.

43. For discussions of corporate conversion efforts, including Kaman, see Robert

DeGrasse Jr., "Corporate Diversification and Conversion Experience", in John Lynch (ed.), *Economic Adjustment and Conversion of Defense Industries* (Bosulder, CO: Westview Press, 1987). For discussion on Frisby Airborne see M. Louise McNeily, "Braving the New World", *Plowshare Press* Vol.15, No.1 (Winter 1990): 1,6.

44. For overview of employee-community efforts see Hill, Deitrick, and Markusen, "Converting the Military Industrial Economy". See also markusen and Yudken, *Dismantling*, Chapter 8.

45. M.G. O'Neil, President of the General Tire and Rubber Company, in response to the letter of inquiry by Senator Abraham Ribicoff (D-CN). Reported in the U.S. Senate, Committee on Government Operations, Subcommittee Executive Reorganization and Government Research, *National Economic Conversion Commission : Responses to Subcommittee Questionnaire*, (September 1970), p.46. Quoted ind the President's Economic Adjustment Committee (PEAC) and the Office of Economic Adjustment (OEA), Office of the Assistant Secretary of Defense (Manpower, Installations and Logistics), "Previous Industrial Conversion Experience", (Prepared by Robert DeGrasse), *Economic Adjustment/Conversion, Appendices* (Washington, D.C. : The Pentagon, July 1985), p.M-4.

46. PEAC and OEA, p.M-20

47. For more detailed treatment of the problem of conversion of defense-dependent engineers and scientists see Joel Yudken and Ann Markusen, "TheLabor Economics of Conversion for Military-Dependent Engineers and Scientists", in Martha Gilliland and Patricia MacCorquodale (eds.), *Engineers and Economic Conversion* (Boulder, CO: Westview Press, in progress). See also Markusen and Yudken, *Dismantling*, Chapter 6, which also includes discussion of the conversion problem of blue-collar workers and managers.

48. U.S. Senate, Committee on Government Operations, *National Economic Conversion Commission ... Questionnaire*, p.162, Cited in PEAC and OEA, p.M-3.

49. Markusen and Yudken, *Dismantling*: p.214.

50. See Markusen and Yudken, *Dismantling*, especially chapter 9.

51. Ibid.

52. See Paul A. Krugman, "Myths and Realities of U.S. Competitiveness". *Science* Vol.254 (November 8, 1991): 811-815.

53. Faux and Schafer also calculate that the U.S. currently suffers an "investment deficit" somewhere between $60-125 billion each year. See Jeff Faux and Todd Schafer, "Increasing Public Investment, New Budget Priorities for Economic Growth in the Post-Cold War Word", Briefing Paper for the Conference on Investing in America's Future, Washington, D.C., October 21, 1991. (Washington, D.C.: Economic Policy Institute, October 1990). See also : Jeff Faux, "Back to the Peace Dividend", *New York Times*, Op-Ed, September 13, 1992; Robert Kuttner, "Savings AloneWon't Jump-Start the Economy", *Business Week*, 21 October, 1991, p.18.

54. For discussion of adjustment problems see Markusen and Yudken, *Dismantling*, Chapters 6-8.

55. For elaboration of this argument see Yudken, "Technology, Defense Conversion and Economic Development".

56. "Critical", "emerging" and "enabling" technologies are terms often used interchangeably. "Critical technologies" are meant to refer to technologies that will lead to new products and industries on a large scale or produce large advances in productivity and quality. They are also thought to be important because they will drive the next generation of technological development. "Critical technology lists" generated by government bodies, such as the Pentagon, Commerce Dept. and Office of Science and Technology, and Private sector groups such as the Council on Competitiveness, are strikingly similar. See *Report of the National Critical Technologies Panel*, (Washington, D.C.: U.S. Government Printing Office, March 1991); U.S. House of Representatives, "Critical Technology: OSTP Report", Hearing before the Subcommittee on Technology and Competitiveness of the Committee on Science, Space and Technology, 102nd Congress, 1st Session, [No.20] (Washington,DC : U.S. Government Printing Office, April 25, 1991); Council on Competitiveness, *Gaining New Ground, Technology Priorities for America's Future*, (Washington, D.C: Council on Competitiveness, March 1991).

57. They defined energy security as "technologies that reduce dependenceon foreign sources, lower energy costs, or improve energy efficiency, and quality of life as the "ability to make strong contributions to health, human welfare, and the environment, both domestically and worldwide", *Report of the National Critical Technologies Panel*, p.122. It is interesting that although the Council on Competitiveness in its report notes the substantial overlap between the U.S. government lists and those put out by Japan's MITI and the European Community, there are also several differences, reflecting the different goals and missions of the respective agencies or institutions which generated the lists. Energy and environmental technologies are listed by the OSTP, the Europeans and Japanese, but are not included in the DOD and Commerce lists. The OSTP is the only one that mentions surface transportation, however. The Japanese list such items as human related technology and disaster prevention technology, which do not appear elsewhere. See Council on Competitiveness, *Gaining New Ground*, pp.65-68.

58. Ken Gieser, "The Greening of America, Making the Transition to a Sustainable Economy", *Technology Review* vol.94, no.6 (August/September 1991): 64-72.

59. See Yudken and Black, "Targeting National Needs" for elaboration of these points.

60. Labor unions such as the International Association of Machinists (IAM), United Autoworkers, and Communication Workers of America have long been concerned with technology and work problems, as well as, in the case of IAM and UAW, economic conversion. In the 1980s the IAM put forth their famous New Technology Bill of Rights, and tried(unsuccessfully) to get legislation passed in support of its provisions. They currently run a large educational center on the outskirts of Washington, D.C. to provide their members in technology and collective bargaining issues. Professional societies have only recently been engaged in trying to promote new research priorities, although groups like the Computer Professionals of Social responsibility,

and unorganized groups in the physics and mathematics disciplines have been active through the 1980s in trying to stop DOD's Strategic Defense Initiative and, more generally have argued for less military dependency in their respective fields. In 1990 a group of representatives of major professional societies, such as the Association for Computing Machinery (ACM), American Physical Society, Institute for Electrical and Electronic Engineers among several others in the physical, natural and social sciences and engineering got together in an attempt to create a new Coalition for Science & Technology in a New Era. The ACM itself has put out a document in 1991 calling for new priorities based on national needs : see Barbara Simons, *On Building a Research agenda for Computer Science,* IBM research report (RJ 8201 (75160) Computer Science) (Yorktown Heights, NY : IBM Research Division, June 28, 1991).

61. See Jaimie Markham and Thomond O'Brien, "Economic Conversion Opportunities at Unisys in Minnesota". Prepared for the Alternative Use Committee (Minneapolis, MN : Markham and Associates, June 16, 1990).

62. See CRT, position papers: Lenny Siegel, Ted Smith, and Rand Wilson, "Sematech, Toxics and U.S. Industrial Policy: Why We Are Concerned", (1990), Rand Wilson, "From Day Care to DARPA" Bargaining for A New Industrial Policy", (1991), Campaign for Responsible Technology, 408 Highland Avenue, Somerville, MA 02144. See also Kirk Ladendorf, "Sematech pressed on environmental safety", *Austin American-Statesman*, 2 June, 1990, p.C1-2.

# APPENDIX

**National Commission for Economic Conversion & Disarmament**
A program for Economic Conversion

**Introduction : Economic Conversion, Employment and Prosperity**

The disengagement of the United States and the former Soviet Union from the Cold War brings with it the prospect of deeper reductions in U.S. military spending. Clearly this is a welcome opportunity for both nations to end the nuclear and conventional arms races. Moreover, reducing military spending could free financial resources for both nations to address their respective budget deficits and many pressing domestic needs.

In order to prevent severe economic dislocation arising from current and proposed military spending reductions, it is critical that the U.S government implement an ambitious economic conversion and adjustment program. While Congress acted last year on a modest defense adjustment amendment to the *Defense Authorization Act of FY 1991,* this three-year $200 million program has failed to deliver timely and well targeted assistance to many regions dependent on the defense industry. Moreover, the Department of Defense has delayed transferring these funds to the Department of Labor and has yet to forward $50 million to the Economic Development Administration of the Commerce Department.

**Why the United States Needs Comprehensive Conversion Policies**

Even limited Pentagon budget cuts passed by Congress to date have produced tremendous economic dislocation. Hundreds of thousands of workers have been affected by these cuts and the nation has lost an opportunity to retain jobs and develop new business opportunities through a morecomprehensive conversion program. Nearly 300,000 defense industry workers have been laid off over the past two years. Under Defense Secretary Cheney's revised budget plan for FY 1992-1997, active duty troops will be reduced by 348,100, civilian DoD personnel will be cut by 148,000 and defense industry workers will be reduced by 813,600. Thus, the government expects to lay off 1,309,700 employees over the next six years.[1] Without a conversion program, cutting the military budget will mean more lay-offs and will exacerbate the recessionary pressures both regionally and nationally. Countering these recessionary pressures will require the implementation of a two fold conversion program.

**Structural Dislocation and Conversion**
The wind-down in military spending has caused economic dislocation in defense-dependent industries, occupations and regions. Advance conversion planning, targeted assistance and incentives are necessary to mitigate and prevent further losses in jobs and business. Incentives for contractors to retook their plants and restructure their firms for civilian production would help to develop new business opportunities and to preserve jobs. In addition, retraining programs are needed to assist managers, engineers and production workers to acquire the knowledge necessary to adapt to the cost and quality control requirements of civilian work. Military-dependent communities already hard-hit by base closures or major contract reductions require economic redevelopment assistance if they are to develop non-military business to reemploy workers and to salvage the small to medium-sized firms so vital to their regional economies.

**New Budget Priorities and Conversion**
Military spending has imposed trade offs on our budget priorities, leading our nation to make fewer productivity-enhancing publicinvestments. This has been especially true for research and development (R & D) where the majority of federal funding has gone to the military, rather than for basic research and civilian-oriented infrastructure. Moreover, military spending has often crowded out public investments in human resources, particularly education and job training. In the 1980s, military spending increases occurred at the expense of public infrastructural investment.[2]

A realignment of federal budget priorities by investing military savings in these critical civilian needs could address fundamental economic and social problems while stimulating key economic sectors to produce jobs. Such public investments should be viewed as wealth-producing assets, rather than as a simple public expense. In the long-run, these expenditures would pay for themselves by generating new economic activity and restoring productivity. Thus it would be short-sighted to exclusively use military savings to reduce the federal budget deficit while ignoring the nation's most pressing economic and social needs.

**Recommended Economic Conversion Policies**

A comprehensive conversion program represents a chance to preserve jobs and to stimulate new business opportunities by creating incentives for firms and workers to move from military production into viable civilian work. The initiatives outlined below form the basis for a structural conversion policy to prevent and mitigate economic dislocation and a fiscal conversion policy to stimulate and redevelop our national economy.

**Structural Conversion Policy**

1. Economic incentives for advance planning by firms to develop newproducts, retook, reorganize production, develop marketing capabilities and retrain their managements, engineers and workforce for commercial work.

   Cost to offset revenue loss = $200 million per year.[3]

2. Community economic redevelopment funding and assistance for adjusting to base closures and major reductions at military plants.

   Community economic development grants = $100 million per year.[4]

3. An adequately funded labor retraining policy which provides job retraining, income maintenance and job relocation benefits.

   Job Training Cost = $600 million per year.
   Income Maintenance & Benefits = $735 million per year.[5]

4. A small business assistance program to help subcontractors and supplier firms who are often the first to be affected by cutbacks.

   Small business loans = $20 million.[6]

5. A.G.I. Bill of Rights for demobilizing military personnel after the Cold War. Such a bill would provide income support and housing stipends for demobilizing troops to go to college or to enter job training programs.

   G.I. Bill Cost = $1.25 billion.[7]

6. Funding the conversion program from the DoD's procurement and RDT&E budgets through a three percent set

aside to a dedicated trust fund for conversion and adjustment.

Estimated Revenue = $3 billion per year.[8]

Two bills pending before Congress form the basis for most of these structural conversion policies. However, neither *The Defense Industry and Worker Realignment Act* (S.2075) sponsored by Senator Joseph Lieberman (D-CT) nor *The Defense Economic Adjustment Act* (H.R.441), sponsored by Congressman Ted Weiss (D-NY) includes all of the components of a comprehensive conversion policy. Indeed, there is currently no legislation pending before Congress that would finance the G.I. bill considered here,[9] although several Congressional offices have expressed interest in considering the problem. If the wind-down in defense spending is to continue there will be increasing need for legislation that addresses all aspects of a comprehensive structural conversion program. Without targeted public policies, we are left with ineffective one-time programs such as the one passed last year.

**Fiscal Conversion Policy**

A compensated shift in federal budget priorities using savings from military cuts to fund critical domestic needs is a complementary fiscal policy which could cushion the economy from the shocks of the military wind-down by stimulating the overall level of economic activity. Such a shift in federal budget priorities could help to maintain the level of employment and output and thereby prevent the current recession from worsening. Perhaps more importantly, public investments in education, health care, job training, infrastructure and civilian research and development could enhance productivity and improve the quality of life.

Numerous economic studies have shown that a shift of federal spending to key infrastructural and human resourceinvestments could generate more jobs and higher levels of economic activities than military spending.[10] Indeed, one study showed that an annual average transfer of $70.5 billion dollars from the military budget to education, infrastructure and other critical needs would generate an annual gain of nearly 477,000 more jobs on average over a four year period; in other words, the study found that every $1 billion transferred from military spending to civilian investment would create a net gain of 6,800 jobs. Moreover, the Gross National Product was shown to grow by an average of $17.6 billion annually.[11]

Additional savings from the Pentagon's R & D budget should be transferred to urgent civilian research priorities, such as alternative energy, transportation and environmental research. Such an approach would maintain the overall level of research effort in the United States, and would also create new conversion opportunities for science-based military serving firms, as well as many civilian firms working on these alternatives.[12]

For over 45 years, Washington has pursued an industrial policy designed to make this nation militarily second to none. It has succeeded brilliantly in this goal. Today, we need industrial policies geared to restoring our nation's economy. Economic conversion policies should be at the forefront of efforts to rebuild the U.S. economy. If our nation is committed to forging a post-Cold War order we must plan for the transition and make the most of the opportunity. Perhaps equally important, we must demonstrate our commitment to peace and economic development by converting our military-oriented economy, rather than merely advocating such a course for our former adversary.

**Endnotes**

1. For 1990 and 1991 estimates on defense worker layoffs, see "National

Defense Industry Job Layoffs for 1990," and "National Defense Industry Job Layoffs for 1991", available from ECD. Information on the Cheney budget taken from the DoD Press Release issued January 29, 1992 outlining the FYDP for FY 1992-1997; estimates based on ECD's Defense Impact Modeling System.

2. See Michael Dee Oden, *A Military Dollar Really is Different* (Lansing MI : Employment Research Associates 1987), pp.20-35.

3. ECD's estimate of offset for revenue loss based on Lieberman S. 2398, Title III formula, 101st Congress.

4. ECD's estimate based on "Military Base Adjustment Experience", in *Economic Adjustment Conversion*, Report prepared by the President's Economic Adjustment Committee and the Office of Economic Adjustment, July 1985. Also, see John Lynch, *Local Economic Development After Military Base Closures* (New York: Praeger Publishers, 1970).

5. See average per worker costs for retraining and income maintenance in *Converting the American Economy* (Lansing, MI: Employment Research Associates, 1990), pp.25-26. Defense Industry lay-offs based on DoD comptroller's Defense Industry Employment to 1997, as calculated by ECD.

6. Based on 8 state surveys of defense contractors obtained by the Commission. The FY 1992 Defense Authorization Act provides $280 million in additional monies for small and medium-sizedbusiness assistance through the Advanced Manufacturing Technology Amendments.

7. Under the Cheney cuts, about 476,000 will be cut by 1996. The annual average cost estimate would cover one-third of those eligible to participate at 60 percent of the weighted average military wage.

8. Projected Defense Budget based on Cheney FYDP 1992.

9. Obviously the G.I bill considered here would provide benefits far greater than those offered under the current Montgomery GI Bill. See *Federal Benefits for Veterans and Dependents,* U.S. Department of Veterans Affairs, January, 1991, pp.11-14 for information about the Montgomery program.

10. See Roger Bezdek, "The Economic Impact-Regional and Occupational -- of Compensated Shifts in Defense Spending" *Journal of Regional Science,* Vol.15, No.2, 1975. Also see Converting *the American Economy : The Economic Effects of an Alternative Security Policy,* Lansing, Michigan, Employment Research Associates, 1991.

11. Ibid, *Converting the American Economy,* pp.26-28.

12. Total potential savings from military R & D cuts for reinvestment could amount to $8.5 billion, according to *Converting the American Economy,* pp.23-24.

**H.R. 441 (102nd Congress) Summary of The Defense Economic Adjustment Act Congressman Ted Weiss**

**ECONOMIC CONVERSION LEGISLATION**

* Paves the way for a smooth transition from a military economy, allowing the United States to address its pressing social needs;

* Creates a job security policy for employees of the military serving sector of the economy.

* Removes the "job pressure" on legislators and communities to save defense plants, leading to a more rational defense policy based on national security and defense needs, not political pressure.

**H.R. 441 INCLUDES :**

**I. ALTERNATIVE USE COMMITTEES**

H.R. 441 calls for the creation of alternative use committees at large defense facilities. These committees, comprised of representatives of both management and labor, would develop a detailed blueprint for the conversion of the facility to civilian-oriented production, in the event that the facility's contract is cancelled or cut back.

If such a blue-print is completed in a timely manner - that is, before the contract reduction is announced, the community would be able to implement the transition of the facility, retrain its workers, and start production without the massive employee layoffs and community disruption which often occurs with contract cancellations.

**II. DEFENCE ECONOMIC ADJUSTMENT COUNCIL**

The bill calls for a Cabinet-level Defense Economic Adjustment Council charged with developing plans for public projects that address human needs. The Council also would serve as a clearing house for information on existing Federal programs relevant to communities affected by military cutbacks.

**III. ADVANCED NOTIFICATION**

The Defense Economic Adjustment Act would also require one year's advance notification of plans to cut back or terminate a defense contract or a military base. The Defense Economic Adjustment Council is responsible for informing the appropriate local officials and their Alternative Use Committees of pending cutbacks.

**IV. ADJUSTMENT ASSISTANCE**

Finally, the bill provides for adjustment assistance for communities and workers while conversion is underway. Communities seriously affected by defense cutbacks would be eligible for Federal planning assistance and individual workers would be eligible for adjustment benefits, including funds for retraining.

**Summary of Budget Proposal CONGRESSMAN SAM GEJDENSON 102nd CONGRESS**

Representative Gejdenson's proposal would reduce both budget authority and outlays for operations and maintenance of overseas military installations by $10 billion each year from FY 1992 through FY 1996. The proposal would reduce the Federal budget deficit by $24.35 billion over five years and would increase, on an annual average, authorizations for the following programs :

* $500 million for international assistance for "tied aid;"

* $800 million for solar and renewable energy research and development;
* $200 million for energy conservation;
* $750 million for pollution control and abatement;
* $250 million for historic preservation;
* $50 million for science and technical research and services of the National Institute of Standards and Technology;
* $750 million average for MAGLEV and high spelled rail R & D;
* $1.5 billion for mass transit;
* $250 million for Community Development grants;
* $250 million for the Economic Development Agency;
* $500 million for job training and employment services block grants to state and dislocated worker assistance under title III of the Job Training and Partnership Act;
* $100 million for Low Income Home Energy Assistance Weatherization program.

Total : $5.9 billion

These funds spend out slowly the first year, hence deficit reduction is greater than it appears.

**S.2075 (102ND congress) Summary of THE DEFENSE INDUSTRIAL DIVERSIFICATION ACT**

**SENATOR JOSEPH I. LIEBERMAN**

**TITLE I/TITLE II - INDUSTRIAL DIVERSIFICATION STUDY**

Requires the Department of Commerce, in consultation with the Department of Defense, the Small Business Administration (S.B.A), and DARPA to perform a study on the extent to which and how diversification of defense industries to non-defense production can be accomplished. Title II establishes the Council on Economic Diversification and Adjustment within the Executive Office of the President. The Council will be responsible for implementing the programs required by this Act.

**TITLE III - DEFENSE INDUSTRIAL DIVERSIFICATION ACCOUNT**

Establishes a Defense Industrial Diversification Account for defense business wishing to become more integrated into the civilian sector. It would function much like an I.R.A.

This title would allow a firm which is at least 35 percent reliant on defense contracts to deposit a portion of its profits into the account, tax free, for the purposes of investing in new non-defense plant and equipment or for the retraining or continued education of its work force.

Deposits would be limited to the amount of a firm's defense-related depreciable assets. These investments could be made for a period of five years. The withdrawals could for 10 years. If the money were withdrawn for purposes other than those specifically stated,then the money would be fully taxed and a 10 percent penalty would be added.

## TITLE IV/TITLE V - SMALL BUSINESS DIVERSIFICATION ASSISTANCE

Establishes, within the S.B.A., a permanent office of Small Business Diversification. The purpose of this office is to identify the problems associated with defense cutbacks for small defense-dependent business. The office would also coordinate the efforts of the S.B.A and its programs for assisting firms adversely affected by defense cutbacks.

Creates a small business diversification guaranteed loan, or grant program, Loans, guaranteed loans, or grants are capped at $40,000 for qualified small firms and require 1:1 match from the firm.

Specific uses for the loan, guaranteed loan, or grant may include, but shall not be limited to, conducting market research or analysis for non-defense products, development of an alternative use or non-defense business plan, basic or applied research and development for non-defense lines of business, strategic planning or non-defense prototype development.

Program is designed to promote diversification of small defense manufacturers and to qualify firms, if needed, for other S.B.A. loan guarantee programs or commercial lending sources.

## S.2133 (102nd Congress) Summary of THE DEFENSE INDUSTRIAL STABILIZATION AND COMMUNITY TRANSITION ACT SENATOR CHRISTOPHER J.DODD

## TITLE I - PRESIDENTIAL COUNCIL ON ECONOMIC DIVERSIFICATION AND ADJUSTMENT

This council is established within the Executive Office of the President. Within the Council, four working groups are established: (1) Technology, (2) Marketing; (3) Small Business; (4) Job Training.

## TITLE VI - ECONOMIC ADJUSTMENT ASSISTANCE FOR EMPLOYEES

Employee notification is required for qualified workers 90 days prior to layoff.

Amends Title III of the Job Training Partnership Act (Displaced Worker Title). Increases Federal set-aside for displaced workers and modifies the formula used todisburse funds to States to factor in defense-related lay-offs.

Allows qualified defense workers, who have been laid-off, to withdraw funds from their Individual Retirement Account to pay the principal or interest on a mortgage of his or her primary residence or for the rental costs of his or her primary residence.

## TITLE VII - COMMUNITY ECONOMIC ADJUSTMENT PLANNING

Identifies communities that will be substantially and seriously affected by defense lay-offs (loss of more than 250 employees).

Allows substantially and seriously impacted communities to be eligible for Title IX Community Planning Grants under the Economic Development Administration.

## TITLE II - GRANT ASSISTANCE FOR ALTERNATIVE USE COMMITTEES

The Council may make a grant of up to $50,000 to any eligible alternative use committee established at a defense facility for the purpose of exploring commercial production. Eligible alternative use committees shall be made up of management, labor and community representatives. Grants may cover no more than one-third of the operating expenses of the committee. They may be used for office space, or for the hiring of marketing and technology consultants.

## TITLE III - DISPLACED WORKERS

The council is authorized to transfer funds from the Defense Economic Adjustment Trust Fund (see below) to existing job training programs under Section 325 of the Job Training Partnership Act.

## TITLE IV - EXPORT OPPORTUNITIES

The State Department is authorized to extend guarantees for the sale of defense articles and services to NATO countries, Japan, Australia, New Zealand and Israel. Theamount guaranteed under this section may not exceed $1 billion.

## TITLE V - DEFENSE PRODUCTION ACT

The Defense Production Act of 1950 is amended to permit the President to make loans, guarantee loans and enter into joint ventures with defense companies for the purpose of converting to commercial production. This provision is available to defense companies critical to the defense industrial base whose existence is threatened by impending defense cuts.

## TITLE VI - SMALL BUSINESS

The Administrator of the Small Business Administration shall take such actions as may be necessary to increase the number of small business development centers (under section 21 of the Small Business Act) in defense-dependent regions of the United States. The allocation for the Small Business Innovative Research Program is increased from 1.25 percent to 1.75 percent.

## TITLE VII - ECONOMIC DEVELOPMENT ASSISTANCE

The Council is authorized to transfer funds from the Defense Economic Adjustment Trust Fund to existing economic development programs under Title IX of the Public Works and Economic Development Act of 1965.

## TITLE VIII - DEFENSE ECONOMIC ADJUSTMENT TRUST FUND

An amount equivalent to 10 percent of the estimated savings from reductions in the defense budget each year shall be placed in this trust fund, for use in carrying out the provisions of this Act.

# ANNEX

## List of authors

**Dr. Arturo Abriani**
FIEEBA
Juncal 2174, 2-25
1125 Buenos Aires
Argentina

**Dr. Torsten Amelung**
Ruhrkohle AG
Rellinghauser Str. 1
4300 Essen
Germany

**Gregory Bischak**
National Commission for Economic
Conversion and Disarmament
Suite 9
1801 18th St. NW
Washington, D.C. 20009
U.S.A.

**Dr. Christoph Butterwegge**
Bremische Stiftung für
Rüstungskonversion und
IFK
Universität Bremen
Sedanstr. 98
2800 Bremen
Germany

**Dr. Anthony DiFilippo**
Department of Sociology
Lincoln University
Lincoln University, PA 19352
U.S.A.

**Vladimir Dmitriev**
Chetek AG
Warwaka-Straße 15
1030012 Moskau
Russia

**Dr. Paul Dunne**
School of Business and
Economics
University of Leeds
Leeds LS2 933
England

**Dr. Jürgen Fortmann**
Ruhrkohle Umwelttechnik GmbH
Rellinghauser Str. 1
4300 Essen 1
Germany

**Dr. Dr. Hans-Joachim Giessmann**
Institut für Friedensforschung
und Sicherheit (IFSH)
Falkenstein 1
2000 Hamburg 55
Germany

**Dr. Ksenia R. Gonchar**
Institute of World Economy
and Intern. Relations
Profsoyuznaya str., 23
Moscow, 117859
Russia

**Evgueni Gouklin**
Uralconversia
ul. Nevskogo 3
Zarechny 624051
Russia

**Alexander Konovalov**
Institute of U.S./Canada Studies
USSR Academy of Sciences
Khlebny per. 2/3
Moscow 121069
Russia

**Prof. Mel Gurtov**
Portland State University
P.O. Box 751
Portland
OR 97207-0751
U.S.A.

**Dr. Fiedrich Kühn**
Bundesanstalt für Geowissenschaften
und Rohstoffe
Aussenstelle Berlin
Invalidenstraße 44
1040 Berlin
Germany

**Burckhardt J. Huck**
SWP
Forschungsinstitut für intern.
Politik und Sicherheit
Haus Eggenberg
8026 Ebenhausen/Isar
Germany

**Dr. Betty G. Lall**
230 E. 81th Street
New York, NY 10028
U.S.A.

**Jean-Paul Hébert**
CIRPES
18 rue beauséjour
76130 Mont-Saint-Aignan
France

**Milton Leitenberg**
CISSM
3106 Morrill Hall
School of Public Affairs
University of Maryland
College Park, Maryland 20742
U.S.A.

**Dr. Dieter Joos**
Industrie-Anlagen
Betriebsgesellschaft mbH
Einsteinstr. 20
8012 Ottobrunn
Germany

**Dr. Peter Lock**
Berghof Stiftung für Friedens-
und Konfliktforschung
Altensteinstraße 48a
1000 Berlin 33
Germany

**Dr. Alfredo Luzuriaga**
3 de Febrero 2191 - 108 B
1428 Buenos Aires
Argentina

**Prof. Dr. Seymour Melman**
Deparment of Ind. Engineering
and Operations Research
304 Mudd Bldg., Dept. IE/OR
Columbia University
New York, N.Y. 10027
U.S.A.

**Peter Menke-Glückert**
Augustastraße 51
5300 Bonn 2
Germany

**Dr. Moustafa A. Mostafa**
Institute of National
Planning
Salah Salem Road
Nasrcity, Cairo
Egypt

**Prof. Dr. R. Mull**
Wilhelm-Raabe-Str. 33
3008 Garbsen 1
Germany

**Dr. Timothy Muzondo**
International Monetary Fund
700 19th Street, N.W.
Washington, D.C. 20431
U.S.A.

**Dr. Bernd Niemeyer**
MBB GmbH
Postfach 1340
8898 Schrobenhausen
Germany

**Vladimir Nikitin**
Central Desine Buro for
Machinebuilding
3, Krasnogvardeyskaya SQ
St. Petersburg
Russia

**Dr. Petra Opitz**
Berghof-Institut für
Friedens- und Konfliktforschung
Altensteinstraße 48 a
1000 Berlin 33
Germany

**Reinhard Peglau**
Umweltbundesamt
Mauerstraße 52
1080 Berlin
Germany

**Anatoli Poljakov**
Scientific and
Research Institute for
Inorganic material
Rogov str. 5
123060 Moscow
Russia

**Alexander Rackitsky**
'Terra Block Soviet Union' Ltd.
10, Petrovka Str., Office 303
103779 Moscow
Russia

**Leonid Raginsky**
Scientific and
Research Institute for
Inorganic material
Rogov str. 5
123060 Moscow
Russia

**Dr. Michael G. Renner**
14-21 31st Drive
Long Island City
New York 11106
U.S.A.

**Theodor Rosendorfer**
MBB GmbH
Postfach 13 40
8898 Schrobenhausen
Germany

**Joachim Schatz**
Treuhandanstalt
Alexanderplatz 6
1020 Berlin
Germany

**Jürgen Scheffran**
IANUS
c/o Institut für Kernphysik
Schloßgartenstr. 9
6100 Darmstadt
Germany

**H. Schröder**
Oberfinanzdirektion
Waterloostr. 4
3000 Hannover
Germany

**Christopher Smart**
Hudson Institute
1015 18th Street N.W.
Suite 200
Washington, D.C. 20036
U.S.A.

**Athar Sultan**
Pakistanische Botschaft
Rheinallee 24
5300 Bonn 2
Germany

**Prof. Piero Tani**
Forum on the Problems
of Peace and War
Viale Matteotti 8
50132 Florenz
Italy

**Amb. Britt Theorin**
Ministry for Foreign Affairs
P.O. Box 16121
10323 Stockholm
Sweden

**Dr. Christian Wellmann**
Projektvbd. Friedenswissenschaften
Kiel
Christian-Albrecht-Universität
Preußerstraße 1-9/III
2300 Kiel 1
Germany

**Horst Wiesebach**
UNIDO
P.O. Box 300
1040 Wien
Austria

**Susan Willett**
Centre for Defence Studies
Kings College
University of London
The Strand
London WC2R 2LS
England

**Dr. Lodewjk Reinaer Wolff**
Energy Conversion Systems B.V.
Neumanlaan 8
5624 KM Eindhoven
Netherlands

**Dr. Joel Yudken**
PRIE
Lucie Stone Hall
Room B119
Rutgers University, Kilmer Campus
New Brunswick, NJ 08903-5078
U.S.A.

# Index